Handbook of

PLANT
and FUNGAL
TOXICANTS

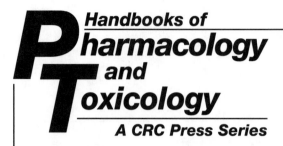

Handbooks of Pharmacology and Toxicology
A CRC Press Series

Mannfred A. Hollinger, Series Editor
University of California, Davis

Published Titles

Handbook of Pharmacokinetic and Pharmacodynamic Correlations with Computer Applications
Hartmut Derendorf

Handbook of Methods in Gastrointestinal Pharmacology
Timothy S. Gaginella

Handbook of Targeted Delivery of Imaging Agents
Vladimir P. Torchilin

Handbook of Pharmacology of Aging, Second Edition
Jay Roberts, David L. Snyder, and Eitan Friedman

Handbook of Plant and Fungal Toxicants
J. P. Felix D'Mello

Forthcoming Titles

Handbook of Theoretical Models in Biomedical Research
David B. Jack

Handbook of Mammalian Models in Biomedical Research
David B. Jack

Handbook of Opportunistic Infections
Vassil St. Georgiev

Handbook of Immunotoxicological Methods
Leonard Ritter

Handbook of

PLANT
and FUNGAL
TOXICANTS

Edited by

J.P. Felix D'Mello

CRC Press
Boca Raton   New York

Acquiring Editor: Paul Petralia
Editorial Assistant: Norina Frabotta
Project Editor: Debbie Didier
Marketing Manager: Susie Carlisle
Direct Marketing Manager: Becky McEldowney
Cover design: Dawn Boyd
PrePress: John Gandour

Library of Congress Cataloging-in-Publication Data

Handbook of plant and fungal toxicants / edited by J. P. Felix D'Mello.
       p.   cm. -- (Handbooks of pharmacology and toxicology)
    Includes bibliographical references and index.
    ISBN 0-8493-8551-2
    1. Plant toxins. 2. Mycotoxins. I. D'Mello, J. P. Felix.
  II. Series.
  RA1250.H36   1997
  615.9'52--dc20
                                       96-27516
                                          CIP

No claim to original U.S. Government works
International Standard Book Number 0-8493-8551-2
Library of Congress Card Number 96-27516
Printed in the United States of America 1 2 3 4 5 6 7 8 9 0
Printed on acid-free paper

Preface

The subject of natural toxicants continues to fascinate many scientists throughout the world. There are those who pursue research in this area purely for academic reasons, but in so doing provide us with a strategic base for applications in the medium- to long-term future. There are those who have already begun the task of exploiting natural toxins in breeding plants for enhanced resistance to disease and insect predators. Again, there are those who seek to use these compounds for therapeutic purposes in medicine. And then there are others whose function it is to ensure the safety of plant products in the interests of public and environmental health. Recent events in Europe concerning the safety of animal products serve to remind those of us entrusted with ensuring quality of plant commodities that on this particular issue, we should never drop our guard.

Commitment to the subject of natural toxicants manifests itself in different ways. Some have felt strongly enough to establish new multidisciplinary journals with the objective of providing an international forum for publication of original results and critical reviews. In my particular case, this interest finds expression in the form of the CRC *Handbook of Plant and Fungal Toxicants*, my second treatise in a related area. In my first book, *Toxic Substances in Crop Plants*, much emphasis was placed on factors which may be regarded as anti-nutritional in their effects or in their mode of action. In *Plant and Fungal Toxicants*, a more diverse range of compounds is being considered. Additionally, the human dimension is being given greater scope in view of widespread disquiet about all matters relating to food safety.

Some topics covered previously have justified expansion into several chapters. Thus, in this Handbook, considerably greater emphasis is being given to alkaloids, both from plant and fungal sources. These include indolizidine alkaloids, glycoalkaloids, *Erythrina* alkaloids, and endophyte alkaloids. The upsurge in research has justified the inclusion of a diverse range of other toxicants and phytochemicals such as bracken fern poisons, polyphenolics, gossypol, flavones, isoflavones, pyrimidine glycosides, fruit and vegetable allergens, linear furanocoumarins, photosensitizing agents, nitrates, oxalates, and *Pinus ponderosa* toxicants. An integrative approach is provided in two chapters which address the important issues of feeding behavior and modeling animal responses to xenobiotics. As in my previous book, I have taken care to emphasize the positive aspects of plant secondary compounds, and examples of beneficial attributes, in the context of environmental protection and human health, will be found throughout this Handbook. However, a specific chapter has also been devoted to the medicinal applications of plant toxicants.

The expansion of the fungal toxin theme is designed to reflect the growing perception of the important role of mycotoxins in determining food and feed safety and in the etiology of certain diseases in humans, particularly esophageal and hepatic cancers. Four main genera of fungi responsible for toxin production are covered in specific chapters on aflatoxins and *Fusarium*, *Penicillium*, and *Alternaria* toxins. The inclusion of this last chapter also makes good on an omission in my earlier book.

Thus, with minimal overlap, these two books provide a comprehensive and authoritative resource for final year undergraduates in a variety of disciplines and for all those engaged in research. I have my contributing authors to thank for providing the necessary expertise and commitment.

The outlook in this Handbook is distinctly global. It has to be, given the diversity and ecological distribution of plant and fungal toxicants. An international authorship was, therefore, inevitable, and, I am grateful to those authors who responded positively to my request for international collaboration in the preparation of their chapters.

Editor

Felix D'Mello, B.Sc., Ph.D., is lecturer at the Scottish Agricultural College, Edinburgh, Scotland.

He received both his degrees from the University of Nottingham, England. His Ph.D. thesis was concerned with the toxicology of, and interactions between, indispensable amino acids. He continued with this theme in Edinburgh, extending the research to the toxicity of the non-protein amino acids of plants. Current interests are more diverse and include the responses of aphids to potato glycoalkaloids and factors affecting the production of *Fusarium* mycotoxins.

His main functions are in education, research, and extension. Dr. D'Mello is currently in charge of a new degree course in Environmental Protection and Management, which he also helped to develop from rudimentary elements. In the course of his research duties, he has published over 80 refereed papers, reviews, and book chapters, solely or with other collaborators. He has also edited three books: *Toxic Substances in Crop Plants* in 1991, *Amino Acids in Farm Animal Nutrition* in 1994, and *Tropical Legumes in Animal Nutrition* in 1995. Dr. D'Mello currently serves on the editorial board of one journal. His extension duties are almost exclusively to do with mycotoxin screening of advisory samples, generally animal feedingstuffs, but on occasion, cereals destined for the food sector. In addition, he is convener of the Mycotoxin Awareness Group which comprises representatives from the main research laboratories in Scotland, including Professor J.E. Smith of the University of Strathclyde, a contributor to this Handbook. As part of its function, the Group has organized a conference on mycotoxins in cereals, for the benefit of industrial clients in the U.K.

Dr. D'Mello is a member of the Nutrition Society and of the World's Poultry Science Association. He is also an honorary member of staff of the University of Edinburgh.

Contributors

David Abramson, Ph.D.
Cereal Research Center
Agriculture and Agri-Food Canada
Winnipeg, Canada

Mahmoud S. Arbid, Ph.D.
Pharmacology Department
National Research Centre
Dokki-Giza, Egypt

Chad C. Chase Jr., Ph.D.
U.S. Department of Agriculture
Agricultural Research Service
Brooksville, Florida

Amrik Singh Chawla, Ph.D.
University Institute of Pharmaceutical
  Sciences
Panjab University
Chandigarh, India

Steven M. Colegate, Ph.D.
Plant Toxins Unit
CSIRO Division of Animal Health
Australian Animal Health Laboratory
Geelong, Victoria, Australia

Moussa M. Diawara, Ph.D.
Department of Biology
University of Southern Colorado
Pueblo, Colorado

Geoffrey R. Dixon, Ph.D.
Department of Horticulture
The Scottish Agricultural College
Auchincruive
Ayr, United Kingdom

Peter R. Dorling, Ph.D.
School of Veterinary Studies
Murdoch University
Murdoch, Western Australia

J. P. Felix D'Mello, B.Sc., Ph.D.
Department of Crop Science and
  Technology
The Scottish Agricultural College
Edinburgh, United Kingdom

Arne Flåøyen, Dr.Med. Vet., Ph.D.
Department of Reproduction and Forensic
  Medicine
Norwegian College of Veterinary Medicine
Oslo, Norway

Stephen P. Ford, Ph.D.
Department of Animal Science
Iowa State University
College of Agriculture
Ames, Iowa

Arne Frøslie, Dr.Med. Vet., Ph.D.
Department of Reproduction and Forensic
  Medicine
Norwegian College of Veterinary Medicine
Oslo, Norway

Emile M. Gaydou, Ph.D.
Laboratoire de Phytochimie
Universite de Droit d'Economie et des
  Sciences d'Aix-Marseille
Marseille, France

Andrew W. Illius, Ph.D.
The Division of Biological Sciences
The University of Edinburgh
Edinburgh, United Kingdom

Neil S. Jessop, Ph.D.
The Division of Biological Sciences
The University of Edinburgh
Edinburgh, United Kingdom

Vijay K. Kapoor, Ph.D.
University Institute of Pharmaceutical
  Sciences
Panjab University
Chandigarh, India

Edward J. Kennelly, Ph.D.
College of Pharmacy
University of Illinois at Chicago
Chicago, Illinois

A. Douglas Kinghorn, Ph.D., D.Sc.
College of Pharmacy
University of Illinois at Chicago
Chicago, Illinois

Shigemitsu Kudo, Ph.D.
Kanesa Company, Ltd.
Tamagawa
Aomori, Japan

Annie McCabe Cookson Macdonald, HNC
Department of Crop Science and
  Technology
The Scottish Agricultural College
Edinburgh, United Kingdom

J. P. Marais, Ph.D.
Biochemistry Section
KwaZulu-Natal Department of Agriculture
Pietermaritzburg, Republic of South Africa

Ronald R. Marquardt, Ph.D.
Department of Animal Science
Faculty of Agricultural and Food Sciences
The University of Manitoba
Winnipeg, Canada

Kazuyoshi K. Okubo, Ph.D.
Department of Food Chemistry and
  Biotechnology
Tohoku University
Amamiyacho
Sendai, Japan

Shantanu Panigrahi, Ph.D.
Natural Resource Management Department
Natural Resources Institute
Chatham Maritime, United Kingdom

Glynn C. Percival, Ph.D.
Department of Horticulture
The Scottish Agricultural College
Auchincruive
Ayr, United Kingdom

Cristina Mihaiela Placinta, Ing.
Institute of Ecology and Resource
  Management
The University of Edinburgh
Edinburgh, United Kingdom

James K. Porter, Ph.D.
Toxicology and Mycotoxin Research Unit
Richard B. Russell Agricultural Research
  Center
U.S. Department of Agriculture
Agricultural Research Service
Athens, Georgia

Frederick D. Provenza, Ph.D.
Department of Rangeland Resources
Utah State University
Logan, Utah

Jess D. Reed, Ph.D.
Department of Animal Science
University of Wisconsin-Madison
Madison, Wisconsin

Carlos A. Risco, DVM
College of Veterinary Medicine
University of Florida
Gainesville, Florida

Gerardo Rodriguez, M.Sc.
Department of Animal Science
University of Wisconsin-Madison
Madison, Wisconsin

Jack P. N. Rosazza, Ph.D.
Division of Medicinal and Natural
  Products Chemistry and Center for
  Biocatalysis and Bioprocessing
University of Iowa
College of Pharmacy
Iowa City, Iowa

Robert E. Short, Ph.D.
U.S. Department of Agriculture
Agricultural Research Service
Fort Keogh Livestock and Range Research
  Laboratory
Miles City, Montana

Barry L. Smith, B.V.Sc., D.Phil.
New Zealand Pastoral Agriculture Research
  Institute
Ruakura Research Centre
Hamilton, New Zealand

John Edward Smith, D.Sc.
Department of Bioscience and Biotechnology
University of Strathclyde
Royal College Building
Glasgow, United Kingdom

John T. Trumble, Ph.D.
Department of Entomology
University of California
Riverside, California

Stefan Vieths, Ph.D.
Department of Allergology
Paul-Ehrlich-Institut
Federal Agency for Sera and Vaccines
Langen, Germany

Na Wang, M.Sc.
Department of Animal Science
Faculty of Agricultural and
  Food Sciences
The University of Manitoba
Winnipeg, Canada

Patricia L. Whitten, Ph.D.
Department of Anthropology
Emory University
Atlanta, Georgia

Table of Contents

Chapter 22

1 Bioactive Indolizidine Alkaloids

S. M. Colegate and P. R. Dorling

CONTENTS

1.1 INTRODUCTION

Complete hydrogen saturation of the indolizine (**1**) system produces the indolizidine (octahydroindolizine) heterocyclic ring system (**2**). This structural unit can be found in alkaloids isolated from plants of various families and from microbial and animal sources.

The indolizidine entity can be the dominant sub-unit from a structural, bioactivity, or biogenetic aspect as in the simple polyhydroxyindolizidine alkaloid swainsonine (**3**, Section 1.2.1), or it can be subordinate to another unit as in the complex, fused ring indoloindolizidine alkaloid 5-methoxycanthinone (**4**, Section 2.9). The potential bioactivity of these alkaloids provides a stimulus for research in the areas of detection, extraction, and isolation, structure determination, synthesis, and biosynthesis.

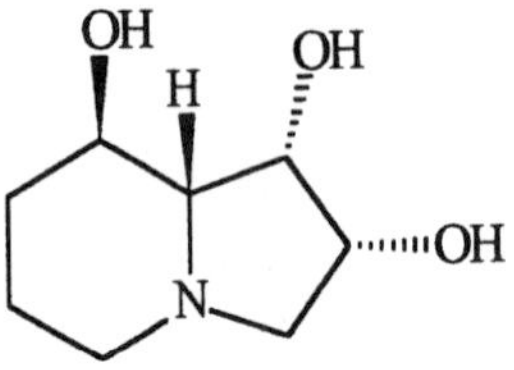

3

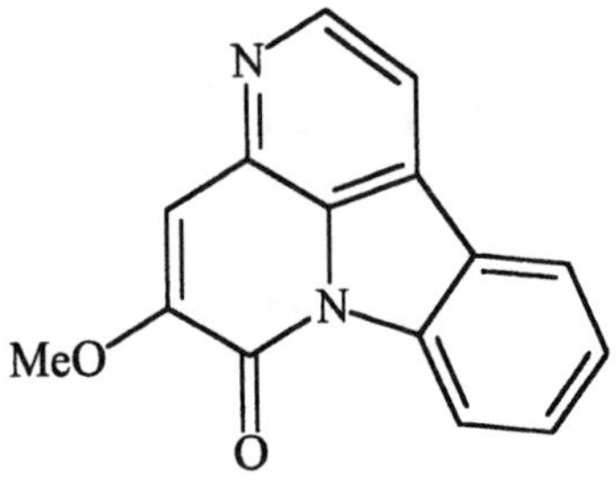

4

Naturally occurring indolizidines have been reviewed on a regular basis.[1] This review, however, is restricted to a discussion of bioactive indolizidine alkaloids isolated from plant or fungal sources. Some of the alkaloids discussed have been isolated from documented toxic plants, while others, although isolated from apparently nontoxic species, possess pharmacological activity. Other indolizidines may simply be associated with plants which have a documented medicinal usage, such as the dendrobium alkaloids dendroprimine (**5**) and crepidamine (**6**) isolated from *Dendrobium teretifolium* (Orchidaceae).[2,3]

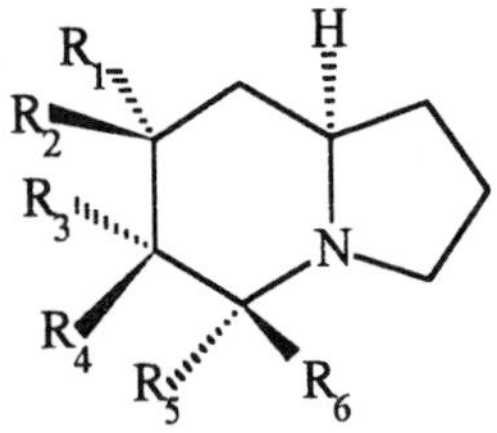

	5	**6**
R_1	CH_3	H
R_2	H	CH_3
R_3	H	Phenyl
R_4	H	OH
R_5	CH_3	CH_2COCH
R_6	H	H

It should be noted that several types of toxic indolizidines have been isolated from animal sources such as the pharaoh ant (e.g., the pheromone, monomorine-I, **7**),[1,2,4] amphibians (e.g., gephyrotoxin, **8**),[1,2] and the marine tunicate *Clavelina picta* (e.g., piclavine A, **9**).[4] Whether these alkaloids are true animal metabolites or are of dietary origin has been under investigation.[5] Indeed some alkaloids from the frog *Dendrobates auratus* have been shown to be derived from insects found in leaf litter on the forest floor.[6] Although there is no reported evidence, it is further conceivable that these insects in turn derive the alkaloids from the plants they consume.

The methods of detection, isolation, structural determination, biosynthesis, and synthesis of the various classes of naturally-occurring indolizidine alkaloids are outside the scope of this review. However, deserving of special mention are the polyhydroxylated indolizidines which are very water soluble and require a different approach to isolation than the normal Stas-Otto, acid/base extraction method for alkaloids. First brought to prominence by the discovery of swainsonine (**3**),[7] Nash et al. suggested that these water-soluble polyhydroxyalkaloids could be more widespread but overlooked by conventional alkaloid detection/isolation

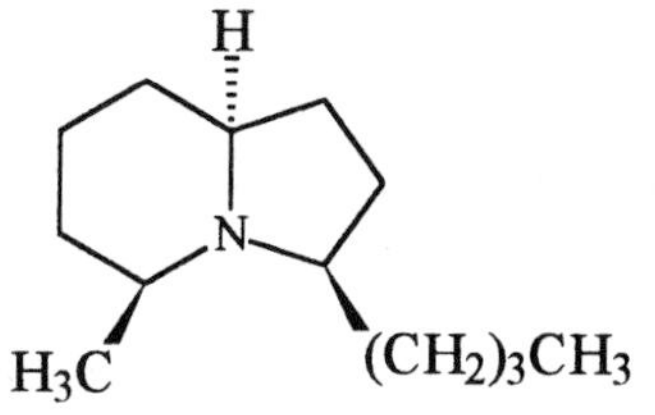

H

H₃C (CH₂)₃CH₃

7

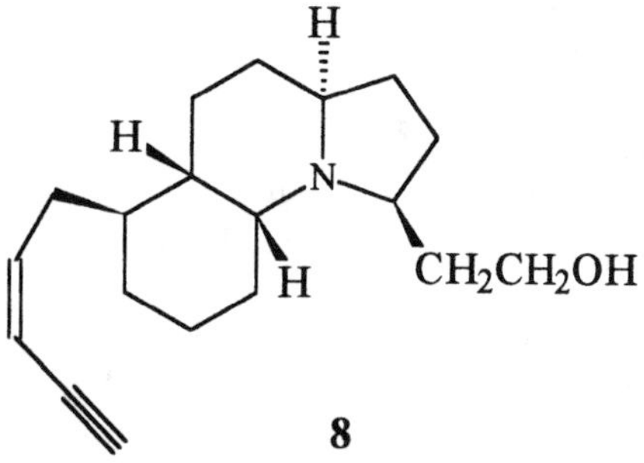

H

H H

CH₂CH₂OH

8

H

(CH₂)₆CH₃

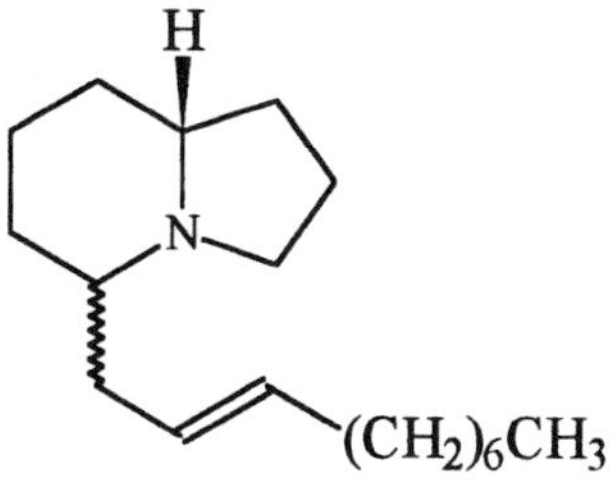

9

methods.[8] More recently, Molyneux discussed these water soluble "cryptic" alkaloids and reviewed successful extraction and isolation procedures used by various workers.[9]

1.2 INDOLIZIDINES FROM PLANTS

1.2.1 SWAINSONINE AND RELATED INDOLIZIDINES

Swainsonine (**3**), the first water-soluble, polyhydroxylated indolizidine to be reported, was first isolated from, and named after, the toxic Australian legume *Swainsona canescens*.[7]

Ingestion of certain species of the *Swainsona* genus by grazing livestock induced a wasting, locomotor syndrome which was shown biochemically and histopathologically to be a consequence of an induced mannosidosis in which the lysosomal glycosidase, α-mannosidase, was inhibited.[10,11] This enzyme inhibition formed the basis of an *in vitro* assay for the presence of the toxin in extracts of *S. canescens* and in subsequent isolation and purification steps. The occurrence, isolation, and biological activities of swainsonine have been discussed at an international symposium and have been extensively reviewed.[12–14] Following the isolation of swainsonine, it was predicted that similar glycomimetic compounds with different stereochemistries may be effective inhibitors of other glycosidases.[7] Indeed, the isolation of swainsonine heralded a new era in the detection and isolation of these water soluble alkaloids, recognition of them as inhibitors of oligosaccharide-processing enzymes, investigations into associated biological and therapeutic effects, and investigations in the chemical synthesis of these compounds, their stereoisomers, and related compounds.

The potential medical uses of swainsonine have been extensively studied and well reviewed.[15,16] Briefly, however, swainsonine has potential as a modulator of the immune system as a consequence of its ability to stimulate interleukin 2 production by T-helper cells, and consequent proliferation of T-lymphocytes. It is also effective as an anti-metastasis agent, inhibiting the spread of some tumors, possibly due to its activity in altering oligosaccharide processing by inhibiting complex glycosidation of glycoproteins. Indeed, swainsonine has achieved success in clinical trials with terminally ill patients with advanced malignancies.[17]

Subsequent to the isolation of swainsonine from *S. canescens*, it was also isolated from *Astragalus lentiginosus* (spotted locoweed) and has been detected in other *Swainsona*, *Astragalus*, and *Oxytropis* spp. from Australia, North and South America, and China.[18–21] More recently, swainsonine has been isolated from *Ipomoea* sp. Q6(aff. *calobra*) by Molyneux et al.[22]

Not restricted to plant sources, swainsonine has been isolated from the fungi *Rhizoctonia leguminicola* (see Section **1.3.1**)[23] and *Metarhizium anisopliae*.[24,25] The potential cancer-related therapeutic effect of swainsonine has stimulated research into increasing production from natural sources by tissue culture[26] as well as trying to optimize production by fungal culture.[25]

The N-oxide of swainsonine (**10**) has been isolated from some *Astragalus* and *Oxytropis* species.[18,27] Two diastereoisomers of 8-dehydroxyswainsonine, lentiginosine (**11**), and 2-*epi*lentiginosine (**12**), have been isolated from *A. lentiginosus*.[28] The latter had previously been synthesized to examine its biological activity,[29] and was isolated as an intermediate on the biosynthetic pathway to swainsonine.[30]

1.2.2 CASTANOSPERMINE AND RELATED INDOLIZIDINES

Ingestion of seeds of an Australian tree, the Moreton Bay Chestnut (*Castanospermum australe*, Leguminosae), by cattle and horses induces an often fatal toxic syndrome characterized by gastroenteritis. Toxicity to humans has also been recorded.[31]

Castanospermine, 1,6,7,8-tetrahydroxyoctahydroindolizine (**13**), was isolated from the seeds of *C. australe* and detected in the leaves and bark.[8,32] It has also been isolated or detected in a number of *Alexa* species, a leguminous tree which is native to South and Central America.[33]

In addition to castanospermine, 7-deoxy-6-*epi*castanospermine (**14**),[34] 6-*epi*castanospermine (**15**),[35] and 6,7-di*epi*castanospermine (**16**)[36] have also been isolated from *C. australe*.

As with the glycomimetic swainsonine and related alkaloids, the biological activities of castanospermine and related indolizidines also involve the inhibition of glycosidases.

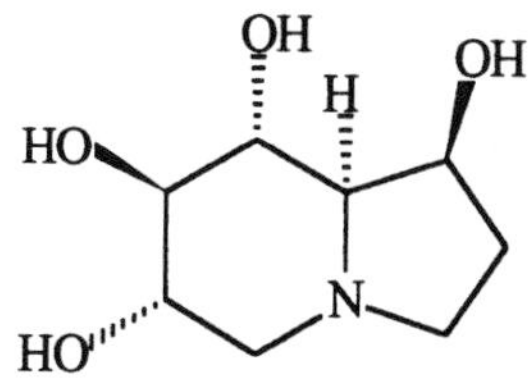

13

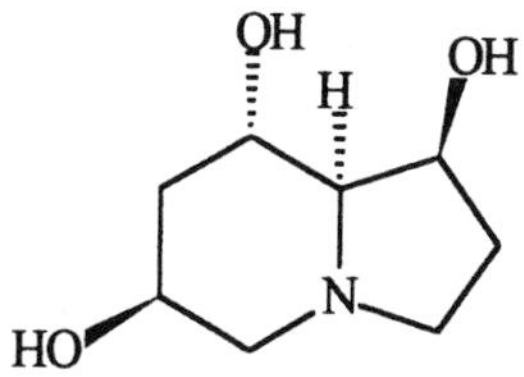

14

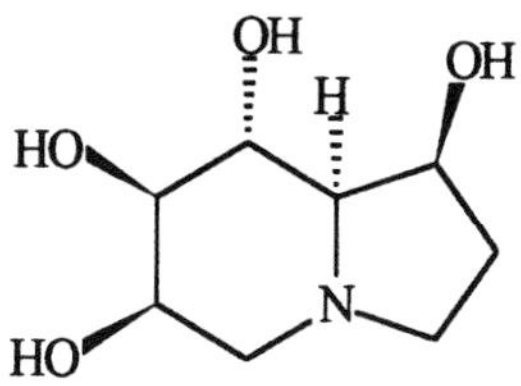

15

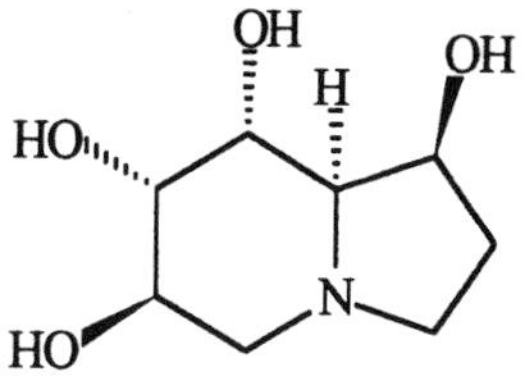

16

Additionally, castanospermine has demonstrated some degree of *in vitro* anti-Human Immu-nodeficiency virus activity. Consequently, there has been extensive effort directed at investi-gating castanospermine and its analogues.[1,37]

1.2.3 IPOMOEA ALKALOIDS

Species of the genus *Ipomoea* (Convolvulaceae) have a long history of use in magico-religious ceremonies due to the presence of hallucinatory alkaloids of the ergoline type (e.g., related to the synthetic LSD). Toxicity problems with livestock grazing species of *Ipomoea* have also been reported.[31] Recently, the trihydroxyindolizidine alkaloid, swainsonine (**3**), was isolated from *Ipomoea* sp. Q6 [aff. *calobra*] (Section **2.1**).[22] The first simple indolizidine alkaloids to be reported from *I. alba* were (+)-ipalbidine (**17**) and (+)-ipalbine (**18**).[38] Ipomine (**19**) was later isolated from *I. muricata*.[39] As can be seen from the structures, both ipalbine and ipomine are glucosides of ipalbidine.

1.2.4 ASTRAGALUS POLYCANTHUS

Indian researchers isolated three indolizidine alkaloids from the leguminous plant *Astragalus polycanthus*. A didehydroindolizidine, 6(H)-(9-isopropylidene-9-hydroxy)-6-methyl-1,2,3,5,7-pentahydroindolizine (polycanthisine, **20**) was isolated and identified by spectro-scopic means.[40] However, acceptance of this proposed structure should be treated with caution since, not only is the structure incorrectly named as a hexahydroindolizine, and evidence for a *trans* ring junction is given for a situation where there is no bridgehead hydrogen, there is insufficient presentation of NMR data in the report to assess the interpretation. Further NMR experiments or unambiguous synthesis are necessary to confirm the proposed structure and to determine relative and absolute stereochemistry.

The arylindolizidine alkaloids polycanthine (**21**) and polycanthidine (**22**) were also iden-tified by NMR spectroscopic and mass spectrometric methods,[41] but again the rationale was reported to be rather perfunctory and thus should be treated with caution until confirmed by

	R
17	H
18	β-glucosyl
19	β-(6-O-p-coumaryl)glucosyl

20

	R$_1$	R$_2$	R$_3$	R$_4$	R$_5$	R$_6$
21	OH	OH	H	H	CH$_3$	H
22	H	H	OH	OH	p-CH$_3$OPh	CH

further NMR experiments or unambiguous synthesis.[1] The stereochemistry of either compound is uncertain and both compounds are readily oxidized to lactams

Species of the genus *Astragalus* have been associated with poisonings around the world.[42,43] Thus, although no biological activity has been reported for any of these *A. polycanthus* alkaloids, they are presented here as indolizidine alkaloids isolated from a species of a documented toxic plant genus.

1.2.5 Lupin Alkaloids

Toxicity problems involving teratogenic effects can occur when stock ingest alkaloid-containing lupins.[44] The lupin alkaloids so far identified as being teratogenic are of the quinolizidine type.[45] Recent studies on lupin alkaloids from *Maackia tashiroi*, a leguminous plant native to Japan, have identified several indolizidine compounds related to the more typical quinolizidine lupin alkaloids.[46,47] Thus, tashiromine (8S-hydroxymethyl-8aR-octahydroindolizine, **23**) and (-)-camoensidine (**24**) are indolizidine analogues of *epi*lupinine (**25**) and lupanine (**26**), respectively. Another alkaloid (**27**) was identified as the N$_{15}$-oxide of (-)-camoensidine by a combination of spectroscopic and chemical methods.

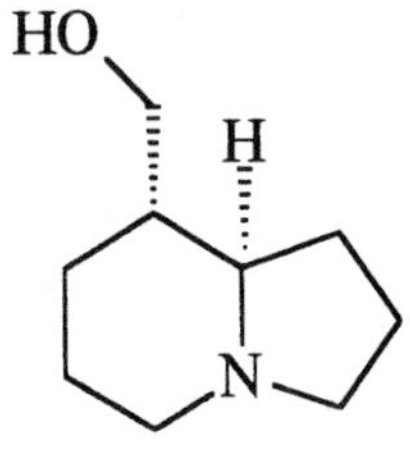

23

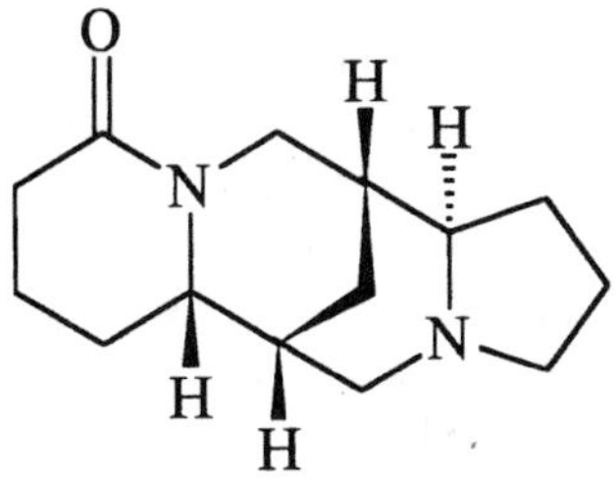

24

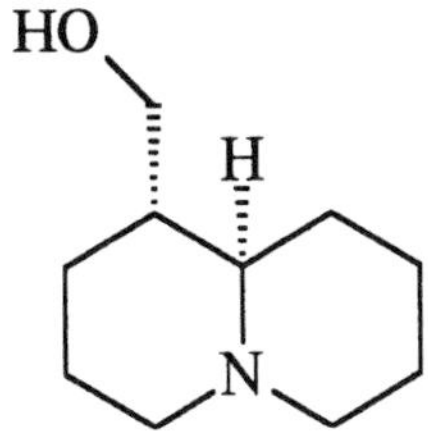

25

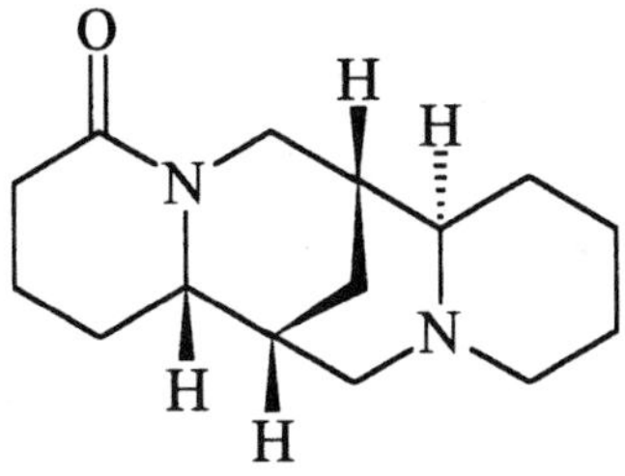

26

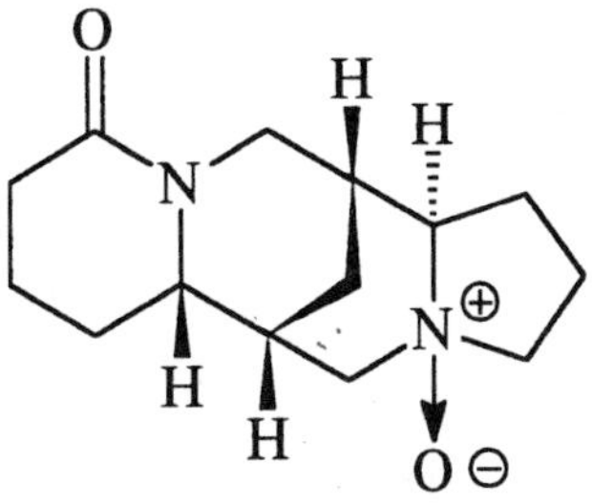

27

1.2.6   *SOLANUM*-DERIVED STEROIDAL INDOLIZIDINES

There are several classes of steroidal alkaloid isolated from *Solanum* spp., such as the solanidanes, spirosolanes, spirostanes, solanocapsines, and epiminocholestanes.[48] The *Solanum*-derived steroidal indolizidines are of the solanidane (**28**) type. The toxic effects of steroidal alkaloids and steroidal glycoalkaloids isolated from *Solanum* species have been well reviewed and include teratogenic effects, necrosis of intestinal and gastric mucosa, acetyl-cholinesterase inhibition, and gastrointestinal irritation resulting in diarrhea, vomiting, nausea, stomach cramps, headache, and dizziness.[48–52]

Over 30 species of *Solanum*, reported to contain alkaloids, are used by humans as food.[52] This fact, in light of the documented toxic effects of *Solanum* species toxicity, has prompted much research into the clinical and pathologic effects of intoxication and into structure-activity relationships. Baker, Gaffield, and Keeler reviewed the toxicology of *Solanum*-derived steroidal alkaloids, the teratogenic components of *Solanum* species, and the mode of action of these glycosteroidal alkaloids.[49–51] A more recent appraisal of the structure requirements for teratogenicity was presented by Gaffield and Keeler at the 4th International Symposium on Poisonous Plants.[52]

Early investigations linked the toxicity of potatoes (*Solanum tuberosum*) with high concentrations of "solanine" in the spouts, leaves, and flowers of the plant, and in the green skins of tubers and in spoiled tubers. "Solanine" was subsequently shown to be a mixture of solanidine (**29**) glycosides, i.e., α-solanine (**30**) and α-chaconine (**31**). In a study with pregnant hamsters, these two alkaloids induced a high incidence of exencephaly and encephalocele in the offspring when the dams were gavaged with the glycoalkaloids on day 8 of gestation.[53] The consequences of the toxicity of steroidal glycoalkaloids for potato consumption and breeding has been reviewed by Van Gelder.[48]

Like α-solanine and α-chaconine, leptine (**32**), a 23-acetoxy derivative of α-chaconine isolated from the foliage of *S. tuberosum*, is also a strong inhibitor of human plasma and eel acetylcholinesterase.[54] It is this effect on the nervous system which may account for the clinical effects such as drowsiness, confusion, apathy, weakness, depression, and unconsciousness.

	R	R$_2$
28	H	H
29	OH	H
30	O-solatriose	H
31	O-chacotriose	H
32	O-chacotriose	OAc

Other *Solanum*-derived steroidal indolizidines include solanthrene (**33**), and various glycosides of solanidine (**29**), rubijervine (**34**), isorubijervine (**35**), leptinidine (**36**), acetylleptinidine (**37**), and demissidine (**38**).[48]

1.2.7   PHENANTHROINDOLIZIDINES

The early investigative work on the isolation, synthesis, and biosynthesis of phenanthroindolizidines has been well reviewed,[2] and the newer developments have been reviewed on a regular basis.[1]

The phenanthroindolizidines vary in stereochemistry and hydroxylation or methoxylation. Thus antofine (**39**) is a demethoxy analogue, and tylophorinine (**40**) is a demethoxy-14S-hydroxy analogue of tylophorine (**41**).[2] N-oxides have also been reported such as the *cis* and *trans* N-oxides of antofine (**39**) isolated in a 4:1 ratio from *Vincetoxicum hirudinaria*,[55] and the N-oxide of 14R-hydroxyantofine (**42**) from the Chinese plant *Cyanchum komarovii*.[56] Quaternary phenanthroindolizidines, dehydrotylophorine (**43**), anhydrodehydrotylophorinine (**44**), and anhydrodehydrotylophorinidine (**45**), have been apparently isolated from *Tylophora indica* but may in fact be oxidative artefacts of the isolation procedure.[57] Septicine (**46**) and the tyloindicines F (**47**) and I (**48**) are *seco*phenanthroindolizidines, varying in the pattern of unsaturation of the indolizidine system, and they have been isolated from *T. indica*.[58]

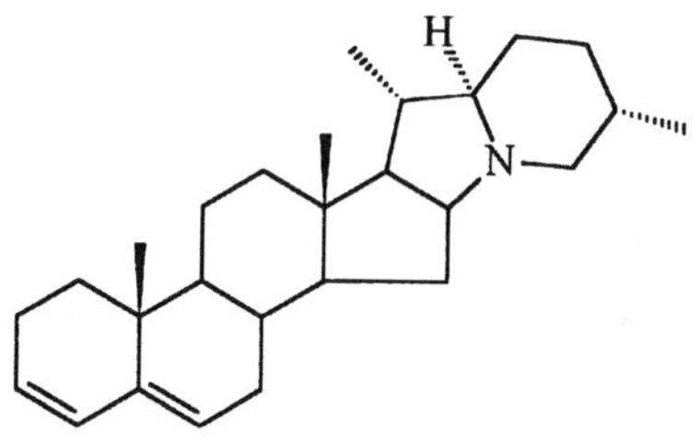

33

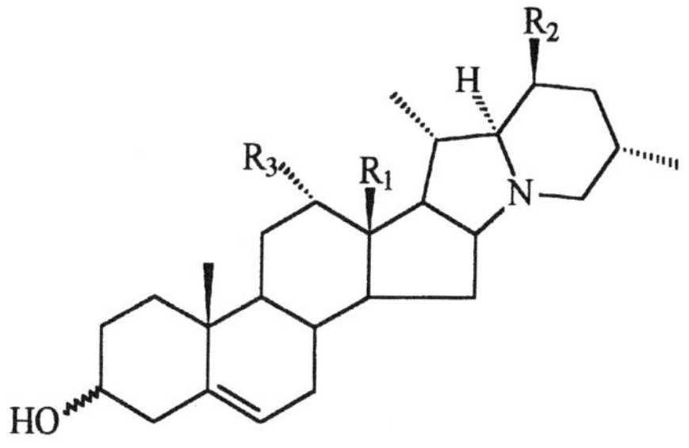

	R_1	R_2	R_3
34	CH$_3$	H	OH
35	CH$_2$OH	H	H
36	CH$_3$	OH	H
37	CH$_3$	OAc	H

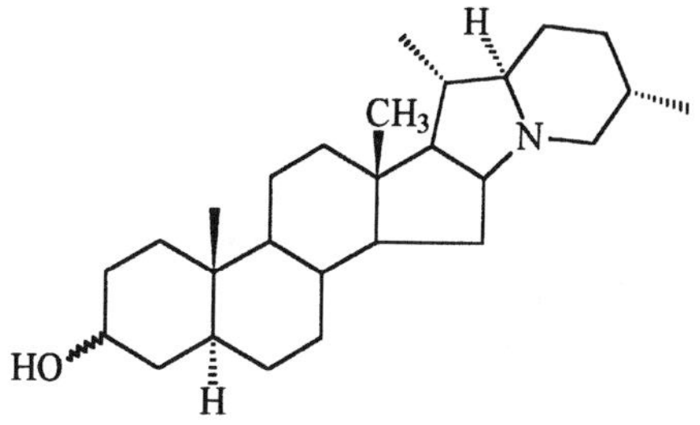

38

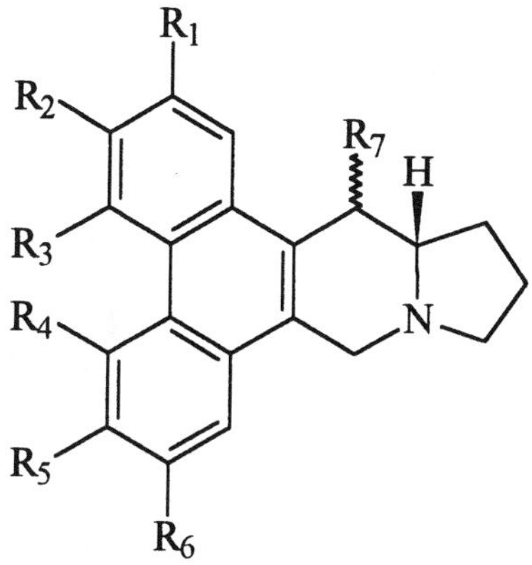

	R_1	R_2	R_3	R_4	R_5	R_6	R_7
39	OCH$_3$	OCH$_3$	H	H	OCH$_3$	H	H
40	H	OCH$_3$	H	H	OCH$_3$	OCH$_3$	OH
41	OCH$_3$	OCH$_3$	H	H	OCH$_3$	OCH$_3$	H
42	OCH$_3$	OCH$_3$	H	H	OCH$_3$	H	OH
49	OCH$_3$	OCH$_3$	H	OCH$_3$	OCH$_3$	H	H
50	H	OCH$_3$	H	H	OH	OCH$_3$	OH

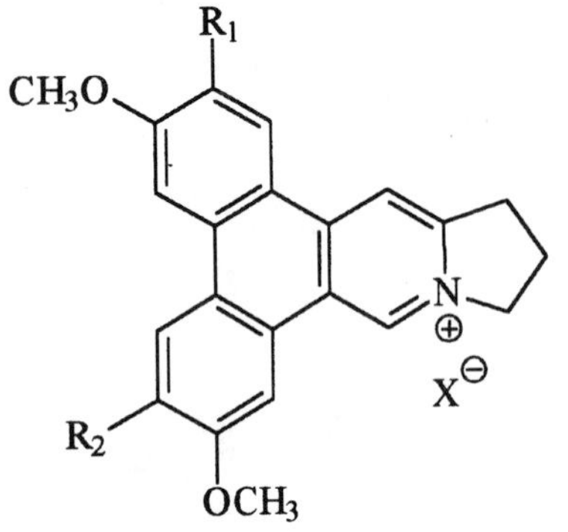

	R_1	R_2
43	OCH$_3$	OCH$_3$
44	H	OCH$_3$
45	H	OH

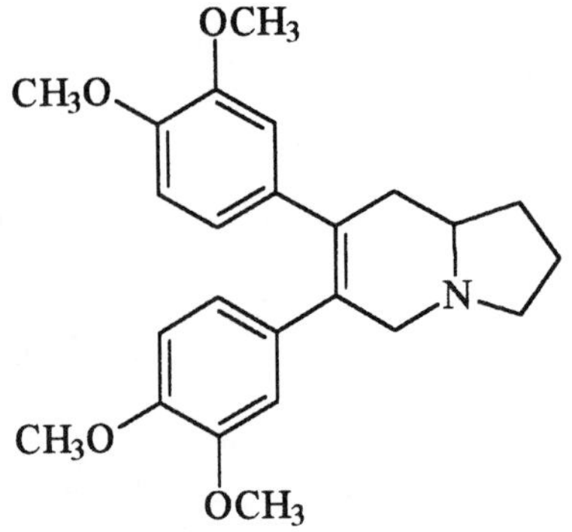

46

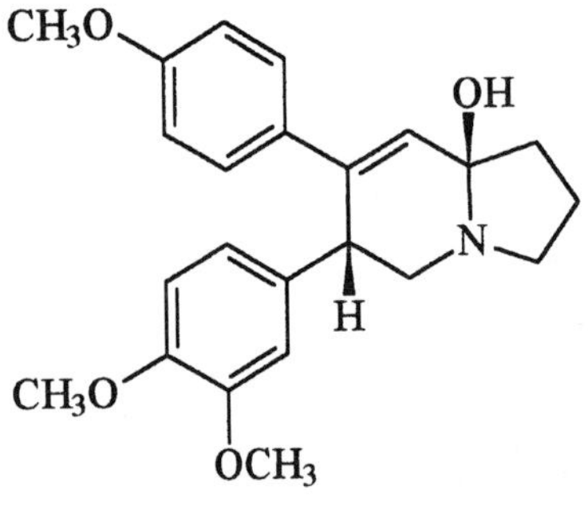

47

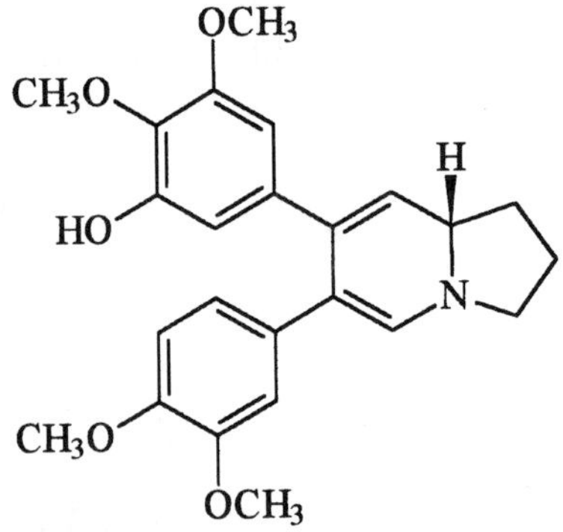

48

Of the many phenanthroindolizidines and secophenanthroindolizidines isolated from species of Asclepiadaceae (*Tylophora*, *Vincetoxicum*, *Cyanchum*, *Pergularia*, and *Antitoxicum*) and Moraceae (*Ficus*), none have been reported as imparting toxicity to the source plant. However, many have been investigated and shown to possess pharmacologic activity which, in overdose, can be detrimental or fatal.

In his review, Gellert indicated that the pharmacologic activity of extracts of phenanthroindolizidine-containing plants, such as *Tylophora* species, has been known for over 100 years.[2] The plants possess emetic and vesicant activity and the alkaloids have been shown to affect the circulatory and respiratory systems and to exert a depressant effect on the heart muscles. Therapeutic benefits of *Tylophora asthmatica* leaves for the treatment of asthma and other allergic disorders have been reported. Extracts of *T. indica* leaves and stems, and the alkaloids (-)-tylocrebine (**49**) from *Tylophora crebifolia* and 14S-tylophorinidine (**50**) from *T. indica*, have been shown to possess antileukemic activity. Antofine (**39**) shows significant antibacterial and antifungal activity.[59]

(+)-Tylocrebine (the enantiomer of **49**), the major alkaloid isolated from *Ficus septica*, caused depression, hypotonia, ataxia, and death in mice at an oral dose of 25 mg kg^{-1} while a twice daily dose of 10 mg kg^{-1} p.o. killed rats within 3 days. Both tylocrebine and a minor alkaloid from *F. septica*, tylophorine (**41**), are active against lymphoid leukemia.[59]

1.2.8 *ELAEOCARPUS* ALKALOIDS

Species of *Elaeocarpus* (Elaeocarpaceae) from Papua, New Guinea contain the simple acylated indolizidines, elaeokanine C (**51**), elaeokanine A (**52**), and isoelaeocarpicine (**53**), the flavanoid-related indolizidines elaeocarpiline (**54**), elaeocarpine (**55**), isoelaeocarpiline (**56**), and isoelaeocarpine (**57**), and the complex indoloindolizidine elaeocarpidine (**58**).[60–62]

	R$_1$	R$_2$
54	αH	βH
56	βH	αH

	R
55	αH
57	βH

1.2.10 Glochidion Alkaloids

The n-hexylated pyridino- and pyrimidinoimidazoles (**67**) and (**68**) were isolated from the Papuan New Guinea plant *Glochidion philippicum* (Euphorbiaceae) which has been associated with antifertility activity. Glochidicine (**67**) and glochidine (**68**) can, at an oral dose of 100–300 mg kg^{-1}, cause depression, low posture, dyspnea, tremors, cyanosis, asphyxial convulsions, and death in mice.[59]

67   **68**

1.3 FUNGAL INDOLIZIDINES

The production of polyhydroxylated indolizidines from fungal sources (*Rhizoctonia leguminicola* and *Metarhizium anisopliae*) has previously been mentioned with respect to swainsonine and related compounds (Section **1.2.1**).

1.3.1 Slaframine

The water soluble, unstable (-)-1S-acetoxy-6S-amino-8aS-indolizidine, slaframine (**69**) was isolated from *Rhizoctonia leguminicola*, a fungus that infects red clover and other leguminous forage crops.[63,64] The infected red clover develops "black patch" disease with consequent loss of seed crop. Ingestion of infected red clover by livestock causes a "slobbering" disease syndrome characterized by excessive salivation, lacrymation, anorexia, diarrhea, and feed refusal.

69   **70**

The structure, initially misassigned on chemical degradation and mass spectrometric considerations, was elucidated by proton nuclear magnetic resonance (^{1}H-NMR) spectroscopy involving spin decoupling.[65] The structure has since been confirmed by several unambiguous syntheses.[4,14]

Studies have indicated that slaframine per se may not be the toxin, but that hepatic oxidation to the ketoimine (**70**) may be necessary to initiate the salivation syndrome. Since administration of atropine prior to dosing with slaframine prevents the excess salivation whereas atropine administration post-dosing does not reverse the salivation effect, it appears as

28. Pastuszak, I., Molyneux, R. J., James, L. F., and Elbein, A. D., Lentiginosine, a dihydroxyindolizidine alkaloid that inhibits amyloglucosidase, *Biochemistry*, 29, 1886, 1990.

29. Colegate, S. M., Dorling, P. R., and Huxtable, C. R., The synthesis and biological activity of (±)-(1α,2α,8aα)-indolizidine-1,2-diol, *Austr. J. Chem.*, 37, 1503, 1984.

30. Harris, T. M., Harris, C. M., Hill, J. E., and Ungemach, F. S., (1S,2R,8aS)-1,2-Dihydroxyindolizidine formation by *Rhizoctonia leguminicola*, the fungus that produces slaframine and swainsonine, *J. Org. Chem.*, 52, 3094, 1987.

31. Everist, S. L., in *Poisonous Plants of Australia, Revised Edition*, Angus and Robertson, Melbourne, 1981.

32. Hohenschutz, L. D., Bell, A. E., Jewess, P. J., Leworthy, D. P., Pryce, R. J., Arnold, E., and Clardy, J., Castanospermine, a 1,6,7,8-tetrahydroxyoctahydroindolizine alkaloid from seeds of *Castanospermum australe*, *Phytochemistry*, 20, 811, 1981.

33. Nash, R. J., Fellows, L. E., Dring, J. V., Stirton, C. H., Carter, D., Hegarty, M. P., and Bell, A. E., Castanospermine in *Alexa* species, *Phytochemistry*, 27, 1403, 1988.

34. Molyneux, R. J., Tropea, J. E., and Elbein, A. D., 7-Deoxy-6*epi*castanospermine, a trihydroxyindolizidine alkaloid glycosidase inhibitor from *Castanospermum australe*, *J. Nat. Prod.*, 53, 609, 1990.

35. Molyneux, R. J., Roitman, J. N., Dunnheim, G., Szumilo, T., and Elbein, A. D., 6-*Epi*castanospermine, a novel indolizidine alkaloid that inhibits α-glucosidase, *Arch. Biochem. Biophys.*, 251, 450, 1986.

36. Molyneux, R. J., Pan, Y. T., Tropea, J. E., Benson, M., Kaushal, G. P., and Elbein, A. D., 6,7-Di*epi*castanospermine, an indolizidine alkaloid glycosidase inhibitor from *Castanospermum australe*, *Biochemistry*, 30, 9981, 1991.

37. Molyneux, R. J., Isolation, characterization and analysis of polyhydroxy alkaloids, *Phytochemical Analysis*, 4, 193, 1993.

38. Gourley, J. M., Heacock, R. A., McInnes, A. G., Nikolin, B., and Smith, D. G., Structure of ipalbine, a new hexahydroindolizine alkaloid isolated from *Ipomoea alba*, *Chem. Comm.*, 709, 1969.

39. Dawidar, A. M., Winternitz, F., and Johns, S. R., Structure of ipomine, a new alkaloid from *Ipomoea muricata* Jacq., *Tetrahedron*, 33, 1733, 1977.

40. Gupta, R. K., Singh, J., and Santani, D. D., Polycanthisine, a new indolizidine alkaloid from *Astragalus polycanthus* Royle (Leguminosae), *Ind. J. Chem.*, 34B, 76, 1995.

41. Gupta, R. K., Singh, J., and Santani, D. D., Polycanthine, a new indolizidine alkaloid from *Astragalus polycanthus*, *Ind. Drugs*, 30, 651, 1993.

42. Various contributors to the 3rd International Symposium on Poisonous Plants, in *Poisonous Plants*, James, L. F., Keeler, R. F., Bailey Jr., E. M., Cheeke, P. R., and Hegarty, M. P., Eds., Iowa State University Press, Ames, 1992.

43. Various contributors to the 4th International Symposium on Poisonous Plants, in *Plant-Associated Toxins: Agricultural, Phytochemical, and Ecological Aspects*, Colegate, S. M. and Dorling, P. R., Eds., CAB International, Wallingford, U.K., 1994.

44. Finnell, R. H., Gay, C. C., and Abbott, L. C., Teratogenicity of rangeland lupines: the crooked calf disease, in *Handbook of Natural Toxins, Volume 6: Toxicology of Plant and Fungal Toxins*, Keeler, R. F. and Tu, A. T., Eds., Marcel Dekker, New York, 1991, Chap. 2.

45. Panter, K. E., Ultrasound imaging: a bioassay technique to monitor fetotoxicity of natural toxicants and teratogens, in *Bioactive Natural Products: Detection, Isolation and Structural Determination*, Colegate, S. M. and Molyneux, R. J., Eds., CRC Press, Boca Raton, 1993, Chap. 20.

46. Ohmiya, S., Kubo, H., Otomasu, K., Saito, K., and Murakoshi, I., Tashiromine: a new alkaloid from *Maakia tashiroi*, *Heterocycles*, 30, 537, 1990.

47. Ohmiya, S., Kubo, H., Nakaaze, Y., Saito, K., Murakoshi, I., and Otomasu, K., (-)-Camoensidine N-oxide; a new alkaloid from *Maakia tashiroi*, *Chemical and Pharmaceutical Bulletin*, 39, 1123, 1991.

48. Van Gelder, W. M., Steroidal glycoalkaloids in *Solanum*: consequences for potato breeding and for food safety, in *Handbook of Natural Toxins, Volume 6: Toxicology of Plant and Fungal Toxins*, Keeler, R. F. and Tu, A. T., Eds., Marcel Dekker, New York, 1991, Chap. 7.

49. Baker, D. C., Keeler, R. F., and Gaffield, W., Toxicosis from steroidal alkaloids of *Solanum* species, in *Handbook of Natural Toxins, Volume 6: Toxicology of Plant and Fungal Toxins*, Keeler, R. F. and Tu, A. T., Eds., Marcel Dekker, New York, 1991, Chap. 5.

50. Keeler, R. F., Baker, D. C., and Gaffield, W., Teratogenic *Solanum* species and the responsible teratogens, in *Handbook of Natural Toxins, Volume 6: Toxicology of Plant and Fungal Toxins*, Keeler, R. F. and Tu, A. T., Eds., Marcel Dekker, New York, 1991, Chap. 6.

51. Gaffield, W., Keeler, R. F., and Baker, D. C., *Solanum* glycoalkaloids: plant toxins possessing disparate physiologically active structural entities, in *Handbook of Natural Toxins, Volume 6: Toxicology of Plant and Fungal Toxins*, Keeler, R. F. and Tu, A. T., Eds., Marcel Dekker, New York, 1991, Chap. 8.

52. Gaffield, W. and Keeler, R. F., Plant steroidal alkaloid teratogens: structure-activity relations and implications, in *Plant-Associated Toxins: Agricultural, Phytochemical and Ecological Aspects*, Colegate, S. M. and Dorling, P. R., Eds., CAB International, Wallingford, U.K., 1994, Chap. 60.

53. Renwick, J. H., Food and malformation, *Practitioner*, 226, 1947, 1982.

54. Bushway, R. J., Savage, S. A., and Ferguson, B. S., Inhibition of acetylcholinesterase by solanaceous glycoalkaloids and alkaloids, *Am. Potato J.*, 64, 409, 1987.

55. Lavault, M., Richomme, P., and Bruneton, J., New phenanthroindolizidine N-oxide alkaloids isolated from *Vincetoxicum hirundinaria* Medic., *Pharm. Acta Helvetica*, 68, 225, 1994.

56. Zhang, R., Fang, S., Chen, Y., and Lu, S., The chemical constituents in *Cyanchum komarovii* Al. Ilsinski, *Chem. Abstr.*, 117, 44583, 1991.

57. Govindachari, T. R., Viswanathan, N., Radhakrishnan, J., Charubala, R., Rao, N. N., and Pai, B. R., Quaternary alkaloids from *Tylophora asthmatica*, *Ind. J. Chem.*, 11, 1215, 1973.

58. Ali, M., Ansari, S. H., and Qadry, J. S., Rare phenanthroindolizidine alkaloids and a substituted phenanthrene, tyloindane from *Tylophora indica*, *J. Nat. Prod.*, 54, 1271, 1991.

59. Collins, D. J., Culvenor, C. C. J., Lamberton, J. A., Loder, J. W., and Price, J. R., in *Plants for Medicines*, and references therein.

60. Johns, S. R., Lamberton, J. A., and Sioumis, A. A., (+)-Elaeocarpiline and (-)-isoelaeocarpiline, new indolizidine alkaloids from *Elaeocarpus dolichostylis*, *Austr. J. Chem.*, 22, 793, 1969.

61. Johns, S. R., Lamberton, J. A., Sioumis, A. A., and Willing, R. I., *Elaeocarpus* alkaloids. I. Structures of (+)-elaeocarpine, (+)-isoelaeocarpine and (+)-isoelaeocarpicine, three new indolizidine alkaloids from *Elaeocarpus polydactus*, *Austr. J. Chem.*, 22, 775, 1969.

62. Johns, S. R., Lamberton, J. A., and Sioumis, A. A., Three new indolizidine alkaloids related to elaeocarpine and isoelaeocarpine, *Chem. Comm.*, 21, 1324, 1968.

63. Rainey, D. P., Smalley, E. B., Crump, M. H., and Strong, F. M., Isolation of a salivation factor from *Rhizoctonia leguminicola* on red clover hay, *Nature (London)*, 205, 203, 1965.

64. Aust, S. D. and Broquist, H. P., Isolation of a parasympathomimetic alkaloid of fungal origin, *Nature (London)*, 205, 204, 1965.

65. Gardiner, R. A., Rinehart, K. L., Jr., Synder, J., and Broquist, H. P., Slaframine. Absolute stereochemistry and a revised structure, *J. Am. Chem. Soc.*, 90, 5639, 1968.

66. Broquist, H. P., The indolizidine alkaloids, slaframine and swainsonine: contaminants in animal forages, *Annu. Rev. Nutr.*, 5, 391, 1985.

67. Freer, A. A., Gardner, D., Greatbanks, D., Poyser, J. P., and Sim, G. A., Structure of cyclizidine (antibiotic M146791): X-ray crystal structure of an indolizidinediol metabolite bearing a unique cyclopropyl side chain, *J. Chem. Soc., Chem. Comm.*, 1160, 1982.

68. Mynderse, J. S., Samlaska, S. K., Fukuda, P. S., Du Bus, R. H., and Baker, P. J., Isolation of A58365A and A58365B, angiotensin-converting enzyme inhibitors produced by *Streptomyces chromofuscus*, *J. Antibiot.*, 38, 1003, 1985.

69. Wong, P. L. and Moeller, K. D., Anodic amide oxidations: total syntheses of (-)-A58365A and (±)-A58365B, *J. Am. Chem. Soc.*, 115, 11434, 1993.

70. Reid, J., Are ACE inhibitors equal in congestive heart failure?, in *ACE Inhibitors, Suppl. S. Afr. Med. J.*, 1, 1993.

71. Leeper, F. J., Shaw, S. E., and Satish, P., Biosynthesis of the indolizidine alkaloid cyclizidine - incorporation of singly and doubly labelled precursors, *Can. J. Chem.*, 72, 131, 1994.

2 Glycoalkaloids

G. C. Percival and G. R. Dixon

CONTENTS

2.1 INTRODUCTION

Potatoes represent one of the major food crops worldwide, with total production increasing in developing and under-developed countries over the last 20 years[1] and a substantial increase in the consumption of processed convenience potato products such as crisps, chips, and tinned potatoes in developed countries recorded.[2] Glycoalkaloids are toxic nitrogenous steroidal glycosides synthesized by members of the Solanaceae, Asclepiadaceae, and Liliaceae.[3] While their families are of interest for their medical, toxicological, and pharmaceutical properties, only work related to glycoalkaloids in cultivated potato (*Solanum tuberosum* L.) will be discussed here since only these compounds are recorded as causing human death and livestock losses[4] and consequently are now regarded as one of the most toxic components in the human diet.[5]

2.2 OCCURRENCE AND COMPOSITION OF GLYCOALKALOIDS

The presence of solanine in potatoes was first reported in 1826[6] and later identified as a glycoside commonly referred to as solanidine.[7] Alpha chaconine was later found in *S. tuberosum* L. leaves and shoots and in foliage of *Solanum chacoense* in 1954.[8] Further work differentiated solanine as a mixture of two glycosides, solanines and chaconines.[9] Alpha solanine and chaconine represent 95% of total glycoalkaloids present in *S. tuberosum* L normally at concentrations of 10–150 mg kg^{-1} fresh weight (FW) of tuber.[10] The remaining 5% consists of β and γ forms of the two glycoalkaloids and the aglycone solanidine; although solanidine and β-chaconine concentrations ≥30% of total glycoalkaloids have been detected in potato peel.[11] Both alkaloids possess a common aglycone (solanidine) but differ in the type of carbohydrate moiety linked via a β-glycoside bond to carbon-3 of the steroid (see Chapter 1). The carbohydrate moiety of α-solanine consists of galactose, glucose, and rhamnose (β-solatriose), while that of α-chaconine is composed of glucose, rhamnose, and rhamnose (β-chacotriose).[12] In general, α-solanine represents 40% of the total glycoalkaloids found in potatoes and α-chaconine 60%[13] although varying ratios of these compounds have been recorded between cultivars.[14]

Because of resistance to biotic and abiotic stress, many wild tuber bearing *Solanum* species are used in potato breeding[15] or screened for potential use.[16] Species such as *S. chacoense* and *S. demissum* contain diverse arrays of glycoalkaloids associated with undesirable flavor, mammalian toxicity, and teratogenicity which may have emerged through hybridization.[10] Incorporation of these species into commercial potato has led to unacceptable increases in tuber glycoalkaloid concentrations,[17] and formation of glycoalkaloids other than α-solanine and α-chaconine.[18] Consequently, an objective of plant breeders is to produce potato genotypes with concentrations <200 mg kg^{-1} tuber FW, a recommended level for food safety. Objectives are hampered since heritability of glycoalkaloid synthesis is high, and is determined by genes under polygenic control.[19] Inheritance studies of tuber glycoalkaloids of segregating families from crosses between accessions of *S. chacoense* indicated that solanine and commersonine were determined by alternative codominant alleles at a single gene locus. A major gene for α-chaconine segregated independently of the solanine/commersonine locus and a recessive gene linked with the commersonine allele was epistatic to the chaconine gene causing the production of β-chaconine instead of α-chaconine.[20]

2.3 DISTRIBUTION IN THE PLANT

Glycoalkaloid concentrations commonly found in potato plants are shown in Table 2.1. Concentrations in the tuber decrease from the epidermis inwards with highest amounts in the periderm and outer parenchyma tissue, negligible concentrations in the pith, and small quantities in intermediate areas.

Higher concentrations are generally present in the bud rich rose end than in the stem tuber end.[21] Glycoalkaloids form in the periderm and cortical parenchyma cells of tubers and areas of high metabolic activity such as bud regions which at high concentrations diffuse throughout the entire tuber.[17]

Sites and cellular organelles responsible for glycoalkaloid synthesis, movement, and accumulation during tuber development are surprisingly unknown. Sectioning of potato and microscopic examination indicated sites of glycoalkaloid synthesis in the peridermal and cortical tuber regions and distribution mainly in and just below compact phellem cells,[31] while use of electron microscopy and cytochemistry localized glycoalkaloid precipitates in sprout tips, unsprouted meristematic tips, and tuber epidermal layers.[32] Subcellular distribution of potato glycoalkaloids following homogenization and centrifugation was highest in the soluble phase, microsomal fraction, mitochondrial phase, and lower fractions, respectively.[33]

TABLE 2.1
Distribution of Glycoalkaloids in Potato Plant

Plant organ	Glycoalkaloid concentration mg kg^{-1} FW	Ref.
Fruits	560–1080	22, 23
Aerial tubers	330–1500	24
Sprouts	2000–10000	23, 25, 26, 27
Flowers	3000–5000	26, 27
Stems	30–300	25, 26, 27
Leaves	400–1400	25, 26, 27
Stolons	150–540	27
Roots	180–850	25, 27
Petals	3060–4970	28
Calyces	4770–5710	28
Skin, 2–3% of tuber	300–600	26, 27, 29
Peel, 10–15% of tuber	150–300	26, 27, 29
Peel and eye, 3mm disk	300–500	26, 27, 29
Peel from bitter tuber	1500–2200	26, 27, 29
Flesh	12–50	26, 27
Whole tuber	10–180	25, 26, 27,
Bitter tubers	250–800	26, 27, 11
In vitro tissue culture	2100–6000	30

Seasonal changes in glycoalkaloid concentrations of whole plants demonstrated light independent *de novo* synthesis of solanine in plant organs, indicating destructive metabolization during growth. Higher glycoalkaloid contents were present in buds and young leaves down to about the eighth node, below which concentrations markedly decreased.[34] At flowering, glycoalkaloid concentrations were low in the tuber, stem, and petiole but high in roots, runners, leaves, and flowers. As the haulm died, glycoalkaloid concentration of the runners, roots, stems, and leaves decreased but altered only slightly in flowers.[35] Studies using reciprocal grafts of potato and tomato suggest that glycoalkaloids are not translocated throughout the plant.[36]

2.4   FACTORS AFFECTING GLYCOALKALOID FORMATION

2.4.1   CULTIVAR

Glycoalkaloid concentrations in commercial potato cultivars range from 10–150 mg kg^{-1} FW, well below the acceptable maximum of 200 mg kg^{-1} FW recommended for food safety.[10,37] Cultivars with naturally high glycoalkaloid concentrations generally produce excessive glycoalkaloids compared with those with naturally low concentrations when subjected to light or poor handling.[38] Tuber concentrations also vary widely among crops of single cultivars produced under various husbandry regimes,[39] and among tubers from a single plant.[34] Immature and small tubers tend to contain higher glycoalkaloid concentrations than mature ones[34] due to a higher surface to volume ratio and generally show greater accumulation rates following exposure to light.[23,40] An inverse relationship has been noted between tuber size and solanine concentration in some cultivars but no correlation between time of maturity and solanine content was observed in others[34]. Relatively little information describing glycoalkaloid concentrations present in plant foliage is available although high correlations have been identified between foliar and tuber concentrations.[41]

2.4.2   LIGHT

Synthesis of glycoalkaloids can be stimulated rapidly after harvest when tubers are exposed to illumination.[42,43] Since tubers are inevitably exposed to light (daylight, fluorescent, and incandescent) in the commercial processing and marketing chains, stimulation of glycoalkaloids presents a major food safety problem for producers, packers, suppliers, and vendors. The blue spectral portion (<500nm, especially ultra violet light <300nm) and infrared light (1300nm) are active elicitors of glycoalkaloid synthesis compared with other wavelengths. Light of 570–700 nm, (yellow-red light) enhances chlorophyll but not glycoalkaloid synthesis.[44,45] Higher light intensities generally result in increased glycoalkaloid synthesis; however, this relationship is not proportional[46] and higher intensities can inhibit synthesis.[47] Tubers exposed to light can synthesize glycoalkaloids when placed into dark storage,[29] and genotypes possessing similar initial concentrations can markedly differ in their rates of accumulation during light exposure. Consequently, selection of cultivars possessing slow glycoalkaloid accumulation rates should now be as essential as aiming for low initial concentrations.[48] Sunlight is a efficient elicitor of solanidine[11] and glycoalkaloid synthesis,[49] regardless of tuber skin pigmentation.[48] Reports of daylight increasing glycoalkaloid concentrations above those recommended for food safety in a matter of hours[50] have recently been disproved.[48,40] Light exposure can induce the synthesis of one glycoalkaloid to a greater degree than the other with the glycoalkaloid present at the lower concentration synthesized more rapidly than the glycoalkaloid present at the higher concentration.[24] These alterations in the proportions of glycoalkaloids may influence toxicity more than absolute concentrations,[51] which in turn may affect the 200 mg kg^{-1} FW food safety limit.

Recent studies indicated initial and subsequent changes in epidermal pigmentation in response to light exposure may absorb or reflect light wavelengths known to be potent elicitors of synthesis, markedly reducing glycoalkaloid accumulation rates.[52] Although exposure to light induces synthesis of nontoxic pigments such as chlorophyll,[53] whether the two biochemical processes are independent is an area of debate. Evidence exists for[42,54] and against a relationship.[12] Indeed, this relationship may be genotype dependent.[43] Associations between glycoalkaloid and other potato pigments such as anthocyanin and carotenoid synthesis have not been investigated. There is evidence that the degraded alkaloid skeleton may be used in pigment biosynthesis suggesting that *Solanum* alkaloids have an indirect role in the attraction of vectors for seed dispersal.[55]

2.4.3   LOCATION, CLIMATE, AND ENVIRONMENT

Significant variation in glycoalkaloid concentration between cultivars grown at different localities in Germany,[56] the U.S.,[39] the U.K.,[57] and Australia[37] have been reported with the recommendation that trials screening new cultivars require more than one season and site to be of any value. Contrary to this, the influence of location may be insignificant[58] and higher concentrations associated with conditions during transport.[49]

High glycoalkaloid concentrations in commercial tubers can be induced by adverse growing conditions which retard maturity such as a cool growing season accompanied by a high number of overcast days.[59] Similarly, tubers subjected to unfavorable climatic conditions such as nutritional imbalance, frost, or hail damage to plant foliage before maturity contained high concentrations.[60] Agronomic practices such as planting depth, level of hilling, date of planting with single and multiple eye pieces, and fertilizer rate were not associated with differences in solanine content.[60] Potatoes grown in hot dry climates contained more glycoalkaloids than those from cooler, higher altitudes or coastal temperate climates in Australia,[37] while higher concentrations were found in summer, autumn, and spring grown potatoes, respectively, in commercial cultivars grown in Pakistan.[61]

Reports of the influence of macronutrients, moisture, and soil organic content are conflicting. Glycoalkaloid concentrations of tubers grown on sludge amended soils were not significantly different from controls.[62] Sodium molybdate applied to the soil one day prior to planting decreased concentrations compared with controls and tubers stored for 5 months with greater decreases at the highest molybdenum concentration.[63] Excess application of nitrogen increased glycoalkaloid concentrations over controls in some cultivars[64] but resulted in lower tuber glycoalkaloid concentrations elsewhere.[65] Exposure of potato vines to ozone significantly reduced tuber glycoalkaloid concentrations of treated plants on a fresh weight but not dry weight basis. In a separate experiment, 18 day old plants exposed to ozone showed no significant differences in leaf glycoalkaloid concentrations compared with nontreated plants.[66]

2.4.4 STORAGE TEMPERATURE AND TIME

Storage temperature and time are significant factors which influence the rate of glycoalkaloid synthesis, but conflicting conclusions have been reached by several workers. In some instances, concentrations increased during storage,[42,64] while elsewhere no accumulation of glycoalkaloids could be detected.[67] Similarly, higher storage temperatures resulted in greater accumulation,[68] while other experimentation suggests the opposite.[69] Concentrations fluctuated widely at different storage temperatures,[70] and in relation to the physiological age of tubers.[71] Light-enhanced glycoalkaloid concentrations are not reduced in the parent tuber during cold storage in the dark.[68]

2.4.5 PHYSIOLOGICAL STRESS

Increased mechanization at harvest and during post-harvest handling can physically damage potato tubers inducing rapid glycoalkaloid accumulation[72] with greater accumulation at higher temperatures,[73,74] in the presence of light,[75,76] and within 15 days after damage.[77] Soaking tuber slices in water lowered concentrations and altered ratios of α-solanine: α-chaconine[78,67] but did not inhibit further glycoalkaloid accumulation during light exposure, with time and at higher temperatures.[79] Wounding tubers by slicing induced the formation of glycoalkaloids atypical of a particular cultivar.[67,80] Severe damage resulting in shatter cracks and splits (external damage) were responsible for the initiation of glycoalkaloid synthesis whereas less severe damage such as bruises and blackspot (internal damage) did not. High correlations between initial glycoalkaloid concentration and increased concentrations after damage have been found, highlighting the importance of selection for low glycoalkaloid clones to reduce accumulation during harvesting and handling.[59]

When compared with light and mechanical injury, physiological disorders such as hollow heart and blackheart are less potent factors stimulating glycoalkaloid synthesis, although tubers affected with these disorders contained higher concentrations in the cortical region than normal tubers with concentrations related to the severity of the disorder.[81]

Glycoalkaloid accumulation in tuber tissue can be suppressed when cut surfaces are inoculated with *Phytophthora infestans*, resulting from alteration of the biogenetic pathway of glycoalkaloids leading to a accumulation of terpenoids.[82] Suppression was greater when compatible races of *P. infestans* were used compared with incompatible races.[83] Accumulation of glycoalkaloid and rishitin in tubers appears due to *de novo* synthesis via the same acetate-mevolanate pathway, with increased accumulation of sesquiterpenoid phytoalexin suggested following diversion of biosynthesis at a branch point beyond mevalonate, probably at farnesyl-pyrophosphate.[84] Glycoalkaloids apparently do not inhibit rishitin accumulation per se since exogenous α-solanine applied to slices had little effect on accumulation rates. Accumulation of solanidine in tuber tissue has also been regarded as an hydrolytic product of glycoalkaloids where cellular disruption or tissue liquification was caused by *Erwinia atroseptica* or *P. infestans*.[85]

2.5   CONTROL OF GLYCOALKALOIDS

2.5.1   PHYSICAL TREATMENTS

Red, green, and violet filters reduced glycoalkaloid accumulation compared with tubers under clear cellophane filters (control), while yellow and blue filters produced slightly, but not significantly, less glycoalkaloids than controls. Only tubers placed beneath orange light contained concentrations higher than controls.[31] Curing tubers at 25°C for 10 days prior to storage reduced the responsiveness of tubers to photo-induced glycoalkaloid synthesis.[86] Mechanically damaged and cut cubes of potato tubers subjected to γ radiation produced less glycoalkaloids than control samples. Maximal inhibitory effects occurred with increasing radiation dosage; however, no effects on existing glycoalkaloids were reported.[87] Vacuum packing with polythene bags[88] and soaking in water prior to illumination effectively inhibited light and wound induced glycoalkaloid formation in some cultivars.[89] Current marketing trends and storage practices, however, limit practical use of these methods.

2.5.2   CHEMICAL TREATMENTS

Several chemical treatments prevent photo and wound induced glycoalkaloid synthesis in potato; however, results can be conflicting. In a glasshouse study, carbofuran in granular and flowable forms applied during tuberization resulted in significantly increased tuber glycoalkaloids.[90] Contrary to this, carbofuran failed to affect glycoalkaloid concentrations in other potato cultivars.[91] Increased glycoalkaloid formation during storage was prevented by a pre-harvest spray of maleic hydrazide[92] in disagreement with conclusions of other authors.[93] Depressed and stimulated glycoalkaloid concentrations in tubers soaked with nicotinic acid and gibberellic acid have been reported.[93] Foliar applications of N6-benzyladenine and 2-chloroethylphosphonic acid applied prior to harvest reduced photo-induced glycoalkaloid synthesis in harvested tubers.[31] The sprout inhibitor *iso*-propyl-N-(3-chlorophenyl)-carbamate (CIPC) inhibited synthesis in mechanically damaged tubers when applied as an emulsified water solution or fumigant, but showed no effect on existing and photo-induced glycoalkaloids. Reasons for this remain unclear.[94]

Treating tubers with paraffin wax at 140–160°C for 0.5 second inhibited light-induced glycoalkaloid synthesis, whereas heating tubers in air alone had no such effect.[95] In later studies, treatments with a range of oils at 22°C were equally effective in inhibiting synthesis, although rancidity of the oils and fats developed over time.[96] Dilutions of oil with acetone also inhibited synthesis without detrimental effects on tuber physiology. Acetone was recovered by passing tubers through warm air and condensing. Further studies on the effectiveness of chemical inhibition of photo-induced glycoalkaloid synthesis include references 97–98. Increased hypobaric pressure[99] and carbon dioxide[54] concentrations >15% during storage had no effect on glycoalkaloid formation.

2.6   GLYCOALKALOIDS IN PROCESSED POTATO

During processing, slices, cubes, mash, strings, strips, and shreds are frequently stored in relatively high light and temperature environments before use, allowing increased glycoalkaloid synthesis and accumulation. Although glycoalkaloids contribute to potato flavor,[56] they possess no known positive role in human nutrition.[100] At concentrations >140 mg kg⁻¹ FW, they impart a bitter taste to tubers while concentrations >220 mg kg⁻¹ FW result in a mild to severe burning in the mouth and throat[101]. Consequently, it has been suggested that

60–70 mg kg^{-1} FW should be considered the acceptable uppermost limit for concentrations in commercial potato cultivars and not 200 mg kg^{-1} FW.[56,23]

The glycoalkaloid concentration of dried potato by-product, pulp meals,[102] and several commercial potato products ranged from 0.4–979 mg kg^{-1} of product;[103] however, analysis of commercial potato peel products demonstrated that many of the cooked peels had glycoalkaloid concentrations two to eight times greater than the upper limit of 200 mg kg^{-1} FW.[104] Glycoalkaloids are not destroyed during food processing treatments such as boiling, baking, frying, and drying, even at high temperatures;[105] indeed, during various modes of cooking, glycoalkaloids migrated from the peel into the cortex.[106]

2.7 PHARMACOLOGY

The first attempt to establish the toxicity of glycoalkaloids using experimental animals failed to isolate either α-solanine or solanidine from the urine of a dog fed α-solanine.[107] Later experiments demonstrated the presence of a human plasma cholinesterase inhibitor in aqueous extracts of potato tissue.[108] Further experiments using dilute aqueous extracts of potato peel demonstrated that this naturally occurring inhibitor differentially inhibited three-serum cholinesterase phenotypes[109] and the anticholinesterase effect of potato juice had a direct relationship to solanine concentration of potato samples alleged to have caused poisoning.[110] This inhibitor was later shown to be α-solanine.[111] Toxicity is now known to result from anticholinesterase activity affecting the central nervous system and membrane disruption activity which affects the digestive system and general body metabolism.

The metabolic fate and distribution of titrated α-solanine and α-chaconine in mammals indicated partial and complete gastrointestinal hydrolysis, poor absorption from the gastrointestinal tract, rapid urinary and fecal excretion of metabolites, and accumulation of a high, but descending order of concentration in tissues such as spleen, kidney, liver, lung, fat, heart, brain, and blood.[112,113,114] Alpha solanine also possessed local anesthetic properties, induced hemolysis, diminished blood catalase, acted as a mitotic poison, and inhibited oxygen uptake in mouse ascites tumor cells.[26] The use of atropine sulfate, paragyline hydrochloride, and amphetamine sulfate to counteract α-solanine toxicity demonstrated that only a prior dose of atropine sulfate reduced the mortality associated with injection into mice.[115]

Several animal species have been tested for sensitivity to a total potato alkaloid extract as well as α-solanine.[26] Results demonstrated α-solanine was more toxic than the aglycone solanidine,[112] total glycoalkaloids were more toxic than α-solanine,[116] and α-chaconine was almost twice as toxic compared with α-solanine.[117] Recent conclusions indicated that the toxicological potency of α-chaconine would be about ten-fold higher than α-solanine.[118]

The cardiotonic activity of glycoalkaloids on frog hearts depended on the aglycone and number of sugars and not by the forms of sugars or their stereochemical configuration,[119] whereas toxicity of glycoalkaloids as evaluated using the Frog Embryo Teratogenesis Assay-Xenopus demonstrated that biological activity was influenced by the number, chemical structure of the carbohydrate, and stereochemical orientation.[114] With the exception of the anticholinesterase activity of leptine I[26], pharmacological and toxicological properties of glycoalkaloids produced by wild *Solanum* spp. remain largely unknown. This could prove to be a critical deficiency in our knowledge as such species are used increasingly in breeding programs.

In vitro studies reported α-solanine, β, and α-chaconine significantly inhibited bovine and human acetylcholinesterase at 100 μm independently of pH. Other solanaceous aglycones produced slight to negligible inhibition and combinations of solanine, chaconine, solasonine, and solamargine produced slightly antagonistic or noninteractive effects. Beta 2 chaconine proved as inhibitory as α-chaconine suggesting that the carbohydrate moiety is less important for inhibition. The slight effect of solanidine, however, suggested that the carbohydrate component makes significant contributions to anticholinesterase activity.[120] So far, no convincing evidence that potato glycoalkaloids pose a significant cancer risk has been presented while studies observing the potential of α-solanine and α-chaconine to induce mutations in *Salmonella typhimurium* also failed to offer convincing evidence of activity.[121] For further detailed reviews, see references 121 and 122.

2.8 TOXICOLOGY

2.8.1 MAMMALS

Consumption of potatoes containing high glycoalkaloid concentrations has caused severe illness and occasionally death.[123] Most cases have occurred in Europe but although poisoning was attributed to potatoes, glycoalkaloid concentrations were not determined.[2] Since then several cases of glycoalkaloid poisoning from potatoes have been reported.[124,125] Complaints from Canadian growers and consumers concerned with unpalatable and bitter tasting potatoes were traced back to high glycoalkaloid contents.[11] Similarly, losses of livestock and poultry by ingestion of potato haulm, sprouted potatoes, and potato peel which had been exposed to light when discarded by a processing plant or left in the field have been reported.[126] The last reported major outbreak of potato poisoning occurred in 1979 in school children.[127] Symptoms included headache, vomiting, diarrhea, and abdominal pain, and various neurological symptoms including apathy, drowsiness, mental confusion, rambling, incoherent stupor, hallucinations, dizziness, trembling and visual disturbances. Potatoes left over from the meal contained significant anticholinesterase activity, equivalent to their glycoalkaloid concentration. Thresholds inducing glycoalkaloid toxicity in humans have been estimated at 1–6 mg kg^{-1} body weight.[4,127,128] It has been suggested that severe birth defects could result from ingestion of imperfect potatoes with the alleged teratogen being possibly fungal synthesized coumarins, tuber synthesized phytoalexins, or steroidal alkaloids.[129] No information is available concerning the long term effects of glycoalkaloids from either animal or human trials involving repeated oral exposure. As potato glycoalkaloids are ingested by large numbers of the population on a routine basis over a lifetime, such fundamental toxicity testing will provide a greater understanding of their potential long term health effects. Such testing is routinely required for the majority of synthetic chemicals whether or not they are required for human consumption.[121]

2.8.2 INVERTEBRATES

Nematicidal activity on the free living saprophytic *Panagrellus redivivus* by α-chaconine was increasingly more toxic with decreasing acidity. Results indicated that the free base is the nematicidal form of the compound as toxic effects were proportional to the release of the free base.[130] Investigations of the resistance of potato cultivars to slug (*Deroceras reticulatum*) demonstrated the presence of glycoalkaloids and phenolics were essential for the expression of resistance. Concentrations of purified glycoalkaloids retarded slug growth at 67.5 mg/100 ml of diet and concentrations of 135mg/100ml of diet produced significant weight loss.[131] Field experiments showed potato tubers low in glycoalkaloids were more susceptible to attack by wireworms, *Agriotes obscurus*, than tubers containing higher concentrations with glycoalkaloids, the key factor predicting larval feeding.[132]

Glycoalkaloids possess repellent properties against the Colorado potato beetle, *Leptinotarsa decemlineata*, eliciting a bursting activity in galeal and tarsal chemosensilla of adult beetles and a similar response on galeal sensilla of larval beetles.[133] The presence of commersonine or dehydrocommersonine rather than α-solanine and α-chaconine as major foliar glycoalkaloids provided a greater degree of resistance demonstrating that the type of glycoalkaloids present in plant foliage may be as important as total glycoalkaloid concentration. Since dehydrocommersonine differs from solanine and chaconine only in the size and composition of the sugar moiety, results suggest that the number of sugar groups or presence/absence of a particular sugar may be more important than inclusion of a particular aglycone.[15] Light intensity during growth can significantly affect Colorado potato beetle resistance of leptine glycoalkaloid synthesizing *S. chacoense* plants[134] since increased light intensity stimulated a 2–4-fold rise in foliar leptines reducing larval development and increasing mortality. An increase in foliar α-solanine and α-chaconine in *S. tuberosum* L. plants was also recorded, however, no enhanced resistance resulted.

Glycoalkaloid concentrations in foliage of wild, tuber bearing *Solanum* species were inversely correlated against nymphal infestation by the potato leaf hopper, *Empoasca fabae* (Harris)[41] and positively effected nymphal survival and feeding behavior,[135] suggesting that foliar concentrations may significantly affect the defence of wild potato species against this pest. Contrary to this, glycoalkaloid concentration of foliage (solasonine and solamargine) and tubers (solamarines) were not correlated to insect resistance against *Myzus persicae*, *Empoasca fabae*, and *Epitrix cucumeris* in *S. berthaultii*. Glandular trichomes rather than glycoalkaloids provided the major element of resistance.[136]

Solanine fed to larvae of the tobacco hornworm, (*Manduca sexta*), indicated solanaceous alkaloids may contribute in the food selection of this pest,[137] whereas no relationship was found between the degree of resistance to the potato cyst nematode (*Globodera pallida*) and glycoalkaloid concentration of roots or tubers of potato clones derived from *S. vernei* × *S. tuberosum*. Nematode infestation neither led to increased glycoalkaloid concentrations of susceptible or resistant potatoes.[138]

2.8.3 PLANT PATHOGENS

Due to their antifungal and antifeedant properties, many workers have attributed the pathogen resistance character of potato cultivars to the presence of glycoalkaloids in the tuber and haulm.[139] Work *in vitro* demonstrated both α-solanine and α-chaconine fungitoxic against the mycelial stages of the pathogens *Helminthosporium carbonum*,[140] *Alternaria solani*,[141] *Fusarium solani*,[142] fungal spores of *Fusarium avenaceum*, *F. culmorum*, and sporangia of *P. infestans*.[143] Alpha solanine, α-chaconine, and solanidine proved toxic to *Fusarium coeruleum* (Lib.) Sacc spores at a LC_{50} of 20–36 mg, 11 mg, and 100 mg l^{-1}, respectively.[143] Toxicity was influenced by sodium:calcium ion ratios but remained unaffected in the presence of the individual ions. Spores of *Streptomyces scabies* and cells of *Bacillus subtilis*, *Micrococcus luteus*, *Erwinia*, and *Pseudomonas* spp., were unaffected by several hours exposure to α-solanine at 2000 mg l^{-1}. Inhibition of radial growth of *Alternaria solani* grown on potato dextrose agar was maximal in the presence of the aglycone, solanidine, followed by α-chaconine and α-solanine, respectively.[144]

The toxicity and biological activity of many glycoalkaloids is pH dependent,[143] with greater effects in alkaline environments due to dissociation into the active unprotonated form.[145,146] Certain pathogens can, however, detoxify glycoalkaloids via the production of extracellular enzymes which hydrolyze the sugar moiety,[147] reducing the sterol binding property[145] and subsequent toxicity.[141]

Work *in vivo* using replicated infected field plots to assess the influence of glycoalkaloids as resistance factors against the pathogens *Alternaria solani*, *Phytophthora infestans*, *Streptomyces scabies*, and *Verticillium albo-atrum* concluded that glycoalkaloids had no direct relationship to resistance.[148] Similarly, no apparent association exists between glycoalkaloid concentrations and levels of multigenic blight resistance,[149] severity of *Rhizoctonia solani* development,[150] resistance to *F. sambucinum*,[151] *F. solani* var. *coeruleum*, and *P. exigua* var. *foveata*.[59] Lack of resistance in some cases may be accounted for by the fact that heavy wounding techniques were employed prior to inoculation.[59,148,150] Since glycoalkaloids are concentrated in the peripheral tuber layers, the production of cracks by dropping or cutting the tuber allows the pathogen to "bypass" the defensive glycoalkaloid barrier.

Lack of correlation between *in vitro* and *in vivo* experimental results in some instances against the same pathogen[152,153] has been discussed in detail elsewhere.[55]

2.9   MODE OF GLYCOALKALOID TOXICITY

Although the antifungal and antifeedant properties of α-solanine and α-chaconine are well documented, the biochemical basis of these effects have not been satisfactorily explained. Evidence that the aglycone solanidine released by membrane bound β-glycosidases is the active form are reported.[154,155,156] There is evidence that toxicity may relate to surfactant effects on cell membranes, but available information points to membrane destabilization resulting primarily from complex formation between the glycoside and membrane bound sterols such as those found in fungal hyphae.[157]

Investigations into the effects of α-solanine and α-chaconine on the permeability of liposome membranes demonstrated the necessity of bound sterols present on the liposome membrane for lysis to occur.[145] There was, however, no close correlation between the extent of sterol binding and liposome disruption, while α-solanine complexed with all sterols *in vitro* but produced no lytic effects except at pH values of eight and above. The importance of the intact carbohydrate moiety was demonstrated by the inability of β-chaconine to complex with sterols or disrupt liposomes. Further research demonstrated a synergistic reaction between α-solanine and α-chaconine enhancing liposome lysis although the order of addition into the test system did not appear crucial[158] and that synergism in relation to membrane disruption is a general biological phenomenon rather than a rare event restricted to particular types of cells.[51]

Infectivity of herpes simplex virus Type 1 in tissue culture was inhibited by prior incubation with aqueous suspensions of glycoalkaloids, the order of activity being α-chaconine, α-tomatine, α-solasonine but not by the corresponding aglycones. Inhibition, however, was not only dependent on the presence of a sugar moiety since α-solanine was inactive under the conditions. It was suggested that inactivation of the virus resulted from insertion of the aglycone into the viral envelope.[159]

2.10   PHYSIOLOGICAL FUNCTIONS

The possible role in tuberization of potato *in vitro*,[160] flowering, species compatibility, male sterility phenomenon, seed formation, sprouting, and post harvest tuber response to environmental factors or cold tolerance are some of the interesting functional properties that arise from the presence of potato glycoalkaloids.[2]

REFERENCES

1. Woolfe, J. A., Glycoalkaloids, proteinase inhibitors and lectins. In *The Potato in the Human Diet*, Ed., Woolfe, J. A., Cambridge University Press, 1987, 162.
2. Jadhav, S. J., Kumar, A., and Chavan, J. K., *Potato Production, Processing and Products*, CRC Press, Boca Raton, 1991.
3. Heftmann, E., Biogenesis of steroids in *Solanaceae. Phytochemistry*, 22(9), 1843, 1983.
4. Morris, S. C. and Lee, T. H., The toxicity and teratogenicity of *Solanaceae* glycoalkaloids, particularly those of the potato (*Solanum tuberosum*): A review, *Food Tech. Aust.*, 36(3), 118, 1984.
5. Hall, R. L., Toxicological burdens and the shifting burden of toxicology. *Food Tech.*, 46(3), 109, 1992.
6. Baup, M., Extrait d'une lettre de M. Baup aux redacteurs sur plusiers nouvelles substances, *Ann. Chim. Phys.*, 31, 108, 1826.

7. Zwenger, C. and Kind, A., Ueber das Solanin und dessen Spaltungsprodukte, *Ann. Chem. Pharm.*, 118, 129, 1861

8. Kuhn, R. and Low, I., New alkaloid glycosides in the leaves of *Solanum chacoense*, *Angew. Chem.*, 69, 236, 1954.

9. Kuhn, R. and Low, I., In *Origins of Resistance to Toxic Agents*, Reid, M.G., and Reynolds, O. E., Eds., Academic Press, New York, 1955.

10. Gregory, P., Glycoalkaloid composition of potatoes: Diversity and biological implications, *Am. Potato J.*, 61, 115, 1984.

11. Zitnak, A., The occurrence and distribution of free alkaloid solanidine in netted gem potatoes, *Can. J. Biochem. Physiol.*, 39, 1257, 1961.

12. Nair, P. M., Behere, A. G., and Ramaswamy, N. K., Glycoalkaloids of *Solanum tuberosum* Linn, *J. Sci. Ind. Res.*, 40, 529, 1981.

13. Guseva, A. R., Borikhina, M. G., and Paseshnichenko, V. A., Utilisation of acetate for the biosynthesis of chaconine and solanine in potato sprouts, *Biokhimiya*, 25(2), 282, 1960.

14. Percival, G. C., Factors influencing accumulation of potato glycoalkaloids and their potential manipulation in tuber pathogen control. Ph.D. Thesis University of Strathclyde/SAC Auchincruive, 1993

15. Sinden, S. L., Sanford, L. L., and Osman, S. F., Glycoalkaloids and resistance to the Colorado Potato Beetle in *Solanum chacoense* Bitter, *Am. Potato J.*, 57, 331, 1980.

16. Gregory, P., Glycoalkaloids of wild tuber bearing *Solanum* species. *J. Agric. Food Chem.*, 29, 1212, 1981.

17. Zitnak, A. and Johnston, G. R., Glycoalkaloid content of B5141-6 potatoes, *Am. Potato J*, 47, 256, 1970.

18. Osman, S. F., Herb, S. F., Fitzpatrick, T. J., and Sinden, S. L., Commersonine, a new glycoalkaloid from two *Solanum* species, *Phytochemistry*, 15, 1065, 1976.

19. Sanford, L. L. and Sinden, S. L., Inheritance of potato glycoalkaloids, *Am. Potato J.*, 49, 209, 1972.

20. McCollum, G. D. and Sinden, S. L., Inheritance study of tuber glycoalkaloids in a wild potato, *Solanum chacoense* Bitter, *Am. Potato J.*, 56, 95, 1979.

21. Reeve, R. M., Hautala, E., and Weaver, M. L., Anatomy and compositional variation within potatoes II. Phenolics, enzymes and other minor components, *Am. Potato J.*, 46, 374, 1969.

22. Coxon, D. T., The glycoalkaloid content of potato berries, *J. Sci. Food Agric.*, 32, 412, 1981.

23. Bomer, A. and Mattis, H., High solanine content of potatoes, *Z. Unters. Nahr. Genussm. Gebrauchsgegenstaende*, 47, 97, 1924.

24. Percival, G. C. and Dixon, G. R., Glycoalkaloid concentrations in aerial tubers of potato (*Solanum tuberosum* L), *J. Sci. Food. Agric.*, 70(4), 439, 1996.

25. Friedman, M. and Dao, L., Distribution of glycoalkaloids in potato plants and commercial potato products, *J. Agric. Food Chem.*, 40, 419, 1992.

26. Jadhav, S. J., Sharma, R. P., and Salunkhe, D. K., Naturally occurring toxic alkaloids in foods, *CRC Critical Reviews in Toxicology*, 21, 21,1981.

27. Lampitt, L. H., Bushill, J. H., Rooke, H. S., and Jackson, E. M., Solanine, glycoside of the potato. 1. Its distribution in the potato plant, *J.Soc. Chem. Ind.*, 62, 48, 1943.

28. Kozukue, N., Kozukue, E., and Mizuno, S., Glycoalkaloids in potato plants and tubers, *HortSci*, 22, 294, 1987.

29. Wood, F. A. and Young, D. A., TGA in potatoes, *Can. Agric. Bull.*, 153, 1974.

30. Khanna, P., Kumar, P., and Singhvi, S., Isolation and characterization of solanine from *in vitro* tissue culture of *Solanum tuberosum* L. and *Solanum dulcamara* L., *Ind. J. Pharm. Sci.*, 50(1), 37, 1988.

31. Jeppsen, R. B., Wu, M. T., Salunkhe, D. K., and Jadhav, S. J., Some observations on the occurrence of chlorophyll and solanine in potato tubers and their control by N6 benzyladenine, Ethephon and filtered lights, *J. Food Sci.*, 39, 1059, 1974.

32. Han, S. R., Campbell, W. F., and Salunkhe, D. K., Ultrastructural localization of solanidine in potato tubers, *J. Food Biochem.*, 13, 377, 1989.

33. Roddick, J. G., Subcellular localisation of steroidal glycoalkaloids in vegetative organs of *Lycoperscion esculentum* and *Solanum tuberosum*, *Phytochemistry*, 16, 805, 1977.

34. Wolf, M. J. and Duggar, B. M., Estimation and physiological role of solanine in the potato, *J. Agric. Res.*, 71, 1, 1946.

35. Morgenstern, F., Uber den solaningehalt der speiseund futterkartoffeln und uber den eenfluss der bodenkultur auf die bildung von solanin in der kartoffelpflanze, *Landwirtsch. Vers. Stn.*, 65, 301, 1907.

36. Roddick, J. G. and Melchers, G., Steroidal glycoalkaloid content of potato, tomato and their somatic hybrids, *Theor. App. Gen.*, 70(6), 655, 1985.

37. Morris, S. C. and Petermann, J. B., Genetic and environmental effects on levels of glycoalkaloids in cultivars of potato (*Solanum tuberosum* L.), *Food Chem.*, 18(4), 271, 1985.

38. Maga, J. A., Potato glycoalkaloids, *CRC Critical Review of Food Science and Nutrition*, 12, 371, 1980.

39. Sinden, S. L. and Webb, R. E., Effect of variety and location on the glycoalkaloid content of potatoes, *Am. Potato J.*, 49, 334, 1972.

40. Griffiths, D. W., Dale, M. F. B., and Bain, H., The effect of cultivar, maturity and storage on photo-induced changes in the total glycoalkaloid and chlorophyll contents of potatoes (*Solanum tuberosum*), *Pl. Sci*, 98, 103, 1994.

41. Tingey, W. M., Mackenzie, D., and Gregory, P., Total foliar glycoalkaloids and resistance of wild potato species to *Empoasca fabae* (Harris), *Am. Potato J.*, 55, 577, 1978.

42. Gull, D. D. and Isenberg, F. M., Chlorophyll and solanine content and distribution in four varieties of potato tubers, *Am. Soc. Hort. Sci.*, 75, 545, 1960.

43. Dale, M. F. B., Griffiths, W., and Bain, H., Glycoalkaloids in potatoes - shedding light on an important problem, *Asp. App. Bio.*, 33, 221, 1992.

44. Conner, H. W., Effect of light on solanine synthesis in the potato tuber, *Plant Physiol.*, 12, 79, 1937.

45. Petermann, J. B. and Morris, S. C., The spectral responses of chlorophyll and glycoalkaloid synthesis in potato tubers (Solanum tuberosum), *Plant Sci.*, 39, 105, 1985.

46. Liljemark, L. E. and Widoff, L. E., Greening and solanine development of white potato in fluorescent light, *Am. Potato J.*, 37, 379, 1960.

47. Percival, G. C. and Dixon, G. R., Effect of light intensity and temperature on glycoalkaloid content of potato tubers, Proceedings of the European Association for Potato Research, 13th Triennial Conference, 1996, pp 43.

48. Percival, G. C., Dixon, G. R., and Sword, A., Glycoalkaloid Concentration of Potato Tubers Following Exposure to Day Light, *J. Sci. Food Agric.*, 71, 59, 1996.

49. Zitnak, A., The influence of certain treatments upon solanine synthesis in potatoes, M.S. thesis, University of Alberta, Edmonton, 1955.

50. Baerug, R., Influence of different rates and intensities of light on solanine content and cooking quality of potato tubers, *Eur. Potato J.*, 5(3), 242, 1962.

51. Roddick, J. G., Rijnenberg, A. L., and Osman, S. F., Synergistic interaction between potato glycoalkaloids α-solanine and α-chaconine in relation to destabilization of cell membranes: Ecological implications, *J. Chem. Ecol.*, 14, 889, 1988.

52. Percival, G. C., Dixon, G. R., and Sword, A., Glycoalkaloid concentration of potato tubers following continuous illumination, *J. Sci. Food Agric.*, 66 139, 1994.

53. Hardenberg, R. E., Greening of potatoes during marketing-A review. *Am. Potato J.*, 41, 215, 1964.

54. Patil, B. C., Salunkhe, D. K., and Singh, B., Metabolism of solanine and chlorophyll in potato tubers as affected by light and specific chemicals, *J. Food Sci.*, 36, 474, 1971.

55. Roddick, J. G., Steroidal alkaloids of the Solanaceae. In *Solanaceae, Biology and Systematics*, D'arcy, W.G., Ed., Columbia University Press, New York, 1986, pp 201.

56. Ross, H., Pasemann, P., and Nitzsche, W., Der glykoalkaloidgehalt von kartoffelsorten in seiner abhangigkeit von anbauort und-jahr und seiner beziehung zum geschmack, *Z. Pflanzenzuchtg*, 80, 64, 1978.

57. Bintcliffe, E. J. B., Clydesdale, A., and Draper, S. R., Effects of genotype, site and season on the glycoalkaloid content of potato tubers, *J. Nat. Inst. Agric. Bot*, 16, 86, 1982.

58. Davis, A. M. C. and Blincow, P. J., Glycoalkaloid content of potatoes and potato products sold in the U.K., *J. Sci. Food Agric.*, 35, 553, 1984.

59. Olsson, K., Anatomical structure and chemical composition of potato tubers, in *Impact Damage, Gangrene and Dry Rot in Potato. Important Biochemical Factors in Screening for Resistance and Quality in Breeding Material*, Olsson, K., Ed., Swedish University of Agricultural Sciences, 1989.

60. Hutchinson, A. and Hilton, R. J., Influence of certain cultural practices on the solanine content and tuber yield in Netted Gem potatoes, *Can. J. Agric. Sci.*, 35, 485, 1955.

61. Rahim, F., Hussain, M., and Khan, F. K., Effect of potato variety and growing season on the total glycoalkaloid (TGA) contents in tuber and its portions, grown in N.W.F.P., Pakistan, *Sar. J. Agric.*, 5(5), 449, 1989.

62. Mondy, N. I., Naylor, L. M., and Phillips, J. C., Total glycoalkaloid and mineral content of potatoes grown in soils amended with sewage sludge, *J. Agric. Food Chem.*, 32, 1256, 1984.

63. Munshi, C. B. and Mondy, N. I., Effect of soil applications of molybdenum on the biochemical composition of Katahdin potatoes: Nitrate nitrogen and total glycoalkaloids, *J. Agric. Food Chem.*, 36, 688, 1988.

64. Cronk, T. C., Kuhn, G. D., and McArdle, F. J., The influence of stage of maturity, level of nitrogen fertility and storage on the concentration of solanine in tubers of three potato cultivars, *Bull. Envir. Cont. Toxic*, 11, 163, 1974.

65. Nowacki, E., Jurzysta, M., and Gorski, P., Fertiliser regimes on tuber quality, *Bull. Acad. Pol. Sci. Serv. Sci. Biol.*, 23, 219, 1975.

66. Speroni, J. J., Pell, E. J., and Weissberger, W. C., Glycoalkaloid levels in potato tubers and leaves after intermittent plant exposure to ozone, *Am. Potato J.*, 58, 407, 1981.

67. Fitzpatrick, T. J., Herb, S. F., Osman, S. F., and McDermott, J. A., Potato glycoalkaloids: Increases and variations of ratios in aged slices over prolonged storage, *Am. Potato J.*, 54, 539, 1977.

68. Percival, G. C., Harrison, J. A. C., Dixon, G. R., The influence of temperature on light enhanced glycoalkaloid synthesis in potato, *Ann. App. Biol.*, 123, 141, 1993 .

69. Hilton, R. J., Factors in relation to tuber quality in potatoes. II.Preliminary trials on bitterness in Netted Gem potatoes, *Sci. Agric.*, 31, 61, 1951.

70. Linnemann, A. R., Andre, V. E., and Hartmans, K. J., Changes in the content of L-ascorbic acid, glucose, fructose and TGA on potatoes (cv.Bintje) stored at 7,16 and 28 C, *Pot. Res.*, 28, 271, 1985.

71. Frydecka-Mazurczyk, A. and Zgorska, K., Changes in chemical composition of tubers during physiological ageing, *European Association of Potato Research, Abstracts of Conference Papers, 11th Triennial Conference*, Edinburgh, Scotland, 152, 1990.

72. Dao, L. and Friedman, M., Chlorophyll, chlorogenic acid, glycoalkaloid and protease inhibitor content of fresh and green potatoes, *J. Agric. Food Chem.*, 42, 633, 1994

73. Mondy, N. I., Leja, M., and Gosselin, B., Changes in total phenolic, total glycoalkaloid and ascorbic acid content of potatoes as a result of bruising, *J. Food Sci.*, 52(3), 631, 1987.

74. Wu, M. T. and Salunkhe, D. K., Changes in glycoalkaloid content following mechanical injuries to potato tubers, *J. Am. Soc. Hort. Sci.*, 101(3), 329, 1976.

75. Salunkhe, D. K., Wu, M. T., and Jadhav, S. J., Effects of light and temperature on the formation of solanine in potato slices, *J. Food Sci.*, 37, 969, 1972.

76. Ahmed, S. S. and Muller, K., Effect of wound damages on the glycoalkaloid content in potato tubers and chips, *Lebensm. Wiss. und Tech.*, 11, 144, 1978.

77. Bergenstrahle, A., Tillberg, E., and Jonsson, L., Regulation of sterol and glycoalkaloid metabolism in ageing potato tuber slices. *Physiol. Plantum*, 79, A117, 1990.

78. Maga, J. A., Total and individual glycoalkaloid composition of stored potato slices, *J. Food Proc. Pres.*, 5, 23, 1981.

79. Mondy, N. I. and Chandra, S., Reduction of glycoalkaloid synthesis in potato slices by water soaking, *HortSci.*, 14(2),173, 1979.

80. Shih, M. and Kuc, J., α-solmarine in Kennebec *Solanum tuberosum* leaves and aged tuber slices, *Phytochemistry*, 13, 997, 1974.

81. Jadhav, S. J., Wu, M. T., and Salunkhe, D. K., Glycoalkaloids of hollow heart and blackheart potato tubers, *HortSci*, 15(2), 147, 1980.

82. Locci, R. and Kuc, J., Steroid alkaloids as compounds produced by potato tubers under stress, *Phytopathology*, 57, 1272, 1967.

83. Shih, M., Kuc, J., and Williams, E. B., Suppression of steroid glycoalkaloid accumulation as related to rishitin accumulation in potato tubers, *Phytopathology*, 63, 821, 1973.

84. Shih, M. and Kuc, J., Incorporation of 14C from acetate and mevalonate into rishitin and steroid glycoalkaloid by potato tuber slices inoculated with *Phytophthora infestans*, *Phytopathology*, 63, 826, 1973.

85. Zacharius, R. M., Kalan, E. B., Osman, S. F., and Herb, S. F., Solanidine in potato tuber tissue disrupted by *Erwinia atroseptica* and by *Phytothphora infestans*, *Physiol. Plant Path.*, 6, 301, 1975.

86. Zitnak, A., Photoinduction of glycoalkaloids in cured potatoes, *Am. Potato J.*, 58, 415, 1981.

87. Wu, M. T. and Salunkhe, D. K., Effect of gamma radiation on wound induced glycoalkaloid formation in potato tubers, *Lebensm. Wiss. Technol.*, 10, 141, 1977b.

88. Wu, M. T. and Salunkhe, D. K., Effects of vacuum packaging on light- induced greening and glycoalkaloid formation of potato-tubers, *Can. Inst. Food Sci. Tech. J.*, 8, 185, 1975.

89. Wu, M. T. and Salunkhe, D. K., After effect of submersion in water on greening and glycoalkaloid formation of potato tubers, *J. Food Sci.*, 43, 1330, 1978.

90. Wilson, J. M. and Frank, J. S., The effect of systemic pesticides on total glycoalkaloid content of potato tubers at harvest, *Am. Potato J.*, 52, 179, 1975.

91. Weingartner, D. P., Shumaker, J. R., and Webb, R. E., Failure of systemic insecticide nematicides to affect total glycoalkaloid content of harvested potato tubers in Northern Florida, *Am. Potato J.*, 53, 247, 1976.

92. Mondy, N. I., Tymiak, A., and Chandra, S., Inhibition of glycoalkaloid formation in potato tubers by the sprout inhibitor maleic hydrazide, *J. Food Sci.*, 43, 1033, 1978.

93. Parups, E. V. and Hoffman, I., Induced alkaloid levels in potato tubers, *Am. Potato J.*, 44, 277, 1967.

94. Wu, M. T. and Salunkhe, D. K., Inhibition of wound induced glycoalkaloid formation in potato tubers (*Solanum tuberosum* L.) by isopropyl-N-(3-chlorophenyl)-carbamate, *J. Food Sci.*, 42(3), 622, 1977.

95. Wu, M. T. and Salunkhe, D. K., Control of chlorophyll and solanine synthesis and spouting of potato tubers by hot paraffin wax, *J. Food. Sci.*, 37, 629, 1972.

96. Wu, M. T. and Salunkhe, D. K., Control of chlorophyll and solanine formation in potato tubers by oil and diluted oil treatments, *HortSci*, 75(5), 466, 1972.

97. Wu, M. T. and Salunkhe, D. K., Use of spray lecithin for control of greening and glycoalkaloid formation of potato tubers, *J. Food Sci.*, 42(5), 1413, 1977.

98. Wu, M. T. and Salunkhe, D. K., Responses of lecithin and hydroxylated lecithin coated potato tubers to light, *J. Agric. Food Chem.*, 26(2), 513, 1978.

99. Jadhav, S. J., Salunkhe. D. K., Wyse, R. E., and Dalvi, R. R., Solanum alkaloids: Biosynthesis and inhibition by chemicals, *J. Food Sci.*, 38, 453, 1973.

100. Sinden, S. L., Sanford, L. L., and Webb, R. E., Genetic and environmental control of potato glycoalkaloids, *Am. Potato J.*, 61, 141, 1984.

101. Sinden, S. L., Deahl, K. L., and Aulenbach, B. B., Effect of glycoalkaloids and phenolics on potato flavor, *J. Food Sci.*, 41, 520, 1976.

102. Bushway, R. J., Barden, E. S., Bushway, A. W., and Bushway, A. A., The mass extraction of potato glycoalkaloids from blossoms, *Am. Potato J.*, 57, 175, 1980.

103. Bushway, R. J. and Ponnampalam, R., α-chaconine and α-solanine content of potato products and their stability during several modes of cooking, *J. Agric. Food Chem.*, 29, 814, 1981.

104. Bushway, R. J., Bureau, J. L., and McGann, D. F., Alpha-chaconine and alpha-solanine content of potato peels and potato peel products, *J. Food Sci.*, 48, 84, 1983.

105. Bushway, A. A., Bushway, A. W., Beltea, P. R., and Bushway, R. J., The proximate composition and glycoalkaloid content of three potato meals, *Am. Potato J.*, 57, 167, 1980.

106. Mondy, N. I. and Gosselin, B., Effect of peeling on total phenols, total glycoalkaloids, discoloration and flavor of cooked potatoes, *J. Food Sci.*, 53(3), 756, 1988.

107. Meyer, G., Ueber vergiftungen durch kartoffeln, I. Ueber den gehalt der kartoffeln an solanin und uber die bildung desselben wahrend der keimung. *Arch. Expl. Path. Pharm.*, 36, 361, 1895.

108. Orgell, H. W., Vaidya, K. A., and Dahm, P. A., Inhibition of human plasma cholinesterase *in vitro* by extracts of Solanaceous plants, *Science*, 128, 1136, 1958.

109. Harris, H. and Whittaker, M., Differential response of human serum cholinesterase types to an inhibitor in potato, *Nature*, 183, 1808, 1959.

110. Abbott, D. C., Observations on the correlation of anticholinesterase effect with solanine content of potatoes, *The Analyst.*, 85, 375, 1960.

111. Pokrovskii, A. A., The effect of the alkaloids of the sprouting potato on cholinesterase, *Biokhimiya*, 21, 705, 1956.

112. Nishie, K., Gumbmann, M. R., and Keyl, A. C., Pharmacology of solanine, *Toxic. App. Pharm.*, 19, 81, 1971.

113. Norred, W. P., Nishie, K., and Osman, S. F., Excretion, distribution, and metabolic fate of (^{3}H-α-chaconine), *Res. Comm. Chem. Path. Pharm.*, 13, 161, 1976.

114. Rayburn, J. R., Bantle, J. A., and Friedman., Role of carbohydrate side chains of potato glycoalkaloids in developmental toxicity, *J. Agric. Food Chem.*, 42, 1511, 1994.

115. Patil, B. C., Sharma, R. P., Salunkhe, D. K., and Salunkhe, K., Evaluation of solanine toxicity, *Food Cosmet. Toxicol.*, 10, 395, 1972.

116. Swinyard, C. A. and Chaube, S., Are potatoes teratogenic for experimental animals, *Teratology*, 8, 349, 1973.

117. Sharma, R. P., Willhite, C. C., Shupe, J. L., and Salunkhe, D. K., Acute toxicity and histo-pathological effects of certain glycoalkaloids and extracts of *Alternaria solani* of *Phytophthora infestans* in mice, *Toxic. Lett.*, 3, 349, 1979.

118. Toyoda, M., Rausch, W. D., Inoue, K., Ohno, Y., Fujiyama, Y., Takagi, K., and Saito, Y., Comparison of solanaeous glycoalkaloids evoked Ca^{2+} influx in different types of cultured cells, *Toxic. in vitro*, 5(4), 347, 1991.

119. Nishie, K., Positive inotropic action of Solanaceae glycoalkaloids, *Res. Comm. Chem. Path. Pharm.*, 15(3), 601, 1976.

120. Roddick, J. G., The acetylcholinesterase-inhibitory activity of steroidal glycoalkaloids and their aglycones, *Phytochemistry*, 28(10), 2631, 1989.

121. Hopkins, J., The glycoalkaloids: Naturally of interest (but a hot potato), *Food Chem. Toxic*, 4, 323, 1995.

122. Van Gelder, W. M. J., Steroidal glycoalkaloids in *Solanum* species: consequences for potato breeding and food safety, Ph.D. thesis, Wageningen Agricultural University, The Netherlands, 1994

123. Damon, S. R., *Food Infections and Food Intoxications*, Williams and Wilkins, Baltimore, 1928

124. Griebel, C., Injurious potatoes rich in solanine, *Z. Unters. Nahr. Genussm. Gebrauchsgegenstaende*, 45, 175, 1923.

125. Rothe, J. C., Illness following eating of potatoes containing solanine, *Z. Hug.*, 88, 1, 1918

126. Jadhav, S. J. and Salunkhe, D. K., Formation and control of chlorophyll and glycoalkaloids in tubers of *Solanum tuberosum* L. and evaluation of glycoalkaloid toxicity, *Adv. Food Res*, 21, 308, 1975.

127. McMillan, M. and Thompson, J. H., An outbreak of suspected solanine poisoning in schoolboys: Examination of criteria of solanine poisoning, *Quart. J. Med.*, 48, 227, 1979.

128. Hellenas, K. E., Determination of potato glycoalkaloids and their aglycone in blood serum by high performance liquid chromatography. Application to pharmacokinetic studies in humans *J. Chromatog. Biomed. Applic.*, 573, 69, 1992

129. Renwick, J. H., Claringbold, W.D.B., Earthy, M. E., Few, J. D., and McLean, A. C. S., Neural tube defects produced in Syrian hamsters by potato glycoalkaloids, *Teratology*, 30, 371, 1984

130. Allen, E. H. and Feldmesser, J., Nematicidal activity of α-chaconine: Effect of hydrogen ion concentration, *J. Nematology*, 3(1), 58, 1971.

131. Johnston, K. A., Kershaw, W. J. S., and Pearce, R. S., Biochemical mechanisms of resistance of potato cultivars to slug attack, *Slugs and Snails in World Agriculture*, BCPC Mono No 41, 281, 1989.

132. Jonasson, T. and Olsson, K., The influence of glycoalkaloids, chlorogenic acid and sugars on the susceptibility of potato tubers to wireworm, *Pot. Res.*, 37, 205, 1995

133. Mitchell, B. K. and Harrison, G. D., Effects of the *Solanum* glycoalkaloids on chemosensilla on the colorado potato beetle. A mechanism of feeding deterrence? *J. Chem. Ecol.*, 11, 73, 1985.

134. Deahl, K. L., Cantelo, W. W., Sinden, S. L., and Sanford, L. L., The effect of light intensity on colorado potato beetle resistance and foliar glycoalkaloid concentration of four *Solanum chaceonse* clones, *Am. Potato J.*, 68, 659, 1991.

135. Raman, K. V., Tingey, W. M., and Gregory, P., Potato glycoalkaloids: Effect on survival and feeding behavior of the potato leafhopper, *J. Econ. Ent.*, 72(3), 337, 1979.

136. Tingey, W. M. and Sinden, S. L., Glandular pubescence, glycoalkaloid composition and resistance to the Green Peach Aphid, Potato Leaf Hopper and Potato Flea Beetle in *Solanum berthaultii*, *Am. Potato J.*, 59, 95, 1982.

137. Boer, D.de. and Hanson, F. E., Feeding response of solanaceous allelochemicals by larvae of the tobacco hornworm, *Maducta sexta*, *Ent. Exper. App.*, 45(2),123, 1987.

138. Forrest, J. M. S. and Coxon, D. T., The relationship between glycoalkaloids and resistance to the white potato cyst nematode, *Globodera pallida* in potato clones derived from *Solanum vernei*, *Ann. App. Biol.*, 94, 265, 1980.

139. Kuc, J., Metabolites accumulating in potato tubers following infection and stress, *Teratology*, 8, 333, 1973.

140. Allen, E. H. and Kuc, J., α-solanine and α-chaconine as fungitoxic compounds in extracts of Irish potato tubers, *Phytopathology*, 58, 776, 1968.

141. Arneson, P. A. and Durbin, R. D., Studies on the mode of action on tomatine as a fungitoxic agent, *Plant Physiol.*, 43, 683, 1968.

142. Ozeretskovskaya, O. L., Vasyukova, N. L., and Davydova, M. A., Investigation of antibiotic substances in potato tubers, *Prikl. Biokhim. I. Microbiol.*, 4, 698, 1968.

143. McKee, R. K., Factors affecting the toxicity of solanine and related alkaloids to *Fusarium coeruleum*, *J. Gen. Micro.*, 20, 686, 1959.

144. Sinden, S. L., Goth, R. W., and O'Brien, M. J., Effect of potato alkaloids on the growth of *Alternaria solani* and their possible role as resistance factors in potatoes, *Phytopathology*, 63, 303, 1972.

145. Roddick, J. G. and Rijnenberg, A. L., Effect of steroidal glycoalkaloids of the potato on the permeability of liposome membranes, *Physiol. Plantarum*, 68, 436, 1986.

146. Roddick, J. G. and Drysdale, R. B., Destabilization of liposome membranes by the steroidal glycoalkaloid α-tomatine, *Phytochemistry*, 23(3), 543, 1984.

147. Holland, T. L. and Taylor, G. J., Transformation of steroids and the steroidal alkaloid, solanine, by *Phytophthora infestans*. *Phytochemistry*, 18, 437, 1979.

148. Frank, J. A., Wilson, J. M., and Webb, R. E., The relationship between glycoalkaloids and disease resistance in potatoes, *Phytopathology*, 65, 1045, 1975.

149. Deahl, K. L., Young, R. J., and Sinden, S. L., A study of the relationship of late blight resistance to glycoalkaloid content in fifteen potato clones, *Am. Potato J.*, 50, 248, 1973.

150. Morrow L. S. and Caruso, F. L., Effect of potato seed tuber glycoalkaloid content on subsequent infection by *Rhizoctonia solani*, *Am. Potato J.*, 60, 403, 1983.

151. Corsini, D. L. and Pavek, J. J., Phenylalanine ammonia lyase activity and fungitoxic metabolites produced by potato cultivars in response to *Fusarium* tuber rot, *Physiol. Plant Path.*, 16, 63, 1980.

152. Paquin, R. and Lachance, R. A., Effects des glycoalcaloides de la pomme de terre sur la croissance de Corynebacterium sepedonicum (Speik and Kott.) Skapt. and Burkh, *Can. J. Micro.*, 10, 115, 1967.

153. Paquin, R., Study on the role of the glycoalkaloids in the resistance of potato to bacterial ring rot, *Am. Potato J.*, 43, 349, 1966.

154. Segal, R. and Milo-Goldzweig, I., On the mechanism of saponin hemolysis - II. Inhibition by aldonolactones, *Biochem. Pharm.*, 24, 77, 1975.

155. Segal, R., Shatkovsky, P., and Milo-Goldzweig, I., On the mechanism of saponin hemolysis - I. Hydrolysis of the glycosidic bond, *Biochem. Pharm.*, 23, 973, 1974.

156. Segal, R. and Schlosser, E., Role of glycosidases in the membranlytic, antifungal action of saponins, *Arch. Micro.*, 104, 147, 1975.

157. Roddick, J. G., Complex formation between solanaceous steroidal glycoalkaloids and free sterols *in vitro*, *Phytochemistry*, 18, 1467, 1979.

158. Roddick, J. G. and Rijnenberg, A. L., Synergistic interaction between the potato glycoalkaloids α-solanine and α-chaconine in relation to lysis of phospholipid/sterol liposomes, *Phytochemistry*, 25(5), 1325, 1987.

159. Thorne, H. V., Clarke, G. F., and Skuce, R., The inactivation of herpes simplex virus by some Solanaceae alkaloids, *Antiviral Res.*, 5, 335, 1985.

160. Taluker, M. M. and Paupardin, C., Effect of solanine on tuberization of potato (*Solanum tuberosum* L) - cultures - *in vitro* at different temperatures, *Bang. J. Agric.*, 11(3), 27, 1986.

3 *Erythrina* Alkaloids

A. S. Chawla and V. K. Kapoor

CONTENTS

3.1   INTRODUCTION

The *Erythrina* genus comprises 108 species of orange or red-flowered trees, shrubs, and herbaceous plants.[1] The plants are found throughout the tropical and semi-tropical regions of the world.[1,2] The genus *Erythrina* (Fabaceae) is a part of the Papilionaceae subfamily of the Leguminosae family. The botanical relationship between the various species within the genus have been classified by the late B. A. Krukoff,[3] for whom they were a life-long study. Besides assuming greater importance as browse legumes for livestock in the tropics, several of the most common species are used for decorative purposes in gardens and city streets. The remarkable curare-like activity of the *Erythrina* alkaloids has provided the stimulus for the extensive chemical studies on the genus. Moreover, the *Erythrina* alkaloids were tertiary bases, whereas other alkaloids with curarizing activity were all quaternary salts.

On the basis of characters of the flowers, fruits, and inflorescence, the genus *Erythrina* has been classified into five subgenera and 26 sections. There are 28 species and 3 subspecies in Mexico, 26 species in Central America, 22 species in South America, 5 species and 2 varieties in the West Indies, 30 species and subspecies in tropical Africa, 6 species in South Africa, 7 species in continental Asia, and 7 species in Malaysia and the Pacific.[3] The external and internal topographic features of about 3500 seeds from 573 samples, representing 101 species have been studied.[4] The close botanical relationship between all the members of the *Erythrina* genus is characterized by the presence of a number of spirocyclic isoquinoline alkaloids which are found in parts of the plants.

The chemistry of the *Erythrina* alkaloids has been extensively reviewed.[5–15] A collective information on the spectral data for more than 90 *Erythrina* alkaloids has been reported.[16] An up-to-date account of the *Erythrina* alkaloids has been published by Chawla and Kapoor.[17] The present chapter reviews the structural features and distribution of the different types of alkaloids in the seeds, leaves, flowers, and other parts of *Erythrina* genus. Spectral features and pharmacological and toxicological aspects are also covered.

3.2   STRUCTURAL FEATURES AND DISTRIBUTION

The characteristic chemical feature of the *Erythrina* alkaloids is the presence of a tetracycline spiroamine system, erythrinane (**1**). The alkaloids occur in the plants both in the "free" base form and also in "combined" form as glycosides or as esters of thioacetic acid. In the early work of these alkaloids, the prefix "erythr-" was used in naming of the free bases, e.g., erythroidine, erythraline, etc., and the prefix "erysothio-" was used to name those alkaloids which contained the thioacetic acid residue, exemplified by erysothiovine and erysothiopine. The alkaloids obtained after mild acid hydrolysis were designated by the prefix "eryso-", e.g., erysodine, erysovine, etc., and termed as "liberated" alkaloids. In the combined form of the alkaloids, the linkage to the thioacetic acid residue or the glucose unit is through a phenolic hydroxyl group.

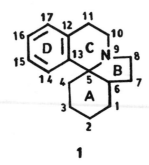

1

Nearly 100 alkaloids have been isolated and characterized from *Erythrina* species. Seeds of the plants are usually the most prolific source of the alkaloids, though their presence in stem, bark, leaves, and flowers has also been reported. Depending upon the location of the double bond(s) within the erythrinan skeleton and nature of the peripheral substituents, and changes in the ring D, the *Erythrina* alkaloids are classified into three main groups. These are the dienoid alkaloids (Table 3.1), the alkenoid alkaloids (Table 3.2), and lactonic alkaloids (Table 3.3). The erythrinan skeleton is almost unique to *Erythrina* genus and the only other examples are in the *Cocculus* genus and in *Pachygone ovata* (Menispermaceae).

In addition to the alkaloids covered under Tables 3.1–3.3, there are several *Erythrina* alkaloids which are structural variants of the dienoid or alkenoid types. Some of the alkaloids occur as quaternary salts such as *N*-methylated alkaloids isolated from pod walls of *Erythrina arborescens*[18-20] and from roots of *Pachygone ovata*,[21] *N*-oxides,[22-25] or dehydrogenated derivatives. These are covered under Table 3.4. A novel alkaloid erymelanthine which has a pyridine ring in the place of benzene ring D in erythrinan skeleton has been isolated by two different research groups[26,27] from the seeds of *Erythrina melanacantha*; the same alkaloid has also been found in *Erythrina merilliana*.

There are some alkaloids isolated from *Erythrina* species which do not have the characteristic erythrinan skeleton but have a benzylisoquinoline, tetrahydrobenzylisoquinoline, or indole system in them. Such alkaloids are given under Table 3.5. Hypaphorine, the indole alkaloid, occurs as a betaine ubiquitously in *Erythrina* species.

Recently, Amer et al.[28] have reported the occurrence of (+)-epierythrinine in an inseparable mixture with its known C-11 epimer (+)-erythrinine and (+)-erysodine in *Erythrina caffra* grown in Egypt. Earlier these authors have also isolated three novel glycodienoid alkaloids from *Erythrina lysistemon*.[29] Occurrence of 8-oxoerythraline, erythraline, erysotrine, and crystamidine in *Erythrina cristagalli*, and of erysotrine as the principal alkaloid in *Erythrina pallida* has been reported.[30] A new alkaloid named as erythrinitol (**77**) has been reported to occur in *Erythrina variegata* flowers.[31,32]

Erythrina alkaloids have also been detected in the milk of goats which had been fed with leaves of these plants. Traces of β-erythroidine, a paralyzing agent, has been detected in the milk of goats fed on foliage of *Erythrina berteroana* and *Erythrina poeppigiana*.[33,34]

The early structural studies on *Erythrina* alkaloids were mainly carried out by Folkers and his group during the late 1930s and early 1940s.[35-37] Subsequently, several other

TABLE 3.1
Dienoid Alkaloids

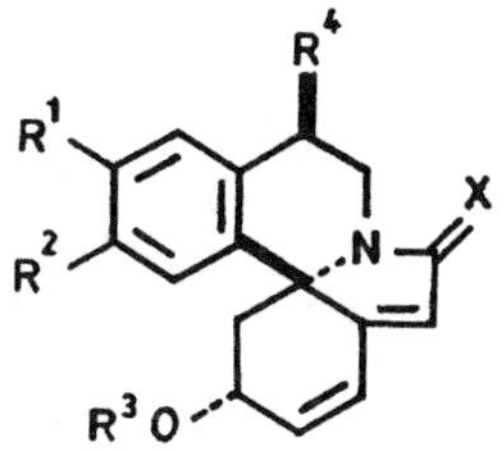

		R^1	R^2	R^3	R^4	X
2.	Erysodine	OH	OCH_3	CH_3	H	H_2
3.	Erysovine	OCH_3	OH	CH_3	H	H_2
4.	Erysopine	OH	OH	CH_3	H	H_2
5.	Erysoline	OCH_3	OH	H	H	H_2
6.	Erysonine	OH	OCH_3	H	H	H_2
7.	Erysotrine	OCH_3	OCH_3	CH_3	H	H_2
8.	Erysotramidine	OCH_3	OCH_3	CH_3	H	O
9.	Erythravine	OCH_3	OCH_3	H	H	H_2
10.	Erythraline	$-OCH_2O-$		CH_3	H	H_2
11.	Erythrinine	$-OCH_2O-$		CH_3	OH	H_2
12.	Erythrocarine	$-OCH_2O-$		H	H	H_2
13.	Erythrascine	OCH_3	OCH_3	CH_3	OAc	H_2
14.	Erythrartine	OCH_3	OCH_3	CH_3	OH	H_2
15.	Erythristemine	OCH_3	OCH_3	CH_3	OCH_3	H_2
16.	11-Hydroxyerysodine	OH	OCH_3	CH_3	OH	H_2
17.	11-Hydroxyerysovine	OCH_3	OH	CH_3	OH	H_2
18.	11-Methoxyerysodine	OH	OCH_3	CH_3	OCH_3	H_2
19.	11-Methoxyerysovine	OCH_3	OH	CH_3	OCH_3	H_2
20.	11-Oxoerysodine	OH	OCH_3	CH_3	O	H_2
21.	11-Oxoerysovine	OCH_3	OH	CH_3	O	H_2
22.	11-Oxoerysopine	OH	OH	CH_3	O	H_2
23.	11-Methoxyerysopine	OH	OH	CH_3	OCH_3	H_2
24.	11-Oxoerythraline	$-OCH_2O-$		CH_3	O	H_2
25.	11-Methoxyerythraline	$-OCH_2O-$		CH_3	OCH_3	H_2
26.	8-Oxoerythraline	$-OCH_2O-$		CH_3	H	O
27.	8-Oxoerythrinine	$-OCH_2O-$		CH_3	OH	O
28.	Glucoerysodine	a	OCH_3	CH_3	H	H_2
29.	11-Methoxyglucoerysodine	a	OCH_3	CH_3	OCH_3	H_2
30.	11-Methoxyglucoerysovine	OCH_3	a	CH_3	OCH_3	H_2
31.	Rhamnoerysodine	b	OCH_3	CH_3	H	H_2
32.	Erysothiovine	OCH_3	c	CH_3	H	H_2
33.	Erysothiopine	OH	c	CH_3	H	H_2
34.	Erysophorine	OCH_3	d	CH_3	H	H_2
35.	Erysodinophorine	d	OCH_3	CH_3	H	H_2
36.	Erysopinophorine	d	OH	CH_3	H	H_2
37.	Iso-erysopinophorine	OH	d	CH_3	H	H_2
38.	Coccuvinine	H	OCH_3	CH_3	H	H_2
39.	Coccuvine	H	OH	CH_3	H	H_2
40.	Coccolinine	H	OH	CH_3	H	O
41.	Coccoline	H	OCH_3	CH_3	H	O

[a] 1-β-glucosyl
[b] 1-β-rhamnosyl
[c] $HOOC-CH_2-SO_3-$
[d] hypaphorine ester

TABLE 3.2
Alkenoid Alkaloids

		R^1	R^2	R^3	R^4	R^5	R^6
42.	Erythratidine	H	OCH$_3$	OCH$_3$	H	H	OH
43.	Epi-erythratidine	H	OCH$_3$	OCH$_3$	H	OH	H
44.	Erythratidinone	H	OCH$_3$	OCH$_3$	H	O	
45.	Erythramine	H	–OCH$_2$O–		H	H	H
46.	Erythratine	H	–OCH$_2$O–		H	OH	H
47.	11-Hydroxyerythratine	H	–OCH$_2$O–		OH	OH	H
48.	Erythratinone	H	–OCH$_2$O–		H	O	
49.	11-Hydroxyerythratidine	H	OCH$_3$	OCH$_3$	OH	H	OH
50.	11-Hydroxy-epierythratidine	H	OCH$_3$	OCH$_3$	OH	OH	H
51.	11-Methoxyerythratidine	H	OCH$_3$	OCH$_3$	OCH$_3$	H	OH
52.	Erysotine	H	OH	OCH$_3$	H	OH	H
53.	Erysotinone	H	OH	OCH$_3$	H	O	
54.	11-Hydroxyerysotinone	H	OH	OCH$_3$	OH	O	
55.	Erysosalvine	H	OCH$_3$	OH	H	OH(H)	
56.	Erysosalvinone	H	OCH$_3$	OH	H	O	
57.	Erysopitine	H	OH	OH	H	OH(H)	
58.	Erysoflorinone	H	OH	OH	H	O	
59.	Dihydroerysodine	H	OH	OCH$_3$	H	H	H
60.	Dihydroerysovine	H	OCH$_3$	OH	H	H	H
61.	Dihydroerysotrine	H	OCH$_3$	OCH$_3$	H	H	H
62.	Coccutrine	OCH$_3$	H	OH	H	H	H
63.	Cocculidine	H	H	OCH$_3$	H	H	H
64.	Cocculine	H	H	OH	H	H	H
65.	Cocculitinine	H	H	OH	H	OH	H
66.	Cocculitine	H	H	OCH$_3$	H	OH	H
67.	Cocculidinone	H	H	OCH$_3$	H	O	
68.	Erythroculine	H	OCH$_3$	COOCH$_3$	H	H	H
69.	Erythlaurine	OH	OCH$_3$	COOCH$_3$	H	H	H
70.	Erythramide	H	OCH$_3$	CONH$_2$	H	H	H

laboratories became involved and, following a series of extensive degradative studies, the structures of the aromatic (dienoid) alkaloids were deduced by Prelog's group in Zurich,[38] and of the erythroidines by Boekelheide's group in Rochester, U.S.[39,40]

In the early work, the alkaloids were isolated from the plant material by methanol extraction, after most of the lipid material had been removed by extraction with light petroleum. The crude product obtained by evaporation of the methanol was then dissolved in dilute hydrochloric acid. Basification of the hydrochloric acid solution with sodium bicarbonate and extraction with chloroform then afforded the crude "free" alkaloid fraction. The aqueous layer from this extraction was then acidified with hydrochloric acid and heated (60–70°) to

TABLE 3.3
Lactonic Alkaloids

71.	X = H$_2$	α-Erythroidine
72.	X = O	8-Oxo-α-erythroidine

73.	X = H$_2$	β-Erythroidine
74.	X = O	8-Oxo-β-erythroidine

75.   Cocculolidine

hydrolyze the combined alkaloids; basifying and extraction with chloroform yielded the corresponding "liberated" alkaloids. Fractional crystallization of the salts was often originally employed to separate the mixtures of the alkaloids obtained, but in most recent years column chromatography and lately semi-preparative high performance liquid chromatography (HPLC) have been predominant methods of isolation and purification.

During the last 20 years or so, several groups have carried out major investigations of the alkaloid content of a variety of *Erythrina* species. The groups worthy of mention are those of Barton[41–44] (then in England), Ito[45–56] (Japan), Singh and Chawla[57–63] and Ghosal[64–68] (India). Rinehart's group in Illinois has carried out systematic screening of seeds of a wide range of species.[69,70] In Cardiff, Jackson's group[71–78] has used gas chromatography-mass spectrometry (GC-MS) as the main technique to facilitate chemotaxonomical studies. Both the "free" and "liberated" fractions were investigated in each case; any free hydroxyl groups in the alkaloids were derivatized as their trimethylsilyl ethers prior to GC-MS. It was observed by Rinehart et al.[69,70] and Singh and Chawla[57] that a significant amount of the alkaloids was also present in the hexane fraction having been co-extracted with the lipids. This observation has resulted in a regular practice of extracting the hexane fraction with acid to obtain the alkaloids present. The various studies have resulted in an observation that individual species are often distinctive in their alkaloid content, although the subgenera and sections are not clearly marked. Erysodine (**2**), erysovine (**3**), and erysopine (**4**) are the most widely distributed alkaloids. American species of *Erythrina* differ from those occurring in other parts of the world in not containing any alkaloid oxygenated at the 11-position of the erythrinan nucleus; occurrence of erythrinine (**11**) in *Erythrina cristagalli*[53] is an exception. The lactonic alkaloids α- and β-erythroidines (**71** and **73**) occur in relatively few species, but when they are found they are usually the major components of the alkaloid fraction. The occurrence of alkenoid alkaloids is fairly wide in the genus; however, their abundance is usually much less in comparison to dienoid or lactonic alkaloids. There is also an increasing number of reports on the presence of 8-oxo-alkaloids of the dienoid and lactonic types.[18,19,74,76,79,80]

While the alkaloid profile of various *Erythrina* species is generally characteristic, significant variations have been found in the alkaloid content of samples collected from different

TABLE 3.4
Alkaloids with Other Structural Modifications

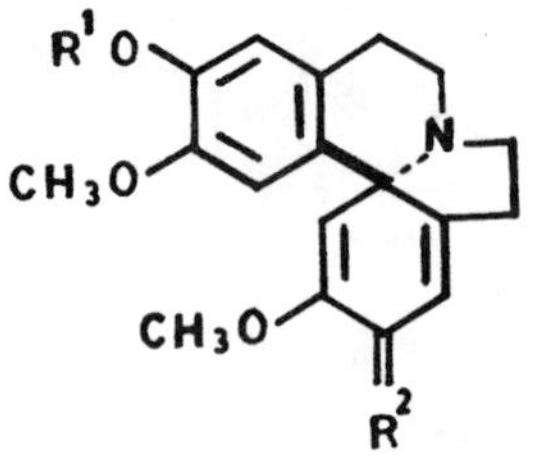

76. $R^1 = H, R^2 = O$ Erysodienone

77. $R^1 = CH_3, R^2 = \begin{smallmatrix}H\\OH\end{smallmatrix}$ Erythrinitol

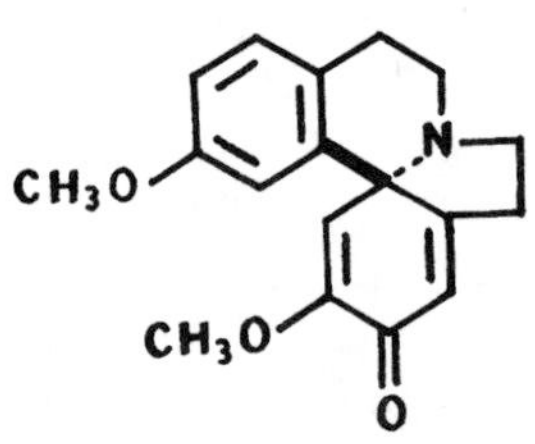

78.   Coccudienone

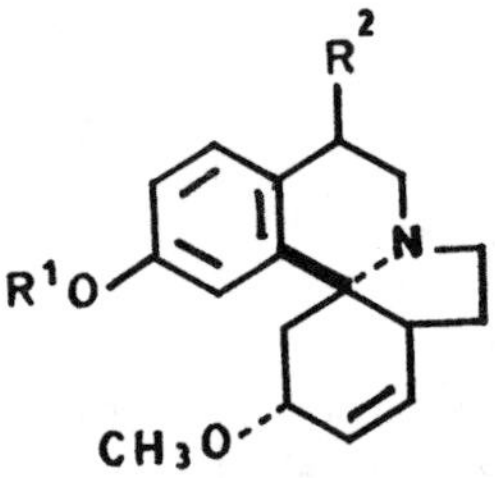

79. $R^1 = H$, $R^2 = OCH_3$ Cocculimine
80. $R^1 = CH_3, R^2 = H$ Isococculidine
81. $R^1 = R^2 = H$ Isococculine

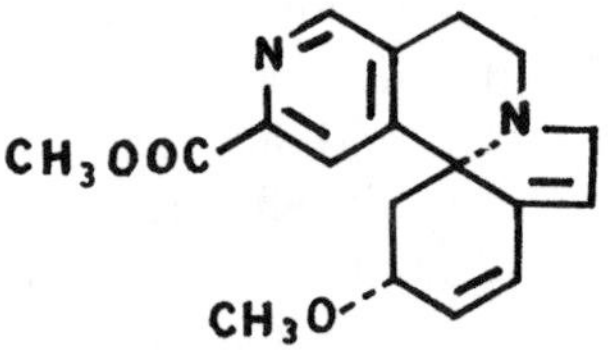

82.   Erymelanthine

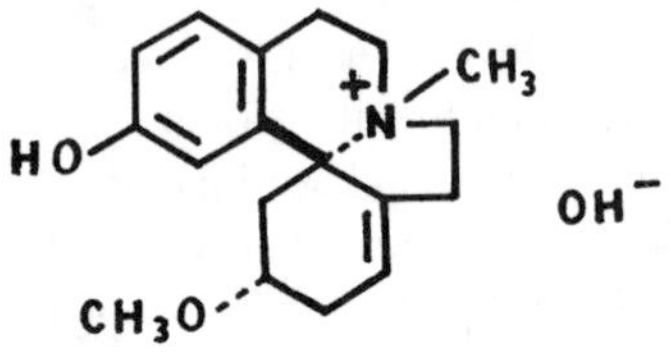

83.   Pachygonine

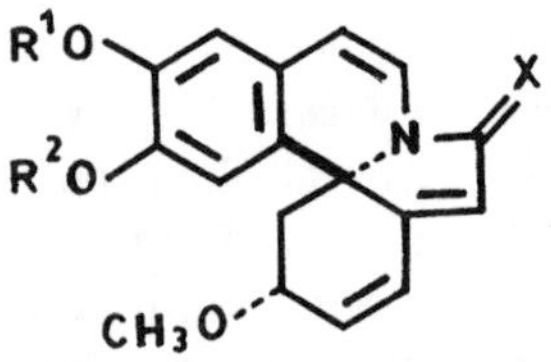

84. $R^1 = R^2 = CH_3$, X = O Erytharbine
85. $R^1R^2 = -CH_2 -$, X = O Crystamidine
86. $R^1 = H, R^2 = CH_3$, X = H$_2$ 10,11-Dehydroerysodine
87. $R^1 = CH_3, R^2 = H$, X = H$_2$ 10,11-Dehydroerysovine

TABLE 3.4 (CONTINUED)
Alkaloids with Other Structural Modifications

88.	$R^1 = R^2 = CH_3$, $R^3 = H$ Erysotrine *N*-oxide	
89.	$R^1 = R^2 = CH_3$, $R^3 = OH$ Erythrartine *N*-oxide	
90.	$R^1 R^2 = -CH_2-$, $R^3 = OCH_3$ 11β-Methoxyerythraline *N*-oxide	
91.	$R^1 = R^2 = R^3 = OCH_3$ Erythristemine *N*-oxide	

places, or at different times form the same place.[73] Differences between the alkaloid content of the seed, bark, leaves, etc., of the plant have also been observed.[64,66,81,82] Some striking variations have been reported; for example, the GC-MS investigation[70] of *Erythrina folkersii* did not reveal the presence of erythraline whereas Folkers and Koniuszy[37] had isolated it from the same plant. The report of occurrence of erysotrine[64] as the major alkaloid in *Erythrina variegata* bark together with minor amounts of erysodine and erysovine was at variance with the observation made by Singh et al.[61] who isolated only erysovine from the same source. Subsequently, Chawla et al.[83] characterized erysotrine, erythratidine, epi-erythratidine, and 11-hydroxy-epi-erythratidine in *E. variegata* bark in addition to those previously reported. A difference in the alkaloidal constituents from two samples of *Erythrina suberosa* seeds collected form different places in India has also been reported.[58] There is also evidence of chemical variation within the species. Letcher[84] inferred that there were two varieties of *Erythrina lysistemon* which yielded either erysotrine or 11-methoxyerythraline, but not both.

3.3   SPECTRAL FEATURES

Various physiochemical techniques such as infrared and ultraviolet spectroscopy, and ^{1}H and ^{13}C nuclear magnetic resonance spectroscopy and mass spectrometry have been extensively employed in determining the structures of *Erythrina* alkaloids. Chiroptical analysis and X-ray crystallography techniques have also been used.

The aromatic dienoid alkaloids show IR absorbances at 1610 cm⁻¹ and UV absorbances around 285–290 nm (dioxygenated aromatic ring) and 235–240 nm (diene component). The alkenoids absorb in the UV around 225 nm, whereas the enone group usually shows absorbance around 230 nm and IR absorbance in the region 1675–1698 cm⁻¹. The lactonic alkaloids, α- and β-erythroidines, exhibit absorption at 1720 cm⁻¹ (lactonic carbonyl group) in the IR spectra and at 224 and 238 nm, respectively, in the UV spectra. The 8-oxo-erythroidines show a lactam absorbance at 1745 cm⁻¹ (five-membered lactam carbonyl group) and an additional UV absorbance at 253 nm arising from the dienone chromophore.[76]

The ^{1}H-NMR readily distinguishes the dienoid alkaloids by the presence of an ABX system corresponding to three olefinic protons, whereas the alkenoid alkaloids only have a single olefinic resonance and the lactonic alkaloids lack the two aromatic singlet resonances. The use of ^{1}H-NMR spectroscopy has also enabled one to distinguish between erysodine (**2**) and erysovine (**3**),[42] and to establish the structures of erythratine (**46**)[42] and 11-methoxyeryth-raline (**25**).[84] INDOR and NOE methods promise to be very useful techniques in structural

TABLE 3.5
Miscellaneous *Erythrina* Alkaloids

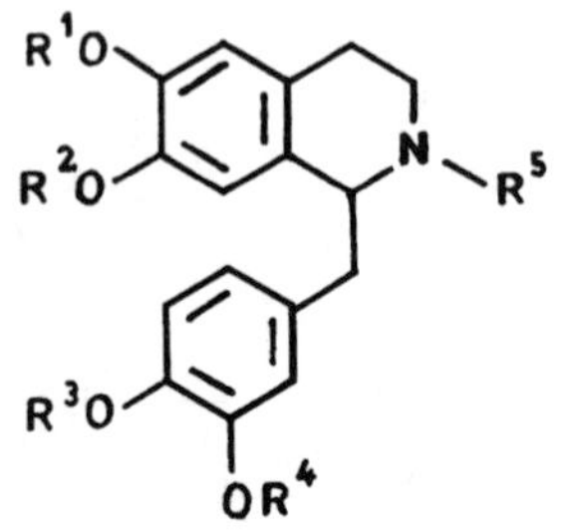

		R¹	R²	R³	R⁴	R⁵
92.	*N*-Norprotosinomenine	H	CH₃	CH₃	H	H
93.	Protosinomenine	H	CH₃	CH₃	H	CH₃
94.	*N*-Nororientaline	CH₃	H	H	CH₃	H
95.	Orientaline	CH₃	H	H	CH₃	CH₃

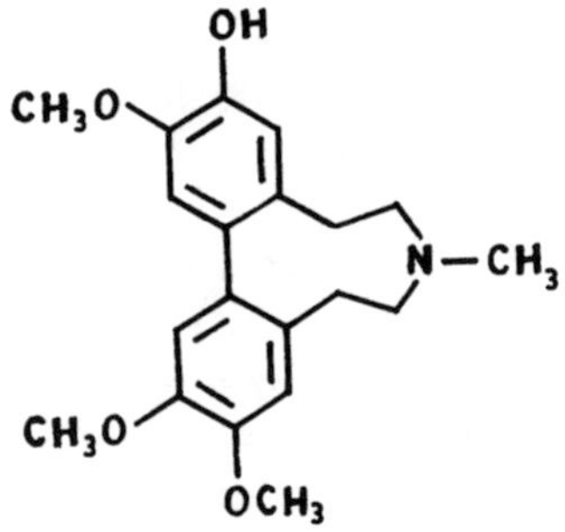

96. Erybidine

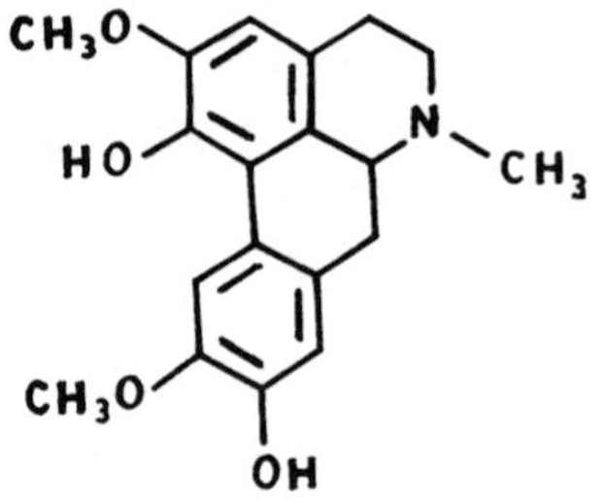

97. Isoboldine

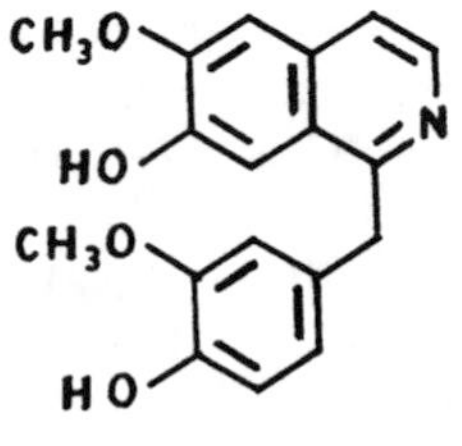

98. Cristadine

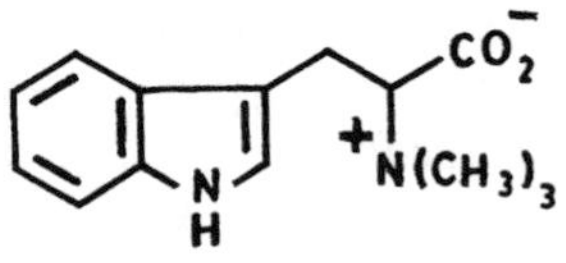

99. Hypahorine

studies, the positions of aromatic hydroxyl and methoxyl groups were shown to be C-15 and C-16, respectively, for dihydroerysovine (**60**).[85] ¹H-NMR spectroscopy has also been employed to establish the stereochemistry at C-2 and C-3 in erythratidine (**42**) and erythraline (**46**).[42,44] With the aid of the INDOR technique, the value of $J_{3,4}$ was found to be 5.5 and 12 Hz for both alkaloids suggesting that the 3-H was axial in both cases. The values $J_{1,2}$ (4.25 Hz) and $J_{2,3}$ (4.25 Hz) in **42** suggested that 2-H is pseudo-equatorial and stereochemistry is that of A (as shown in Figure 3.1). In **46** and its C-2 epimer (epi-erythratine) the values of $J_{2,3}$ were found to be 7.5 and 3–4 Hz, respectively, thereby suggesting the stereochemistry

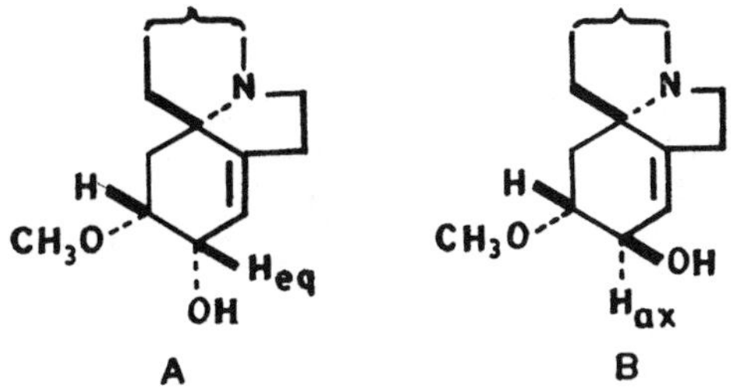

FIGURE 3.1 Stereochemistry of ring A.

B and A at C-2 for erythratine and epi-erythratine, respectively. Thus **42** and **46** have opposite stereochemistry at C-2.

The ^{13}C-NMR chemical shifts of a number of dienoid, alkenoid, and lactonic alkaloids have been reported;[77] these have been assigned by internal comparisons within the series of related compounds and also by the use of model compounds.

Mass spectrometry has played a vital role in the identification of *Erythrina* alkaloids, particularly when present in very small quantities. Mass spectrometry, especially in conjunction with gas chromatography, has proved particularly useful in preliminary screening of *Erythrina* species for their alkaloid content.[69–71] A detailed analysis of the mass spectra of a range of *Erythrina* alkaloids showed[86] that the predominant fragmentation of the dienoid and lactonic alkaloids corresponded to cleavage of methyl or methoxyl from the C-3 methoxyl unit (Figure 3.2).

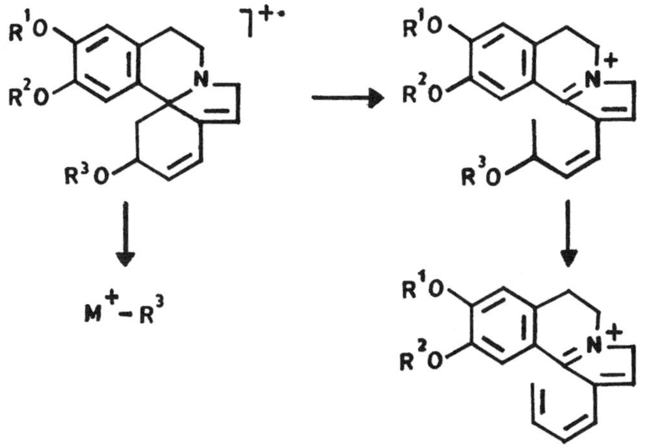

FIGURE 3.2 Mass spectral fragmentation of dienoid *Erythrina* alkaloids.

The alkenoid alkaloids also showed similar fragmentations; another prominent peak at M^+-58 corresponding to retro-Diels-Alder fragmentation, and loss of C-3, C-4, and C-3 methoxyl group also occurred (Figure 3.3). Mass spectrometry has also been used to distinguish between epimers, e.g., erythratidine and epi-erythratidine.[27]

Chiroptical analysis studies have shown that dienoid alkaloids show a positive Cotton effect. A similar effect was shown by the alkenoid alkaloids.

X-Ray crystallography has been used to determine stereochemistry of *Erythrina* alkaloids. The absolute configuration of lactonic and aromatic alkaloids has been established to be (3R,5S).[87–89] An X-ray analysis of 2-bromo-4,6-dinitrophenolate salt of erythristemine (**15**) showed the absolute configuration at C-11 as *S*. The preparation of this derivative constituted a new method and may be applicable in other cases.[43,90] On the basis of optical rotation and ^{1}H-NMR data, the absolute configurations of erythratidine (**42**) and erythratine (**46**) have been determined to be 2*S*, 3*R*, 5*S* and 2*R*, 3*R*, 5*S*, respectively.[44]

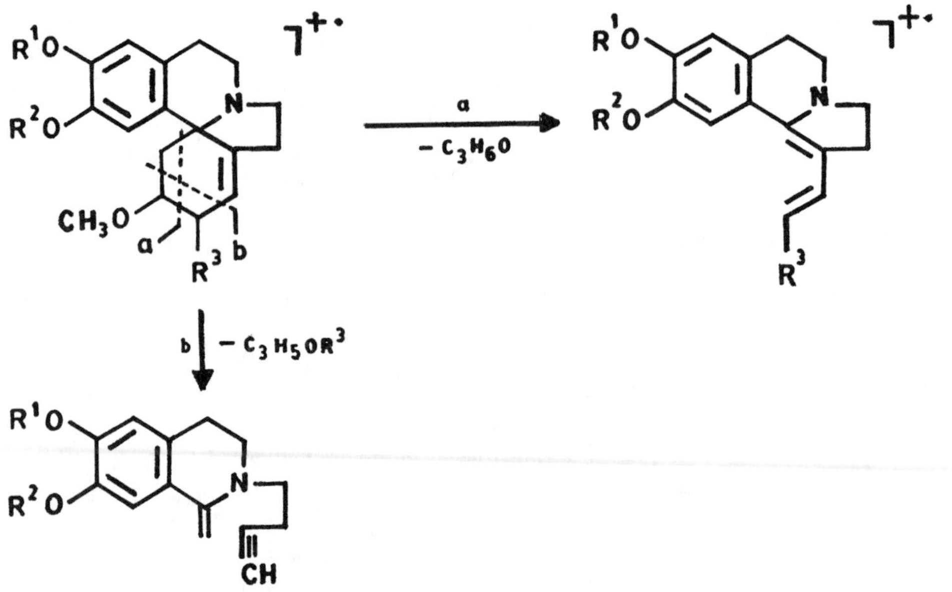

FIGURE 3.3 Mass spectral fragmentation of alkenoid *Erythrina* alkaloids.

3.4   PHARMACOLOGY AND TOXICOLOGY

Erythrina alkaloids in their tertiary basic form have relatively high paralyzing activity.[91,92] Quaternization greatly diminishes the potency. The most active of the alkaloids in this respect is β-erythroidine (**73**). Dihydro-β-erythroidine, obtained by hydrogenation of β-erythroidine, is about five times more active than **73**. The erythroidines are effective by mouth. Depression of blood pressure and respiration are the side effects of these alkaloids. Respiratory depression, decrease in blood pressure, and skeletal muscle relaxation consistent with those of a competitive neuromuscular blocking agent have been reported for erysotrine (**7**) in anesthetized dogs.[93] Muscle relaxant activity has also been reported for the total alkaloids of *E. variegata*.[94] The alkaloids also caused an increase in the sedative effects of hexobarbital.

The total alkaloids obtained from the trunk bark of *E. variegata* are reported to exhibit different pharmacological effects such as neuromuscular blocking, smooth muscle relaxation, central nervous system depression, anticonvulsant, potentiation of phenobarbital hypnosis, inhibition of acetylcholine-induced spasm, and moderate negative inotropic and chronotropic effects.[66,95] The alkaloids did not have any analgesic, antipyretic, antiinflammatory, laxative, and diuretic effects. A recent report,[96] however, makes a mention of antiinflammatory effect of the total alkaloids extracted from the bark of *E. variegata*. Spasmolytic activity has been reported for the ethanolic extract of *Erythrina velutina*.[97] It has been observed that 50% ethanol extractive of the leaves of *E. suberosa* has antineoplastic activity against Sarcoma 180 in the mouse.[98]

REFERENCES

1. Krukoff, B. A. and Earneby, R. C., *Lloydia*, 37, 332, 1974.
2. Krukoff, B. A., *Lloydia*, 40, 407, 1977.
3. Krukoff, B. A., *Allertonia*, 3, 121, 1982.
4. Gunn, C. R. and Barnes, D. E., *Lloydia*, 40, 454, 1977.
5. Marion, I., in *The Alkaloids*, Vol. 2, Manske, R. H. F. and Holmes, H. L., Eds., Academic Press, New York, 1952, 499.

6. Boekelheide, V., in *The Alkaloids*, Vol. 7, Manske, R. H. F., Eds., Academic Press, New York, 1960, 201.

7. Hill, R. K., in *The Alkaloids*, Vol. 9, Manske, R. H. F., Ed., Academic Press, New York, 1967, 483.

8. Mondon, A., in *Chemistry of the Alkaloids*, Pelletier, S. W., Ed., Van Nostrand Reinhold Company, New York, 1970, 173.

9. Kametani, T. and Fukumoto, K., *Synthesis*, 657, 1972.

10. Dyke, S. F. and Quessy, S. N. in *The Alkaloids*, Vol. 18, Rodrigo, R. G. A., Ed., Academic Press, New York, 1981, 1.

11. Snieckus, V. A., *The Alkaloids* (London), 1, 145, 1971; 2, 199, 1972; 3, 180, 1973; 4, 273, 1974; 5, 176, 1975; 7, 176, 1977.

12. De Silva, S. O. and Snieckus, V. A., *The Alkaloids* (London), 8, 144, 1978.

13. Jackson, A. H., *The Alkaloids* (London), 9, 144, 1979.

14. Chawla, A. S. and Jackson, A. H., *The Alkaloids* (London), 11, 137, 1981; 12, 155, 1982; 13, 196, 1983.

15. Chawla, A. S. and Jackson, A. H., *Nat. Prod. Reports*, 1, 371, 1984; 3, 555, 1986; 6, 55, 1989; 7, 565, 1990.

16. Amer, M. E., Shamma, M., and Freyer, A. J., *J. Nat. Prod.*, 54, 329, 1991.

17. Chawla, A. S. and Kapoor, V. K., in *Alkaloids: Chemical & Biological Perspectives*, Vol. 9, Pelletier, S. W., Ed., Elsevier Science Ltd., Oxford, 1995, 85.

18. Tiwari, K. P. and Masood, M., *Phytochemistry*, 18, 704, 1979.

19. Tiwari, K. P. and Masood, M., *Phytochemistry*, 18, 2069, 1979.

20. Masood, M. and Tiwari, K. P., *Phytochemistry*, 19, 490, 1980.

21. Bhat, S. V., Dornauer, H., and De Souza, N. J., *J. Nat. Prod.*, 43, 588, 1980.

22. Chawla, A. S., Gupta, M. P., and Jackson, A. H., *J. Nat. Prod.*, 50, 1146, 1987.

23. Chawla, A. S., Raja Reddy, R., and Jackson, A. H., *Indian J. Pharm. Sci.*, 51, 189, 1989.

24. Sarragiotto, M. H., Filho, H. L., and Marsaioli, A. J., *Can. J. Chem.*, 59, 2771, 1981.

25. Chawla, A. S., Sood, A., Kumar, M., and Jackson, A. H., *Phytochemistry*, 31, 372, 1992.

26. Dagne, E. and Steglich, W., *Tetrahedron Lett.*, 24, 567, 1983.

27. Jackson, A. H., in *The Chemistry and Biology of Isoquinoline Alkaloids*, Phillipson, J. D., Roberts, M. F., and Zenk, M. H., Eds., Springer-Verlag, Berlin, 1985, 62.

28. Amer, M. E., Kassem, F. F., El-Masry, S., Shamma, M., and Freyer, A. J., *Alexandria J. Pharm. Sci.*, 7, 28, 1993.

29. Amer, M. E., El-Masry, S., Shamma, M., and Freyer, A. J., *J. Nat. Prod.*, 54, 161, 1991.

30. Mantle, P. G., *Phytochemistry*, 38, 1315, 1995.

31. Chawla, H. M. and Sharma, S. K., *Fitoterapia*, 64, 15, 1993.

32. Chawla, H. M. and Sharma, S. K., *Fitoterapia*, 64, 383, 1993.

33. Soto-Hernandez, M. and Jackson, A. H., *Phytochem. Anal.*, 4, 97, 1993.

34. Payne, L. D. and Foley, J. P., Anal. Antibiot./Drug Residue Food Prod. Anim. Orig [Proc. Am. Chem. Soc. Agric. Food Chem. Div. Symp. 1991], Agarwal, V. K., Ed., Plenum, New York, 1992, 211; *Chem. Abstr.*, 118, 58330d, 1993.

35. Folkers, K. and Koniuszy, F., *J. Am. Chem. Soc.*, 61, 1232, 1939.

36. Folkers, K. and Koniuszy, F., *J. Am. Chem. Soc.*, 61, 3053, 1939.

37. Folkers, K. and Koniuszy, F., *J. Am. Chem. Soc.*, 62, 436, 1940.

38. Kenner, G. W., Khorana, H. G., and Prelog, V., *Helv. Chim. Acta*, 34, 1969, 1951.

39. Boekelheide, V. and Grundon, M. F., *J. Am. Chem. Soc.*, 75, 2563, 1953.

40. Boekelheide, V., Weinstock, J., Grundon, M. F., Sauvaga, G. L., and Agnello, E. J., *J. Am. Chem. Soc.*, 75, 2550, 1953.

41. Barton, D. H. R., James, R., Kriby, G. W., Turner, D. W., and Widdowson, D. A., *Chem. Commun.*, 294, 1966.

42. Barton, D. H. R., James. R., Kriby, G. W., Turner, D. W., and Widdowson, D. A., *J. Chem. Soc.*, (C), 1529, 1968.

43. Barton, D. H. R., Jenkins, P. N., Letcher, R., Widdowson, D. A., Hough, E., and Rogers, D., *Chem. Commun.*, 391, 1970.

44. Barton, D. H. R., Gunatilaka, A. A. L., Letcher, R. M., Lobo, A. M. F. T., and Widdowson, D. A., *J. Chem. Soc. Perkin Trans. I*, 874, 1973.

45. Ito, K., Furukawa, H., and Tanaka, H., *Chem. Commun.*, 1076, 1970.

46. Ito, K., Furukawa, H., and Tanaka, H., *Chem. Pharm. Bull.*, 19, 1509, 1971.
47. Ito, K., Furukawa, H., and Tanaka, H., *Yakugaku Zasshi*, 93, 1211, 1973; *Chem. Abstr.*, 79, 146713, 1973.
48. Ito, K., Furukawa, H., and Tanaka, H., *Yakugaku Zasshi*, 93, 1215, 1973; *Chem. Abstr.*, 79, 146715, 1973.
49. Ito, K., Furukawa, H., Tanaka, H., and Rai, T., *Yakugaku Zasshi*, 93, 1218, 1973; *Chem. Abstr.*, 79, 146714, 1973.
50. Ito, K., Furukawa, H., and Haruna, M., *Yakugaku Zasshi*, 93, 1611, 1973; *Chem. Abstr.*, 80, 68387, 1974.
51. Ito, K., Furukawa, H., and Haruna, M., *Yakugaku Zasshi*, 93, 1617, 1973; *Chem. Abstr.*, 80, 48212, 1974.
52. Ito, K., Furukawa, H., Haruna, M., and Lu, S. T., *Yakugaku Zasshi*, 93, 1671, 1973; *Chem. Abstr.*, 80, 68390, 1974.
53. Ito, K., Furukawa, H., Haruna, M., and Ito, M., *Yakugaku Zasshi*, 93, 1674, 1973; *Chem. Abstr.*, 80, 68391, 1974.
54. Ito, K., Haruna, M., and Furukawa, H., *Yakugaku Zasshi*, 95, 358, 1975; *Chem. Abstr.*, 82, 167515, 1975.
55. Ito, K., Haruna, M., Jinno, Y., and Furukawa, H., *Chem. Pharm. Bull.*, 24, 52, 1976.
56. Ito, K., Suzuki, F., and Haruna, M., *J. Chem. Soc. Chem. Commun.*, 733, 1978.
57. Singh, H. and Chawla, A. S., *Experientia*, 25, 785, 1969.
58. Singh, H. and Chawla, A. S., *J. Pharm. Sci.*, 59, 1179, 1970.
59. Singh, H. and Chawla, A. S., *Planta Med.*, 19, 71, 1970.
60. Singh, H. and Chawla, A. S., *Planta Med.*, 19, 378, 1971.
61. Singh, H., Chawla, A. S., Jindal, A. K., Conner, A. H., and Rowe, J. W., *Lloydia*, 38, 97, 1975.
62. Singh, H., Chawla, A. S., Kapoor, V. K., Kumar, N., Piatak, D. M., and Nowicki, W., *J. Nat. Prod.*, 44, 526, 1981.
63. Singh, H., Chawla, A. S., Kapoor, V. K., and Kumar, J., *Planta Med.*, 41, 101, 1981.
64. Ghosal, S., Ghosh, D. K., and Dutta, S. K., *Phytochemistry*, 9, 2397, 1970.
65. Ghosal, S., Majumdar, S. K., and Chakraborti, A., *Aust. J. Chem.*, 24, 2733, 1971.
66. Ghosal, S., Dutta, S. K., and Bhattacharya, S. K., *J. Pharm. Sci.*, 61, 1274, 1972.
67. Ghosal, S., Chakraborti, A., and Srivastava, R. S., *Phytochemistry*, 11, 2101, 1972.
68. Ghosal, S. and Srivastava, R. S., *Phytochemistry*, 13, 2603, 1974.
69. Hargreaves, R. T., Johnson, R. D., Millington, D. S., Mondal, M. N., Beavers, W., Becker, L., Young, C., and Rinehart, K. L., Jr., *Lloydia*, 37, 569, 1974.
70. Millington, D. S., Steinman, D. H., and Rinehart, K. L., Jr., *J. Am. Chem. Soc.*, 96, 1909, 1974.
71. Games, D. E., Jackson, A. H., Khan, N. A., and Millington, D. S., *Lloydia*, 37, 581, 1974.
72. Barakat, I., Jackson, A. H., and Abdullah, M. I., *Lloydia*, 40, 471, 1977.
73. Abdullah, M. I., Barakat, I. E., Games, D. E., Ludgate, P., Mavraganis, V. G., Ratnayake, V. U., and Jackson, A. H., *Ann. Missouri Bot. Gard.*, 66, 533, 1979.
74. Jackson, A. H. and Chawla, A. S., *Allertonia*, 3, 39, 1982.
75. Jackson, A. H., Ludgate, P., Mavraganis, V., and Redha, F., *Allertonia*, 3, 47, 1982.
76. Chawla, A. S., Jackson, A. H., and Ludgate, P., *J. Chem. Soc. Perkin I*, 2903, 1982.
77. Chawla, A. S., Chunchatprasert, S., and Jackson, A. H., *Org. Magn. Reson.*, 21, 39, 1983.
78. Chawla, A. S., Redha, F. M. J., and Jackson, A. H., *Phytochemistry*, 24, 1821, 1985.
79. Dagne, E. and Steglich, W., *Phytochemistry*, 23, 449, 1984.
80. Mantle, P. G., Laws, I., and Widdowson, D. A., *Phytochemistry*, 23, 1336, 1984.
81. El-Olmey, M. M., Ali, A. A., and El-Mottaleb, M. A., *Lloydia*, 41, 342, 1978.
82. Ghosh, D. K. and Majumdar, D. N., *Curr. Sci.*, 41, 578, 1972.
83. Chawla, A. S., Krishnan, T. R., Jackson, A. H., and Scalabrin, D. A., *Planta Med.*, 54, 526, 1988.
84. Letcher, R. M., *J. Chem. Soc.* (C), 652, 1971.
85. Ju-ichi, M., Ando, Y., Yoshida, Y., Kunitomo, K., Shingu, T., and Furikawa, H., *Chem. Pharm. Bull.*, 25, 533, 1977.
86. Boar, R. B. and Widdowson, D. A., *J. Chem. Soc.* (B), 1591, 1970.
87. Hanson, A. W., *Proc. Chem. Soc.*, 52, 1963; *Acta Cryst.*, 16, 939, 1963.
88. Wenzinger, G. R. and Boekelheide, V., *Proc. Chem. Soc.*, 53, 1963.
89. Boekelheide, V. and Wenzinger, G. R., *J. Org. Chem.*, 29, 1307, 1964.

90. Hough, E., *Acta Cryst.*, Sect. B, 32, 1154, 1976.
91. Paton, W. D. M., *J. Pharm. Pharmacol.*, 1, 273, 1949.
92. Craig. L. E., in *The Alkaloids*, Vol. 5, Manske, R. H. F., Ed., Academic Press, New York, 1955, 265.
93. Qayum, A., Khanum, K., and Miana, G. A., *Pak Med. Forum.*, 6, 35, 1971; *Chem. Abstr.*, 77, 148526, 1972.
94. Nguyen Van Tuu, Pham Thanh Ky, Pho Duc Thuan, and Do Cong Huynh, *Tap Chi Duoc Hoc*, (6), 13, 26, 1991; *Chem. Abstr.*, 117, 40226u, 1992.
95. Bhattacharya, S. K., Debnath, P. K., Sanyal, A. K., and Ghosal, S., *J. Res. Indian Med.*, 6, 135, 1971.
96. Nguyen Van Tuu, Pham Thanh Ky, Hoang Thi Bach Yen, and Pho Duc Thuan, *Tap Chi Duoc Hoc*, (1), 25, 1992; *Chem. Abstr.*, 117, 62543x, 1992.
97. Barros, S. G., Matos, F. J. A., Vieira, J. E. V., Sousa, M. P., and Medeiros, M. C., *J. Pharm. Pharmacol.*, 22, 116, 1970.
98. Dhar, M. L., Dhar, M. M., Dhawan, B. N., Mehrotra, B. N., and Ray, C., *Indian J. Exp. Biol.*, 6, 232, 1968.

4 Endophyte Alkaloids

J. K. Porter

CONTENTS

4.1 INTRODUCTION

Food and feed toxicity problems associated with humans and livestock have led to the isolation and identification of numerous biologically active plant and fungal alkaloids. Economic factors associated with reproduction problems in livestock, low birth weights, and poor animal performances (i.e., reduced weight gains) on pasture grasses also have catalyzed numerous studies into defining toxic substituents. One of the oldest food and feed borne toxicity problems associated with animal and human health is the *Story of Ergot*.[1] The toxic syndrome, referred to as ergotism (St. Anthony's Fire), has plagued both animals and humans since before the black sclerotia (i.e., ergot-like bodies; cockspur; wolf's tooth, etc.) of *Claviceps purpurea* (Fries) Tulasne and *Claviceps paspali* Stevens and Hall on wheat and rye were associated with the gangrenous and convulsive forms of ergotism that swept all of Europe during the Middle Ages.[1] The subsequent isolation and identification of the active constituents produced by these fungi began to unravel one of the most potent and biologically significant group of compounds in the medicinal sciences: *the ergot alkaloids* (or, *Die Mutterkornalkaloide* = alkaloids of the mother grain; the reader is referred to the detailed reviews of Bove, Hofmann, Floss, and Berde and Schild).[1-4] Currently, ergotism is not an immediate threat to humans since grain regulations limit *Claviceps*-(ergot)-contaminated products from entering foods used for human consumption. Nevertheless, there is a prevalent and constant threat to domestic livestock and an indirect risk to humans.[5] In addition to grains infected with *Claviceps*, livestock grazed on clavicipitaceous endophyte (fungal)-infected pasture grasses suffer devastating production losses due to the same ergot alkaloids.[6-10]

Throughout the modern era of agriculture, classical ergot toxicity, not related to *Claviceps*, has been described in livestock grazed on pastures which later were identified as endophyte-infected: e.g., tall fescue (*Festuca arundinaceae*. Scrib.) toxicity in cattle ("gangrenous fescue foot") and horses (agalactia in pregnant mares), and perennial ryegrass (*Lolium perenne* L.)

51

staggers in sheep and other livestock ("convulsive rye grass staggers").[11-15] Although *Acremonium spp.*, the ergopeptine alkaloids, and the indole-diterpenoid lolitrem alkaloids, respectively, have been defined as the endophytes and compounds associated with these two toxic syndromes in livestock,[12,13,15,16] other genera of endophytic fungi (i.e., *Epichloe, Balansia, Myriogenospora*) also have been related to atypical or idiopathic ergotism (that not caused by *Claviceps*) in livestock on pasture grasses.[13,14,16,17] Most interestingly, *Acremonium, Epichloe, Balansia, Myriogenospora,* and *Claviceps* are taxonomically aligned within the Clavicipitaceae;[18,19] and whereas ergot alkaloid biosynthesis was once thought specific for the genus *Claviceps*, because of their corresponding evolutionary relationships, it is not surprising to find that the *Acremonium, Epichloe,* and *Balansia* endophytes produce the same compounds.

This chapter will address the known ergot and other alkaloids produced by selected endophytes with evolutionary and biosynthetic interrelationships to alkaloid production (i.e., the lolitrem alkaloids) and those alkaloids believed to result from the endophyte-grass host associations (i.e., the pyrrollizidine loline alkaloids). Although primary emphasis will be placed on the ergot and lolitrem alkaloids produced by *Acremonium spp.*, because of their historical significance and genetic relationships, alkaloid production and identification by *Epichloe spp.* and *Balansia spp.* also will be considered. The chemical structures, biosynthesis, and precursor relationships (i.e., beginning with anthranilic acid and tryptophan) for production of the simple ergot alkaloids (clavine, lysergic acid, simple lysergic acid amides); the tricyclic ergopeptine (and ergopeptam) alkaloids; and the lolitrem alkaloids have been presented in earlier reviews along with the lolines and the pyrolopyrazine alkaloid peramine.[16] Livestock toxicities, reduced reproduction efficiency, reduced weight gains, neuroendocrine mechanisms, physiological parameters, and toxic syndromes ("gangrenous" fescue foot; "convulsive" rye grass staggers) in livestock (cattle, sheep, horses, and other ungulates) grazed on infected pastures also have been presented in reviews with major considerations directed at these compound classes.[5,8-17]

(**Note added in proof:** based on DNA sequence analyses, *Acremonium coenophialum* (Morgan-Jones & W. Gams), *A. typhinum* (Morgan-Jones & W. Gams), *A. lolii* (Latch, Christensen & Samuels), *A. chisosum* (J.F. White & Morgan-Jones), *A. starrii* (J.F. White & Morgan-Jones), and *A. unicinatum* (W. Gams, Petrini & D. Schmidt) have been reclassified as: *Neotyphodium coenophialum-, N. typhinum-, N. lolii-, N. chisosum-, N. starrii-,* and *N. unicinatum-* Glenn, Bacon, & Hanlin comb. nov., respectively. A dual system of nomenclature differentiates *Neotyphodium* and the sexually reproducing species classified in *Epichloe*).[89]

4.2 NATURAL OCCURRENCE

4.2.1 ERGOT ALKALOIDS

One of the first correlations of endophytic fungi with ergot-like syndromes in cattle on pasture grasses stemmed from investigations of a disease known as "bermuda grass (*Cynodon dactylon*) tremors" which affected some 50,000 animals in Louisiana (U.S.) in the early 1970s.[20,21] Almost simultaneously, *Balansia epichloe, Balansia henningsiana, Balansia strangulans,* and *Epichloe typhina* were associated with toxic pastures of bermuda grass and toxic tall fescue pastures throughout the southeastern United States.[22-31] The indoleglycerols, [*erythro-* and *threo*-1-(3-indoly)propane-1,2,3-triol; 4-(3-indoly)butane-1,2,3-triol); and 3-(3,3-diindolyl)propane-1,2-diol],[24,26] the ergot clavine (chanoclavines; agroclavine; elymoclavine; penniclavine), and simple lysergic acid amide alkaloids (ergonovine; ergonovinine) were isolated from the *Balansia spp.*,[25,27-29] and the ergot peptide alkaloids (ergovaline, ergosine) were subsequently identified from cultures of *E. typhina*;[25,30,31] these studies were the *prima facie* linkage between the endophytes and atypical ergotism in cattle on pasture grasses and the production of lysergic acid amide derivatives in fungi outside of the genus *Claviceps*. However, it was not until ergovaline was isolated and identified as the major ergot peptide alkaloid from toxic *E. typhina*-infected fescue [32,33] that the ergot alkaloid-endophyte associations became recognized as significant (For a historical perspectives, the reader is directed to previous reviews).[11,16,17,34,35] Predicated on *in vitro* culture studies, a taxonomic reclassification of the endophyte of toxic tall fescue from *E. typhina* to *Acremonium coenophialum* Morgan-Jones and Gams was initiated.[36] Subsequent investigations linked the endophyte *A. lolii* Latch, Christensen, Samuels, and the tremorgenic lolitrem alkloids with perennial ryegrass staggers in sheep.[37,38] Although the taxonomy of *Acremonium/Epichloe* is still unsettled,[34,39] current relationships according to both DNA characterizations and biogenesis of the ergot peptide and lolitrem alkaloids would suggest these two are synonymous.[18,19] Then too, the teleomorphic state of *A. typhinum* Morgan-Jones and Gams is *E. typhina*.[37] (The reader is referred to the **note added in proof** after the Introduction and Reference 89 for current taxonomy on *Acremonium* endophytes.)

Our primary knowledge of the endophyte alkaloids (i.e., from *Acremonium*-infected grasses) stems from attempts to identify the constituents related to fescue toxicity in cattle and horses and ryegrass staggers in sheep (and other livestock) and attempts to identify those compounds which impart insect resistance and drought tolerance to the infected plants.[40,41] Limited research has been conducted within the *Acremonium*, *Epichloe*, and *Balansia* for the commercial production of these compounds, but an evaluation of these fungi, their plant host, alkaloid production, and their relationships to livestock toxicities and plant defense mechanisms underscores how important the endophyte-grass associations play within the ecological survival of both fungi and grass species. These evaluations also demonstrate the economic significance to livestock production and that deficiencies in animal performances (i.e., weight gains, reproduction efficiency) on endophyte-infected pasture and range grasses is a ubiquitous problem. Concomitantly, utilization of both pasture and range grasses for livestock production is dependant on the mutualistic grass-endophyte symbionts.

Tables 4.1 and 4.2 outline the known ergot alkaloids produced *in vitro* by the grass endophytes with their current (*in vivo*) implications from the endophyte-infected grasses. Ergovaline, ergosine, lysergic acid amide, and chanoclavine are the major naturally occurring ergot alkaloids isolated from *A. coenophialum*, *E. typhina*, and infected grass and seeds of fescue.[32,33,42–45] Perennial ryegrass infected with *A. lolii*, *E. typhina,* and artificially infected with *A. coenophialum* also produced ergovaline as the primary ergot peptide alkaloid.[42,43,46] Similar results have been reported with undefined endophyte-infected fescue and perennial ryegrasses.[47] Although minor quantities of clavine, lysergic acid amides, and ergot peptide alkaloids (Table 4.2) have been isolated from endophyte-infected fescue, attention has focused on ergovaline because it is the primary ergopeptine produced in culture and infected plants and associated with toxicity in domestic animals (see below).

TABLE 4.1
Principal Endophytic Fungi of the Clavicipitaceae that Produce either Ergot and/or Lolitrem Alkaloids *In Vitro* and/or *In Planta*

Epichloe typhina

Acremonium coenophialum; A. lolii; A. typhinum; A. uncinatum

Balansia epichloe; B. henningsiana; B. claviceps; B. strangulans; B. obtecta; B. cyperi

For taxonomic cosanguinity, *in vitro* approaches, and procedures in *Epichloe/Acremonium* refer to references 36, 37, 43, 71, 86, 87; *Balansia* refer to references 71, 88.

The isolation and identification of ergobalansine and its C-8 epimer ergobalansinine from cultures of *Balansia cyperi, Balansia obtecta,* and also *Balansia obtecta*-infected sanbur grass (*Cenchrus echinatus*) has introduced yet another novel and unique class of ergot peptide

TABLE 4.2
Ergot Alkaloids from Grass Endophytes (Culture and/or Infected Plants)

Acremonium and *Epichloe* spp.[25,30–33,44–47,49, 72–75]	*Balansia* spp.[25,27–29,31,48]
Ergovaline[a]	Ergonovine[a]
Ergovalinine	Ergonovinine
Ergosine	Ergobalansine
Ergosinine	Ergobalansinine
beta-Ergosine	6,7,-Secoagroclavine
Ergotamine	Penniclavine
Ergotaminine	Elymoclavine
Ergonine	Agroclavine
Ergocrystine	Isochanoclavine I
alpha- and *beta*-Ergokryptine	Chanoclavine I
Ergocornine	Dihydroelymoclavine
Ergonovine	Isodihydrolysergamide[b]
Ergonovinine	
Lysergic acid amide (Ergine)	
Isolysergic acid amide (Erginine)	
Lysergylmethylcarbinolamide[c]	
Lysergic acid	
Isolysergic acid	
8-Hydroxylysergic acid amide	
Penniclavine	
Elymoclavine	
Agroclavine	
6,7-Secoagroclavine	
Festuclavine	
Chanoclavine I, II	

Note: -*inine* suffix and *iso*- prefix denotes the C-8 epimers (not considered naturally occurring alkaloids but an epimerization product from isolation procedures)

[a] major alkaloid produced *in vitro* and *in planta*;
[b] tentative identification;
[c] by inference from the isolation and identification of lysergic acid amide.[16,44,49]

alkaloids.[48] This class begins a new series devoid of the proline ring in the tricyclic peptide portion of the molecule and truly opens avenues so far unrecognized in ergoline biosynthesis. Yet in another study, *Acremonium*-infected *Stipa robusta* (Sleepygrass) was shown to contain lysergic acid amide, 8-hydroxylysergic acid amide, ergonovine, and chanoclavine I. Lysergic acid amide was the major alkaloid (20 ug/g) and was suggested as the primary agent responsible for the sedative effects in horses.[49] These studies suggest *Acremonium*-endophytes have the biogenetic capability to produce an entire spectrum of known and novel ergot alkaloids.

4.2.2 LOLITREM ALKALOIDS

The indole-diterpenoid alkaloids and related biosynthetic precursors produced by *A. lolii* are paxilline, paxitriol(s), lolitriol, and lolitrems A through E(F) (Table 4.3).[38,40–54] Although the lolitrem alkaloids A and B are primarily associated with the toxic *A. lolii*-infected perennial ryegrass (see biological activity),[54] paxilline and lolitrem B also have been identified in

TABLE 4.3

Lolitrem Alkaloids and Biosynthetic Precursors Isolated from *Acremonium lolii*-infected Perennial Ryegrass and Cultures or other *Acremonium spp.*-Infected Grasses[13,14,38,42,50–55]

Paxilline[a,b]; *alpha*-Paxitrol; Lolitriol
Lolitrem A(A1 and A2)[b,c]; Lolitrem B[a,b,c]; Lolitrem C,
Lolitrem D; Lolitrem E[c]; (Lolitrem F)

[a] Major alkaloids produced *in vitro* and *in planta* (B, 77%; E, 15%; A, 4%);[54]
[b] Related to neurological tremors syndrome;
[c] Also reported as having insect resistance activity.[54]

cultures of *A. coenophialum* and *E. typhina*-infected plants.[42,55] *A. lolii* isolated from varieties of ryegrass and artificially introduced into varieties of fescue still produced predominately lolitrem B.[42]

4.2.3 PERAMINE

The pyrrolopyrazine alkaloid peramine is the primary alkaloid associated with insect deterrences in *A. lolii*-infected perennial ryegrass.[42,56–60] The alkaloid also is produced by *A. coenophialum, E. typhina,* and *A. starrii*; it has a wide distribution within the endophyte infected grasses, and it occupies a major role in the resistance of ryegrass to the Argentine stem weevil, sod web worm, and certain species of aphids.[42,60] Currently, no other natural analogues of peramine have been identified or associated with the endophyte/grass hosts which possibly suggests another specificity within this genera.

4.2.4 LOLINE ALKALOIDS

Even though there does not appear to be a direct biosynthetic relationship between the ergot and pyrrollizidine alkaloids N-acetylloline, N-formylloline, and the other lolines (Table 4.4),

TABLE 4.4

Alkaloids Isolated from Endophyte-Infected Grasses Primarily Associated with Insect Resistance[a] and/or Identified as Compounds Resulting from the Endophyte-Grass Associations[b]

Peramine[a]
Loline (festucine) [a,b]; N-acetylloline[a,b]; N-formylloline[a,b]; N-methylloline
Norloline; N-acetylnorloline

Note: Peramine is produced by the endophytes, *A. lolii , A. coenophialum, E. typhina;*[56–60] the lolines are related to the endophytes-(*A. coenophialum/E. typhina*)-host grass associations;[47,61–63,65,66,70,73] N-acetylloline, N-formylloline = major alkaloids isolated from infected plants.[47,62]

these compounds are referenced because of their important associations with *A. coenophi-alum*-grass host symbionts, their contribution to insect deterrence, alleopathy, and their contributions to livestock toxicities.[61–70] Production of these alkaloids within the grass-host symbiont associations is not well defined but may be indirectly related with available nitrogen, the free amino acid pool, and stress mechanisms on the grass and fungus, (see below); thus, these important definitions await future investigations.Quantitatively, the loline alkaloids (primarily N-acetylloline and N-formylloline) by far exceed (100-1000 x the concentrations) either the ergot, peramine, and/or lolitrem alkaloids produced within the *A. coenophialum*-hosts associations.[47,62] The natural occurrence of these alkaloids depends on the interdependent associations of *A. coenophialum* with their grass hosts for they have not been isolated from either fungal cultures or uninfected plants.[42,62] However, when *A. coenophialum* from its natural host of tall fescue was artificially introduced into ryegrass, production of the loline alkaloids equivalent to its natural state was observed.[42,62] In addition to *A. coenophialum*-infected fescue, loline alkaloid accumulation also has been reported to be extremely high (mg/g) in meadow fescue (*F. pratensis* Hudson) infected with *A. uncinatum*.[62]

4.3   FACTORS AFFECTING ALKALOID PRODUCTION BY ENDOPHYTES: BIOLOGICAL, PHYSICAL, CHEMICAL

Alkaloid production by most endophyte-host associations is predominately related to defense and species survival and therefore is contingent on dynamic, environmental influences. Consequently, culturing of these endophytes has not reached commercialization and generally resides within laboratory investigations. Except for the unique production of ergobalansine, the novel 6,7-secoagroclavine, dihydroelymoclavine, and festuclavine by certain species of *Balansia*,[31,48] and the use of *A. coenophialum* and *A. lolii* to produce confirmatory standards of ergovaline and the lolitrems, utilization of *Acremonium* vs. *Claviceps* for the commercial production of the ergot peptide (and/or the lolitrem) alkaloids has met with limited interest.

Bacon and White have reviewed stains, media, and procedures for analyzing *Balansia* and *Acremonium* endophytes and for their production of ergot alkaloids.[71] Apparently, growth is dependent on fresh isolates and "undefined growth factors" in a sorbitol medium. Mantle and Weedon[53] have referenced media for the production of the lolitrems in cultures of *A. lolii* which apparently follow growth typical of the species.[37] The *in vitro* alkaloid production by the endophytes may depend on whether the fungus has a parasitic, a mutualistic, or epiphytic relationship with its host. Siegle et al. have defined *in planta* alkaloid production by *Acremonium* and *Epichloe* as an endophyte-plant response to herbivorous insects (and possibly livestock grazing pressure).[42] Bush et al. have outlined a similar relation to production of the loline alkaloids in *A. coenophialum*-infected fescue.[62] Then too, alkaloid production may depend on: the endophyte's growth within the plant, regulatory mechanisms of the host on the fungus, and (obviously) the endophyte's genetic ability for alkaloid biosynthesis.[43] Although the ergot alkaloids and peramine biosyntheses is by *A. coenophialum* within infected tall fescue, Hill et al.[72] indicate ergot alkaloid concentration (i.e., ergovaline) is predominantly regulated by the plant while Roylance et al.[73] indicate peramine production *in planta* is controlled by the endophyte. Other studies have depicted a relationship between alkaloid production and environmental stress (temperature/drought/available water), available nitrogen and free amino acids; carbohydrate (sugar alcohols) levels and osmotic pressure related to plant viability.[74–77] Carbohydrate levels, in this regard, are also needed as a carbon source for alkaloid biosynthesis. The isolation of the intermediate clavines along with the tricyclic peptide alkaloids[25,28,30,32,33,45] from cultures and endophyte-infected plants suggest similar biosynthetic pathways as those produced by *Claviceps*; their interrelationships with tryptophan metabolism and amino acid and carbohydrate availability have been discussed.[16] Mantle and Weedon [53] have outlined a hypothetical metabolic grid for the biosynthesis of the lolitrems alkaloids by *A. lolii* with interrelated considerations for similar alkaloids produced by *Claviceps, Penicillium, and Aspergillus*. Monday-Finch et al. [54] have related stereochemical considerations in the biosynthesis of lolitrem A (A1 and A2) with previous suggestions outlined by Miles et al.[51,52]

The natural production of ergot alkaloids by *Acremonium* and *Epichloe* and those that have been isolated from *Balansia spp.*, respectively, fall into two distinct classes somewhat analogous to observations with certain species of *Claviceps* (e.g., *C. purpurea* and *C. paspali*). Except for the unique ergot peptide alkaloids produced by *B. obtecta* and *B. cyperi* (i.e., ergobalansine), [48] production of the tricyclic-ergot peptide alkaloids appears to reside with *Acremonium* and *Epichloe* (Table 4.2). Apparently, *A. lolii, A. coenophialum,* and other *Acremonium* biotypes have developed the capability to produce either and/or the ergopeptine and lolitrem alkaloids *in planta* and *in vitro*. From an evolutionary standpoint, *Acremonium* and *Epichloe* are the most advanced group of grass endophytes which may have significance to their biogenesis and survival among grass species.[18] Although not an alkaloid, production of 5-*alpha*-ergosta-7,22E-dien-3*beta*-ol by *A. lolii* has been suggested as a diagnostic character to differentiate this endophyte[53] from *A. coenophialum* in which ergosta-4,6,8,(14),22-tetraene-3-one has been defined as the principal steroid.[78] The heteroannular triene-3-one steroid was originally associated with bermuda grass tremors in cattle, toxic endophyte infected fescue (i.e., *E. typhina* pre *A. coenophialum*), cultures of *B. epichloe* and infected smut grass, and cultures of a *Claviceps* isolated from the toxic bermuda grass; this steroid was suggested as a field indicator for ergot-type toxicities in cattle.[16,21]

4.4   BIOLOGICAL ACTIVITY

A plethora of literature exists on the biological activities of the ergot alkaloids in mammalian systems (humans, nonhuman primates, rats, mice, other laboratory animals) and also on their associated toxicities to both humans and livestock.[1,4] Because of their historical significance to both gangrenous and convulsive ergotism, their therapeutic applications in migraines and obstetrics, and the infamous semisynthetic hallucinogenic lysergic acid diethylamide (LSD), the ergot alkaloids are one of the most pharmacologically defined group of compounds in the medicinal sciences.[4] Nevertheless, it was not until the isolation and identification of ergovaline from *Acremonium*-infected plants that concerted studies were initiated to more clearly define their activities in ruminants and other ungulates. In cattle, decreased prolactin and melatonin, increased internal body temperatures and vasoconstriction, reduced weight gains, reproduction problems, and agalactia in pregnant mares have been related to the known effects of the ergot alkaloids in infected fescue pastures.[5,8–12,17,79–81]

Production problems in animals on *A. coenophialum*-infected fescue are defined more toward an epiphenomenal synergistic effect of the co-occurring alkaloids along with the environmental stress conditions of temperature (summer heat/winter cold). There is little doubt that environmental temperatures and stress on animals play a major role with enhanced activities of these compounds and the manifestation of toxic syndromes in livestock.[8–12,17,79–81] Also, the convulsive or neurological tremors syndrome associated with *A. lolii*-infected perennial ryegrass has been related to both lolitrems B and A.[13,14,54] The above relationships are consistent with the most efficient survival mechanisms involving plant resistance to herbivorous insects, which have been described in terms of the co-occurrence of peramine, the lolines, and the ergot alkaloids.[42,43,56,60,62,64–67,82] Moreover, insect deterrence has been reported for the lolitrems A, B, and E.[54]

Some avenues for combating these problems in livestock have been outlined and involve therapeutic practices; treatment with compounds that inhibit the effects of the ergot alkaloids (antidopaminergics); general vaccines against the toxins; breed selection for resistance or tolerance; toxin dilution with supplemental feeding; and the genetic exploitation of the grass and fungus to produce a specific host-endophyte symbiont with all the qualities needed in pasture grasses without the livestock toxicity problems.[8–12,17,19,60,79–81,83] The latter approach may be more practical for the production of turf and certain ornamental grasses not utilized for livestock. But genetic strategies that exploit the positive aspects of grass and fungal symbionts should be approached parsimoniously. Grazing animals along with wildlife (rabbits, birds, deer, bison, rodents, and insects) represent vectors in the transmission and dissemination of endophyte-grass symbionts. A priori, with the evolutionary probabilities for species survival among the grass and fungal endophytes, prudence dictates this discipline be approached under guarded management practices to prevent more problems from occurring than it is designed to solve. For example, new cultivars of *A. lolii* from perennial ryegrass that demonstrated negligible ryegrass staggers toxicity to livestock and retained resistance to the Argentine stem weevil, produced ergovaline at levels too high to consider for commercial use.[60] Other parameters within these considerations have been outlined.[54] Therefore, a grass-fungus symbiont combination that aberrantly evolves either similar or more draconian protective mechanisms could and would turn into a major environmental bane for domestic livestock, wildlife, and subsequently human populations (i.e., the *Pandora* effect). Release of new genetically-designed endophyte-grass symbionts should be considered only after comprehensive chemical analyses and the implementation of controlled agronomic and protected animal trials.

4.5   CONCLUSIONS

Endophytic fungi that live their entire life cycle within their plant host have survived by developing a close interrelationship beneficial to both. Throughout the evolutionary processes, subsistence for both fungus and plant host became dependent on a reciprocal exchange of biochemical mechanisms and the subsequent production of compounds needed for their mutual survival; that is, production of those compounds associated with defense (the ergot alkaloids, the pyrrollizidine loline alkaloids, the pyrrolopyrazine alkaloid peramine, and to some extent the lolitrem alkaloids[54] provide protection against herbivorous insects); production of compounds associated with growth (auxin production; indole acetic acid, indole acetamide, indole ethanol, indole glycerols, etc.,[16,84,85]); production of compounds associated with dissemination (alleochemical production; possibly the alleopathic activities of the loline alkaloids[63]); and the development of mechanisms associated with proliferation of the species (via sexual or asexual reproduction: *Epichloe* vs. *Acremonium*[19,86]). Understanding the interrelationships with the endophyte-host plants associations and alkaloid production promises to be one of our most unique natural resources for controlling and developing agriculture on numerous levels and one of the most prolific sources of novel biologically active compounds.

REFERENCES

1. Bove, F. J., *The Story of Ergot*, S. Karger, New York., 1970.
2. Hofmann, A., *Die Mutterkornalkaloide*, Enke Verlag, Stuttgart, 1964.
3. Floss, H. G., Biosynthesis of ergot alkaloids and related compounds, *Tetrahedron*, 32, 873-912, 1976.
4. Berde, B. and Schild, H. O., Ergot Alkaloids and Related Compounds, *Handbook Exp. Pharm.*, Vol 49, Springer-Verlag, New York, 1978.
5. Thompson, F. N. and Porter, J. K., Tall fescue toxicoxes in cattle: could there be a public health problem, *Vet. Human Toxicol.*, 32, 51-57, 1991.
6. Stuedemann, J. A. and Hoveland, C. S., Fescue Toxicity: History and impact on animal agriculture, *J. Prod. Agric.*, 1, 39-44, 1988.

7. Hoveland, C. S., Importance and economic significance of *Acremonium* endophytes to performance of animals and grass plant, in: *Acremonium/Grass Interactions*, Joost, R. E. and Quisenberry, S. S., Eds., Elsevier Scientific Publishers, Amsterdam, Netherlands, *Agric., Ecosys. Envir.*, 44, 3-12, 1993.

8. Paterson, J., Forcherio, C., Larson, B., Samford, M., and Kerley, M., The effects of fescue toxicosis in beef cattle productivity, *J. Anim. Sci.,* 73, 889-898, 1995.

9. Cheeke, P. R., Endogenous toxins and mycotoxins in forage grasses and their effects on livestock, *J. Anim. Sci.*, 73, 909-918, 1995.

10. Cross, D. L., Redmond, L. M., and Strickland, J. R., Equine fescue toxicosis: signs and solutions, *J. Anim. Sci.*, 73, 899-908, 1995.

11. Robbins, J. D., Porter, J. K., and Bacon, C. W., Occurrence and clinical manifestation of ergot and fescue toxicoses, in: *Diagnosis of Mycotoxicoses*, Richards, J. L. and Thurston, J. R., Eds., Martinus Nijhoff Publishers, Dordrecht, Netherlands, pp 61-74, 1986.

12. Bacon, C. W., Lyons, P. C., Porter, J. K., and Robbins, J. D., Ergot toxicities from endophyte-infected grasses: a review, *Agronomy J.*, 78, 106-116, 1986.

13. Prestidge, R. A., Causes and control of perennial ryegrass staggers in New Zealand, in: *Acremonium/Grass Interactions*, Joost, R. E. and Quisenberry, S. S., Eds., Elsevier Scientific Publishers, Amsterdam, Netherlands, *Agric., Ecosys. Envir.*, 44, 283-300, 1993.

14. Cunningham, P. J., Foot, J. Z., and Reed, K. F. M., Perennial ryegrass (*Lolium perenne*) endophyte (*Acremonium lolii*) relationships: the Australian experience, in: *Acremonium/Grass Interactions*, Joost, R. E. and Quisenberry, S. S., Eds., Elsevier Scientific Publishers, Amsterdam, Netherlands, *Agric., Ecosys. Envir.*, 44, 157-168, 1993.

15. Yates, S. G., Tall fescue toxins, in: *Handbook of Naturally Occurring Food Toxicants*, M. Recheigl, Ed., CRC Press Inc., Boca Raton, FL, pp. 249-273, 1983.

16. Porter, J. K., Chemical constituents of grass endophytes, Chapter 8, in: *Biotechnology of Endophytic Fungi of Grasses*, Bacon, C. W. and White, J. F., Jr., Eds., CRC Press, Boca Raton, pp. 103-123, 1994.

17. Porter, J. K. and Thompson, F.N. Jr., Effects of fescue toxicosis on reproduction in livestock, *J. Anim. Sci.,* 70, 1594-1603, 1992.

18. Bacon, C. W., Fungal endophytes, other fungi, and their metabolites as extrinsic factors of grass quality, Chapter 8, in: *Forage Quality, Evaluation and Utilization,* Fahey, G. C., Jr., Ed., American Society of Agronomy, Madison, Wisconsin, pp. 318-366, 1994.

19. Schardle, C. L. and Siegel, M. R., Molecular genetics of *Epichloe typhina* and *Acremonium coenophialum*, in: *Acremonium/Grass Interactions*, Joost, R. and Quisenberry, S., Eds., Elsevier Scientific Publishers, Amsterdam, Netherlands, *Agric. Ecosys. Envir.*, 44, 169-185, 1993.

20. Porter, J. K., Bacon, C. W., and Robbins, J. D., Major alkaloids of a *Claviceps* isolated from toxic bermuda grass, *J. Agric. Food Chem.*, 22, 838-841, 1974.

21. Porter, J. K., Bacon, C. W., Robbins, J. D., and Higman, H. C., A field indicator in plants associated with ergot-type toxicities in cattle, *J. Agric. Food Chem.* 23, 771-775, 1975.

22. Bacon, C. W., Porter, J. K., and Robbins, J. D., Toxicity and occurrence of *Balansia* on grasses from toxic fescue pastures, *Applied Microbiol.*, 29, 553-556, 1975.

23. Bacon C. W., Porter, J. K., Robbins, J. D., and Luttrell, E. S., *Epichloe typhina* from toxic tall fescue, *Applied Environ. Microbiol.*, 34, 576-581,1977.

24. Porter, J. K., Bacon, C. W., Robbins, J. D., Himmelsbach, D. S., and Higman, H. C., Indole alkaloids from *Balansia epichloe* (Weese), *J. Agric. Food Chem.*, 25, 88-93, 1977.

25. Porter, J. K., Bacon, C. W., and Robbins, J. D., Clavicipitaceae: *Claviceps* related fungi and their production of ergot alkaloids, *Lloydia* (*J. Nat. Prod.*) 41, 654-655, 1978.

26. Porter, J. K., Robbins, J. D., Bacon, C.W., Himmelsbach, D. S., and Haeberer, A. F., Determination of epimeric 1-(3-indolyl)-propane-1,2,3-triol isolated from *Balansia epichloe*, *Lloydia* (*J. Nat. Prod.*) 41, 43-49, 1978.

27. Bacon, C. W., Porter, J. K., and Robbins, J. D., Laboratory production of ergot alkaloids by species of *Balansia*, *J. Gen. Microbiol.*, 113, 119-126, 1979.

28. Porter, J. K., Bacon, C. W., and Robbins, J. D, Lysergic acid amide derivatives from *Balansia epichloe* and *Balansia claviceps* (Clavicipitaceae), *J. Nat. Prod.* 42, 309-314, 1979.

29. Bacon, C. W., Porter, J. K., and Robbins, J. D., Ergot alkaloid biosynthesis by isolates of *Balansia epichloe* and *Balansia henningsiana*, *Can. J. Bot.* 59, 2534-2538, 1981.

30. Porter, J. K., Bacon, C. W., and Robbins, J. D., Ergosine, ergosinine and chanoclavine I from *Epichole typhina*, *J. Agric. Food Chem.*, 27, 595-598, 1979.

31. Porter, J. K., Bacon, C. W., Robbins, J. D., and Betowski, D., Ergot alkaloid identification in clavicipitaceae systemic fungi of pasture grasses, *J. Agric. Food Chem.*, 29, 653-657, 1981.

32. Lyons, P. C., Plattner, R. D., and Bacon, C. W., Occurrence of peptide and clavine alkaloids in tall fescue grass, *Science*, 232, 487-489, 1986.

33. Yates, S. G., Plattner, R. D., and Garner, G. B., Detection of ergopeptine alkaloids in endophyte-infected, toxic K-31 fescue by mass spectrometry/mass spectrometry, *J. Agric. Food Chem.*, 33, 719-722, 1985.

34. Garner, G. B., Rottinghaus, G. E., Cornell, C. N., and Testereci, H., Chemistry of compounds associated with endophyte/grass interaction: ergovaline and ergopeptine related alkaloids, in: *Acremonium/Grass Interactions*, Joost, R., Quisenberry S., Eds., Elsevier Scientific Publishers, Amsterdam, Netherlands, *Agric. Ecosys. Envir.*, 44, 65-80, 1993.

35. Bacon, C. W., Toxic endophyte-infected tall fescue and range grasses: historical perspectives, *J.Anim.Sci.*,73, 861-870.

36. Morgan-Jones, G. and Gams, W., *Notes on Hypomycetes*, XLI, An endophyte of *Festuca arundinacea* and the anamorph of *Epichloe typhina*, new taxa in one of two new sections of *Acremonium*, *Mycotaxon*, 15, 311, 1982

37. Latch, G. C. M., Christensen, M. J., and Samuels, G. J., Five endophytes of *Lolium and Festuca* in New Zealand, *Mycotaxon*, 20, 535-550, 1984.

38. Gallagher, R. T., Hawkes, A. D., Steyn, P. S., and Vlleggaar, R., Tremorgenic neurotoxins from perennial ryegrass causing ryegrass staggers disorders of livestock: structure and elucidation of lolitrem B., *J. Chem. Soc., Chem Commun.*, 614-616, 1984.

39. Rykard, D. M., Bacon, C. W., and Lutrell, E. S., Host relations of Myriogenospora atrementosa and *Balansia epichloe* (Clavicipitaceae), *Phytopathology* 75, 950-956, 1985.

40. *Biotechnology of Endophytic Fungi of Grasses*, Bacon, C. W. and White, J. F., Jr., Eds., 1994, CRC Press, Boca Raton.

41. *Acremonium*/Grass Interactions, Joost, R. and Quisenberry, S., Eds., Elsevier Scientific Publishers, Amsterdam, Netherlands, 1993.

42. Siegel, M. R., Latch, G. C. M., Bush, L .B., Fannin, F. F., Rowan, D .D., Tapper, B. A., Bacon, C. W., and Johnson, M. C., Fungal endophyte-infected grasses: alkaloid accumulation and aphid response, *J. Chem. Ecol.*, 16, 3301-3315, 1990.

43. Siegel, M. R. and Bush, L. P., Importance of endophytes in forage grasses, a statement of problems and selection of endophytes, Chapter 10, in: *Biotechnology of Endophytic Fungi of Grasses*, Bacon, C. W. and White, J. F., Jr., Eds., CRC Press, Boca Raton, pp 135-150, 1994.

44. Shelby, R. A., Improved HPLC method for the detection of ergot alkaloids in endophyte-infected tall fescue, p.3, in: Proc. Tall Fescue Toxicoses Work Shop, Southern Extension and Research Activity Information Exchange-8, Nov. 16-17, Memphis, TN, 1992.

45. Yates, S. G. and Powell, R. G., Analysis of ergopeptine alkaloids in endophyte-infected tall fescue. *J. Agric. Food Chem.* 36:337-340, 1988.

46. Rowan, D. D. and Shaw, G. J., Detection of ergopeptine alkaloids in endophyte-infected perennial ryegrass by tandem mass spectrometry, *New Zealand Vet. J.*, 35, 197, 1987.

47. TePaske, M. R., Powell, R. G., and Clement, S. L., Analyses of selected endophyte-infected grasses for the presence of loline-type and ergot-type alkaloids, *J. Agric. Food Chem.*, 41, 229-2303, 1993.

48. Powell, R. G., Plattner, R. D., Yates, S. G., Clay, K., and Leuchtmann., A., Ergobalansine, a new ergot-type peptide alkaloid isolated from *Cenchrus echinatus* (sandbur grass) infected with *Balansia obtecta*, and produced in liquid cultures of *B. obtecta* and *Balansia cyperi*, *J. Nat. Prod.*, 53, 1272-1279, 1990.

49. Petroski, R. J., Powell, R. G., and Clay, K., 1992. Alkaloids of *Stipa robusta* (sleepy grass) infected with an *Acremonium* endophyte. *Nat. Toxins* 1:84-88.

50. Weedon, C. M. and Mantle, P. G., Paxilline biosynthesis by *Acremonium loliae*; a step toward defining the origin of lolitrems neurotoxins, *Phytochem.*, 26, 969, 1987.

51. Miles, C. O., Wilkins, A. L., Gallagher, R. T., Hawkes, A. D., Munday, S. C., and Towers, N. R., Synthesis and tremorgenicity of paxitrols and lolitriol: possible biosynthetic precursors of lolitrem B, *J. Agric. Food Chem.*, 40, 234-238, 1992.

52. Miles, C. O., Munday, S. C., Wilkins, A. L., Ede, R. M., and Towers, N. R., Large scale isolation of lolitrem B and structure determination of lolitrem E, *J. Agric. Food Chem.*, 42, 1488-1492, 1994.

53. Mantle, P. G. and Weedon, C. M., Biosynthesis and transformation of tremorgenic indolediterenoids by *Pennicillium paxilli* and *Acremonium lolii*, *Phytochem.* 36, 1209-1217, 1994.

54. Munday-Finch, S. C., Miles, C. O., Wilkins, A. L., and Hawkes, A. D., Isolation and structural elucidation of lolitrem A, a tremorgenic mycotoxin from perennial ryegrass infected with *Acremonium lolii*, *J. Agric. Food Chem.*, 43, 1283-1288, 1995.

55. Penn, J., Garthwaite, I., Christensen, M. J., Johnson, C. M., and Towers, N. R., The importance of paxilline in screening for potentially tremorgenic *Acremonium* isolates, pp. 88-92, in: Hume, D. E., Latch, G. C. M., and Easton, H. S., Eds., Proc. 2nd Int. Sym. on *Acremonium*/Grass Interactions, Feb. 4-6, 1993, Massey University, AgResearch, Palmerston North, New Zealand.

56. Rowan, D. D., Hunt, M. B., and Gaynor, D. L., Peramine, a novel insect feeding deterrent from rye grass infected with *Acremonium loliae*, *J. Chem. Soc. Chem. Commun.*, 935, 1986.

57. Rowan, D. D. and Tapper, B. A., An efficient method for the isolation of peramine, an insect feeding deterrent produced by the fungus *Acremonium lolii*, *J. Nat. Prod.*, 52, 193-195, 1989.

58. Tapper, B. A, Rowan, D. D., and Latch, G. C. M., Detection and measurement of alkaloid peramine in endophyte-infected grasses, *J. Chromatog.*, 463, 133, 1989.

59. Fannin, F. F., Bush, L. P, Siegel, M. R., and Rowan, D. D., Analysis of peramine in fungal endophyte-infected grasses by reversed-phase thin-layer chromatography, *J. Chromatog.*, 503, 288-292, 1990.

60. Rowan, D. D. and Latch, G. C. M., Utilization of endophyte-infected perennial ryegrasses for increased insect resistance, Chap. 12, in: *Biotechnology of Endophytic Fungi of Grasses*, Bacon, C. W. and White, J. F., Jr., Eds., CRC Press, Boca Raton, pp. 169-183, 1994.

61. Bush, L. P., Cornelius, P. C., Buckner, R. C., Varney, D. R., Chapman, R. A., Burns, P. B., Kennedy, C. W., Jones, T. A., and Saunders, M. J., Association of N-acetylloline and N-formylloline with *Epichloe typhina* in tall fescue, *Crop. Sci.*, 22, 941, 1982.

62. Bush, L. P., Fannin, F. F., Siegel, M. R., Dahlman, D. L., and Burton, H. R., Chemistry, occurrence and biological effects of saturated pyrrollizidinerolizidine alkaloids associated with endophyte-grass interactions, in: *Acremonium*/Grass Interactions, Joost, R. and Quisenberry, S., Eds., Elsevier Scientific Publishers, Amsterdam, Netherlands, *Agric., Ecosystems Environ.*, 44, 81-102, 1993.

63. Petroski, R. J., Dornbos, D. L., Jr., and Powell, R. G., Germination and growth inhibition of annual ryegrass (*Lolium multiflorum* L.) and alfalfa (*Medicago sativa* L.) by loline alkaloids and synthetic N-acetylloline derivatives, *J. Agric. Food Chem.* 38, 1716-1718, 1990.

64. Clay, K., Fungal endophytes of grasses. A defensive mutualism between plants and fungi, *Ecology*, 69, 10-16, 1988.

65. Petroski, R. J. and Powell, R. G., Naturally occurring pest bioregulators, in: *A.C.S. Symposium Series* 449, Hedin, P. A., Ed., American Chemical Society, pp 426-433, 1991.

66. Reidell, W. E., Kieckhefer, R. E., Petroski, R. J., and Powell, R. G, Naturally-occurring and synthetic loline alkaloid derivatives: Insect feeding behavior modification and toxicity, *J. Entomol. Sci.*, 26, 122-129, 1991.

67. Patterson, C. G., Potter, D. A., and Fannin, F. F., 1991, Feeding deterrency of alkaloids from endophyte infected grasses to Japanese beetle grups, *Entomol. Exp. Appl.*, 61, 285-289, 1991.

68. Oliver, J. W., Powell, R. G., Abney, L. K., Linnabary, R. D., and Petroski, R. J., N-acetylloline-induced vasoconstriction of the lateral saphenous vein (cranial branch) of cattle, in: Quisenberry, S. S. and Joost, R. E., Eds., Proc. International Symposium *Acremonium*/Grass Interactions, New Orleans, LA, pp. 239-243, Nov. 5-7, 1990.

69. Oliver, J. W., Linnabary, R. D., Abney, L. K., and Strickland, J. R., Response of blood vessels to N-acetylloline and serotonin, Proceedings Tall Toxicoses Fescue Workshop, SERAIEG-8, Atlanta GA, Oct. 25-26, p. 51, 1993.

70. Powell, R. G. and Petroski, R. J., The loline group of pyrrollizidinerolizidine alkaloids, in: *The Alkaloids: Chemical and Biological Perspectives*, Pelletier, S.W., Ed., Vol. 8., Springer-Verlag, New York, pp. 320-338, 1992.

71. Bacon, C. B. and White, J. F., Jr., Stains, media, and procedures for analyzing endophytes, Chapter 4, in: *Biotechnology of Endophytic Fungi of Grasses*, Bacon, C.W. and White, J. F., Jr., Eds., CRC Press, Boca Raton, pp 47-56, 1994.

72. Hill, N. S., Parrott, W. A., and Pope, D. D., Ergopeptine alkaloid production by endophytes in a common tall fescue genotype, *Crop Sci.*, 31,1545-1547, 1991.

73. Roylance, J. T., Hill, N. S., and Agee, C. S., Ergovaline and peramine production in endophyte-infected tall fescue-independent regulation and effects of plant and endophyte genotype, *J. Chem. Ecology*, 20, 2171-2183, 1994.

74. Arachevaleta, M., Bacon, C. W., Hoveland, C. S., and Radcliffe, D. E., Effects of tall fescue on plant response to environmental stress, *Agron. J.*, 81, 83-90, 1989.

75. Arachevaleta, M., Bacon, C. W., Plattner, R. D., Hoveland C. S., and Radcliffe, D. E., Accumulation of ergopeptide alkaloids in symbiotic tall fescue grown under deficits of soil water and nitrogen fertilizer, *Appl. Environ. Microbiol.*, 58, 857-861, 1992.

76. Richardson, M. D., Chapman, G. W., Jr., Hoveland, C. S., and Bacon, C. W., Sugar alcohols in endophyte-infected tall fescue under drought, *Crop Sci.*, 32, 1060-1061, 1992.

77. Lyons, P. C., Evans, J. J., and Bacon, C. W., Effects of fungal endophyte*Acremonium coenophialum* on nitrogen accumulation and metabolism in tall fescue,*Plant Physiol.*, 92, 726-732, 1990.

78. Davis, N. D., Cole, R. J., Dorner, J. W., Weete, J. D., Backman, P. A., Clark, E. M., King, C. C., Schmidt, S. P., and Diener, U. L., Steroid metabolites of*Acremonium coenophialum*, an endophyte of tall fescue, *J. Agric. Food Chem.*, 34, 105-108, 1986.

79. Schmidt, S. P. and Osborn, T. G., Effects of endophyte-infected tall fescue on animal performance, in: *Acremonium/*Grass Interactions, Joost, R., Quisenberry S., Eds., Elsevier Scientific Publishers, Amsterdam, Netherlands, *Agric. Ecosys. Envir.*, 44, 233-262, 1993.

80. Thompson, F. N. and Stuedemann, J. A., Pathophysiology of fescue toxicoses, in: *Acremonium/*Grass Interactions, Joost, R. and Quisenberry, S., Eds., Elsevier Scientific Publishers, Amsterdam, Netherlands, *Agric. Ecosys. Envir.*, 44, 263-281, 1993.

81. Thompson, F. N. and Garrner, G. B., Vaccines and pharmacological agents to alleviate fescue toxicoses, Chapter 9, in: *Biotechnology of Endophytic Fungi of Grasses*, Bacon, C. W. and White, J. F., Jr., Eds., CRC Press, Boca Raton, pp. 125-131, 1994.

82. Yates, S. G., Fenster, J. C., and Bartlet, R. J., Assay of tall fescue seed extracts, fractions, and alkaloids using the large milkweed bug, *J. Agric. Food Chem.*, 37, 354-357, 1989.

83. Clay, K., The potential role of endophytes in ecosystems, Chapter 6, in: *Biotechnology of Endophytic Fungi of Grasses*, Bacon, C. W. and White, J. F., Jr., Eds., CRC Press, Boca Raton, pp. 73-86, 1994,

84. Porter, J. K., Bacon, C. W., Cutler, H. G., Arrendale, R. F., and Robbins, J. D.,*In vitro* auxin production by *Balansia epichloe*, *Phytochemistry*, 24, 1429-1431, 1985.

85. DeBattista, J. P., Bacon, C. W., Severson, R. F., Plattner, R. D., and Bouton, J. H., Indole acetic acid production by the fungal endophyte of tall fescue,*Agron. J.,* 82, 878-880, 1990.

86. Schardl, C. L., Molecular and genetic methodologies and transformation of grass endophytes, Chapt. 11, in: *Biotechnology of Endophytic Fungi of Grasses*, Bacon, C.W. and White, J. F., Jr., Eds., CRC press, Boca Raton, pp. 151-165, 1994.

87. Parrott, W. A., *In vitro* approaches for the study of*Aremonium-Festuca* Biology, Chapt. 3, in: *Biotechnology of Endophytic Fungi of Grasses*, Bacon, C. W. and White, J. F., Jr., Eds., CRC press, Boca Raton, pp. 37-46, 1994.

88. White, J. F., Taxonomic relationships among the members of the Balansieae (Clavicipitales), Chapt. 1, in: *Biotechnology of Endophytic Fungi of Grasses*, Bacon, C. W. and White, J. F., Jr., Eds., CRC Press, Boca Raton, 3-20, 1994.

89. Glenn, A. E., Bacon, C. W., Price, R., and Hamlin, R. T., Molecular phylogeny of*Acremonium* and its taxonomic implications, *Mycologia*, 88, 369–383, 1996.

5   The Toxicity of Bracken Fern (genus *Pteridium*) to Animals and its Relevance to Man

B. L. Smith

CONTENTS

5.1   INTRODUCTION

Bracken fern (genus *Pteridium*), one of the most ubiquitous plants on our planet, has been described[1] as the most locally intensive and globally extensive of all plants. Most common in temperate areas at forest margins, on recently deforested areas, and on regressing farmland, its distribution appears to be limited by only the extremes of latitude, altitude, or heat. In the presence of adequate moisture, it aggressively colonizes open areas especially from forest margins and becomes part of a natural progression towards reforestation. Bracken is considered to be encroaching worldwide with encroachment rates for U.K. bracken being calculated to be between 1–3%.[2]

5.2   TAXONOMY AND GEOGRAPHICAL DISTRIBUTION

Traditionally, bracken fern has been regarded as a single species, *Pteridium aquilinum*, but with different subspecies and varieties recognized.[3] In this scheme, two subspecies were proposed. The largely northern hemisphere subspecies, *aquilinum*, contains the varieties *aquilinum, decompositum, pubescens, feei, latiusculum, wrightianum, pseudocaudatum*, and *africanum*. The mainly southern hemisphere subspecies, *caudatum*, contains the varieties *caudatum, esculentum, arachnoideum, revolutum*, and *yarrabense*. The varieties are widely distributed on a geographical basis[1] with toxicity or the toxin ptaquiloside having been recorded from almost all subspecies and varieties.[4]

More recently, the monotypic view of bracken has been reevaluated and proposals have been made for separate species (e.g., *Pteridium esculentum* in Australia[5] and *Pteridium aquilinum* and *Pteridium pinetorum* for Scottish brackens[6]). Increasingly, chromosomal ploidy, isoenzyme polymorphism, DNA and RNA nucleic acid base sequence analysis, and DNA homologies[6–9] are being used in conjunction with cladistic analysis to reassess taxonomic and evolutionary relationships within the genus *Pteridium*.

5.3   SYNDROMES OF BRACKEN POISONING

At times of restricted available nutritious feed, domestic herbivores and pigs will readily consume bracken fern. Depending on the species and age of the animals, the quantity and quality of the bracken available, and the length of time and consumption rate, a number of disease syndromes may occur. Bracken toxicity to animals and the nature of some of these syndromes,[10–15] as well as the implications for man,[14,16] have been reviewed. The isolation of bracken components responsible for the different syndromes, especially those involving the carcinogenesis and the isolation of ptaquiloside, will be detailed in a later section.

Most recent interest has been in the carcinogenic nature of bracken fern. It is the only higher plant proven to cause cancer naturally in animals.

5.3.1   NATURAL SYNDROMES

5.3.1.1   Cyanogenic Potential

Bracken has been reported to contain prunasin, a cyanogenic glycoside,[17] but cyanogenesis has not been reported as a cause of death in herbivores. Cyanogenic phenotypes of bracken have been identified in the U.K., New Zealand, and Australia.[18–20] An interesting southward increase in frequency of cyanogenic plants was noted in Australia.[20] The possible role of cyanogenic potential for protecting bracken from herbivory has been discussed.[19]

5.3.1.2   Induced Thiamine Deficiency

Bracken contains antithiamine factors,[21] in particular a thiaminase type 1,[22] which splits thiamine into its two component ring structures. Unlike ruminants who make their own, monogastric animals require thiamine in their diet and thus they are more susceptible to the effects of thiaminase from a variety of sources, including bracken. The condition is seen clinically as mild to severe nervous signs. At its worst, bilateral cerebrocortical necrosis of the brain is observed. In its milder forms (inappetence and mild to moderate nervous symptoms), the condition responds well to thiamine administration. Although ruminants are more resistant to this manifestation of acute bracken poisoning, it has been produced experimentally in mature sheep fed 15–30% bracken rhizome in their diet.[23] The natural condition occurs often in the horse[24,25] and pig[26] and has been recorded after the experimental feeding of bracken to rodents[27–29] and the horse.[24]

5.3.1.3   Acute Hemorrhagic Syndrome

Acute bracken poisoning in ruminants is the clinical manifestation of a degenerative change in the more rapidly dividing cells of the animal. Sheep are more resistant to the disease than cattle.[30] In the gut epithelium, there is necrosis of the larynx, pharynx, and small intestine, especially of young animals. The bone marrow aplasia which also occurs has severe and dramatic effects. The megakaryocytes cease to produce platelets and when profound thrombocytopenia is reached, a hemorrhagic crisis occurs with multiple hemorrhages developing throughout the body. Other dividing precursor stem cells of the hemopoietic system, the granulocytes and lymphocytes, are affected, leading to severe leukopenia. In cattle, the most severe effect is on granulocytes,[31] whereas in sheep a profound lymphocytopenia has been reported.[32] Symptoms and death may occur some weeks after animals are removed from bracken. This syndrome, or at least the leukopenia, thrombocytopenia, or the marrow aplasia has been reproduced by feeding bracken to sheep[32] and cattle.[33,34]

5.3.1.4   Bright Blindness

In the U.K., sheep feeding on bracken develop a progressive degeneration of the neuroepithelium of the retina which results in blindness.[35,36] Because the retinal atrophy causes an increased reflectance of the tapetum lucidum in semi-dark conditions, the disease is known as "bright blindness." A narrowing of the retinal vessels has also been recorded. This progressive retinal degeneration has been reproduced in sheep by feeding bracken fern.[37] In another experiment, bracken fern known to contain ptaquiloside caused the condition.[38] The narrowing of the retinal vessels also occurred.

5.3.1.5   Enzootic Hematuria

Enzootic hematuria is named for the clinical expression of the presence of multiple mixed tumors in the urinary bladder (Figure 5.1). Some of the tumors are of epithelial tissue origin, but many are of vascular origin and these, in particular, give rise to intracystic hemorrhage. The condition occurs in sheep,[39] but more commonly in cattle,[11,40,41] after prolonged ingestion of bracken. In cattle, the disease has an extremely wide geographical distribution. It occurs throughout Europe, Africa, North, Central, and South America, India, Nepal, China, Indonesia, the Philippines, Japan, Australia, New Zealand, Turkey, and many other regions. The lesions have been reproduced in cattle[42,43] by feeding them bracken fern.

 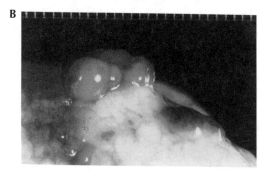

FIGURE 5.1A AND B   Urinary bladder of cow with numerous raised areas of epithelial neoplasia and dark hemorrhagic areas of hemangiomata. The cow had bovine enzootic hematuria and was from an area of New Zealand where the disease is endemic. (From Smith, B.L. and Beatson, N.S., *N. Z. Vet. J.*, 18(6), 115–120, 1970. With permission.)

5.3.1.6   Upper Alimentary Carcinoma

Epithelial carcinomata of the nasopharynx, esophagus, and the forestomachs of cattle have been reported from the U.K., Kenya, and Brazil[44–46] when the cattle have been exposed to bracken in their pastures. It has been suggested[45,47] that these tumors are caused by the malignant transformation of bovine papilloma virus (BPV) type 4.[48,49] An association between urinary bladder carcinoma and BPV2 has also been suggested.[50] The immunosuppressive effects of the bracken carcinogen and other toxins on latent papilloma virus infections of cattle and other species[51] have been examined.

5.3.2   EXPERIMENTAL SYNDROMES

While several of the natural syndromes of bracken toxicity have been reproduced experimentally in the species of concern or in laboratory animals, experimental administration of bracken or its fractions have resulted in hitherto unreported lesions, especially neoplastic lesions.

Thiamine deficiency has been reproduced in the horse[24] and in rodents[27,28] by feeding diets containing known amounts of bracken fern. Bracken feeding has reproduced the bone marrow aplasia in cattle,[33,34] and its toxicity to bovine intestinal crypt cells.[52] The progressive retinal neuroepithelial degeneration and the narrowing of retinal vessels has been reproduced by feeding bracken to rats.[38]

The experimental feeding of bracken fern has reproduced the urinary bladder tumors in rats[28,53–55] and guinea pigs,[56] but has also caused neoplasia in other organs. Neoplasia has been reproduced in the gut, especially the ileum, of the rat,[28,29,54,57–59] guinea pig,[56] quail,[52] and Egyptian toad.[60] Lung adenoma was caused in the mouse by feeding pellets made from bracken fronds[52] or spores.[61,62] The feeding of bracken fern or its parts has also caused mammary adenocarcinoma in mice and in rats.[61,63] Interestingly, the feeding of *P. revolutum* bracken for long periods of time reproduced urinary bladder neoplasia in cattle but not in rabbits or goats.[64]

5.4   OTHER FERNS PRODUCING SAME SYNDROMES OR BRACKEN TOXINS

Thiamine deficiency has been recorded in livestock consuming *Cheilanthes sieberi* and *Marsilea drummondi* (Nardoo) in Australia.[65] Also within Australia, both the acute hemorrhagic and enzootic hematuria syndromes have been recorded in cattle both inside and outside the area of distribution of bracken fern. These cases could not be accounted for by stock movement and were associated with the ingestion of the rock fern, *Cheilanthes sieberi*.[41] *Cheilanthes sieberi* has been shown to contain the carcinogen ptaquiloside[66] and to cause the acute hemorrhagic disease experimentally.[67] It appears that *Cheilanthes*, growing in abundance, as occurs after rain in the drier areas of Australia, is capable of causing at least two of the syndromes caused by bracken fern.

Other fern genera which have been shown to contain ptaquiloside or ptaquiloside-like mutagenic compounds with TLC, HPLC, or a modified Ames test,[68,69] are *Histiopteris incisa*, *Cheilanthes myriophylla*, *Cibotium barometz*, *Dennstaedtia hirsta*, *Dennstaedtia scabra*, *Hypolepis punctata*, *Hypolepis bamleriana*, *Hypolepis tennifolia*, *Pteris cretica*, *Pteris nipponica*, *Pteris oshimesis*, *Pteris tremula*, *Pteris wallichiana*, *Pityrogramma calome*, and *P. sulphurea*.[70,71] Although some of these may be of local concern, they are insignificant as causes of stock ill health. Bracken, however, because of its ubiquity and its known association with the disease syndromes, remains the fern of primary concern.

5.5 PTAQUILOSIDE AND THE BRACKEN SYNDROMES

Various compounds have been isolated from bracken, many showing negative or uncertain results when tested for carcinogenicity. These include flavonoids, indanones, pterolactams, pterosins, and tannins. Shikimic acid from bracken was reported to be carcinogenic,[72] but subsequent work did not support this.[73] One experiment strongly suggested that quercetin, which is found in bracken, caused intestinal and urinary bladder tumors in rats.[74] Again, subsequent work has failed to substantiate this finding.[75] The carcinogenicity of quercetin has been specifically discussed and the evidence found to be insufficient.[76,77] None of the foregoing work has given results as consistently unequivocal as those associated with ptaquiloside, especially when associated with carcinogenicity.

Almost simultaneously, two groups of workers isolated and determined the structure of a major bracken compound shown to be carcinogenic to rats by oral and parenteral dosing[78–80] and mutagenic by a modified Ames test using *Salmonella typhimurium*.[81–83] One group named the compound, a norsesquiterpene glycoside, ptaquiloside and the second group aquilide A. The term ptaquiloside is now more widely used. Animal studies have shown ptaquiloside to cause ileal, urinary bladder, and mammary gland carcinoma in rodents.[53,84] In addition, it has been shown that bracken fern from an area with high ptaquiloside concentration caused tumors (mainly ileal and urinary bladder), in a high percentage of rats, whereas bracken with a low ptaquiloside concentration did not cause many tumors.[28] The administration of ptaquiloside alone to a Suffolk lamb by means of jejunal catheterization caused retinal degeneration and narrowing of the retinal vessels, both arteries and veins.[38] Prolonged oral dosing with ptaquiloside[38] caused thrombocytopenia and neutropenia in a calf but the acute hemorrhagic syndrome had not occurred when dosing was discontinued due to depletion of the available ptaquiloside. At autopsy, myeloid aplasia was present.

Ptaquiloside, whose reactivity is pH dependent, has been shown to be mutagenic,[83,85] clastogenic,[86,87] and to react with DNA.[88,89] Ptaquiloside is believed to form covalent adducts with DNA through the N3 of adenine or the N2 of guanine through opening of the cyclo-propane ring of the reactive dienone. The sequence selectivity for adduct formation and DNA cleavage has been demonstrated.[88,89] The most favorable sequence has been shown to be 5'-AAAT. It is proposed that ptaquiloside under weakly basic conditions forms an aglycone, ptaquilosin, which then rearranges to form the ultimate carcinogen, the dienone, which is highly reactive with DNA, its principal biological target (Figure 5.2). The natural aglycone, (-)-ptaquilosin, has been synthesized and the DNA cleaving activities of its dienone demonstrated.[90] Its enantiomer (+)-ptaquilosin when converted to its dienone was shown to have minimal DNA cleaving activity. Analogues of the ptaquiloside dienone have been synthesized and shown to possess DNA cleaving activities also.[90] Structure activity relationships of the ptaquiloside bioactive moiety have been examined by comparing the mutagenic, clastogenic, and DNA reactivities of ptaquiliside, ptaquilosin, and closely related compounds.[85,87,91,92] An intact cyclopropane ring together with the proximity of the hydroxyl or ketone group on the adjacent carbon of the six membered dienone ring together with the presence of the (-)-ptaquilosin enantiomer configuration[90] appears to be essential for the highest activity of the ultimate carcinogen.

Ptaquiloside has been assayed by a mutagenicity bioassay,[69] thin-layer chromatography,[71] and by high performance liquid chromatography.[4,93,94] The methods used for its early extraction and isolation[78,79,83] have been subsequently modified.[95]

Although different taxa of *Pteridium* have been compared[4] for differences in ptaquiloside concentration, no clear-cut differences between taxa have been determined. Regional differences in the carcinogenicity of bracken have been identified.[96] Very high and variable concentrations of ptaquiloside (20–12,945 mg kg^{-1} dried bracken) were recorded during a survey of eastern Australian bracken.[4] The same survey showed a southward increase in ptaquiloside concentrations.[4] In New Zealand, very high concentrations of ptaquiloside

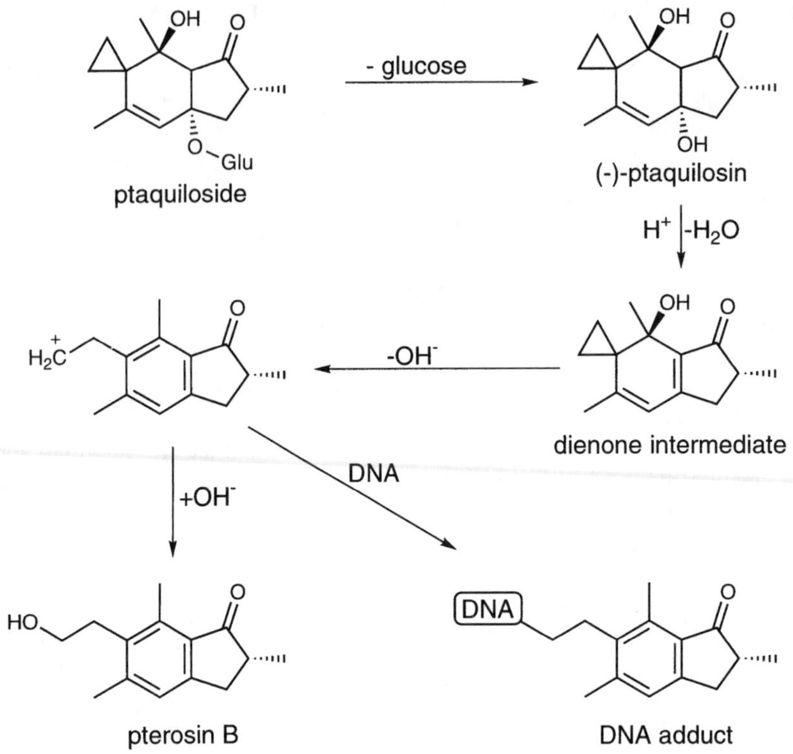

FIGURE 5.2 Formation of either inactive pterosin B or DNA adducts from ptaquiloside via its intermediates. The DNA adducts result in DNA cleavage and base sequence transversions.

(>5000 mg kg⁻¹ dried bracken) occurred in bracken taken from a farm on which bovine enzootic hematuria occurred compared with bracken from another region (<100 mg kg⁻¹ dried bracken).[97,98] This difference was maintained when rhizomes from the two areas were transplanted and grown in common environments, suggesting either a genetic component or the transfer of some other factor such as an endophyte with the rhizome as was suggested earlier.[99]

In the New Zealand studies, the highest concentrations of ptaquiloside were found to occur in the crozier with concentrations lessening as the frond aged (Table 5.1).[98,100] The highest concentrations of ptaquiloside occurred in the spring with another peak occurring in the autumn (Figure 5.1).[97] No ptaquiloside was found in rhizomes, even in the small pre-emergent shoot apices. However, ptaquiloside was found in the young croziers very soon after emergence.[101] Ptaquiloside has not been detected in the spores of bracken[71] or in those of *Cheilanthes sieberi*, another ptaquiloside-containing fern.[66] Bracken spores, however, have been considered as a source of bracken toxins particularly in view of the human exposure to bracken spores in recreational and occupational areas where they may be inhaled. Bracken (var. *aquilinum*) spores were shown[62] to cause gastric tumors, mammary tumors, pulmonary adenomas, and leukemias when given to mice for up to 27 months. Tumors occurred in 2 of 49 control animals whereas tumors developed in 53 of 98 treated animals. Recently,[61] significantly greater numbers of pulmonary and mammary tumors occurred in mice given a diet containing bracken *caudatum* spores for up to 20 months. Therefore, in view of the twice demonstrated carcinogenicity of bracken spores, it would appear that either the extraction of ptaquiloside from the spores of *Pteridium* or *Cheilanthes* has been unsuccessful, that other carcinogenic compounds exist in the spores, or that the spores were collected from geographical areas where the bracken did not contain ptaquiloside.

TABLE 5.1
Concentrations of Ptaquiloside in Different Parts and Growth Stages of Bracken Fern Taken at the Same Time from a Site where Australian *Pteridium esculentum* was Growing Uniformly. The Pterosin B was Analyzed after Base Acid Reaction to Convert Ptaquiloside to Pterosin B. Pterosin B Already Present in the Fern had been Removed during the Cleanup Procedure.

	Ptaquiloside (μg g^{-1})	Pterosin B (μg g^{-1})
Croziers		
Tips	905	1535
Stems	246	443
Immature Fern		
Apical tips of unfolding laminae	975	1916
Next two pinnae of laminae	330	744
Basal two pinnae of laminae	223	454
Fresh mature fern		
Basal pinnae (light green and soft)	66	99
Older mature fern		
Basal pinnae (dark green and hard)	82	163

Little is known about the disposition of ptaquiloside. It is an unstable compound degraded by light and heat especially at high pH.[71] A high proportion is lost during concentration of aqueous solutions particularly at high temperatures.[95] Very little has been determined of the metabolism of ptaquiloside *in vivo* except that the observation has been made[83] that the alkalinity of the ileal section of the small intestine and the alkaline nature of herbivore urine may explain the presence of tumors in these sites. Ptaquiloside has been detected either directly or indirectly in urine,[102] meat,[102] and milk[102,103] of cattle given a high proportion of bracken in their diet. Despite its reported instability it does withstand cooking temperatures.[16]

5.6 VIRAL COMPONENT?

In 1978, Jarret and colleagues[45] noted a high prevalence of upper alimentary cancer of cattle in the northern U.K. and suggested an interaction between experimental carcinogens and bovine papilloma virus. Similar conditions have been recorded in Kenya and Brazil and in all three areas strong circumstantial evidence points to bracken fern as the interacting factor. The papilloma virus was later identified as BPV-4.[48] Other bovine papilloma types (BPV-2) have been associated with urinary bladder papillomatosis[50] and it has been suggested that bracken carcinogen may cause activation or malignant transformation of either the overt or latent relevant papilloma infection. The identification and role of ptaquiloside or of other bracken fern constituents is still relatively unknown. Recent research[49] has established the presence of latent papilloma virus infection as full genomes in cattle, not just in epithelial tissue but also in lymphocytes. Interestingly, BPV-4 has been described recently in Australia.[104] The isolated DNA was obtained from sessile bovine papillomata from the mouth but carcinomata of the digestive tract of cattle are not a common finding there.

It is becoming increasingly apparent that past, present, and future experiments need interpretation in the light of a whole range of causal, interacting, and modifying agents or influences. These include numerous bracken constituents, of which ptaquiloside must be the major active agent, and the presence or absence of viral infection. The roles of these as initiators, promoters, cocarcinogens, or immunosuppressors must also be considered.

While it has been established that ptaquiloside is at least a major carcinogen in bracken with proven mutagenicity and clastogenicity and with established reactions with DNA forming adducts, there are also other bracken constituents which may play a role in some aspects of the different syndromes. These include the various analogues of ptaquiloside, the known constituents of uncertain significance and other compounds as yet undiscovered.

5.7   BRACKEN AND HUMAN HEALTH

Bracken fern has been considered to be a risk to man either through direct or indirect consumption of the toxic principles. The problem has been recently reviewed.[16] In Japan and some other countries, bracken croziers are consumed directly. Despite the customary treatments of the croziers in Japan with boiling water and soda ash which reduces the carcinogenicity,[96] a significantly greater prevalence of upper esophageal carcinoma was associated with the consumption of bracken croziers. Bracken rhizomes were consumed by the aboriginal peoples of Australia and New Zealand,[105,106] but no records exist of disease syndromes associated with this. Although pigs consuming rhizomes have developed the acute hemorrhagic syndrome,[26] neither neoplasia in the pigs nor the presence of significant quantities of ptaquiloside[93,97] in rhizomes have been recorded.

As already mentioned, spores of neither bracken fern nor *C. sieberi*, a species containing ptaquiloside, have not given detectable quantities of ptaquiloside. However, in the two experiments in which bracken spores have been included in the diet of mice, pulmonary adenomas or mammary carcinomas have been detected in significant numbers. The presence of sporing bracken in recreational and working areas has given some cause for concern and warnings to the public have been issued.[107] However, the risk is not well defined.

The indirect consumption of bracken toxins by man, especially through the milk of cattle, has caused more concern.[16] In a survey of patients with gastric cancer in North Wales,[108] a significantly increased risk (x 2.34) was found among people who had spent their childhood in bracken infested areas when compared with matched controls. The authors concluded, however, that, with the advent of bulked milk supplies in these areas, the risk may no longer exist. However, in Costa Rica many small villages exist in the mountainous areas, and the local population consumes the milk of cows grazing the local bracken infested pastures. Here epidemiological studies have shown a correlation between the prevalence of enzootic hematuria in the cattle and gastric and esophageal cancer in man.[109] The same situation is believed to occur in areas of northern South America.

The milk of cattle both experimentally and naturally exposed to bracken fern has also been shown to be carcinogenic to laboratory animals[105,110,111] and the milk of lactating cows fed bracken fern has been shown to contain ptaquiloside.[103]

The risks to man from the direct consumption of croziers can readily be avoided or, in any event, the risk reduced by the use of treatments referred to above. The risk from the indirect consumption of carcinogens from bracken is not well defined. Some risk may be likely where a stable human population regularly consumes the milk of local cows grazing bracken infested pastures. In this respect, attention has been focused on some villages in central American areas.[109] Doubtless, several other such areas exist in other developing countries. In developed countries, similar situations may also exist in regions where locally produced milk is consumed. With the bulk collection of milk and subsequent redistribution, the dilution effects become great and the significance of the bracken risk to man subsequently reduced. However, species susceptibilities are known to vary, and may change the risks substantially. Whereas laboratory animal studies may demonstrate cancer over a year to two year exposure period at most, the consumption of agricultural products by man may be over lifetimes of 60 or more years. The possibility that ptaquiloside or other carcinogens from bracken may occur as chemical residues in agricultural products other than milk may be significant, particularly since ptaquiloside withstands boiling and cooking temperatures.[112] We have found that ptaquiloside both alone and in a meat gruel is still present at 60–70% of original concentrations after 80°C for 3 h.[102]

5.8   CONCLUSIONS

The ubiquitous bracken fern is the only higher plant known to cause cancer in animals. It is also responsible for other disease syndromes in animals. These diseases are important to livestock owners in some regions. The potent norsesquiterpene glycoside ptaquiloside is probably responsible for most of the mutagenicity and carcinogenicity of this plant and can occur at very high concentrations in bracken. Other carcinogens or immunosuppressants may also exist in the plant. The plant may also be of significance as a human carcinogen through the direct or indirect consumption of carcinogen(s). Ptaquiloside and possibly other bracken constituents present an opportunity for establishing models of many aspects of carcinogenesis both in animals and in man.[56] The illudins, compounds closely related to ptaquiloside, are also being investigated as possible chemotherapeutic agents.[113]

REFERENCES

1. Taylor, J. A., The bracken problem: a global perspective, in *Bracken Biology and Management*, Thomson, J. A. and Smith, R. T., Eds., Australian Institute of Agricultural Science, Sydney, 1990, 3.
2. Taylor, J. A., The Bracken Problem: A Local Hazard and Global Issue, in *Bracken, Ecology, Land Use and Control Technology*, Smith, R. T. and Taylor, J. A., Eds., Parthenon, Carnforth, 1986, 21.
3. Tryon, R. M., A revision of the genus *Pteridium*, *J. N. Engl. Bot. Club*, 43, 1, 1941.
4. Smith, B. L., Seawright, A. A., Ng, J. C., Hertle, A. T., Thomson, J. A., and Bostock, P. D., Concentration of ptaquiloside, a major carcinogen in bracken fern (*Pteridium spp.*), from eastern Australia and from a cultivated worldwide collection held in Sydney, Australia, *Nat. Toxins*, 2, 347, 1994.
5. Brownsey, P. J., The taxonomy of bracken (*Pteridium:* Dennstaedtiaceae) in Australia., *Austr. Sys. Botany*, 2, 113, 1989.
6. Page, C. N., Structural variation in western European bracken: An updated taxonomic perspective., in *Bracken: An Environmental Issue.*, Smith, R. T. and Taylor, J. A., Eds., International Bracken Group Special Publication No. 2, August 1995, Aberystwyth, 1995, 13.
7. Wolf, P. G., Sheffield, E., Thomson, J. A., and Sinclair, R. B., Bracken taxa in Britain: a molecular analysis, in *Bracken: An Environmental Issue.*, Smith, R. T. and Taylor, J. A., Eds., International Bracken Group Special Publication No. 2, August, 1995, Aberystwyth, 1995, 16.
8. Sheffield, E., Wolf, P. G., and Ranker, T. A., Genetic analysis of bracken in the Hawaiian Islands, in *Bracken: An Environmental Issue.*, Smith, R. T. and Taylor, J. A., Eds., International Bracken Group Special Publication No. 2, August, 1995, Aberystwyth, 1995, 29.
9. Thomson, J. A., Weston, P. H., and Tan, M. K., A molecular approach to tracing major lineages in *Pteridium.*, in *Bracken: An Environmental Issue*, Smith, R. T. and Taylor., J. A., Eds., International Bracken Group Special Publication No. 2, August 1995, Aberystwyth, 1995, 21.
10. Anon., Some naturally occurring and synthetic food components, furocoumarins and ultraradiation, WHO International Agency for Research on Cancer, Lyon, 40, 1986.
11. Evans, W. C., Patel, M. C., and Koohy, Y., Acute bracken poisoning in homogastric and ruminant animals, *Proc. R. Soc., Edinburgh*, 81B, 29, 1982.
12. Fenwick, G. R., Bracken (*Pteridium aquilinum*)-toxic effects and toxic constituents, *J. Sci. Food Agric.*, 46, 147, 1989.

13. Hopkins, A., Bracken (*Pteridium aquilinum*): its distribution and animal health implications, *Br. Vet. J.*, 146, 316, 1990.

14. Trotter, W. R., Is bracken a health hazard?, *Lancet*, 336, 1563, 1990.

15. Smith, B. L., Bracken fern and animal health in Australia and New Zealand, in *Bracken Biology and Management,* Thomson, J. A. and Smith, R. T., Eds., Australian Institute of Agricultural Science, Sydney, 1990, 227.

16. Smith, B. L. and Seawright, A. A., Bracken fern (*Pteridium spp.*) carcinogenicity and human health: A brief review, *Nat. Toxins*, 3, 1, 1995.

17. Cooper-Driver, G., Finch, S., Swain, T., and Bernays, E., Seasonal variation in secondary plant compounds in relation to the palatability of *Pteridium aquilinum.*, *Biochem. Syst. Ecol.*, 5, 177, 1977.

18. Hadfield, P. R. and Dyer, A. F., Polymorphism of cyanogenesis in British populations of bracken (*Pteridium aquilinum* L. Kuhn)., in *Bracken: Ecology, Land Use and Control Technology,* Smith, R. T. and Taylor, J. A., Eds., Parthenon, Carnforth, 1986, 293.

19. Jones, C. G., Phytochemical variation, colonization and insect communities: the case of bracken fern (*Pteridium aquilinum*)., in *Variable Plants and Herbivores in Natural and Managed Systems,* Denno, R. F. and McClure, M. S., Eds., Academic Press, 1983, 513.

20. Low, V. H. K. and Thomson, J. A., Cyanogenesis in Australian bracken (*Pteridium esculentum*): distribution of cyanogenic phenotypes and factors influencing activity of the cyanogenic glucosidase, in *Bracken Biology and Management,* Thomson, J. A. and Smith, R. T., Eds., Australian Institute of agricultural Science, Sydney, 1989, 105.

21. Somogyi, J. C., On antithiamine factors of fern, *J. Vitaminol.*, 17, 165, 1971.

22. Evans, W. C., Bracken thiaminase-mediated neurotoxic syndromes, *Bot. J. Linnean Soc.*, 73, 113, 1976.

23. Evans, W. C., Evans, I. A., Humphreys, D. J., Lewin, B., and Davies, W. E. J., Induction of thiamine deficiency in sheep, with lesions similar to those of cerebrocortical necrosis., *J. Comp. Pathol.*, 85, 253, 1975.

24. Evans, E. T. R., Evans, W. C., and Roberts, H. E., Studies on bracken poisoning in the horse, *Br. Vet. J.*, 107, 364, 1951.

25. Fernandes, W. R., Garcia, R. C. M., Medeiros, R. M. A., and Birgel, E. H., Experimental *Pteridium aquilinum* intoxication of horses, *Arq. Escola Med. Vet. Univ. Fed. Bahia,* 13, 112, 1990.

26. Evans, I. A., Humphreys, D. J., Goulden, L., Thomas, A. J., and Evans, W. C., Effects of bracken rhizomes on the pig, *J. Comp. Pathol. Ther.*, 73, 229, 1963.

27. Evans, W. C. and Evans, E. T. R., The effects of the inclusion of bracken (*Pteris aquilina*) in the diet of rats, and the problem of bracken poisoning in farm animals, *Br. Vet. J.*, 105, 175, 1949.

28. Smith, B. L., Embling, P. P., Agnew, M. P., Lauren, D. R., and Holland, P. T., Carcinogenicity of bracken fern (*Pteridium esculentum*) in New Zealand, *N. Z. Vet. J.*, 36, 56, 1988.

29. Pamukcu, A. M., Yalciner, S., Price, J. M., and Byran, G. T., Effects of the coadministration of thiamine on the incidence of urinary bladder carcinomas in rats fed bracken fern, *Cancer Res.*, 30, 2671, 1970.

30. Moon, F. E. and McKeand, J. M., Bracken poisoning in sheep, *Br. Vet. J.*, 109, 321, 1953.

31. Evans, W. C., Evans, I. A., Thomas, A. J., Watkins, J. E., and Chamberlain, A. G., Studies on bracken poisoning in cattle, *Br. Vet. J.*, 114, 180, 1958.

32. Sunderman, F. M., Bracken poisoning in sheep, *Austr. Vet. J.*, 64, 25, 1987.

33. Stockman, S., Bracken poisoning of cattle in Great Britain, *J. Comp. Pathol. Ther.*, 35, 273, 1922.

34. Naftalin, J. M. and Cushnie, G. H., Experimental bracken poisoning in calves, *J. Comp. Pathol. Ther.*, 64, 75, 1954.

35. Watson, W. A., Barlow, R. M., and Barnett, K. C., Bright blindness- A condition prevalent in Yorkshire hill sheep, *Vet. Rec.*, 77, 1060, 1965.

36. Watson, W. A., Barnett, K. C., and Terlecki, S., Progressive retinal degeneration (Bright Blindness) in sheep: A review, *Vet. Rec.*, 91, 665, 1972.

37. Watson, W. A., Terlecki, S., Patterson, D. S. P., Sweasey, D., Herbert, C. N., and Done, J. T., Experimentally produced progressive retinal degeneration (Bright Blindness) in sheep, *Br. Vet. J.*, 128, 457, 1972.

38. Hirono, I., Ito, M., Yagyu, S., Haga, M., Wakamatsu, K., Kishikawa, T., Nishikawa, O., Yamada, K., Ojika, M., and Kigoshi, H., Reproduction of progressive retinal degeneration (bright blindness) in sheep by administration of ptaquiloside contained in bracken, *J. Vet. Med. Sci.*, 55, 979, 1993.

39. Harbutt, P. R. and Leaver, D. D., Carcinoma of the Bladder of Sheep, *Austr. Vet. J.*, 45, 473, 1969.

40. Smith, B. L. and Beatson, N. S., Bovine enzootic haematuria in New Zealand, *N. Z. Vet. J.*, 18, 115, 1970.

41. McKenzie, R. A., Bovine enzootic haematuria in Queensland, *Austr. Vet. J.*, 54, 61, 1978.

42. Rosenberger, G. and Heeschen, W., Alderfarn (*Pteris aquilina*) die ursache des sog. Stallrotes der rinder (haematuria vesicalis bovis chronica), *Dtsch. tierarzliche Wochenschz.*, 67, 201, 1960.

43. Pamukcu, A. M., Goksoy, S. K., and Price, J. M., Urinary bladder neoplasms induced by feeding bracken fern (*Pteris aquilina*) to cows, *Cancer Res.*, 27, 917, 1967.

44. Dobereiner, J., Tokarnia, C. H., and Canella, C. F. C., The occurrence of enzootic hematuria and epidermoid carcinomas in the upper digestive tract of cattle in Brazil, *Pesq. agrop. Brazil.*, 2, 489, 1967.

45. Jarrett, W. H. F., McNeil, P. E., Grimshaw, W. T. R., Selman, I. E., and McIntyre, W. I. M., High incidence area of cattle cancer with a possible interaction between an environmental carcinogen and a papilloma virus, *Nature*, 274, 215, 1978.

46. Plowright, W., Linsell, C. A., and Peers, F. C. A., A focus of rumen cancer in Kenyan cattle, *Br. J. Cancer*, 25, 72, 1971.

47. Jarrett, W. F. H., Environmental carcinogens and papillomaviruses in the pathogenesis of cancer, *Proc. R. Soc. (Lond.)*, 231, 1, 1987.

48. Campo, M. S., Moar, M. H., Jarrett, W. F. H., and Laird, H. M., A new papillomavirus associated with alimentary cancer in cattle, *Nature*, 286, 180, 1980.

49. Campo, M. S., Jarrett, W. F. H., O'Neil, W., and Barron, R. J., Latent papillomavirus infection in cattle, *Res. Vet. Sci.*, 56, 151, 1994.

50. Campo, M. S., Jarrett, W. F. H., Barron, R., O'Neil, B. W., and Smith, K. T., Association of bovine papillomavirus type 2 and bracken fern with bladder cancer in cattle, *Cancer Res.*, 52, 6898, 1992.

51. Campo, M. S., O'Neil, B. W., Barron, R. J., and Jarrett, W. F. H., Experimental reproduction of the papilloma-carcinoma complex of the alimentary canal in cattle, *Carcinogenesis*, 15, 1597, 1994.

52. Evans, I. A., The radiomimetic nature of bracken toxin, *Cancer Res.*, 28, 2252, 1968.

53. Hirono, I., Ogino, H., Fujimoto, M., Yamada, K., Yoshida, Y., Ikagawa, M., and Okumura, M., Induction of tumors in ACI rats given a diet containing ptaquiloside, a bracken carcinogen, *J. Natl. Cancer Inst.*, 79, 1143, 1987.

54. Pamukcu, A. M. and Price, J. M., Induction of intestinal and urinary bladder cancer in rats by feeding bracken fern (*Pteris aqulina*), *J. Natl. Cancer Inst.*, 43, 275, 1969.

55. Price, J. M. and Pamukcu, A. M., The induction of neoplasms of the urinary bladder of the cow and the small intestine of the rat by feeding bracken fern (*Pteris aquilina*)., *Cancer Res.*, 28, 2247, 1968.

56. Bringuier, P.-P., Piaton, E., Berger, N., Dubruyne, F., Perrin, P., Schalken, J., and Devonec, M., Bracken fern-induced bladder tumors in guinea pigs. A model for human neoplasia., *Am. J. Pathol.*, 147, 858, 1995.

57. Schacham, P., Philip, R. B., and Gowdey, C. W., Antihematopoietic and carcinogenic effects of bracken fern (*Pteridium aquilinum*) in rats, *Am. J. Vet. Res.*, 31, 191, 1970.

58. Pamukcu, A. M., Wattenberg, L. W., Price, J. M., and Bryan, G. T., Phenothiazine inhibition of intestinal and urinary bladder tumors induced in rats by bracken fern, *J. Natl. Cancer Inst.*, 47, 155, 1971.

59. Evans, I. A. and Mason, J., Carcinogenic Activity of Bracken, *Nature*, 208, 913, 1965.

60. El-Mofty, M. M., Sadek, I. A., and Bayoumi, S., Improvement in detecting the carcinogenicity of bracken fern using an Egyptian toad, *Oncology*, 37, 424, 1980.

61. Villalobos-Salazar, J., Mora, J., Meneses, A., and Pashov, B., The carcinogenic effects of bracken spores, in *Bracken: An Environmental Issue*, Smith, R. T. and Taylor, J. A., Eds., International Bracken Group, Aberystwyth, 1995, 102.

62. Evans, I. A., Smith, R. T., and Taylor, J. A., The carcinogenic, mutagenic and teratogenic toxicity of bracken, in *Bracken: Ecology, Land Use and Control Technology,* Smith, R. T. and Taylor, J. A., Eds., Parthenon, Carnforth, 1986, 139.

63. Hirono, I., Aiso, S., Hosaka, S., Yamaji, T., and Haga, M., Induction of mammary cancer in CD rats fed bracken (*Pteridium aquilinum*) diet, *Carcinogenesis*, 4, 885, 1983.

64. Zhao, S. Z., Hu, W. Y., Ni, B. Y., Zhu, T., Meng, X. Q., Li, Z. C., Liu, N., Liu, L. X., and Yang, H. S., Experimental studies on bladder tumors induced by *Pteridium revolutum* Nakai in yellow cattle, *Chin. J. Vet. Med.*, 14, 5, 1988.

65. Chick, B. F., Quinn, C., and McCleary, B. V., Pteridophyte intoxication of livestock in Australia., in *Plant Toxicity,* Seawright, A. A., Hegarty, M. P., James, L. F., and Keeler, R. F., Eds., Queensland Poisonous Plant Committee, Brisbane, 1985, 453.

66. Smith, B. L., Embling, P. P., Lauren, D. R., Agnew, M. P., Ross, A. D., and Greentree, P. L., Carcinogen in rock fern (*Cheilanthes sieberi*) from New Zealand and Australia, *Austr. Vet. J.*, 66, 154, 1989.

67. Clark, I. A. and Dimmock, C. K., The toxicity of *Cheilanthes sieberi* to cattle and sheep, *Austr. Vet. J.*, 47, 149, 1971.

68. Ames, B. N., McCann, J., and Yamasaki, E., Methods for detecting carcinogens and mutagens with the Salmonella/mammalian-microsome mutagenicity test, *Mut. Res.*, 31, 347, 1975.

69. Matoba, M., Saito, E., Saito, K., Koyama, K., Natori, S., Matsushima, T., and Takimoto, M., Assay of ptaquiloside, the carcinogenic principle of bracken, *Pteridium aquilinum*, by mutagenicity testing in *Salmonella typhimurium*, *Mutagenesis*, 2, 419, 1987.

70. Saito, K., Nagao, T., Takasuki, S., Koyama, K., and Natori, S., The sesquiterpenoid carcinogen of bracken fern, and some analogs, from the Pteridaceae, *Phytochemistry*, 29, 1475, 1990.

71. Saito, K., Nagao, T., Matoba, M., Koyama, K., Natori, S., Murakami, T., and Saiki, Y., Chemical assay of ptaquiloside, the carcinogen of *Pteridium aquilinum*, and the distribution of related compounds in the *Pteridaceae*, *Phytochemistry*, 28, 1605, 1989.

72. Evans, I. A. and Osman, M. A., Carcinogenicity of bracken and shikimic acid, *Nature*, 250, 348, 1974.

73. Hirono, I., Fushimi, K., and Matsubara, N., Carcinogenicity test of shikimic acid in rats, *Toxicol. Lett.*, 1, 9, 1977.

74. Pamukcu, A. M., Yalciner, S., Hatcher, J. F., and Bryan, G. T., Quercetin, a rat intestinal and bladder carcinogen present in bracken fern (*Pteridium aquilinum*), *Cancer Res.*, 40, 3468, 1980.

75. Stoewsand, G. S., Anderson, J. L., Boyd, J. N., Hrazdina, G., Babish, J. G., Walsh, K. M., and Losco, P., Quercetin: a mutagen, not a carcinogen, in Fischer rats, *J. Toxicol. Environ. Health*, 14, 105, 1984.

76. Ito, N., Letter to the editor: Is quercetin carcinogenic?, *Jpn. J. Cancer Res.*, 83, 312, 1992.

77. Hirono, I., Letter to the editor: Is quercetin carcinogenic?, *Jpn. J. Cancer Res.*, 83, 313, 1992.

78. Hirono, I., Yamada, K., Niwa, H., Shizuri, Y., Ojika, M., Hosaka, S., Yamaji, T., Wakamatsu, K., Kigoshi, H., Niiyama, K., and Uosaki, Y., Separation of carcinogenic fraction of bracken fern, *Cancer Lett.*, 21, 239, 1984.

79. Niwa, H., Ojika, M., Wakamatsu, K., Yamada, K., Hirono, I., and Matsushita, K., Ptaquiloside, a novel norsesquiterpene glucoside from bracken, *Pteridium aquilinum* var. *latiusculum*, *Tetrahedron Lett.*, 24, 4117, 1983.

80. Niwa, H., Ojika, M., Wakamatsu, K., Yamada, K., Ohba, S., Saito, Y., Hirono, I., and Matsushita, K., Stereochemistry of ptaquiloside, a novel norsesquiterpene glucoside from bracken, Pteridium aquilinum var. latiusculum, *Tetrahedron Lett.*, 24, 5371, 1983.

81. Van der Hoeven, J. C. M., Lagerweij, W. J., Meeuwissen, C. A. J. M., Hauwert, P. C. M., Voragen, A. G. J., and Koeman, J. H., Mutagens in food products of plant origin, in *Mutagens in Our Environment,* Sorsa, M. and Vaino, H., Eds., Alan R Liss, New York, 1982, 327.

82. Van der Hoeven, J. C. M. and Van Leewen, F. E., Isolation of a Mutagenic fraction from Bracken (*Pteridium aquilinum*), *Mut. Res.*, 79, 377, 1980.

83. Van der Hoeven, J. C. M., Lagerweij, W. J., Posthumus, M. A., Van Veldhuizen, A., and Holterman, H. A. J., Aquilide A, a new mutagenic compound isolated from bracken fern (*Pteridium aquilinum* (L.) Kuhn), *Carcinogenesis*, 4, 1587, 1983.

84. Hirono, I., Aiso, S., Yamaji, T., Mori, H., Yamada, K., Niwa, H., Ojika, M., Wukamatsu, K., Kigoshi, H., Niiyama, K., and Uosaki, Y., Carcinogenicity in rats of ptaquiloside isolated from bracken, *Gann*, 75, 833, 1984.

85. Nagao, T., Saito, K., Hirayama, E., Uchikoshi, K., Koyama, K., Natori, S., Morisaki, N., Iwasaki, S., and Matsushima, T., Mutagenicity of ptaquiloside, the carcinogen in bracken, and its related illudane-type sesquiterpenes. I. Mutagenicity in *Salmonella typhimurium*, *Mut. Res.*, 215, 173, 1989.

86. Mori, H., Sugie, S., Hirono, I., Yamada, K., Niwa, H., and Ojika, M., Genotoxicity of ptaquiloside, a bracken carcinogen, in the primary culture/DNA-repair test, *Mut. Res.*, 143, 75, 1985.

87. Matsuoka, A., Hirosawa, A., Natori, S., Iwasaki, S., Sofuni, T., and Ishidate, M., Jr., Mutagenicity of ptaquiloside, the carcinogen in bracken, and its related illudane-type sesquiterpenes. II. Chromosomal aberration tests with cultured mammalian cells, *Mut. Res.*, 215, 179, 1989.

88. Smith, B. L., Shaw, G., Prakash, A., and Seawright, A. A., Studies on DNA adduct formation by ptaquiloside, the carcinogen of bracken ferns (*Pteridium spp.*), in *Plant-Associated Toxins*, Colegate, S. M. and Dorling, P. R., Eds., CAB International, Wallingford, 1994, 167.

89. Kushida, T., Uesugi, M., Sugiura, Y., Kigoshi, H., Tanaka, H., Hirokawa, J., Ojika, M., and Yamada, K., DNA damage by ptaquiloside, a potent bracken carcinogen: Detection of selective strand breaks and identification of DNA cleavage products, *J. Am. Chem. Soc.*, 116, 479, 1994.

90. Kigoshi, H., Imamura, Y., Mizuta, K., Niwa, H., and Yamada, K., Total synthesis of natural (levo)-ptaquilosin, the aglycon of a potent bracken carcinogen ptaquiloside, and the (dextro)-enantiomer and their DNA cleaving activities, *J. Am. Chem. Soc.*, 115, 3056, 1993.

91. Ojika, M., Wakamatsu, K., Niwa, H., and Yamada, K., Ptaquiloside, a potent carcinogen isolated from bracken fern *Pteridium aquilinum var. latiusculum*: Structure elucidation based on chemical and spectral evidence, and reactions with amino acids, nucleosides and nucleotides, *Tetrahedron*, 43, 5261, 1987.

92. Kigoshi, H., Tanaka, H., Hirokawa, J., Mizuta, K., and Yamada, K., Synthesis of analogues of a bracken ultimate carcinogen and their DNA cleaving activities, *Tetrahedron Lett.*, 33, 6647, 1992.

93. Alonso Amelot, M. E., Perez Mena, M., Calcagno, M. P., and Jaimes Espinoza, R., Quantitation of pterosins A and B, and ptaquiloside, the main carcinogen of *Pteridium aquilinum* (L. Kuhn), by high pressure liquid chromatography, *Phytochem. Anal.*, 3, 160, 1992.

94. Agnew, M. P. and Lauren, D. R., Determination of ptaquiloside in bracken fern (*Pteridium esculentum*), *J. Chromatogr.*, 538, 462, 1991.

95. Oelrichs, P. B., Ng, J. C., and Bartley, J., Purification of ptaquiloside, a carcinogen from *Pteridium aquilinum*, *Phytochemistry*, 40, 53, 1995.

96. Hirono, I., Shibuya, C., Shizmu, M., and Fushmi, K., Carcinogenic activity of processed bracken as human food, *J. Natl. Cancer Inst.*, 48, 1245, 1972.

97. Smith, B. L., Embling, P. P., Lauren, D. R., and Agnew, M. P., Carcinogenicity of *Pteridium esculentum* and *Cheilanthes sieberi* in Australia and New Zealand, in *Poisonous Plants. Proceedings of the Third International Symposium*, James, L. F., Keeler, R. F., Bailey, E. M., Cheeke, P. R., and Hegarty, M. P., Eds., Iowa State University Press, 1992, 448.

98. Smith, B. L., Lauren, D. R., Embling, P. P., and Agnew, M., Carcinogenicity of bracken fern (*Pteridium spp.*) in relationship to grazing ruminants, in *Proceedings of the 17th International Grassland Congress*, Broughan, J., Eds., New Zealand Grassland Association, 1993, 1396.

99. Schoental, R., Bracken toxicity and soil mycotoxins, *Vet. Rec.*, 115, 500, 1984.

100. Smith, B. L., Lauren, D. R., Embling, P. P., and Agnew, M. P., Ptaquiloside in Australian and New Zealand ferns as a cause of neoplasia, in *Bracken biology and management*, Thomson, J. A. and Smith, R. T., Eds., Australian Institute of Agricultural Science, Sydney, 1990, 241.

101. Smith, B. L. and Agnew, M. P., (unpublished results), 1995.

102. Smith, B. L. and Lauren, D. R., (unpublished results), 1995.

103. Alonso Amelot, M. E., Castillo, U., and De Jongh, F., Passage of the bracken fern carcinogen ptaquiloside into bovine milk, *Lait*, 73, 323, 1993.

104. Bloch, N., Sutton, R. H., Breen, M., and Spradbrow, P. B., Bovine papillomavirus type 4 in Australia, *Austr. Vet. J.*, 72, 273, 1995.

105. Villalobos Salazar, J., Meneses, A., and Salas, J., Carcinogenic effects in mice of milk from cows fed on bracken fern *Pteridium aquilinum*, in *Bracken Biology and Management*, Thomson, J. A. and Smith, R. T., Eds., Australian Institute of Agricultural Science, Sydney, 1990, 247.

106. Veitch, B., Aspects of Aboriginal use and manipulation of bracken fern, in *Bracken Biology and Management*, Thomson, J. A. and Smith, R. T., Eds., Australian Institute of Agricultural Science, Sydney, 1990, 215.

107. Anon., Heathlands of England harbour cancer spores, *N. Sci.*, 14 April, 23, 1988.

108. Galpin, O. P., Whitaker, C. J., Whitaker, R., and Kassab, J. Y., Gastric cancer in Gwynedd. Possible links with bracken, *Br. J. Cancer*, 61, 737, 1990.

109. Villalobos-Salazar, J., Meneses, A., Rojas, J. L., Mora, J., Porras, R. E., and Herrero, M. V., Bracken derived carcinogens as affecting animal and human health in Costa Rica, in *Bracken Toxicity and Carcinogenicity as Related to Animal and Human Health*, Taylor, J. A., Ed., International Bracken Group, 1989, 40.

110. Evans, I. A., Jones, R. S., and Mainwaring-Burton, R., Passage of bracken fern toxicity into milk, *Nature*, 237, 107, 1972.

111. Pamukcu, A. M., Erturk, E., Yalciner, S., Milli, U., and Bryan, G. T., Carcinogenic and mutagenic activities of milk from cows fed bracken fern (*Pteridium aquilinum*), *Cancer Res.*, 38, 1556, 1978.

112. Hirono, I., Ushimaru, Y., Kato, K., Mori, H., and Sasaoka, I., Carcinogenicity of boiling water extract of bracken, *Pteridium aquilinum*, *Gann*, 69, 383, 1978.

113. Kelner, M. J., McMorris, T. C., and Taetle, R., Preclinical evaluation of illudins as anticancer agents: basis for selective cytotoxicity, *J. Natl. Cancer Inst.*, 82, 1562, 1990.

6 Analysis of Proanthocyanidins and Related Polyphenolic Compounds in Nutritional Ecology

G. Rodriguez and J. D. Reed

CONTENTS

6.1 INTRODUCTION

Phenolic compounds such as proanthocyanidins (PA), hydrolyzable tannins, phenolic acids, and flavonoids are secondary metabolites that are widely distributed in the plant kingdom. Polyphenolic compounds are of general interest because of their wide-ranging ecological effects from the organism to the ecosystem level.[1]

Research in nutritional ecology has demonstrated that polyphenols affect plant/animal interactions through their involvement in feeding stimulation, feeding deterrence, digestion inhibition, digestion stimulation, toxicity, toxicity amelioration, disease resistance, signal inhibition, signal transduction, and regulation of nutrient cycles.[2-7] Food of plant origin provides higher animals with nearly all of their requirements for phenolic nutrients (i.e., tyrosine and phenylalanine) because higher animals are unable to synthesize compounds with benzenoid rings from their aliphatic precursors.[8]

The two major types of tannin, PA and hydrolyzable tannins, are chemically quite different. Hydrolyzable tannins are gallic or hexahydroxydiphenic acid esters of glucose or other polyols. The ester bonds are hydrolyzed by acid, base, and enzymes, yielding gallic acid subunits and the core polyol.[9] On the other hand, proanthocyanidins (PA), more commonly referred to as condensed tannins, are chains of flavan-3-ols units with 4–6 and 4–8 carbon-carbon bonds joining the individual flavonoids , and are sometime branched. Proanthocyanidins are not susceptible to hydrolysis but can be oxidatively degraded in strong acid to yield anthocyanidins.[10,11] Heating PA in acidic alcohol solutions (i.e., butanol) produces the red anthocyanidins. The traditional nomenclature for PA is base on this acid-catalyzed oxidation.[12]

Quantification of phenolic compounds is necessary in order to explore hypotheses related to the effect of phenolic compounds on plant/animal interactions. Numerous researchers have investigated the use of phenolic assays in ecological and nutritional research.[8,11,13–19] Studies of the ecological or nutritional significance of polyphenols are often confounded by the use of inappropriate assays and standards that lead to incorrect conclusions.[16,17] A good understanding of the chemistry of polyphenolic compounds is essential to using and developing methods for ecological investigations.

6.2   BIOSYNTHESIS OF PROANTHOCYANIDINS AND RELATED POLYPHENOLS

The biosynthesis of anthocyanidins, proanthocyanidins (condensed tannins), and other flavonoids starts with the synthesis of chalcones from precursors that originate in both the "Shikimate" and "Acetate-Malonate" pathways of primary metabolism. The condensation of 1 molecule of 4-coumaroyl-CoA and three molecules of malonyl-CoA is carried out by the enzyme clalcone synthase (CHS) yielding naringenin chalcone. The coumaroyl-CoA is synthesized from the amino acid phenylalanine by three enzymatic steps (phenylalanine ammonia-lyase (PAL), cinamic 4-hydroxylase, and 4-coumarate-CoA ligase), collectively called the general phenylpropanoid pathway. This compound is also the precursor of a variety of compounds such as lignin , coumarins, stilbenes, and esters. The enzymes responsible for the biosynthesis of the precursors of anthocyanidins (flavan-3,4-diols) and proanthocyanidins (flavan-3,4-diols and flavan-3-diols) have been demonstrated in cell-free preparations from several plants species.[20,21] However, the enzymes involved in the final step in the biosynthesis of PA (condensing enzyme) and anthocyanidins have not been isolated.[22] The condensing enzyme joins flavan-3-ols with an intermediate arising from flavan-3,4-diols to form oligomeric PA. However, until enzyme synthetase systems for the polymerization have been isolated and described, the details will remain unknown.

6.3   AFFINITY OF PROANTHOCYANIDINS WITH PROTEINS AND CARBOHYDRATES

A broad definition of tannin is a polyphenolic compound with sufficient molecular weight and substitution with suitable groups (i.e., carboxyl) that enable them to form effectively strong complexes with proteins and other macromolecules such as certain types of polysaccharides and carbohydrates.[23] The binding mechanism involved in the complexation between tannins and proteins needs to be better understood in order to predict the nutritive value of plant protein in relation to the content of tannin in feeds.[24] This is because some tannin-containing feeds have pro-nutritional and others have anti-nutritional effects.

As with proteins, similar parameters are important in complex formation with polysaccharides, where increased molecular weight, low water solubility, and conformational flexibility lead to stronger association. However, in contrast to proteins, changes in pH does not markedly alter the binding of proanthocyanidins with carbohydrates. As pH rises, there is only a slightly greater attachment of the substrates to dextran gels. As in protein-polyphenol interaction, the size and flexibility of the polyphenol are also important in the affinity between carbohydrate and PA, especially in those situations where the interaction between the polymers is a surface event. However, additional interaction between polyphenols and carbohydrates is observed when the carbohydrate structure allows the occlusion of phenolic compounds into cavities in the carbohydrate. These properties influence the binding of PA to distinct forms of starch and other carbohydrates.[25,26] Deshpande and Salunkhe[27] showed that tannic acid (hydrolysable tannin) and catechin have a significant extent of association with amorphous forms of starch such as amylose and amylopectin; however, the interaction with compact and crystalline structures of native legumes and potato starches were poorer. Haslam[28] suggested that the type of binding and the driving force in the formation of inclusion complexes is unclear, but the Van der Waals forces, hydrogen bonding, release of solvent water molecules from the cavity during complexation, and the release of strain energy in the macrocycle are probably the most important ways of interaction.

The carbohydrate component of PA may result from sugar moieties in the monomeric units of the PA polymer (glycosides) and the linkages of PA with the fiber component of the cell wall. This last group of PA are insoluble in the common solvents that are used to extract PA (i.e., aqueous acetone).

6.4   PROANTHOCYANIDINS AND LIGNIN

Proanthocyanidins and lignin are polymers formed by phenolic compounds. The first group can be found in two different forms; those that are soluble in aqueous-organic solvents and those that are insoluble. The insoluble forms may be associated with carbohydrates, fiber, and others polyphenols such as lignin. The insoluble forms of PA are easily confused with lignin, especially in analytical systems for plant phenolic analysis.[29] The similarities between insoluble PA and lignin has led to the conclusion that these polymers have analogous roles in the cell wall.[28]

However, some important distinctions need to be considered. Proanthocyanidins combine two different biosynthetic pathway, the Shikimate pathway with the formation of one of the precursors, phenylpropane-CoA, and the Acetate-Malonate pathway with the formation of the second precursor, the malonate-CoA, while the only precursor for the biosynthesis of lignin is phenylpropane produced by the same pathway. The cellular localization of lignin is very restricted to the cell wall and the final biosynthetic steps are catalyzed by peroxidase. Proanthocyanidins also can be found in the cell wall in some plants. However, the biosynthesis is basically located in vesicles budded off from the endoplasmic reticulum and stored in the large central vacuole.[30]

Characterization of tannins from the South American tree, Quebracho colorado (*Schinopsis balansae*), showed that the purified PA yielded sugars and gallic acid after acid hydrolysis and lignin derived degradation products upon $KMnO_4$ oxidation. Both moieties are covalently linked to the matrix of the PA.[31]

6.5   ANALYSIS OF PROANTHOCYANIDINS AND POLYPHENOLS

There are several methods of analyses for tannins and related polyphenols.[32] However the relationship between quantitative analysis of polyphenols and their nutritional or toxic effects are often poor because the physical and chemical properties of polyphenolics that determine their reactivity in analytical techniques are often different than the properties that determine their ecological effects.

The methods for polyphenol analysis are divide into chemical assays and protein-binding assays. The first one is important in the quantification and elucidation of structure; the second type of assay can be used either to determine the amount of tannin in a sample or to evaluate biological activity. The chemical methods are also divided into assays for general polyphenol and assays that depend on specific functional groups.[11]

The Folin-Dennis and Folin-Ciocalteu's reagents are commonly used assays for general classes of polyphenols. The Folin reaction is based on the reduction of phosphomolybdic acid by phenols in aqueous alkali. These methods do not distinguish between tannin and other types of polyphenol and also react with easily oxidized substances such as ascorbic acid.[11,29]

The vanillin-HCl reagent reacts with flavan-3-ols and proanthocyanidins. The reaction is based on the condensation of the phenolic aldehyde (vanillin) with the phloroglucinol structure of flavan-3-ols and PA under acidic condition in an organic solvent (methanol or ethanol). The correction for "blank" color is mandatory, and the modification of the solvent for this method may provide a measure of the degree of polymerization of proanthocyanidins.[33] The sensitivity of this method may be increased by the use of analogs of vanillin such as 4-dimethylaminocinnamaldehyde (DMACA).[34]

The butanol-HCl assay is specific for proanthocyanidins. The flavonoid subunits of the PA polymer are oxidatively cleaved to yield anthocyanidin. Some features of this method are that the assay does not involve hydrolysis, the yield of anthocyanidin is critically altered by impurities of transition metal ions, and the color of the reaction is solvent-dependent.[11]

The lack of appropriate standards is a problem that is common to all of the colorimetric assays, and limits their usefulness for quantitative analysis. The most frequently used standards are tannic acid for the Folin-Ciocalteu reagent, catechin for the vanillin-HCl reaction, and quebracho tannins for the butanol-HCl reaction. These standards lead to considerable error in estimates of the content of polyphenols because the extinction coefficients produced by the chromophores in the standards differ from those of the compounds in the plant extract.[29]

The protein-binding methods are considered to be more useful for determining the amount of tannin in plants because these assays may be closely related to ecological effects.[35,36] Assays that rely on protein precipitation usually quantify the amount of protein precipitated or the amount of polyphenolic compound that remains in solution. There are several factors involved in the formation of complexes between protein and phenolic polymers: molecular weight and structural heterogeneity of phenolics and proteins; degree of glycosylation, amino acid composition, and molecular weight of the proteins; and reaction conditions such as pH, temperature, time, and relative concentration of reactants.[11] The radial diffusion assay[36] is a recently developed protein precipitation method. The amount of precipitated complex is proportional to the content of tannin in the sample. This method is suitable for a large number of samples and uses tannic acid as a standard. However, this procedure is less effective than colorimetric assays for quantification.

A gravimetric method for estimating total tannin and related phenolic compounds that is based on their precipitation by trivalent ytterbium has been described.[18] This procedure can be use for quantitative analysis and isolation of phenolic compounds. The advantages of this method are that standards are not required and the precipitate can be easily dissolved with oxalic acid to yield a solution of the tannin that can be use for further experiments.

6.6 INSOLUBLE PROANTHOCYANIDINS AND ANALYTICAL METHODS

In most plant samples that contain PA, there is a portion that is not extracted by aqueous organic solvents. This group of insoluble proanthocyanidins may be similar to lignin and covalently bound to carbohydrates or other polymers within the plant cell wall,[28] or an artifact of sample preparation.

Bate-Smith[37] showed that in sainfoin (*Onobrychis viciaefolia*) a notable percentage of the PA was not extracted by aqueous organic solvents. The fibrous residue after extraction contained a significant amount of PA, even if the sample is fine milled using hot methanol for extraction. The same problem occurred with procyanadin from species such as *Aesculus hippocastanum* and *Pilea cadieri*.[38]

The treatment of plant samples prior to extraction has a major effect on tannin analysis. The tissue of interest must be maintained so that the tannins are not altered or destroyed during the sampling process. Extraction of tannins with aqueous organic solvents from fresh samples are likely to give the highest yield because complex formation and oxidative poly-merization are avoided.[29] Steaming and autoclaving fresh samples of oak leaves decreased the amount of tannin detected by more than 50%. The reduction was likely caused by destruction of polyphenols and/or reduced solubility through complexation with other mac-romolecules.[39] Hagerman[19] recommends using fresh samples if possible or freeze-drying the sample before tannin extraction, because air-drying leaves is known to result in the modifi-cation of relatively stable condensed tannin. Orians[40] showed that air-dried leaves from members of the Salicaceae family have reduced concentration of PA, while vacuum-dried fresh leaves have high concentration. Thus, the accuracy with which PA can be quantified depends upon such factors as species, growing season, extracting solvent, method of analysis, etc. In general, there is no agreement on the best method for preserving plant material, the most effective aqueous organic solvent for extracting polyphenols, or the best method of analysis.[19] Torti et al. (1995) showed that the method of solvent extraction is also important because a homogenizer was more efficient and consistent in comparison to extraction in a sonicator/shaker bath which only reduced the plant material into smaller material while the homogenizer also destroyed cell walls, and therefore increased yield and efficiency.

The analysis of PA that were bound to the fiber fraction or present as tannin -protein complexes in leaf samples of *Quercus incane*, *Prosopis cineraria*, and *Robinia pseudoacacia* was more accurate when the samples were freeze-dried.[42] These authors also found that the reactivity of PA to butanol-HCl in tannin-protein complexes that were dried at 50°C or 100°C for 2 h, 4 h, and 24 h was significantly lower when compared to complexes that were freeze-dried or undried. There were no statistical differences between samples that were freeze-dried and heated at 50°C in the content of PA in fiber fractions, neutral detergent fiber (NDF), and acid detergent fiber (ADF). However, lower results in the content of PA were obtained in the fiber fractions when samples were dried at 100°C as compared to freeze-dried samples.

Proanthocyanidins have a large effect on the analysis of NDF in eucalyptus (*Eucalyptus ovaca*). The interference of tannin in the NDF was reduced by dividing the samples into 3 particle size fractions (small, medium, and large) and analyzing each fraction separately.[43] Another alternative suggested by McArthur[43] is presoaking the sample in polyethylene glycol (PEG). The content of PA in small particles was the lowest, but the variation of NDF was higher in small and medium particles. The variation in NDF decreased when PEG was used which indicated that PA affects the accuracy of the NDF measurement. PEG may have decreased the variation in the determination of NDF by disrupting the tannin-protein com-plexes. The complexation of proanthocyanidins by PEG has also been used for quantification of tannin.[44,45] However, Makkar et al.[45] showed that the PEG-complexes are insoluble in the detergent solution and appear in NDF and ADF fractions.

Terrill et al.[46] showed that the treatment of sample prior to the analysis of PA is important if accurate results are to be obtained. Aqueous acetone (70%) had the maximum extraction of PA from *Sericea Lespedeza* samples for all drying treatment. Using the vanillin-HCl method for PA, the color developed by the reaction decreased dramatically when water was added to the solution. The concentration of proanthocyanidins in these samples was almost twice when catechin was used as a standard compared with purified tannin. The sample preservation method determined the concentration of PA because polymerization or complex-ation with other cellular components during sample preparation reduced extractability. For *Sericea* samples, freeze drying appeared to be the most efficient method of preservation. Terrill et al.[47] also reported a decrease in analyzable tannin in *Sericea* samples that had been dried in the field.

A method using boiling sodium dodecyl sulphate containing 3-mercaptoethanol (SDS) was developed by Terrill et al.[48] for the extraction of protein-bound and fiber-bound condensed tannin. The procedure uses a modified butanol-HCl assay for PA and purified PA in SDS solution as a standard.

A considerable amount of PA in several tree species remained in the neutral detergent fiber (NDF) after sequential extraction with aqueous organic solvent (7:3, acetone:water) followed by neutral detergent.[49,50] Cassava (*Manihot esculenta*, Crantz), has a good potential as a forage because leaf blades normally contain more than 20% crude protein (CP); however, a high content of PA present in the NDF may be an important factor that limits the nutritive value.

The presence of PA in fiber fractions may interfere with the interpretation of results from detergent system of forage analysis. For instance, in 17 East African browse species, the content of insoluble proanthocyanidins was positively correlated with neutral detergent fiber (NDF). Insoluble PA can be solubilized with the formation of red solution of anthocyanidins by refluxing with acid detergent. In *Acacia* species, insoluble PA were associated with high levels of detergent insoluble nitrogen which may result from formation of tannin -protein complexes. In the detergent system of forage analysis, hemicellulose is calculated by the difference between NDF and ADF; therefore, the insoluble PA that are solubilized by acid detergent will contribute to increased hemicellulose values. The interpretation of NDF in these type of samples is misleading because PA in the complex invalidates the supposition that NDF represents cell wall carbohydrates and lignin. A complete estimation of both soluble and insoluble fractions is required in order to apply the detergent system of analysis to forages that contain PA.[29]

Bartolome et al.,[51] working with three quite different foods (pears, lentils, and cocoa) showed that the average molecular size of PA that is present in the dietary fiber fraction is larger than the average size of PA that is present in the whole food.

The PA in seed coats of cowpeas interfered with the analysis of lignin using the acetyl bromide method.[52] The interference is due to the absorption at 280 nm by both polymers when the cell wall is dissolved in acetyl bromide in glacial acetic acid (1:3, v:v). In addition to PA and lignin, proteins also absorb at 280 nm. This situation can be alleviated by destroying the protein using perchloric acid in the acetyl bromide reagent. Nevertheless, other components of the cell wall such as ferulic and p-coumaric acid can also cause interference. Morrison et al.[52] showed that by removing the soluble fraction of proanthocyanidins with aqueous acetone before the lignin analysis it is possible to obtain a good estimation of lignin by using the acetyl bromide method. However, the determination will also be influenced by the ratio of soluble to insoluble PA in cultivars of cowpea.

Makkar et al.[53] also showed that PA interferes with fiber determination in *Acacia saligna*. In order to figure out the effect of PA on the fiber values of this species, three different methods were followed: the fiber fraction was analyzed by the standard method which uses separate samples for NDF and ADF; it was analyzed by the sequential method where the NDF is treated with acid detergent to obtain ADF; and the fiber fraction of the feed was determined by the standard method and feces by the sequential method. In the leaves of *A. saligna*, the sequential analysis of ADF and acid detergent lignin (ADL) were much higher than the standard determination. The opposite result was observed for fecal samples from sheep and goats. These results indicate that the fiber in *A. saligna* leaves differs from the fiber in feces from ruminants that consume the leaves. Differences of more than 10 and 5% were found in the content of hemicellulose (NDF-ADF) in the original leaf samples and feces, respectively, depending if the sample was analyzed by the direct or sequential method. Similar results were obtained when the content of cellulose (ADF-ADL) was assayed. The higher result found in the sequentially determined ADF and ADL could be explained by the higher content of PA in this fiber fraction.

Makkar and Singh[54] studied the variation of PA in the fiber fraction of five different species of oak and showed that PA were associated with NDF and ADF in samples of mature leaves. There was a significantly positive correlation between the content of PA in the fiber fractions and the total PA. These relationships are comparable to those describe for *Acacia* species.[50]

Reed[55] showed differences between bird resistant (BR) and non bird resistant (NBR) sorghum in fiber composition that were related to differences in content of PA. NDF content was analyzed by two different methods; the amylase procedure gave higher NDF values than the urea/amylase method. All estimates of PA were significantly and negatively correlated with fiber digestibility. Reduction in fiber digestibility in these samples can be attributed to the inhibition of microbial digestion by strong covalent linkage between insoluble PA and lignin to fiber fractions. Estimation of soluble PA may give incomplete determination of inhibitory capacity of PA, because insoluble PA may increase the ADF and lignin fraction and decrease digestibility of NDF.

Wiegand et al.,[57] demonstrated the effects of PA in the utilization of the detergent system of forage analysis. In this research, three accessions of *Sesbania sesban* and one accession of *Sesbania goetzei* were used as sources of protein in diets for sheep. The results showed that the accession with the highest content of PA had negative digestion coefficients for ADL and neutral detergent insoluble nitrogen (NDIN). Linear regression of the apparent digestible amount of N and the amount of N in the diet can be used for the estimation of true digestibility. However, as the level of PA increased, forage N became nutritionally nonuniform and true digestibility of N could not be predicted by regression analysis.[29] This effect was observed in *S. Sesban* accession 15036 and *S. goetzei*. Plants containing high levels of PA such as *Acacia cyanophyla*, *A. sieberiana*, and *A. seyal* showed similar effects when fed to sheep.[56]

Wiegand et al.,[57] also showed that the N in samples with high content of PA is digested in a nutritionally non-uniform manner; and fecal excretion of NDIN was highly correlated to intake of PA. Both observations indicated that PA caused the formation of detergent-insoluble complexes in the digestive tract, elevating the fecal excretion of NDIN and ADL. According to Reed (1995), the basic principles of applying the detergent analysis for estimating the nutritive value of forages are violated by the formation of these complexes. These principles are that neutral detergent soluble and crude protein are nutritionally uniform and have high true digestibility, and fiber fractions cannot be formed in the digestive tract (it has to be from plants origin). However, the violation of these principles in forages containing PA does not invalidate the use of the detergent system of analysis because the deviation from the theoretical behavior of forage components can be used to explain the effects of PA on protein and fiber digestion.

6.7 CONCLUSION

Proanthocyanidins are secondary metabolites that have a large effect on the nutritive value and toxicology of plants. Proanthocyanidins are very reactive and precautions are needed in handling samples to avoid reactions that affect quantitative and qualitative analysis. A better understanding of the chemical structure of PA will improve the quality and interpretation of the results of analysis. None of the available assays are completely effective. Researchers need to understand the principles behind each method of analysis in order to choose those methods that are most suitable to the samples and hypotheses being tested. The information in this chapter emphasizes the necessity to determine both soluble and insoluble proanthocyanidins.

REFERENCES

1. Appel, H. M., Phenolic in ecological interactions: The importance of oxidation, *J. Chem. Ecol.*, 19, 1521, 1993.
2. Bernays, E. A., Cooper Driver, G., and Bilgener, M., Herbivores and plant tannins, *Adv. Ecol. Res.*, 19, 263, 1989.

3. Schultz, J. C., Tannin-insect interactions, in *Chemistry and Significance of Condensed Tannins,* Hemingway, R. W. and Karchesy, J. J., Eds., Plenum Press, New York, 1989, 417.

4. Peters, N. K. and Verma, D. P. S., Phenolics compounds as regulators of gene expression in plant-microbe interaction, *Mol. Plant-Microbe Interact.,* 3, 4, 1990.

5. Friend, J., Plants phenolics, lignification and plant disease, *Prog. Phytochem.,* 7, 197, 1981

6. Schlesinger, W. H., *Biogeochemistry: An Analysis of Global Change.* Academic Press, San Diego, California, 1991, 78, 91, 185.

7. Schultz, J. C., Hunter, M. D., and Appel, H. M., Antimicrobial activity of polyphenol mediates plant-herbivore interactions, in *Plant Polyphenols: Biogenesis, Chemical Properties, and Significance,* Hemingway R. W., Ed., Plenum Press. 1992, 621.

8. Deshpande, S. S., Cheryan, M., and Salunkle, D. K., Tannin analysis of food products, *CRC Crit. Rev. Food Sci. Nutr.,* 244, 401, 1986.

9. Haslam, E., Symmetry and promiscuity in procyanidin biochemistry, *Phytochemistry,* 17, 1625, 1977.

10. Porter, L. J., Hrstich, L. N., and Chan, B. C., The conversion of procyanidins and prodelphinidins to cyanidin and delphinidin, *Phytochemistry,* 25, 223 ,1986.

11. Hagerman, A. E. and Butler, L. G., Choosing appropriate methods and standards for assaying tannin , *J. Chem. Ecol.,* 15, 1795, 1989.

12. Haslam, E., Proanthocyanidins, in *The Flavonoids: Advances in Research,* Harborne, J. B. and Mabrey, T. J., Eds., Chapman and Hall, London, 1982, Chap. 7.

13. Peri, C. and Pompei, C., Estimation of different phenolic groups in vegetable extracts, *Phytochemistry,* 10, 2187, 1971.

14. Hagerman, A. E. and Butler, L. G., Condensed tannin purification and characterization of tannin-associated proteins, *J. Agric. Food Chem.,* 28, 947, 1980.

15. Mole, S. and Waterman, P. G., A critical analysis of techniques for measuring tannins in ecological studies, *Oecologia,* 78, 93, 1987.

16. Wisdom, C. S., Gonzales-Coloma, A., and Rundel, P. W., Ecological tannin assays, *Oecologia,* 72, 395, 1987.

17. Mole, S., Butter, L. G., Hagerman, A. E., and Waterman, P. G., Ecological tannin assays: A critique, *Oecologia,* 78, 93, 1989.

18. Reed, J. D., Horvath, P. J., Allen, M. S., and Van Soest, P. J., Gravimetric determination of soluble phenolics including tannin from leaves by precipitation with trivalent ytterbium, *J. Sci. Food Agric.,* 36, 255, 1985.

19. Hagerman, A. E., Extraction of tannin from fresh and preserved leaves, *J. Chem. Ecol.,* 14, 453, 1988.

20. Stafford, H. A., Lester, H. H., and Porter. L. J., Chemical and enzymatic synthesis of monomeric procyanidins (leucocyanidins or 3′,4′,5,7-tetrahydroxy-flavan-3,4-diols) from (2R, 3R)-dihydroquercetin, *Phytochemistry,* 24, 333, 1985.

21. Ruhnau, B. and Forkmann, G., Flavan-3,4-diols in anthocyanin biosynthesis, enzymatic formation with flower extracts from Callistephus chinensis, *Phytochemistry,* 27, 1035, 1988.

22. Stafford, H. A., *Flavonoid Metabolism,* CRC Press, Boca Raton, FL, 1990, Chap. 2, 3.

23. Horvath, P. J., *The Nutritional and Ecological Significance of Acer-Tannins and Related Polyphenols.* M. S. Thesis, Cornell University, Ithaca, New York, 1981.

24. Mueller-Harvey, I. and McAllan, A. B., Tannins: Their biochemistry and nutritional properties, *Adv. Plant Cell Biochem. Biotechnol.,* 1, 151, 1992.

25. Haslam, E. and Lilley, T. H., Interaction of natural phenol with macromolecules, in *Plant Flavonoids in Biology and Medicine: Biochemical, Pharmacological, and Structure-Activity Relationships,* Middleton, C. V. and Harborne, J. B., Eds., Alan R. Liss, New York, 1985, 53.

26. Davis, A. B. and Hoseney, R. C., Grain sorghum condensed tannin. I. Isolation, estimation, and selective adsorption by starch, *Cereal Chem.,* 56, 310, 1979.

27. Deshpande, S. S. and Salunkhe, D. K., Interactions of tannic acid and catechin with legume starches, *J. Food Sci.,* 47, 2080, 1982.

28. Haslam, E., *Plant Polyphenols-Vegetable Tannins Revised,* Cambridge University Press, Cambridge, U. K. 1989.

29. Reed, J. D., Nutritional Toxicology of tannin and related polyphenol in forage legumes, *J. Anim. Sci.,* 73, 1516, 1995.

30. Stafford, H. A., Proanthocyanidins and the lignin connection, *Phytochemistry*, 27, 1, 1988.
31. Streit, W. and Fengel, D., Purified tannins from quebracho colorado, *Phytochemistry*, 36, 481. 1994.
32. Waterman, P. G. and Mole, S., *Analysis of Phenolic Plant Metabolites*, Blackwell Scientific Publications, Oxford, U. K. 1994, Chap. 4, 5, 7.
33. Butler, L. G., Price, M. P., and Brotherton, J. E., Vainillin assay for proanthocyanidins (condensed tannins): Modification of the solvent for estimation of the degree of polymerization, *J. Agric. Food Chem.*, 30, 1087, 1982.
34. Li, Y-G, Tanner, G., and Larkin, P., The DMACA-HCl Protocol and the threshold proanthocyanidin content for bloat safety in forage legumes, *J. Sci. Food Agric.*, 70, 89, 1996.
35. Martin, J. S. and Martin, M. M., Tannin assays in ecological studies: Lack of correlation between phenolics, proanthocyanidins and protein-precipitating constituents in mature foliage of six oak species, *Oecology*, 54, 205, 1982.
36. Hagerman, A. E., Radial diffusion method for determining tannin in plant extract, *J. Chem. Ecol.*, 13, 437, 1987.
37. Bate-Smith, E. C., Tannins of herbaceous leguminosae, *Phytochemistry*, 12, 1809, 1973.
38. Bate-Smith, E. C., Phytochemistry of proanthocyanadins, *Phytochemistry*, 14, 1107, 1975.
39. Makkar, H. P. S. and Singh, B., Effect of steaming and autoclaving Oak (Quercus incana) leaves on levels of tannins, fiber and In-sacco dry matter digestibility, *J. Sci. Food Agric.*, 59, 469, 1992.
40. Orians, C. M., Preserving leaves for tannin and phenolic glycoside analyses: A comparison of methods using three Willow Taxa, *J. Chem. Ecol.*, 21, 1235, 1995.
41. Torti, S. D., Dearing, M. D., and Kursar, T. A., Extraction of phenolic compounds from fresh leaves: A comparison of methods, *J. Chem. Ecol.*, 21, 117, 1995.
42. Makkar, H. P. S. and Singh, B., Determination of condensed tannins in complexes with fiber and proteins, *J. Sci. Food Agric.*, 69, 129, 1995.
43. McArthur, C., Variation in neutral detergent fiber analysis of tannin -rich foliage, *J. Wildl. Manage.*, 52, 374, 1988.
44. Makkar, H. P. S. and Becker, K., Behavior of tannic acid from various commercial sources towards redox, metal complexing and protein precipitation assay of tannins, *J. Sci. Food Agric.*, 62, 295, 1993.
45. Makkar, H. P. S., Blummel, M., and Becker, K., Formation of complexes between polyvinyl pyrrolidones or polyethylene glycols and tannins, and their implication in gas production and true digestibility in *in vitro* techniques, *Br. J. Nutri.*, 73, 897, 1995.
46. Terrill, T. H., Windham, W. R., Evans, J. J., and Hoveland, C. S., Condensed tannin concentration in *Sericea lespedeza* as influenced by preservation method, *Crop Sci.*, 30, 219, 1990.
47. Terrill, T. H., Windham, W. R., Hoveland, C. S., and Amos, H. E., Forage preservation method influences on tannin concentration, intake, and digestibility of *Sericea Lespedeza* by sheep, *Agron. J.*, 81, 435, 1989.
48. Terrill, T. H., Rowan, A. M., Douglas, G. B., and. Barry. T. N, Determination of extractable and bound condensed tannin concentrations in forage plants, protein concentrate meals and cereal grains, *J. Sci. Food Agric.*, 58, 321, 1992.
49. Reed, J. D., McDowell, R. E., Van Soest, P. J., and Horvath, P. J. Condensed tannins: A factor limiting the use of cassava forage, *J. Sci. Food Agric.*, 33, 213, 1982.
50. Reed, J. D., Relationships among soluble phenolics, insoluble proanthocyanidins and fiber in east African browse species, *J. Range Manage.*, 39, 5, 1986.
51. Bartolomé, B., Jiménes-Ransey, L. M., and Butler, L. G., Nature of the condensed tannin present in the dietary fiber fractions in foods, *Food Chem.*, 53, 357, 1995.
52. Morrison, M., Asiedu, E. A., Stuchbury, T., and Powell, A. A., Determination of lignin and tannin contents of cowpea seed coats, Ann. Bot., 76, 287, 1995.
53. Makkar, H. P. S., Borowy, N. K., Becker, K., and Degen, A., Some problems in fiber determination of a tannin-rich forage (*Acacia saligna* leaves) and their implications in *in vivo* studies, *Anim. Feed Sci. Technol.*, 55, 67, 1995.
54. Makkar, H. P. S. and Singh, B., Distribution of condensed tannins (proanthocyanidins) in various fiber fractions in young and mature leaves of some oak species, *Anim. Feed Sci. Technol.*, 32, 253, 1991.

55. Reed, J. D., Phenolics, fiber, and fiber digestibility in bird resistant and non bird resistant sorghum grain, *J. Agric. Food Chem.*, 35, 461, 1987.

56. Reed, J. D., Soller, H., and Woodward, A., Fodder tree and straw diets for sheep: Intake, growth, digestibility and the effects of phenolics on nitrogen utilization, *Anim. Feed Sci. Tech.*, 30, 39, 1990.

57. Wiegand, R. O., Reed, J. D., Said, A. N., and Ummuna, V. N., Proanthocyanidins (condensed tannins) and the use of leaves from *Sesbania sesban* and *Sesbania goetzei* as protein supplements, *Anim. Feed Sci. Technol.*, 54, 175, 1995.

7 Gossypol

C. A. Risco and C. C. Chase, Jr.

CONTENTS

7.1 INTRODUCTION

Gossypol is a naturally occurring compound in the cotton plant (*Gossypium* spp.). Cotton fibers are primarily produced from *Gossypium hirsutum* (Upland cotton) but also from *Gossypium barbadense* (Egyptian, Tanguis, or Pima). Because gossypol has been recognized since the turn of the century to be toxic to livestock, a limiting factor of whole cottonseed and cottonseed meal as feed sources is their gossypol content.[1] Whole cottonseed and cottonseed meal are popular by-products of the cottonseed processing industry. Whole cottonseed is a high source of energy, fiber, and protein in diets of lactating dairy cows.[2] As much as 3.5 kg per cow per day may be fed with excellent results in maintaining both high milk production and milk fat percentages. Cottonseed hulls are commonly used to increase the fiber source in dairy and beef rations. After extraction of oil from prepared cottonseed, the resulting cottonseed meal is available for livestock feeding (Figure 7.1). Cottonseed meal is a popular protein supplement in beef cow diets. Cottonseed meal is also an excellent and economical source of protein in rations for poultry, swine, lamb, and beef and dairy calf rations when properly managed.

Nonruminant species such as swine and poultry do exhibit susceptibility to gossypol toxicity under certain feeding regimens. Also, despite years of practical experience in feeding cottonseed meal to preruminant livestock in limited amounts, gossypol toxicity has been demonstrated in young calves as well as lambs.[3,4] These reports have suggested that in calves and lambs the preruminant, with an undeveloped rumen, functions essentially as a nonruminant and is unable to detoxify gossypol. Ruminants reportedly detoxify gossypol in the rumen by binding the compound to soluble proteins.[5] However, reports of toxicity in mature cattle suggests that the capacity of the rumen to detoxify gossypol can be exceeded.[6,7]

The chemical properties and toxicity of gossypol are covered in this chapter. In addition, current accepted recommendations for safe levels of gossypol consumption and prevention of toxicity are also reviewed.

0-8493-8551-2/97/$0.00+$.50
© 1997 by CRC Press, Inc.

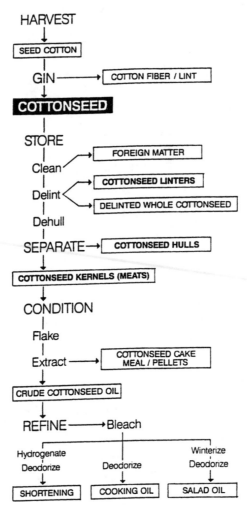

FIGURE 7.1 Processing steps from harvest to products of cottonseed. (From Jones, L. A., *Cattle Research with Gossypol Containing Feeds*, 1991, 1. With permission.)

7.2   GOSSYPOL PIGMENT

Gossypol, $C_{30}H_{30}O_8$, a binaphthyl polyphenolic pigment, is found in the glands, stems, and roots of the cotton plant belonging to the genus *Gossypium* as well as the plants in the Malvaceae family.[1] It is a chiral molecule due to its steric hindrance to rotation about the internaphthyl bond. Gossypol is isolated from the cotton plant as a racemic mixture of (+) and (-) enantiomers, by extraction followed by crystallization from solutions containing acetic acid, which produces the complex (+) or (-) gossypol-acetic acid. The enantiomeric ratio of gossypol appears to be related to the *Gossypium* species. An enantiomeric excess of (+) gossypol was found in each variety of *Gossypium arboreum*, *Gossypium herbaceum*, and *G. hirsutum*, whereas (-) gossypol was in excess in each variety of *G. barbadense* or pima cotton investigated.[8] While gossypol is the major phenolic in cottonseed, there are at least 15 closely related compounds. The phenolic groups on the molecule are chemically reactive, so it can bind readily with minerals and amino acids.

Whole cottonseed is composed of seed, linters, and hull. The major source of gossypol in whole cottonseed is found in the seed glands, which appear as small black dots on the cut surface of the seed.[9] The amount of gossypol in the various whole cottonseed components obtained from samples collected at cottonseed mills during 1993 to 1994 is shown in Table 7.1.[10] The level of gossypol in cottonseed is dependent on plant species as well as climatic and soil condition, water supply, and amount and composition of fertilizers used.[11] These factors may explain the variation of gossypol content in whole cottonseed reported in the literature during the past 30 years.[10] Gossypol in the plant is related to insect resistance.[12] Although there are varieties of glandless cotton plants, they make up less than a tenth of 1% of the acreage of the U.S. cotton crop.[9]

TABLE 7.1
Gossypol Content in Cottonseed Products from Samples Collected at U.S. Cottonseed Oil Mills During 1993–1994

Product	Gossypol[a]		Gossypol isomers[b]	
	Total	Free	(+)	(−)
	mg kg^{-1}	mg kg^{-1}	Percent of total	
Cottonseed hulls	1070	490	58.4	40.6
Decorticated seed[c]	12,000	12,400	61.2	38.8
Whole linted seed[d]	6600	6800	61.2	38.8

[a] Total and free gossypol determined by *Official and Tentative Methods of Analysis*, 3rd ed., Amer. Oil Chem. Soc., Champaign, IL, 1988, 1.
[b] Determined by high performance liquid chromatography, Hron, R. J., Kim, H. L., and Calhoun M. C., presented at Annual Meeting Amer. Oil Chem. Soc., San Antonio, TX, May 7–21, 1995.
[c] Seed without hulls and lint.
[d] Determined by multiplying percentage of gossypol in decorticated seed by 0.55.

Source: From Forster, L. A., Jr. and Calhoun, M. C., *Feedstuffs*, 67, 16, 1995. With permission.

7.3   SOURCE OF GOSSYPOL IN FEEDS

Gossypol exists in two forms, free and bound. The free form is toxic, whereas the bound form is considered nontoxic because it is not released in the rumen.[13] Officially, gossypol content in feeds is measured and reported in terms of total and free. In whole seed, virtually all the gossypol is in the free form. The processing of whole cottonseed into meal complexes gossypol with proteins, primarily the epsilon amino group of lysine and becomes bound, the biologically inactive form. Although researchers are concerned about the effect of bound gossypol, evidence is lacking to consider it along with free gossypol in terms of toxicity. This subject is currently under investigation.[14,15,16]

The total gossypol content of cottonseed meal is not affected by the method used for oil extraction because total gossypol levels equals free gossypol plus bound gossypol.[9] Table 7.2 shows that the final level of free gossypol content in the cottonseed meal is dependent on the initial seed content and the method used to extract the oil.[10] The direct solvent extraction of cottonseed oil yields the highest content of gossypol while the screw press method yields the lowest.[17] Essentially, all of the cottonseed mills that employ the solvent method now also include the use of expanders, which is a modified extruder, used prior to solvent treatment. Expanders introduce some heat and pressure into the process of solvent extraction of cottonseed oil.[9] The production of heat in this process binds some of the gossypol (bound gossypol). The different processes used in the U.S. to extract oil have been described.[9] They are listed in the order of most common use:

TABLE 7.2
Gossypol Content in Cottonseed Meal Processed by Different Methods from Samples Collected at U.S. Cottonseed Oil Mills During 1993–1994

Product	Gossypol[a] Total mg kg⁻¹	Gossypol[a] Free mg kg⁻¹	Gossypol isomers[b] (+) Percent of total	Gossypol isomers[b] (−) Percent of total
Mechanical	10,900	600	59.9	40.1
Prepress Solvent	10,600	700	58.1	41.9
Direct Solvent	12,100	1500	60.9	39.1
Expander Solvent	11,600	1400	58.3	41.7

[a] Total and free gossypol determined by *Official and Tentative Methods of Analysis*, 3rd ed., Amer. Oil Chem. Soc., Champaign, IL, 1988, 1.
[b] Determined by high performance liquid chromatography, Hron, R. J., Kim, H. L., and Calhoun M. C., presented at Annual Meeting Amer. Oil Chem. Soc., San Antonio, TX, May 7–21, 1995.

Source: From Forster, L. A., Jr. and Calhoun, M. C., *Feedstuffs*, 67, 16, 1995. With permission.

1. EXPANDER-SOLVENT PROCESS. Prepared seed kernels are run through an expander before using hexane to extract the oil. Currently, the majority of cottonseed meal produced in the U.S. is produced by the expander-solvent process.
2. EXPELLER OR SCREW PRESS PROCESS (OLD PROCESS). Large screw presses are used alone without the use of solvent extraction. This method yields the lowest levels of free gossypol.
3. PREPRESS SOLVENT PROCESS. Oil is removed using the large screw presses before using hexane to extract the oil.
4. DIRECT SOLVENT PROCESS. Uses no expanders and results in the highest levels of free gossypol.

7.4   TOXICITY

Ingestion of excessive levels of whole cottonseed, cottonseed meal, or occasionally cottonseed hulls containing sufficient levels of gossypol is the route of exposure known to cause toxicity. The level predisposing the animal to toxicity depends on the following factors: species, duration of ingestion, rumen function, and protein, mineral, and fiber content of the ration. It is known that the toxic effect is cumulative in nature, where toxicity occurs after several weeks to months of ingestion. A quantity of gossypol must accumulate to a certain level in order to exert its effects and cause clinical signs or death. Levels of gossypol in the liver and kidney, but not muscle, are elevated in proportion to the amount of free gossypol consumed.[4]

In swine,[18] lambs,[4] and calves,[3,19] gossypol appears to be cardiotoxic, and death and clinical signs are attributable to heart failure. This accounts for the chronically labored breathing (dyspnea), pendulous abdomen, intermandibular swelling, and jugular vein distention that are commonly observed in animals with gossypol toxicity prior to death. On post mortem examination, the following findings are usually present: severe effusion of a straw-colored fluid into the body cavities (ascites), edema and icteric discoloration of the mesentery, an enlarged liver (hepatomegaly) with a brown to golden color (nutmeg) on the cut surface of the liver, and an enlarged heart (cardiomegaly) that appears pale, streaked, and flabby.[3,4,18,19]

The above clinical and post mortem findings are compatible with heart failure due to passive congestion and venous stasis. The most consistent histopathological finding is centrilobular hepatic congestion and necrosis. Swollen cells, congestion of portal triads, pyknosis, karyorrhexis, and karyolysis of hepatocytes, perivascular edema, and severe fatty changes are seen with the varying degrees of necrosis. The changes present in the liver may be secondary to heart failure; however, a direct effect of gossypol on the liver cannot be ruled out.

Chemistry panels designed to monitor serum proteins, energy status, serum electrolytes, renal function, liver function, and necrosis generally are not useful in demonstrating impending gossypol toxicity when used as a group monitoring test.[19] Clinical chemistry changes may only be characteristic of liver failure during the terminal stages in affected calves. Alterations in measures of erythroid cell population such as hematocrit and hemoglobin occurs in calves ingesting gossypol and may be of a diagnostic value. Hematocrit and hemoglobin concentrations have been reported to be decreased in calves with gossypol toxicoses.[19] Gossypol binds to iron when fed to swine and is associated with decreased hematocrit.[20] When ferrous ions were added to the diet containing gossypol, normal erythroid cell measures resulted, but toxicity also resulted[20]. It has been suggested that iron bound to gossypol is unavailable for normal biosynthesis of hemoglobin and other processes in the animal.[21]

A decrease in hematocrit could also result from a direct effect of gossypol on the erythrocytes. An increase in erythrocyte osmotic fragility appears to be a consistent finding in various gossypol studies.[22,23,24,25,26] Changes in erythrocyte indices such as mean corpuscular volume, mean corpuscular hemoglobin, and mean corpuscular hemoglobin concentration are not common in gossypol toxicoses indicating normal hematopoietic response is maintained.[19] Therefore, the degree of anemia is mild. In the above study,[19] it was observed that the overall changes in group means for hematology and serum chemistry variables are modest and insufficient to be used to distinguish, on a diagnostic basis, between safe and unsafe levels of gossypol in the diet of young calves.

In swine and dogs, gossypol content of the liver is shown to be related directly to changes in electrical patterns of the heart.[27] Affected changes in the electrocardiogram were the T wave amplitude, T wave duration, and the isoelectric S-T segment. Gossypol may bind to phospholipids or the epsilon amino group of lysine present in the cytoplasm of cardiac cells, altering cellular permeability of potassium. Changes in intracellular potassium concentration may affect cardiac conductivity and function. Therefore, sudden death caused by gossypol toxicity may be attributed to an acute alteration of intra- and extracellular potassium content interfering with normal myocardial conduction.[12]

The relationship between gossypol toxicity and α-tocopherol (vitamin E) exists. Gossypol can interact directly with biological membranes and promote formation of highly reactive oxygen containing free radicals.[28] Free radicals provoke oxidative injury and compromise the oxidant system of living organisms through lipid peroxidation of membranes, protein cross-linkage, nonperoxidative mitochondrial damage, and DNA damage. The inhibition and scavenging systems for free radicals include glutathione peroxidase, vitamin E, and ascorbate (vitamin C). Bender et al.,[28] found that vitamin E, ascorbate, glutathione peroxidase, and other antioxidants were reduced by feeding rats high levels of gossypol. In dairy cattle, Lane and Stuart[29] found that feeding high amounts of gossypol decreased plasma vitamin E levels, implicating a possible relationship between vitamin E and gossypol toxicity. Pre- and post-partum consumption of free gossypol impaired aspects of calf skeletal development with lowered serum vitamin E and B carotene.[26]

7.4.1 NONRUMINANTS

Gossypol toxicity has been reported in rats, mice, rabbits, guinea pigs, dogs, chickens, and swine.[12] In chickens, increasing levels of gossypol have been correlated with increased mortality and reduction in weight gains.[30] In addition, cottonseed meal in the ration of laying hens results in a discoloration of the yolks and whites of stored and fresh eggs. Gossypol reacting with yolk iron results in an olive yolk discoloration. The pink discoloration in the egg white has been attributed to the cyclopropenoids fatty acids present in the residual oil in cottonseed meal.[30] In chickens, tolerable levels of gossypol reported in the literature vary from a low of 160 mg kg^{-1} to a high of 1000 mg kg^{-1}.[30] In 1957, a committee on the status of cottonseed meal in poultry rations concluded that any cottonseed meal containing more than 400 mg kg^{-1} free gossypol should be used with caution.

In swine, the recommended limit for free gossypol is no more than than 100 mg kg^{-1} in the diet.[31] In poultry and swine diets, iron is added to rations in the form of ferrous sulfate at a ratio of 1:1 iron to free gossypol in order to help prevent toxicity. An insoluble complex is formed between the iron and gossypol which prevents absorption. In addition, high intake of protein and calcium appears to have a protective effect. Recommended levels for free gossypol in poultry and swine diets are depicted in Table 7.3.[31,32,33]

TABLE 7.3
Currently Recommended Safe Levels of Free Gossypol in Poultry and Swine Diets

Class	Free gossypol intake, mg kg^{-1}	Maximum free gossypol intake with iron, mg kg^{-1}
Broiler[a]	100–150	400[b]
Layers[a]	50	150[c]
Swine[d]	100	400[e]

[a] Waldroup, P. W., *Feedstuffs*, 53, 21, 1981.
[b] 1 to 2 mg kg^{-1} of Fe to 1 mg kg^{-1} of free gossypol.
[c] 4 mg kg^{-1} of Fe to 1 mg kg^{-1} of free gossypol.
[d] Tanksley, T. D. and Knabe, D. A., *Feedstuffs*, 53, 24, 1981.
[e] 1 mg kg^{-1} of Fe to mg kg^{-1} of free gossypol.

Source: From Martin, S. D., *Feedstuffs*, 62, 89, 1990. With permission.

7.4.2   YOUNG RUMINANTS

Gossypol is thought to be detoxified in the rumen by binding to soluble proteins (epsilon amino group of lysine) produced by ruminal microbes and by dilution and slowed absorption.[5] Dairy calves and lambs are often fed rations that contain cottonseed meal preweaning (less than 2 months of age) or after weaning. It is not known when the rumen becomes functional and cannot be entirely determined by age. Therefore, young ruminants with undeveloped rumens appear to function as nonruminants and the rumen cannot be expected to effectively detoxify gossypol. At birth and while the young ruminant is consuming milk, the safe level of free gossypol in diets approximates that for nonruminant species, 100 mg kg^{-1}.[14] The period from 3 to 8 weeks of age is considered a transitional period which, with adequate dry feed intake the rumen becomes functional. During this period of transition, the ability of the rumen to detoxify dietary free gossypol increases and 200 mg kg^{-1} is considered safe.[14]

A 100% mortality occurred prior to 30 days of age in lambs administered 900 mg kg^{-1} of gossypol acetic acid, and myocardial lesions were evident in euthanatized healthy lambs consuming 100 mg kg^{-1} of gossypol acetic acid.[4] The free gossypol in the study reported above was administered orally via gelatin capsules, which resulted in a more concentrated, undiluted dose of free gossypol being made available for intestinal absorption. Because death or clinical signs of gossypol toxicity did not occur to the group consuming 100 mg kg^{-1}, Morgan et al.[4] suggests that up to 100 mg kg^{-1} of free gossypol in lambs is safe when gossypol is administered in this form. However, when a diet composed of cottonseed meal that contained 800 mg kg^{-1} was fed to lambs, no adverse effects were noted.[34]

Death losses from gossypol toxicosis in calves have been reported to range from 350 mg kg^{-1} to 850 mg kg^{-1} of free gossypol.[3,35,36] In the above reports, the variability in calf responses to various levels of gossypol consumption was due to different undefined dietary and management conditions, raising questions about the actual safe concentration of free gossypol in diets. In an attempt to help define at what levels free gossypol is safe in young calves, the effects of feeding diets containing 0, 100, 200, 400, or 800 mg kg^{-1} of free gossypol to Holstein calves from 1 to 120 days of age was measured.[19] Clinical evidence of disease was limited to calves fed 400 or 800 mg kg^{-1} free gossypol. It was concluded from that study that a ration containing up to 200 mg kg^{-1} of free gossypol is safe, 400 mg kg^{-1} of free gossypol approaches toxicity and 800 mg kg^{-1} causes death losses. In terms of mean daily free gossypol consumption, 2 grams approaches toxicity.

7.4.3 Mature Ruminants

In ruminants, the protein binding detoxifying mechanism can be overwhelmed if the free gossypol content is excessive or a low protein content is present in the rumen. The symptoms of gossypol toxicity seen in functional ruminants include sudden death as in nonruminants, and suspected physiological and reproductive effects without death.[37]

Physiological changes occurred and gossypol was found in tissues from 24 lactating dairy cattle fed a diet containing 15% direct solvent extracted cottonseed meal for 14 weeks.[6] Free gossypol intake averaged 24.2 g day^{-1}. The clinical signs observed by Lindsey et al.[6] were limited to reduced milk production, respiratory stress, and one death. Physiological changes were a decrease in hemoglobin content and an increase in erythrocyte fragility, which appeared to be detrimental only in periods of heat or nutritional stress.

A 10% mortality rate was reported in lactating dairy cattle fed 2.7 to 4.54 kg day^{-1} of whole cottonseed for a 6 to 8 month period.[37] The level of free gossypol in the whole cottonseed ranged from 1,700 to 6,800 mg kg^{-1}. Milk production was not affected and death was preceded by minimal signs of toxicity. Lesions found on post mortem examination resembled those seen in swine with gossypol toxicity. Gossypol was not found in the milk. In the report cited above, the whole cottonseed was ammoniated.[37] It has been suggested that the ammoniation of whole cottonseed may increase the availability of free gossypol.[38]

It should be kept in mind that in the reports cited above cows were fed extremely high levels of cottonseed meal or whole cottonseed which contained high concentrations of free gossypol. There has been a paucity of research evaluating at what level free gossypol is safe for functional ruminants. As pointed out in the review by Rogers and Poore,[13] defining safe recommendations based upon commercial feeding practices has received little attention. Table 7.4 shows current recommended safe levels of free gossypol in the total diets of non-functional and functional ruminants.[13,14] As indicated in Table 7.4, it is suggested that a higher level of free gossypol from whole cottonseed can be included in the total diet than free gossypol from cottonseed meal.[13] The reason for this difference, however, is not clearly understood.

Because gossypol intake has been associated with infertility in men in China, considerable interest in the role of this compound as a male contraceptive has occurred.[39] Research in this area over the past years has created some concern about the safety of cottonseed products when fed to breeding age cattle. The reproductive effects of gossypol in mammals has been reviewed by Randel et al.[7] In relation to female reproduction, gossypol has been shown to inhibit implantation,[40] delay embryo development,[41] and decrease ovarian steroidogenesis.[42] In contrast, in a study involving post pubertal beef heifers, direct solvent extracted cottonseed meal and whole cottonseed were fed to provide intakes of 0, .4, 1.7, 3.3, and 8.2 g day^{-1} of

TABLE 7.4
Currently Recommended Safe Levels of Free Gossypol in the Total Diet of Ruminants

Stage of rumen development	Age	Cottonseed meal (mg kg^{-1})	Whole cottonseed (mg kg^{-1})
Preruminant	0–3 weeks	100	100
Transitional	3–8 weeks	200	900
Functional post-weaning	8–24 weeks	200	600
Mature			
Females	>24 weeks	600	1200
Males	>24 weeks	200	900

Source: From Rogers, G. M. and Poore, M. H., V*et. Med.*, 995, 1995. With permission.

free gossypol for 62 days. Cumulative 30-day pregnancy rates after the feeding of gossyopol ceased were not different among treatment group.[25]

Gossypol appears to exert unique and selective effects upon the male reproductive system. These include reduced spermatogenesis and impaired sperm motility; the latter was associated with morphological aberrations of the sperm midpiece.[43] Feeding gossypol containing feedstuffs to bulls and rams has been shown to damage the spermatogenic epithelium.[7,44] An increase in sperm cell midpiece abnormality and erythrocyte fragility in Brahman bulls fed 2.75 kg day^{-1} of cottonseed meal (8.2 g day^{-1} of free gossypol) has been reported.[24]

The type of cottonseed product (meal vs. whole seed), and gossypol enantiomer (+ or -) may determine the extent of the toxicological effect that will occur. It has been suggested that detoxification of gossypol in the rumen is more efficient with whole seed diets than with cottonseed meal diets.[13] In the study by Chase et al.,[44] Brahman bulls fed 1.8 g day^{-1} of free gossypol from cottonseed meal had similar damage to seminiferous epithelium when compared to bulls fed 16 g day^{-1} of free gossypol from whole cottonseed. The spermicidal effect of gossypol may also depend on the predominant + or - gossypol enantiomer present in the cottonseed product. Due to its stereospecific binding properties, the (-) gossypol enantiomer is less bound to plasma proteins and appears better able to cross the blood-testis barrier *in vivo* and inhibit the biological activity of some proteins.[45]

Collectively, the studies cited above suggest that gossypol can escape complete detoxification in the rumen and have an effect on erythrocyte fragility and testicular tissue. However, the relevance of these studies to commercial cattle operations needs to be brought into its proper perspective. The free gossypol content in the cottonseed meal study rations were obtained from solvent extraction methods, which accounts for less than 2% of the oil extraction method used today. In addition, males in the above cited studies were not subjected to a fertility study.

7.5 GOSSYPOL ANALYSIS

The standard analytical method used to determine gossypol levels in feedstuffs employs an aniline reaction procedure according to the American Oil Chemists Society methods Ba 8-78 and Ba 7-58 for total gossypol and free gossypol, respectively.[46] However, the standard method is unsatisfactory when applied to mixed feed. The problem with this method is incomplete recovery of free gossypol from feed mixtures and the extraction of other feed constituents interfering in the subsequent colorimetric determinations. A procedure has been described, using mixed isopropyl alcohol hexane water solvent containing a gossypol complexing agent, 3-amino-1-propanol, which prevents interference by feed constituents and stabilizes gossypol during extraction.[47] In addition, gossypol levels can also be determined by means of high performance liquid chromatography (HPLC), this method is not affected by interference of feed constituents.[48]

Tissue assays for gossypol can be performed by aniline reaction procedures.[3,49] Liver gossypol concentrations of calves fed toxic levels of gossypol (800 mg kg^{-1}) were greater than 160 μg g^{-1} of wet liver. In contrast, concentrations in calves fed nontoxic concentrations of gossypol were less than 50 μg ml^{-1}.[19] The use of HPLC methodology in plasma is also used in some laboratories in the U.S.[16] The HPLC procedure for measuring plasma gossypol is sensitive and separates (+) and (-) gossypol enantiomers. Plasma gossypol reflects the level and availability of gossypol in livestock diets; it plateaus in 4 to 6 weeks after feeding and thereafter remains relatively constant.[16] Furthermore, plasma and liver levels of gossypol are highly correlated.

7.6   DIAGNOSTIC PROCEDURES AND SUMMARY

Much of the gossypol research published over the past years employed very high levels of gossypol to demonstrate some form of gossypol toxicity. Under practical conditions, when the appropriate processing and feeding guidelines are followed for cottonseed products, the risk for gossypol toxicity is low. In the U.S., cottonseed products are commonly used in swine, poultry, beef, and dairy cattle rations without adverse effects being reported. Nevertheless, because of the potential for toxicity due to feeding errors that may occur when feeding cottonseed products, it is important for nutritionists and veterinarians to be familiar with the diagnosis of gossypol toxicity.

When suspecting or investigating a gossypol toxicity problem the following factors as described by Morgan[12] should be considered. Determine if cottonseed products such as whole cottonseed, cottonseed meal, or cottonseed hulls are being fed. Have the feed tested for free and total gossypol by an official gossypol testing laboratory.[13] Analyze the cottonseed products (whole cottonseed, cottonseed meal, or cottonseed hulls) separately and not the complete ration which includes the cottonseed products mixed with the other feed constituents. Analysis for total and free gossypol concentration in mixed rations is unsatisfactory if the standard analytical method used is the aniline reaction procedure. If the level of free gossypol is reported on a kernel basis, to estimate the free gossypol level in the whole cottonseed multiply the free gossypol level in the kernel by 0.55%.[16,38] From the levels of free gossypol in the whole cottonseed, cottonseed meal, or cottonseed hulls, calculate the total content of free gossypol that would be present in the complete mixed ration. Compare the estimated level of free gossypol obtained to the recommended levels that appear in Tables 7.3 and 7.4, taking into consideration species, class, and stage of rumen development.

There must be a clinical history which is suggestive of gossypol toxicosis. From published results involving field cases, these include multiple animals, sudden death syndrome, chronic respiratory problems that are unresponsive to antibiotics, animals that are doing poorly, and a history of infertility in adult cattle. On necropsy, gross lesions that are compatible with failure of the cardiovascular and respiratory system are suggestive of gossypol toxicity. On histopathology, the most common finding reported is centrolobular hepatic congestion and necrosis related to hepatic anoxia.

Gossypol levels in porcine liver can be determined as the aniline reaction product, with minced tissue residue (described as bound gossypol) following extraction with aqueous acidic ethanol-ether solutions (described as free gossypol).[49] The range for total gossypol in pigs consuming 600 mg kg^{-1} was 240 to 338 μg gossypol g^{-1} of tissue.[49] Using a procedure similar to the one described above, Holmberg et al.[3] reported mortalities in calves with liver gossypol levels from 34 to 56 μg g^{-1} of wet liver tissue. These levels were ten times that of control calves which had no cotton products in their diet, and two to four times that of adult cattle consuming various levels of whole cottonseed and cottonseed meal in their diet. Plasma gossypol levels determined by HPLC method reflect the availability of gossypol in a diet. Currently, ≤ 5 µg ml^{-1} is considered the safe upper limit for plasma gossypol in dairy cattle fed cotton products for extended periods.[16] Gossypol toxicity has been suspected in a herd with plasma gossypol levels ≥ 10 µg ml^{-1}.[16] In dairy herds fed a total mixed ration with 15% whole cottonseed, plasma gossypol levels averaged 1.5 to 3.5 µg ml^{-1}.[16] The relationship of plasma gossypol level to gossypol content in the diet of other species has not been evaluated and warrants further research.

There are minimal changes in blood chemistry and erythrocytes parameters during toxicity. Changes that have been consistently reported are an increase in erythrocyte osmotic fragility, hemoglobinuria, anemia, and a decrease in hemoglobin content.

The differential diagnosis for gossypol toxicity includes monensin toxicity, vitamin E and selenium toxicity, and toxicity to certain plants (coffee senna, bracken fern, white snakeroot, Lantana, and milk weed).[12]

REFERENCES

1. Adams, R. and Geissman, T. J., Gossypol, pigment of cottonseed, *Chem. Rev.*, 60, 555, 1977.
2. Coppock, C.E., Lanham, J. K., and Horner, J.I., A review of the nutritive value and utilization of whole cottonseed, cottonseed meal and associated by-products by dairy cattle, *Anim. Feed Sci. Technol.*, 18, 89, 1987.
3. Holmberg, C. A., Weaver, L. D., Guterbock, W. M., Genes, J., and Montgomery, P., Pathological and toxicological studies of calves fed a high concentration cottonseed meal diet, *Vet. Pathol.*, 25, 147, 1988.
4. Morgan, S. E., Stair, E. L., Martin, T., Edwards, W. C., and Morgan, G. L., Clinical, clinicopathologic, pathologic, and toxicologic alterations associated with gossypol toxicosis in feeder lambs, *Am. J. Vet. Res.*, 49, 493, 1988.
5. Reiser, R. and Fu, H. C., The mechanisms of gossypol detoxification by ruminant animals, *J. Nutr.*, 76, 215, 1962.
6. Lindsey, T. O., Hawkins, G. E., and Guthrie, L. D., Physiological response of lactating cows to gossypol from cottonseed meal rations, *J. Dairy Sci.*, 63, 562, 1980.
7. Randel, R. D., Chase, C. C., Jr., and Wyse, S. J., Effects of gossypol and cottonseed products on reproduction of mammals, *J. Anim. Sci.*, 70, 1628, 1992.
8. Cass, Q. E., Tiritan, E., Matlin, S.A., and Freire, E.C. Gossypol enantiomer ratios in cotton seeds, *Phytochemistry*, 30, 2655, 1991.
9. Jones, L.A., Definition of gossypol and its prevalence in cottonseed products, in *Cattle Research with Gossypol Containing Feeds*, Jones, L. A., Kinard, D. H., and Mills, J. S., Eds., The National Cottonseed Products Association, Memphis, Tennessee, 1991, 1.
10. Forster, L. A., Jr. and Calhoun, M. C., Nutrient values for cottonseed products deserve a new look, *Feedstuffs*, 67, 44, 1995.
11. Altschul, A. M., Lyman, C. M., and Thurber, F. H., Cottonseed meal, in *Processed Plant Protein Foodstuffs*, Altschul, A. M., Ed., Academic Press, New York, NY., 1958, 469.
12. Morgan, S. E., Gossypol as a toxicant in livestock. *Vet. Clin. North Amer. Food Anim. Prac.*, 5, 251, 1989.
13. Rogers, G. M. and Poore, M. H., Optimal feeding management of gossypol-containing diets for beef cattle, *Vet. Med.*, 995, 1995.
14. Calhoun, M. C. and Holmberg C. A., Safe use of cotton by- products as feed ingredients for ruminants: a review, in *Cattle Research with Gossypol Containing Feeds*, Jones, L.A., Kinard, D. H., and Mills, J. S., Eds., The National Cottonseed Products Association, Memphis, Tennessee, 1991, 97.
15. Calhoun, M. C., Cottonseed meal processing and gossypol toxicity for early-weaned lambs, presented at Sheep and Goat Field Day, Texas A&M University, San Angelo TX, 1989.
16. Calhoun, M. C., Kuhlman, S. W., and Baldwin, B. C., Assessing the gossypol status of cattle fed cotton feed products, in *Proceedings from the Pacific Northwest Nutrition Conference*, Seattle, WA, 1995, 1.

17. Berardi, L. C. and Goldblatt, L. A., Gossypol, in *Toxic Constituents of Plant Foodstuffs*, Liener, I. E., Ed., Academic Press, New York, NY., 1980, 184.

18. Smith, F. H., The pathology of gossypol poisoning, *Am. J. Pathol.*, 33, 353, 1957.

19. Risco, C. A., Holmberg, C. A., and Kutches, A., Effect of graded concentrations of gossypol on calf performance: toxicological and pathological considerations, *J. Dairy Sci.*, 75, 2787, 1992.

20. Brahan, J. E., Jarquin, R., Bressani, R., Gonzales, J. M., and Bressani, R., Effect of gossypol on the iron binding capacity of serum in swine, *J. Nutr.*, 93, 2415, 1967.

21. Barraza, M. L., Coppock, C. E., Brooks, K. N., Wilks, D. L., Saunders, R. G., and Latimer, G.L., Jr., Iron sulfate and feed pelleting to detoxify free gossypol in cottonseed diets for dairy cattle, *J. Dairy Sci.*, 74, 3457, 1991.

22. Kuhlman, S. W., Calhoun, M. C., Huston, J. E., Baldwin, B. C., Engdahl, B. S., and Ueckert, D. N., Increased erythrocyte fragility in cattle, sheep and goats fed whole cottonseed, *J. Anim. Sci.*, 69 (Suppl.1), 63, 1991.

23. Wyse, S. J., Velez, J. A., Stahringer, R. C., Greene, L. W., and Randel, R. D., Effects of diets containing free gossypol on erythrocyte fragility and packed cell volume in cattle, *J. Anim. Sci*, 69 (Suppl. 1), 43, 1991.

24. Risco, C. A., Chenoweth, P. J., Larsen, R. E., Velez, J., Shaw, N., Tran, T., and Chase, C. C., Jr., The effect of gossypol in cottonseed meal on performance and on hematological and semen traits in postpubertal Brahman bulls, *Theriogenology*, 40, 629, 1993.

25. Gray, M. L., Greene, L. W., and Williams G. L., Effects of dietary gossypol consumption on metabolic homeostasis and reproductive endocrine function in postpubertal beef heifers and cows, *J. Anim. Sci.*, 71, 3052, 1993.

26. Willard, S. T., Neuendorff, D. A., Lewis, A. W., and Randel, R. D., Effects of free gossypol in the diet of pregnant and postpartum Brahman cows on calf development and cow performance, *J. Anim. Sci.*, 73, 496, 1995.

27. Albrecht, J. E., Clawson, A. J., Ulberg, L. C., and Smith, F. H., Effect of high gossypol cottonseed meal on the electrocardiogram of swine, *J. Anim. Sci.*, 27, 976, 1968.

28. Bender, H. S., Derolf, S. Z., and Misra, H. P., Effects of gossypol on the antioxidant defense system of the rat testis, *Arch. Androl.*, 21, 59, 1988.

29. Lane, A. G. and Stuart, R. L., Gossypol intake may effect vitamin status of dairy cattle, *Feedstuffs*, 62, 13, 1990.

30. Ensminger, M. E. and Olentine, C. G., *Feeds and Nutrition*, Ensminger Publishing, Clovis, CA, 1978, 357.

31. Martin, S. D., Gossypol effects in animal feeding can be controlled, *Feedstuffs*, 62, 89, 1990.

32. Waldroup, P. W., Cottonseed meal in poultry diets, *Feedstuffs*, 53, 21, 1981.

33. Tanksley, T. D. and Knabe, D. A., Use of cottonseed meal in swine rations, *Feedstuffs*, 53, 24, 1981.

34. Danke, R. J., Panciera, R. J., and Tillman, A. D., Gossypol toxicity studies with sheep, *J. Anim. Sci.*, 24, 1199, 1965.

35. Hollon, B. F., Waugh, R. K., Wise, G. H., and Smith, F. H., Cottonseed meals as the primary protein supplement in concentrate feeds for young calves, *J. Dairy Sci.*, 41, 286, 1958.

36. Rogers, P. A. M., Henaghen, T. D., and Sheeler, B., Gossypol poisoning in young calves, *Ir. Vet. J.*, 29, 9, 1975.

37. Smalley, S. A. and Bicknell, E. J., Gossypol toxicity in dairy cattle, *Comp. Cont. Educ. Pract. Vet.*, 4, 378, 1982.

38. Calhoun, M. C., Understanding and managing gossypol in cattle diets, in *Proceedings of the Southwest Nutrition and Management Conference*, Ahwatukee, AZ, 1995, 17.

39. Liu, G. Z. and Segal, S. J., Introduction and history of gossypol, in *Gossypol. A Potential Contraceptive for Men*, Segal, S. J., Ed., Plenum Press, New York, 1985, 1.

40. Lagerlof, R. K. and Tone, J. M., The effect of gossypol acetic acid on female reproduction, *Drug Chem. Toxicol.*, 2, 469, 1985.

41. Zirkle, S. M., Lin, Y. C., Gwazdauskas, F. C., and Canseco, R. S., Effect of gossypol on bovine embryo development during the preimplantation period, *Theriogenology*, 30, 75, 1988.

42. Gu, Y., Chang, C. J., Rikihisa, Y., and Lin, Y. C., Inhibitory effect of gossypol on human gonadotropin (hCG)- induced progesterone secretion in cultured bovine luteal cells, *Life Sci.*, 47, 407, 1990.

43. Chenoweth, P. J, Risco, C. A., Larsen, R. E., Velez, J., Tran, T., and Chase, C. C., Jr., Effects of dietary gossypol on aspects of semen quality, sperm morphology and sperm production in young Brahman bulls, *Theriogenology*, 42, 1, 1994.

44. Chase, C. C., Jr., Bastidas, P., Ruttle, J. L., Long, C. R., and Randel R. D., Growth and reproductive development in Brahman bulls fed diets containing gossypol, *J. Anim. Sci.*, 72, 445, 1994.

45. Wang, J. M., Tao, L., Wu, X. L., Lin, L. X., Wu, J., Wang, M., and Zhang, G. Y., Differential binding of (+) and (-) gossypol to plasma protein and their entry into rat testis, *J. Reprod. Fertil.*, 95, 277, 1992.

46. American Oil Chemists Society, Determination of total gossypol, official method Ba 8-78, determination of free gossypol, official method Ba 7-58, *Official and Tentative Methods of Analysis*, 3rd ed., Amer. Oil Chem. Soc., Champaign, IL, 1988, 1.

47. Pons, W. A., Jr. and Hoffpauir, C. L., Determination of free and total gossypol in mixed feeds containing cottonseed meals, *J. Offic. Assoc. Agric. Chem.*, 40, 1068, 1957.

48. Hron, R. J., Kim, H. L., and Calhoun M. C., Determination of gossypol in cottonseed products, presented at a Symp. Available Gossypol in Cottonseed Products, Annual Meeting Amer. Oil Chem. Soc., San Antonio, TX, May 7-21, 1995.

49. Smith, F. H., Determination of free and bound gossypol in swine tissues, *J. Amer. Oil Chem. Soc.*, 42, 145, 1965.

8 Flavones

E. M. Gaydou

CONTENTS

8.1 INTRODUCTION

Flavones comprise a large group of common plant metabolites. It appears that they are secondary metabolites involved in the plant growth hormone and growth regulators in defense against infection. They exhibit a wide range of biochemical and physiological activities toward other life forms such as viruses, fungi, bacteria, and insects. Lipophilic flavones and hydroxyflavones with specific glycosidic moieties could be used in integrated pest strategies in the growth and development of important crop pests. Flavones are regular components of the human diet; it has been estimated that the average daily intake of flavonoids varies from 50 mg[1] to about 1 g per person.[2] The amounts normally consumed by humans are not acutely toxic since these compounds contain, for most of them, neither nitrogen nor sulfur elements often associated with toxicity. Flavone sulfates and in particular alkaloid flavones are uncommon in plants. Some flavonoids are mutagenic in bacterial test systems, but are not mutagenic *in vivo*. They possess the capacity to inhibit chemical carcinogenesis in experimental animals.

    This review on the chemistry and toxic effects of flavones is not exhaustive for several reasons: 1) the number of flavones isolated from plants increase considerably every year;

2) the toxic effects of flavones on different enzymes and biochemical processes are too numerous to be outlined; and 3) numerous reviews have been published on the chemistry and distribution of flavones[3-8] and their biological properties.[9-12]

8.2 CHEMISTRY AND LOCALIZATION IN PLANTS

8.2.1 GENERAL ASPECTS

The basic structural feature of the flavone compounds is the 2-phenyl-4H-1-benzopyran-4-one or 2-phenyl-γ-benzopyrone, comprising 2 benzene rings (commonly named A and B, Figure 8.1). Flavonols are 3-hydroxyflavones.

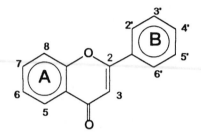

FIGURE 8.1 Structure and carbon numbering of flavone.

Substantial progress on the flavonoid biosynthesis has been reviewed.[13] The B-ring and part of the heterocyclic ring of the flavonic skeleton are provided by a suitable hydroxycinnamic acid CoA ester, whereas the A-ring originates from three acetate units via malonyl-CoA. Both precursors are derived from carbohydrates. 2,3-Dihydroflavones (or flavanones) are the direct precursors of the class of flavones. Flavanones can be hydroxylated to C-3 yielding dihydroflavonols which are biosynthetic intermediates in the formation of flavonols.[13] Flavones are present mainly as glycosides and, being water-soluble, are present in all plants as vacuolar constituents. Besides their taxonomic implications, flavones are biologically active natural products. Flavonols were shown to play an important role in plant reproduction[14] and are essential for pollen germination.[15] The lipophilic nature of low hydroxylated and O-methylated flavones leads to extracellular accumulation in oil glands or in bud excretions. It has been shown that the lipophilicity is correlated with fungal and bacterial toxicities.[16] Besides the signalization in plant-microbe interactions, they serve as phytoalexins.[17] Flavones, with complex substitution like C-alkyl or C-glycosylflavones, biflavones, and heterocyclic-flavones occur in most cases in a single plant species.

8.2.2 MAIN CLASSES OF FLAVONES

8.2.2.1 Flavones and Flavonols

Theoretically, the number of flavones and flavonols substituted with hydroxy substituents and/or with O-methyl substituents is about 40,000.[18] The number of naturally known flavonols is about 400; among them, kaempferol, quercetin, and myricetin are very common (number and position of substituents of flavones and flavonols having a trivial name are given in Figure 8.2 and/or Appendix 8.1). Flavones may occur in free state or in per-O-methylated form such as tangeretin found in *Citrus* peel oil. Among the flavones (about 300 are known[19]), apigenin and luteolin (corresponding to the hydroxylation pattern of kaempferol and quercetin respectively) are widely distributed in plants.

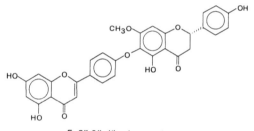

A: Rutin

B: Echinacin

C: Baohuoside - 1

D: Ficine

E: 4',7''-di-*O*-methylamentoflavone

F: 2'',3''-dihydrocryptomerin

FIGURE 8.2 Structure of some selected flavones.

8.2.2.2 Flavone and Flavonol Glycosides

As mentioned above, a large number of flavones are present in plants in association with a sugar. Ten monosaccharides (D-apiose, L-arabinose, L-rhamnose, D-xylose, D-allose, D-galactose, D-glucose, D-mannose, D-galacturonic and D-glucuronic acids) are known to occur in pyranose form. For L-arabinose, pyranose and furanose forms are encountered.[20] Among the disaccharides associated with flavones, rutinose (*O*-α-L-rhamnosyl-(1 → 6)-glucose) is widespread. Quercetin-3-rutinoside or rutin (Figure 8.2A), first isolated from rue, *Ruta graveolens*, is used for capillary fragility and shows antiviral and antibacterial activity.[10] Neohesperidose (*O*-α-L-rhamnosyl-(1 → 2)-glucose) and sophorose (*O*-β-D-glucosyl-(1 → 2)-glucose) are also often encountered. Trisaccharides (linear or branched) are sometimes components of flavonol glycosides and a branched tetrasaccharide has been found attached to acacetin.[21] Various acylated derivatives have been reported, among them, echinacin (Figure 8.2B), isolated from *Echinops echinatus* which shows antifungal activity.[22]

8.2.2.3   Flavones and Flavonols with Various Substituents

The number of flavones with various substituents is considerable. In *C*-glycosylflavones, the sugar moieties (14 glycosyl residues[23]) are generally linked at C-6 and/or C-8 position of the flavonic skeleton, apigenin and luteolin being by far the most common. Vitexin, 8-glycosylapigenin, present in pearl millet (*Pennisetum millet*) and the main source of food energy for rural poor people living under semi-arid tropics, inhibits thyroid peroxidase activity and contributes to the genesis of endemic goiter.[24] *C*-Methylflavones have been found in several sources.[23] Among the flavones with isoprene substitution, the linkage is also generally observed at C-6 and/or C-8 position of flavone. A prenylated flavonol glycoside, baohuoside-1 (Figure 8.2C), isolated from *Epidemium davidii*, was shown to suppress antibody and delayed-type hypersensitivity responses in mice in a dose-dependent fashion.[25] Gomphrenol, 3,5,4′-trihydroxy-6,7-methylenedioxyflavone, was synthesized during the hypersensitive reaction of *Gomphrena globosa* leaves to tomato bushy stunt virus infection.[26] Pyranoflavones and furanoflavones occur in few families, *Fabaceae, Rutaceae, and Moraceae*. About 100 flavone and flavonol sulfates (mono up to tetrasulfates) have been reviewed.[20] Some flavone alkaloids such as ficine (Figure 8.2D), have been reported[27] from *Ficus pantanonia*. Biflavonoids have been proved to be biologically active compounds. Bioactivity-guided fractionation of the leaves of *Selaginella willldenowii* afforded 4′,7″-di-*O*-methyl amentoflavone (Figure 8.2E), which is significantly cytotoxic against a panel of human cancer lines. On the other hand, 2″,3″-dihydroisocryptomerin (Figure 8.2F), a recently discovered dihydrobiflavone,[28] was noncytotoxic.

8.2.3   Isolation, Identification, and Synthesis

8.2.3.1   Analytical Procedures for Flavone and Flavonol Isolation

Isolation, purification, and preliminary identification of flavones using paper chromatography, column chromatography (CC), and thin layer chromatography (TLC) have been described.[9] Flavonol glycosides are often purified by gel filtration.[29] Polymethoxylated flavones and other hydroxyflavones and flavonols have been separated using straight-phase high-performance liquid chromatography (HPLC),[30] reversed-phase RP-HPLC,[31,32] superfluid critical chromatography,[33] gas chromatography,[34] and ion exchange chromatography.[35] Centrifugal partition chromatography (CPC) has been proven to be an indispensable tool in the separation of flavonols[36] and flavonol glycosides,[37] but in most cases, primary partition by classical CC (cellulose or polyamide) has to be performed. Isolation of *C*-glycosylflavones was achieved using CC filled with polystyrene unpolar adsorbents or diethylaminoethyl (DEAE) cellulose.[38]

8.2.3.2   Spectroscopic Determination of Flavones and Flavonols

The literature on ultraviolet (UV) spectra of flavones and flavonols was examined.[9,39] The methanol spectra of these compounds exhibit two major absorption peaks called Band I (300–380 nm), associated with absorption due to the B ring cinnamoyl system, and Band II (240–280 nm), associated with absorption due to the A ring benzoyl system. The effects of sodium methylate, sodium acetate or $AlCl_3$, and $AlCl_3/HCl$ on the UV spectrum have been reviewed. Differentiation of 5-hydroxy- and 5-hydroxy-3-methoxyflavones with or without B-ring substitutions[40,41] has been published. RP-HPLC, in combination with photodiode-array detection and post-column derivatization, led to structural identification of compounds as they were separated.[42]

^{1}H-nuclear magnetic resonance (NMR) including the newer 1D and 2D NMR techniques of flavone aglycones and/or glycosides has been well documented.[43] A wide range of reference spectra under identical conditions, together with tabulated data from the literature have been given.[43] Comprehensive compilations on ^{13}C-NMR have also appeared.[44,45] Reinvestigations of signal assignment in ^{13}C NMR spectra of highly methoxylated flavones using 2D NMR have been recently reported.[46] Natural abundance ^{17}O-NMR spectra of various methoxyflavones was discussed for steric and conformational changes in order to understand their biochemical activity.[47]

Hydroxy- and polymethoxyflavones have been studied with electron ionization (EI) mass spectra.[48–50] Fast atom bombardment (FAB) mass spectrometry provided useful information for the structural elucidation of flavonoid glycosides.[51] Negative ion desorption chemical ionization (DCI) of underivatized flavonoid *O*- and *C*-glycosides[52] and monotrimethylsilylated flavonoid methyl ethers[53] was shown to be a good approach for the structural characterization of flavonoid glycosides.

8.2.3.3   Syntheses of Flavones and Flavonols

Structural determination of pure natural flavones using some spectral data is sometimes difficult, in particular between 5,6-dihydroxy-7,8-dimethoxyflavones and 5,8-dimethoxyflavones, using NMR techniques. Reference data could be obtained starting from synthesized molecules, particularly when small amounts of pure natural products are isolated. Among the various ways available for the synthesis of flavones and their derivatives,[54] the Kostanecki-Robinson (or Allan-Robinson) reaction has been largely used. This synthesis consists of a condensation of substituted benzoic anhydrides and the corresponding sodium salt, on an *o*-hydroxyacetophenone, followed by an alkaline hydrolysis. The yields are increased using an anhydrous tertiary amine.[55] Another method, the Algar-Flynn-Oyamada (or AFO) reaction, involves the oxidation of 2′-hydroxychalcones with alkaline hydrogen peroxide,[56] but the Baker-Venkataraman rearrangement remains the preferred synthetic method for chemists. Flavones are thus obtained in three steps, starting from substituted acetophenones without isolation of any intermediates. The *o*-hydroxybenzoates of acetophenones are converted into the corresponding oily diaroylmethanes with potassium hydroxide in an aprotic polar solvent such as pyridine or dimethylsulfoxide.[57,58] In the case of *o,o*′-dihydroxyacetophenones, the ring closure of the corresponding *o,o*′-dihydroxydibenzoylmethanes is sometimes ambiguous: the structural identification, on a spectral basis, of pure natural products is difficult, and if molecular model syntheses fails, some erroneous structures need to be revised. Therefore, some reported structures have been proved wrong after unambiguous synthesis methods, recently developed.[59] This is the case, for example, of candirone, isolated from *Tephrosia candida* which was assumed to be 5,4′-dihydroxy-3,6,8-trimethoxyflavone.[60] This was revised to 5,4′-dihydroxy-3,6,7-trimethoxyflavone, penduletin, on the basis of spectral data of the synthetic isomers.[61]

8.2.4   Localization in Plants

Flavones occur practically in all parts of plants, most frequently in glycosidic combination. In bud exudates and on leaf and fruit surfaces they occur in the free state (aglycone) and/or in *O*-methylated form. The flavonoid patterns of mosses, ferns, gymnosperms, and angiosperms have been reviewed up to 1991.[20] Comparisons between the lipophilic flavone composition at the leaf surface (epicuticular) within the wax, and the hydrophilic flavone glycosides within the cell vacuole (intracellular) have been made. Flavone aglycones are most often accumulated on plants externally, in buds and in leaf and fruit waxes. They are produced by plants having secretory glands and/or living in arid or semi-arid areas. As a result, aglycones predominate in the *Lamiaceae*, the majority of them being substituted not only in the 5- and 7-positions of the A-ring but also in 6- and/or 8-positions.[62] *Rutaceae* produces a large number of highly methylated flavonoids. In *Citrus* fruits and leaves, seven C-glycosylflavones and 22 aglycones have been reviewed, the highly substituted being 3,5,6,7,8,3′,4′-heptamethoxyflavone. Few flavone glycosides have been identified in *Citrus* juices.[63] Wollenweber has done an extensive survey of aglycone-accumulating families.[19] If water-soluble flavone glycosides are generally cell vacuolar constituents, they may be obtained in some case from lipophilic leaf fractions when they are highly acylated.[20] For example, in *Picea abies* needles, kaempferol 3-*O*-glucoside showed a metabolism indicating rapid turnover and/or translocation from the soluble pool into an ester-bound insoluble pool.[64]

Various protecting roles against environmental conditions or attack on plant tissues by predators induced biosynthesis of flavone derivatives. Higher concentrations in flavones on leaf surface and intracellular plant tissues for protection against damaging UV radiation has been observed.[20] Among the wide range of organic compounds appearing in the sapwood of trees after wounding, injury, or fungal attack, flavonic derivatives were identified. [65]

Flavone and flavonol glycosides are found in food plants; in particular, quercetin, kaempferol, myricetin, apigenin, and luteolin *O*-glycosides are constituents of fruits,[66] vegetables, cereals, tea, cocoa, and spices. The contents of these five flavonic compounds in various freeze-dried vegetables and fruits, after extraction and acid hydrolysis of *O*-glycosides, were determined.[67] Quercetin levels in the edible parts of most vegetables were generally below 10 mg kg^{-1} except for onions with 284–486 mg kg^{-1}, and in most fruits the content averaged 15 mg kg^{-1}. Kaempferol could only be detected in kale (211 mg kg^{-1}), endive (15–91 mg kg^{-1}), leeks (11–56 mg kg^{-1}), and turnip tops (31–64 mg kg^{-1}).[67] The content of myricetin, luteolin, and apigenin was below the limit of detection (<1 mg kg^{-1}). Seasonal variability was observed for leafy vegetables with the highest flavonoid content in summer.[67]

8.3   TOXICITY

8.3.1   VIRUSES

Naturally occurring flavones; quercetin, morin, rutin, apigenin, and other flavonoids, have been reported to possess toxicity against various types of viruses.[11,68] The antiviral potency of flavones against poliomyelitivirus type 1 and human rhinovirus type 15 has been determined.[69] The antiviral activity seems to be in relation to a 3-hydroxy or a 3-methoxy group on the flavonic skeleton being present. Natural 3-*O*-methylquercetin isolated from *Euphorbia grantii* was active against picornaviruses and visicolar stomatitis virus.[70] Synthetic 3-methoxyflavones showed the role of substitution at C-4′ and C-5.[71] Synergistic antiviral effects between quercetin and ascorbate were observed.[72] The inhibitory effect of baicalin, 7-D-glucuronic acid-5,6-dihydroxyflavone, isolated from the Chinese medicinal plant *Scutellaria baicalensis* against human immunodeficiency virus (HIV-1) infection has been studied *in vitro*.[73] Another anti-HIV principle, acacetin-7-*O*-β-D-galactopyranoside, was isolated from *Chrysanthemum moriflorum*.[74] A structure-activity correlation with some related flavonoids showed that flavonoids with hydroxyl group at C-5 and C-7 were more potent inhibitors of HIV growth.[74] Data obtained suggested that baicalin might serve as a useful drug for treatment and prevention of HIV-1 infections since this plant is extensively used in China for treatment of many diseases without detrimental side effects.[73]

8.3.2   FUNGI

The various levels of interaction between plants and fungi[75] and the characterization of various antifungal metabolites in higher plants has been reviewed.[76] Among the plant substances showing antifungal activity, phenolic hydroxyl groups which have a high affinity to proteins, may act as inhibitors to fungal enzymes necessary to infect plants.[77] For example, mycelial extraction of *Aspergillus oryzae* grown on starch media was fractionated and among the fraction having a toxic effect against *Cladosporium cucumerinum*, methylenedioxy-3′4′-5-hydroxy-3,6-dimethoxyfuranflavone were isolated.[78] The relative acidity and the number of hydroxyl groups appear to be the main factors contributing to fungal toxicity.[16] On the other hand, stimulating effects of myricetin and quercetin on germination and growth of vesicular-arbuscular mycorrhiza fungi have been found.[79] Quercetin has been proven to be toxic for *Candida tropicalis*[80] and *Candida albicans*[81] in a concentration between 50 and 100 µg l[-1]. Kumatakenin, 5,4'-dihydroxy-3,7-dimethoxyflavone, was active against *C. tropicalis* and *Fusarium solani*.[82] Xanthomicrol, 5,4'-dihydroxy-6,7,8-trimethoxyflavone and 3,3'-di-*O*-methylquercetin, isolated from *Varthemia iphionoides*, were very active against *F. solani*, *C. tropicalis*, and *Aspergillus parasiticus*[82] and 7,3'-*O*-dimethylquercetin isolated from *Wedelia biflora* also has antifungal properties.[83]

Various compounds, including apigenin, apigenin-7-*O*-glucoside, and echinacin (Figure 8.2B), showed high efficiency against spore germination of *Alternaria tennissima* at 25–50 µg l[-1].[22] Kaempferol-3-(2,3-diacetoxy-4-*p*-coumaroyl) rhamnoside isolated from the leaves of *Myrica gale* was shown to have a varying inhibitor activity against five species of fungi found on these leaves.[84] The fungal toxicity of epicuticular methoxylated flavones of the aerial parts of *Helichrysum nitens*[85] and *Psiadia trinervia*[86] was ascertained by TLC assay using spores of *C. cucumerinum*. The permethylated flavonols, 3,5,6,7,8,4'-hexamethoxy- and 3,5,6,7,8,3',4'-heptamethoxyflavones obtained by methylation of naturally occurring compounds and isolated from *Bellardia trixago,* inhibited the growth of *Cladosporium herbarum*.[87] By contrast, 5-hydroxy-6,7-dimethoxyflavone and almetin, 5-hydroxy-6,7,8-trimethoxyflavone were inactive and topazolin, 5,7,4'-trihydroxy-3-methoxy-6-(3,3-dimethylallyl)-flavone, isolated from the roots of yellow lupine, *Lupinus luteus*, has only weak activity against *C. herbarum*.[88] Apigenin-7-*O*-glucoside and luteolin-7-*O*-glucoside, isolated from *Veratrum grandiflorum*, did not show antifungal activity against *C. herbarum*.[89] Although the fungal toxicity of flavone decreased dramatically when the methoxy group in position 5 was replaced by a hydroxyl,[90] umuhengerin, 5-hydroxy-6,7,3',4',5'-pentamethoxyflavone exhibited antifungal properties against *Aspergillus niger*, *Aspergillus fumigatus*, *Trichophyton mentagrophytes*, and *Mucrosporum canis*.[91]

Apigenin, luteolin, five flavone *C*-glycosides (vitexin, isoorientin, isoswertisin, molludistin, and schaftoside), and luteolin-7-*O*-glucoside inhibit *in vitro* growth and asexual reproduction of *Phytophora parasitica* as well as that of *Verticillium albo atrum* and *Colletotrichum musae*. *C*-Glycosylflavones were more toxic than the corresponding aglycones. The *O*-glucoside and *C*-glycosides similarly inhibited both micromycetes lacking β-glucosidase, but inhibited pectinolytic enzymes differently.[92] Isorhamnetin was inhibitory on the growth and development of *Phytophthora sojae* at a concentration of 60–120 µmol l[-1] and was fungicidal at 240 µmol l[-1].[93] Quercetin and its 3-*O*-β-D-glucoside, isoquercetin, caused significantly prolonged lags in *P. sojae* growth at 60–240 µmol l[-1], but were not fungicidal at any of these concentrations. Kaempferol, apigenin, chrysin, and rutin were not inhibitory over this range.[93]

Bioassays compared mycelial growth of *Cytospora persoonii*, which causes perennial canker on the bark of peach, plum, and sweet cherry, on tectochrysin 5-glucoside, chrysin, and tectochrysin from resistant sour cherries *Prunus cerasus*.[94] Chrysin stopped mycelial growth at 1.0–2.5 mmol l[-1] and techtochrysin was more effective than its 5-glucoside.[94] The protector role of kaempferol, quercetin, and myricetin to *Mangifera indica* from the ingress of two pathogenic fungi, *A. niger* and *Fusarium moniliformae* was indicated.[95] The postinfectional formation of the flavonols in flowers of *M. indica* suggests that they might function as phytoalexins. In healthy leaves, the concentration of the 3-flavonols increased with maturity.

Flavonoid mixtures always containing two different flavonoid compounds have been tested against grain contaminating fungi, *A. tennissima*, *Cladosporium cladosporioides*, *Spicellum roseum*, and *Trichoderma hamatum*. Highest activity up to 100% inhibition of mycelial growth was observed when the mixture contained flavone and/or flavanone, flavonol or 7-hydroxyflavone.[96,97] Flavones show generally higher activity (over the concentration range of 1–8 10^{-4} mol l^{-1}) against four *Deuteromycotina* than flavanone.[98] Similar results were obtained in malt extract broth against five storage fungi of the genus *Aspergillus (A. repens, A. amstelodami, A. chevalieri, A. flavus* and *A. petrakii)*.[99] Some flavones and flavonols were tested against various pathogens commonly found during the storage of fruits and vegetables; all the flavonoids tested, with the exception of kaempferol-3-rutinoside, showed an appreciable activity against *Penicillium* sp.[100] The highest inhibition was observed with apigenin-*O*-glucoside and kaempferol-3-*O*-rutinoside. This latter compound was inactive against *Penicillium digitatum*.[100] These results suggest that a right combination of flavone/fungus/plant is of fundamental importance for improving crop storage.

8.3.3 BACTERIA

It has been shown that lipophilic flavonoids display antimicrobial activity, a property due to their ability to penetrate biological membranes. With bacteria, it seems that there is not a simple correlation between lipophilicity and antibacterial activity.[16] The resistance of fresh onion, *Allium cepa*, to bacterial infection with *Pseudomonas cepacia* may be ascribed to the increasing accumulation of isorhamnetin 3-glucoside and quercetin 3'-glucoside, following inoculation.[101] Quercetin 3-α-L-arabinopyranoside-2″-gallate appeared to have a potent antibacterial activity in nutritional agar medium at a concentration of 100 µg ml^{-1} against *Escherichia coli*.[102] Isokaempferide and 5,7,4'-trihydroxy-3,8-dimethoxyflavone were shown to be active against *Bacillus cereus*.[86] Quercetin, kaempferol, kaempferide, and myricetin isolated from various plants showed antibacterial activity.[103,104] The effectiveness of various antibiotics was about 20-fold higher than quercetin against *Staphylococcus aureus, Enterobacter aerogenes*, and *E. coli*.[80] Similar results were obtained for quercetin and rutin, the 3-rutinosyl derivative.[80] Galangin was the most active flavonol tested against a skin bacterium, *Staphylococcus epidermis*.[105] Umuhengerin, exhibited *in vitro* antibacterial properties in concentration up to 200 µg ml^{-1} against various bacteria including *S. aureus* and *Salmonella typhimurium*.[91] Among the various aglycones and glycosides of apigenin and luteolin isolated from the leaves of *Salvia palaestina*, cirsimaritin, 4',5-dihydroxy-6,7-dimethoxyflavone showed a high activity against *S. aureus, S. epidermitis, E. coli, Klebsiella pneumoniae, Proteus vulgaris*, and *Pseudomonas aeruginosa*.[106]

More than 40 flavone aglycones and/or glycosides were tested[107] for antibacterial activity against *Pseudomonas maltophilia* and *Enterobacter cloacae*, which occur in the gut of cotton bollworm *Heliothis zea* and tobacco budworm *H. virescens*. A general trend towards higher activity as the number of hydroxyl groups was shown. Growth inhibition of both species was observed for *ortho*-dihydroxyl groups. Methylation of a hydroxyl decreased activity.[107] Various flavonol-*O*-glycosides including quercetin-*O*-galactosyl-(1→6)-glucoside[108] and gossypetin-8-*O*-rhamnoside[109] were also tested.

In a structure-activity study of 16 active flavones, free 3',4',5'-trihydroxy B-ring and a free 3-hydroxy were shown to be necessary against *S. aureus* and *P. vulgaris*.[110] Chrysoeriol and luteolin have been reported to activate bacterial nodulation genes involved in control of nitrogen fixation.[111] Luteolin and chrysin promoted growth of *Bradyrhizobium* sp., a symbiotic bacterial culture with peanut, *Arachis hypogea*.[112] The flavone decomposition product, 5,7-dihydroxychromone (DHC), was found to inhibit the radial growth cultures of the soil pathogenic fungi *Rhizoctonia solani* and *Sclerotium rolfsii* with I_{50} (the concentration of DHC required to inhibit growth was 50%) values 18 and 26 µmol l^{-1} respectively.[112]

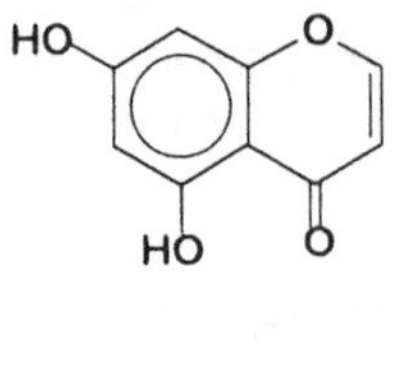

FIGURE 8.3 Structure of 5,7-dihydroxychromone (DHC).

8.3.4 Insects

Flavones, like other secondary metabolites, have been shown to affect the feeding behavior of insects.[113–115] Although flavones and flavone glycosides are not very toxic to insects, some of them can act as a growth inhibitor on phytophagous insects at relatively low concentrations. For example, Matsuda [116] showed that the feeding of various flavones (morin, myricetin, quercetin, quercitrin, and rutin) on various leaf beetles, among which *Plagiodera versicolora* in the adult stage, was stimulated by all compounds. *Altica oleracea* and *A. nipponica* were stimulated to feed by rutin, quercetin, myricetin, and myricitrin. The feeding of the daikon leaf beetle *Phaedon brassicae* and the rice leaf beetle *Oulema oryzae* was inhibited by all the flavones tested, suggesting that flavonoids can play a significant role in the host selection of phytophagous insects.[116] Ermanin, 7,4'-di-*O*-methylkaempferol, isolated from *Passiflora foetida* is a strong feeding deterrent to *Dione juno* larvae[117] and unsubstituted flavone was the better inhibitoty flavonic compound on the growth of the navel orangeworm *Amyelois transitella*.[125] Maysin, a 6-*C*-glycosyl luteolin, retards growth of the bollworm *Heliothis zea*.[119] Vitexin, a 8-*C*-glycosyl apigenin, has a deterrent activity on the homoptera *Myzus persicae*.[120]

The sugar moiety seems to play a selective role in the behavior of insects, as observed using quercetin glycosides derivatives: rutin (3-*O*-rutinosyl quercetin), quercitrin (3-*O*-rhamnosyl quercetin), and quercimeritrin (quercetin 7-*O*-β-glucoside).[121–125] Rutin is toxic toward the larvae of *H. zea* development,[125] the European corn borer *Ostrinia nubilalis*,[123] and the southern armyworm *Spodotera eridania*.[124] For *O. nubilalis*,[123] the quercetin glycoside derivatives have a variable antifeedant effect. The substituent influence on the flavonic skeleton against the development of *H. zea*, using more than 40 flavones having a varied hydroxylation *C*-glycosyl and *O*-glycoside pattern was investigated.[125] Luteolin and iso-orientin were active, but luteolin 3',4'-dimethyl ether and isovitexin were inactive.[125]

It seems that flavonol rhamnosides such as quercitrin or gossypetin-8-*O*-rhamnoside are more toxic than the corresponding glucosides and development of cotton lines with higher levels of such rhamnosides, either by breeding or by genetic strategies, are research objectives.[121] Therefore, Asiatic cottons (*Gossypium arboreum*) have been investigated as a source of resistance to the tobacco budworm *H. virescens*, since gossypetin 8-*O*-glucoside and gossypetin 8-*O*-rhamnoside, the more toxic flavonol derivatives, are present in this species.[121]

8.3.5 Mammals

Mammalian toxicity of flavones is very low as demonstrated by pharmacological studies. They are subjected to hydrolysis by intestinal microorganisms and will only be absorbed to a limited extent as intact flavones and flavonols. As a result, although the average amount of flavonoid in diet is about 1 g, the reasonable estimated daily intake of flavonol *O*-glycosides is perhaps 50 mg (quercetin equivalent).[1] Flavones are metabolized by a variety of cells, especially those in the liver.[126] Since the metabolites are excreted into the urine, there is no metabolite accumulated into the body.[127] A regular intake is not considered to present a significant acute toxicity and, for human usage, doses lower than 1 g each day per adult had been recommended.[128] In the 1970s, polymethoxylated flavones from *Citrus* such as sinensetin and nobiletin have been proven to increase resistance of erythrocytes to aggregation by shifts in plasma constituents.[129] Some of these flavonoid compounds formed citroflavonoids and were considered to possess vitamin-like activity and defined as *Flavonoids with a vitamin P effect*.[130]

The impact of plant flavonoids on mammalian biology has been reviewed.[11] Some flavones, among them quercetin and kaempferol, are mutagenic in bacterial test systems.[131,132] A moderate to low cytotoxicity of various flavones in a human cell lung carcinoma and colorectal cancer cell lines has been proven.[133] Quercetin and kaempferol suppressed cytotoxicity of active oxygen species on Chinese hamster V79 cells.[134] However, many questions are still unresolved regarding their long-term toxicity, since the carcinogenic potential of some flavones is equivocal. 5,3'-Dihydroxy-3,6,7,8,4'-pentamethoxyflavone, isolated from *Polanisia dodecandra*, showed remakable cytotoxicity *in vitro* against panels of cancers.[135] The potential therapeutic application of quercetin as an anticancer drug either alone or in combination with an antitumor antibiotic such as Adriamycin, in multidrug-resistant breast tumor cells has been investigated.[136] The pro-oxidant properties of flavones, including myricetin, which are generally considered to be antioxidants and anticarcinogens, were demonstrated and a dual role for these compounds in mutagenesis and carcinogenesis suggested.[137]

ACKNOWLEDGMENTS

Thanks go to L. Kalaiji and V. Masotti for their help with the references and data compilation.

REFERENCES

1. Brown, J. P., A review of the genetic effects of naturally occurring flavonoids, anthraquinones and related compounds, *Mutat. Res.*, 75, 243, 1980.
2. Kühnau, J., The flavonoids, a class of semi-essential food components: their role in human nutrition, *World Rev. Nutr. Diet.*, 24, 117, 1976.
3. Geissman, T. A., *The Chemistry of Flavonoid Compounds*. Pergamon Press, Oxford, 1962.
4. Harborne, J. B., Mabry, T. J., and Mabry, H., *The Flavonoids*. Chapman and Hall, London, 1975.
5. Harborne, J. B. and Mabry, T. J., *The Flavonoids: Advances in Research*, Chapman and Hall, London, 1982.
6. Harborne, J. B., *The Flavonoids: Advances in Research since 1980*, Chapman and Hall, London, 1988.
7. Harborne, J. B., *The Flavonoids: Advances in Research since 1986*, Chapman and Hall, London, 1993.
8. Harborne, J. B., *Methods in Plant Biochemistry*, Volume 1, *Plant Phenolics*. Academic Press, London, 1989.
9. Mabry, T. J., Markham, K. R., and Thomas, M. B., *The Systematic Identification of Flavonoids*. Springer Verlag, New York, 1970.
10. Harborne, J. B. and Baxter, H., *Phytochemical Dictionary. A Handbook of Bioactive Compounds from Plants*, Taylor and Francis, London, 1993.
11. Middleton, E., Jr. and Kadaswami, C., The impact of plant flavonoids on mammalian biology: implications for immunity, inflammation and cancer, in *The Flavonoids: Advances in Research since 1986*. Harborne, J. B., Ed., Chapman and Hall, London, 1993, chap. 15.
12. Pathak, D., Pathak, K., and Singla, A. K., Flavonoids as medicinal agents. Recents advances, *Fitoterapia*, 63, 371, 1991.
13. Heller W. and Forkmann, G., Biosynthesis of flavonoids, in *The Flavonoids: Advances in Research since 1986*, Harborne, J. B., Ed., Chapman and Hall, London, 1993, chap. 11.
14. Vogt, T., Pollack, P., Tarlyn, N., and Taylor, L. P., Pollination -or wound- induced kaempferol accumulation in petunia stigmas enhances seed production, *Plant Cell.*, 6, 11, 1994.
15. Vogt, T., Wollenweber E., and Taylor, L.P., The structural requirements of flavonols that induce pollen germination of conditionally male fertile Petunia, *Phytochemistry*, 38, 589, 1995.
16. Laks, P. E. and Pruner, M. S., Flavonoid biocides: structure/activity relations of flavonoid phytoalexin analogs, *Phytochemistry*, 28, 87, 1988.
17. Koes, R. E., Quattrocchio, F., and Mol, J. N. M., The flavonoid biosynthetic pathway in plants: functions and evolution, *BioEssays*, 16, 123, 1994.
18. Iinuma, M. and Mizuno M., Natural occurrence and synthesis of 2'-oxygenated flavones, flavonols, flavanones and chalcones, *Phytochemistry*, 28, 681, 1989.
19. Wollenweber, E., Flavones and flavonols, in *The Flavonoids: Advances in Research since 1986*. Harborne, J. B., Ed., Chapman and Hall, London, 1993, chap.7.

20. Williams, C. A. and Harborne, J. B., Flavone and flavonol glycosides, in *The Flavonoids: Advances in Research since 1986.* Harborne, J. B., Ed., Chapman and Hall, London, 1993, chap.8.

21. Ahmed, A. A. and Saleh, N. A. M., Peganetin, a new branched acetylated tetraglycoside of acacetin from *Peganum armala*, *J. Nat. Prod.*, 50, 256, 1987.

22. Singh, U. P., Pandey, V. B., Singh, K. N., and Singh, R. D. N., Antifungal activity of some flavones and flavone glycosides of *Echinops echinatus*, *Can. J. Bot.*, 66, 1901, 1988.

23. Jay, M., *C*-Glycosylflavonoids, in *The Flavonoids: Advances in Research since 1986.* Harborne, J. B., Ed., Chapman and Hall, London, 1993, chap. 3.

24. Gaitan, E., Lindsay, R. H., Reichert, R. O., Ingbar, S. H., Cookseey, R. C., Legan, J., Meydrech, E. F., Hill, J., and Kubota, K., Antithyroid and goitrogenic effects of millet: role of *C*-glycosylflavones, *J. Clin. Endocrinol. Metab.*, 68, 707, 1989.

25. Li, S. Y., Ping, G., Geng, L., Seow, W. K., and Thong, Y. H., Immunopharmacology and toxicology of the plant flavonoid baohuoside-1 in mice, *Int. J. Immunopharmac.*, 16, 227, 1994.

26. Redolfi, P., Cantisani, A., Matta, A., and Pennazio, S., Some properties of a phenolic compound synthesized in *Gomphrena globosa* during the hypersensitive response to Tomato bushy stunt virus, *Phytopath. medit.*, 18, 107, 1979.

27. Johns, S. R. and Russel, J. H., Ficine, a novel flavonoidal alkaloid from *Ficus pantanonia*, *Tetrahedron Lett.*, 24, 1987, 1965.

28. Silva, G. L., Chai, H., Gupta, M. P., Farnsworth, N. R., Cordell, G. L., Pezzuto, J. M., Bheecher, C. W. W., and Kinghorn, A. D., Cytotoxic biflavonoids from *Selaginella willdenowii*, *Phytochemistry*, 40, 129, 1995.

29. Hiermann, A., Gel filtration of flavonoids on Fractogel PGM 2000, *J. Chromatogr.*, 362, 152, 1986.

30. Bianchini, J. P. and Gaydou, E. M., Substituent effect of polymethoxylated flavones in high-performance liquid chromatography, *J. Chromatogr.*, 259, 150, 1983.

31. Barberan, F. A. T., Tomas, F., Hernandez, L., and Ferreres, F., Reversed-phase high-performance liquid chromatography of 5-hydroxyflavones bearing tri- or tetrasubstituted A-rings, *J. Chromatogr.*, 347, 443, 1985.

32. Vande Casteele, K., Geiger, H., and Van Sumere, C. F., Separation of flavonoids by reversed-phase high-performance liquid chromatography, *J. Chromatogr.*, 240, 81, 1982.

33. Morin, P., Gallois, A., Richard, H., and Gaydou, E. M., Fast separation of polymethoxylated flavones by carbon dioxide supercritical fluid chromatography, *J. Chromatogr.*, 586, 171, 1991.

34. Gaydou, E. M., Berahia, T., Wallet, J. C., and Bianchini, J. P., Gas chromatography of some polymethoxylated flavones and their determination in orange peel oils, *J. Chromatogr.*, 549, 440, 1991.

35. Archambault, J. C., Morin, P., Gaydou, E. M., Biesse, J., Andre, P., and Dreux, M., Optimization of pH and SDS concentration factors in MECC separation of flavonoids, *Polyphenols 94*, 69, 437, 1994.

36. Oka, H., Oka, F., and Ito, Y., Multilayer coil planet centrifuge for analytical high-speed counter-current chromatography, *J. Chromatogr.*, 479, 53, 1989.

37. Vanhaelen, M. and Vanhaelen-Fastre, R., Counter-current chromatography for isolation of flavonol glycosides from *Gingko biloba* leaves, *J. Liquid Chromatogr.*, 11, 2969, 1988.

38. Maggi, L., Stella, R., Ganzerli-Valentini, M. T., and Pietta, P., Model compound sorption by the resin XAD-2, XAD-8 and diethylaminocellulose. An useful application to flavonoid isolation, *J. Chromatogr.*, 478, 225, 1989.

39. Jurd, L. M. , *The Chemistry of Flavonoid Compounds.* T. A. Geissman, Ed., Pergamon Press, Oxford, 1962.

40. Voirin, B., UV spectral differentiation of 5-hydroxy- and 5-hydroxy-3-methoxyflavones with mono-(4'), di-(3',4') or tri-(3',4',5')-substituted B-rings, *Phytochemistry*, 22, 2107, 1983.

41. Tomas-Barberan, F. A. Iniesta-Sanmartin, E., Ferreres, F., and Tomas-Lorente, F., High Performance Liquid Chromatography, thin layer chromatography and ultraviolet behaviour of flavone aglycones with unsubstituted B-rings, *Phytochemical Analysis*, 1, 44, 1990.

42. Schaufelberger, D. and Hostettmann, K., High performance liquid chromatographic analysis of secoiridoid and flavone glycosides in closely related *Gentiana* species, *J. Chromatogr.*, 389, 450, 1987.

43. Markham, K. R. and Geiger H., ^{1}H nuclear magnetic resonance spectroscopy of flavonoids and their glycosides in hexadeuterodimethylsulfoxide, in *The Flavonoids: Advances in Research since 1986*. Harborne, J. B., Ed., Chapman and Hall, London, 1993, chap. 10.

44. Markham, K. R., *Techniques of Flavonoid Identification*. Academic Press, London, 1982.

45. Agrawal, P. K., Studies in organic chemistry. 39. Carbon-13 NMR of flavonoids. Agrawal, P. K., Ed., Elsevier, Amsterdam, 1989.

46. Wallet, J. C. and Gaydou, E. M., ^{13}C NMR Spectra of a Series of 2′,3′,4′-trimethoxyflavones. Reinvestigation of Signal Assignment in ^{13}C NMR Spectra of Flavones, *Magn. Reson. Chem.*, 31, 518, 1993.

47. Wallet, J. C., Gaydou, E. M., and Cung, M. T., ^{17}O NMR Study of Steric Hindrance in Methoxylated Flavones. Substitution on the Phenyl ring and the Heterocycle, *Magn. Reson. Chem.*, 28, 557, 1990.

48. Rizzi, G. P. and Boeing, S. S., Mass spectral analysis of some naturally occurring polymethoxyflavones, *J. Agric. Food Chem.*, 32, 551, 1984.

49. Hedin, P. A. and Phillips, V. A., Electron impact mass spectral analysis of flavonoids, *J. Agric. Food Chem.*, 40, 607, 1992.

50. Berahia, T., Gaydou, E. M., Cerrati, C., and Wallet, J. C., Mass Spectrometry of Polymethoxyflavones, *J. Agric. Food Chem.*, 42, 1697, 1994.

51. Sakushima, A., Nishibe, S., Takeda, T., and Ogihara, Y., Positive and negative ion mass spectra of flavonoid glycosides by fast atom bombardment, *Mass Spectrosc.*, 36, 71, 1988.

52. Sakushima, A., Nishibe, S., and Brandenberger, H., Negative ion desorption chemical ionization mass spectrometry of Flavonoid glycosides, *Biomed. Environ. Mass Spectrom.*, 18, 809, 1989.

53. Sakushima, A. and Nishibe, S., Mass spectrometry of monotrimethylsilylated flavonoid methyl ethers, *Biomed. Environ. Mass Spectrom.*, 19, 319, 1990.

54. Gripenberg, J., Flavones, in *The Chemistry of Flavonoid Compounds*. Geissman, T. A., Ed., Pergamon Press, Oxford, 1962, chap.13.

55. Looker, J. H., McMechan, J. H., and Mader, J. W., An amine solvent modification of the Kostanecki-Robinson reaction. Application to the synthesis of flavonols, *J. Org. Chem.*, 43, 2344, 1978.

56. Cummins, B., Donnelly, D. M. X., Eades, J. F., Fletcher, H., Cinneide, F. O., Philbin, E. M., Swirski, J., Wheeler, T. S., and Wilson, R. K., Oxidation of chalcones (AFO reaction), *Tetrahedron*, 19, 499, 1963.

57. Gaydou, E. M. and Bianchini, J. P., Etudes de composés flavoniques. I. Synthèses et propriétés (UV, RMN, du ^{13}C) de quelques flavones, *Bull. Soc. Chim. Fr.*, II, 43, 1978.

58. Gaydou, E. M. and Bianchini, J. P., Etudes de composés flavoniques. II. Synthèses et propriétés (UV, RMN, du ^{13}C) de quelques dihydroxy-5,7 flavonols, *Ann. Chim.*, 2, 303, 1977.

59. Horie, T., Kawamura, Y., Yamamoto, H., Kitou, T., and Yamashita, K., Synthesis of 5,8-dihydroxy-6,7-dimethoxyflavones and revised structures for some natural flavones, *Phytochemistry*, 39, 1201, 1995.

60. Parmar, V. S., Jain, R., Simonsen, O., and Boll, P. M., Isolation of candirone: a novel pentaoxygenation pattern in a naturally occurring 2-phenyl-4*H*-1-benzopyran-4-one from *Tephrosia candida*, *Tetrahedron*, 43, 4241, 1987.

61. Horie, T., Kawamura, Y., Kobayashi, T., and Yamashita, K., Revised structure of a natural flavone from *Tephrosia candida*, *Phytochemistry*, 37, 1189, 1994.

62. Wollenweber, E. and Dietz, V. H., Occurrence and distribution of free flavonoid aglycones in plants, *Phytochemistry*, 20, 869, 1981.

63. Horowitz R. M. and Gentili, B., Flavonoids constituents of *Citrus*, in *Citrus Science and Technology*. Vol. 1, Nutrition, anatomy, chemical composition and bioregulation. Nagy, S., Shaw, P. E. and Veldhuis, M. K., Eds., Avi Publishing, Wesport, 1977, chap. 10.

64. Strack, D., Heilemann, J., Wray, V., and Dirks, H., Structures and accumulation patterns of soluble and insoluble phenolics from Norway spruce needles, *Phytochemistry*, 28, 2071, 1989.

65. Kemp, M. S. and Burden, R. S., Phytoalexins and stress metabolites in the sapwood of trees., *Phytochemistry*, 25, 1261, 1986.

66. Macheix, J. J., Fleuriet, A., and Billot, J., *Fruit phenolics*, CRC Press, Inc., Boca Raton, 1990, 60.

67. Hertog, M. G. L., Hollman, P. C. H., and Katan, M. B., Content of potentially anticarcinogenic flavonoids of 28 vegetables and 9 commonly consumed in the Netherlands, *J. Agric. Food Chem.*, 40, 2379, 1992.

68. Selway, J. W. T., Biochemical, Pharmacological and Structure-Activity Relationships in *Plant Flavonoids in Biology and Medicine*, Cody, V., Middleton, E., and Harborne, J. B., Eds., Alan R. Liss, New York, 1986, 521.

69. Vanden Berghe, D. A., Vlietinck, A. J., and Van Hoof, L., Plant products as potential antiviral agents, *Bull. Inst. Pasteur*, 84, 101, 1986.

70. Van Hoof, L., Vanden Berghe, D. A. R., Hatfield, G. M., and Vlietinck, A. J., Plant antiviral agents. V.3- Methoxyflavones as potent inhibitors of viral-induced block of cell synthesis,*Planta Med.*, 50, 513, 1984.

71. De Meyer, N., Vlietinck, A. J., Pandey, H. K., Mishra, L., Pieters, L. A. C. L., Vanden Berghe, D. A. and Haemers, A., Synthesis and antiviral properties of 3-methoxyflavones, in *Flavonoids in Biology and Medicine III*: Current issues in Flavonoid Research, Das, N. P., Ed., National University of Singapore, 1990, 403.

72. Vrijsen, R., Everaert, L., and Boeye, A., Antiviral activity of flavones and potentiation by ascorbate, *J. Gen., Virol.*, 69, 1749, 1988.

73. Li, B.-Q., Fu, T., Yan, Y.-D., Baylor, N. W., Ruscetti, F. W., and Kung, H.-F., Inhibition of HIV infection by baicalin - a flavonoid compound purified from chinese herbal medicine,*Cel. Mol. Biol. Res.*, 39, 119, 1993.

74. Hu, C.-Q., Chen, K., Shi, Q., Kilkuskie, R. E., Cheng, Y.-C., and Lee, K.-H., Anti-AIDS agents, 10. Acacetin-7-*O*-β-D-galactopyranoside, an anti-HIV principle from *Chrysanthemum moriflorum* and a structure-activity correlation with some related flavonoids,*J. Nat. Prod.*, 57, 42, 1994.

75. Mayer, A. M., Plant fungal interactions: a plant physiologist's viewpoint, *Phytochemistry*, 28, 311, 1989.

76. Grayer, R. J. and Harborne, J. B., A survey antifungal compounds from higher plants, 1982-1993, *Phytochemistry*, 37, 19, 1994.

77. Prusky, D. and Keen, N. T., Involvement of preformed antifungal compounds in the resistance of subtropical fruits to fungal decay, *Plant Disease*, 77, 114, 1993.

78. Dahiya, J. S., Isolation and characterization of*Aspergillus oryzae* pigments, *Indian J. Microbiol.*, 27, 12, 1987.

79. Werner, D., Bernard, S., Goerge, E., Jacobi, A., Kape, R., Kosch, K., Mueller, P., Parniske, M., Schenk, S., Shmidt, P., and Streit, W., Competitiveness and communication for effective inoculation by *Rhizobium, Bradyrhizobium* and vesicular-arbuscular mycorrhiza fungi, *Experientia*, 50, 884, 1994.

80. Ghazal, S. A., Abuzarqa, M., and Mahasneh, A. M., Antimicrobial activity of *Polygonum equisetiforme* extracts and flavonoids, *Phytother. Res.*, 6, 265, 1992.

81. Wafaa, A., Aziza, M. M., Amer, M. M., Hassan, A. B., and Soliman, G. A., Antimicrobial activity of *Roemeria hybrida* flavonoid-glycosides, *J. Drug. Res.* , 19, 199, 1990.

82. Afifi, F. U., Al-Khalil, S., Abdul-Haq, B. K., Mahasneh, A., Al-Eisawi, D. W., Sharaf, M., Wong, L. K., and Schiff, P. L., Jr., Antifungal flavonoids from *Varthemia iphionoides*, *Phytother. Res.*, 5, 173, 1991.

83. Miles, D. H., Chittawong, V., Hedin, P. A., and Kokpol, U., Potential agrochemicals from leaves of *Wedelia biflora*, *Phytochemistry*, 32, 1427, 1993.

84. Carlton, R. R., Deans, S. G., Gray, A. I., and Waterman, P. G., Antifungal activity of a flavonol glycoside from the leaves of bog myrtle (*Myrica gale*), *Chemoecology*, 2, 69, 1991.

85. Tomas-Barberan, F. A., Msonthi, J. D., and Hostettmann, K., Antifungal epicuticular methylated flavonoids from *Helichrysum nitens*, *Phytochemistry*, 27, 753, 1988.

86. Wang, Y., Hamburger, M., Gueho, J., and Hostettmann, K., Antimicrobial flavonoids from *Psiadia trinervia* and their methylated and acetylated derivatives, *Phytochemistry*, 28, 2323, 1989.

87. Tomas-Barberan, F. A., Cole, M. D., Garcia-Viguera, C., Tomas-Lorente, F., and Guirado, A., Epicuticular flavonoids from *Bellardia trixago* and their antifungal fully methylated derivatives, *Int. J. Crude Drug. Res.*, 28, 57, 1990.

88. Tahara, S., Hashidoko, Y., and Mizutani, J., New 3-methoxyflavones in the roots of yellow lupine (*Lupinus luteus* L. cv Topaz), *Agric. Biol. Chem.*, 51, 1039, 1987.

89. Hanawa, F., Tahara, S., Mizutani, J., Antifungal stress compounds from *Veratrum grandiflorum* leaves treated with cupric chloride, *Phytochemistry*, 31, 3005, 1992.

90. Harborne, J. B., in *Plant and Fungal Toxins*. Kuler, R. F., and Tu, A. T. Ed., Marcel Dekker, New York, 1983.

91. Rwangabo, P. C., Claeys, M., Pieters, L., Corthout, J., Vanden Berghe, D. A., and Vlietinck, A. J., Umuhengerin, a new antimicrobially active flavonoid from *Lantana trifolia*, *J. Nat. Prod.*, 51, 966, 1988.

92. Ravise, A., Kirkachrian, B. S., Chopin, J., and Kunesch, G., Phenolic compounds and structural analogs of phytoalexins. Influence of structure and substituents on the *in vitro* inhibition of micromycetes and some lytic enzymes, *Ann. Phytopathol.*, 12, 335, 1980.

93. Rivera-Vargas, L. I., Schmitthenner, A. F., and Graham, T. L., Soybean flavonoid effects on and metabolism by *Phytophthora sojae*, *Phytochemistry*, 32, 851, 1993.

94. Geibel, M., Sensitivity of the fungus *Cytospora persoonii* to the flavonoids of *Prunus cerasus*, *Phytochemistry*, 38, 599, 1995.

95. Ghosal, S., Biswas, K., and Chattopadhyay, B. K., Differences in the chemical constituents of *Mangifera indica*, infected with *Aspergillus niger* and *Fusarium moniliformae*, *Phytochemistry*, 17, 689, 1978.

96. Weidenboerner, M. and Jha, H. C., Fungicidal activity of flavonoid mixtures against grain contaminating fungi, *Meded. Fac. Landbouwkd. Toegepaste Biol. Wet.*, 59, 1017, 1994.

97. Weidenboerner, M. and Jha, H. C., Antifungal activity of flavonoids and their mixtures against different fungi occurring on grain, *Pestic. Sci.* , 38, 347, 1993.

98. Weidenboerner, M. and Jha, H. C., Optimization of the concentrations of flavone and flavanone regard to their antifungal activities on four *Deuteromycotina*, *Z. Pflanzenkrankh. Pflanzenschutz*, 101, 662, 1994.

99. Weidenboerner, M., Hindorf, H., Jha, H. C., and Tsotsonos, P., Antifungal activity of flavonoids against storage fungi of the genus *Aspergillus*, *Phytochemistry*, 29, 1103, 1990.

100. Attanzo, V., De Cicco, V., Di Venere, D., Lima, G., and Salerno, M., Antifungal activity of phenolics against fungi commonly encountered during storage, *Ital. J. Food Sci.*, 1, 23, 1994.

101. Omidiji, O. and Ehimidu, J., Changes in the content of antibacterial isorhamnetin 3-glucoside and quercetin 3'-glucoside following inoculation of onion (*Allium cepa* L. cv. Red Creole) with *Pseudomonas cepacia*, *Physiol. Mol. Plant Pathol.*, 37, 281, 1990.

102. Iwagawa, T., Kawasaki, J., Hase, T., Sako, S., Okubo, T., Ishida, M., and Kim, M., An acylated flavonol glycoside from *Lasiobema japonica*, *Phytochemistry*, 29, 1013, 1990.

103. Aumente Rubio, M. D., Ayuso Gonzalez, M. J., Garcia Gimenez, M. D., and Toro Sainz, M. V., Flavonols isolated from *Erica andevalensis* Cabezudo-Ribera: antimicrobial activity of the species, *Planta Med. Phytother.*, 22, 113, 1988.

104. Meresta, L. and Meresta, T., Antibacterial activity of flavonoid compounds of propolis, occurring in flora in Poland, *Bull. Vet. Inst. Pulawy*, volume date 1985 28-29, 61, 1986.

105. Nishino, C., Enoki, N., Tawata, S., Mori, A., Kobayashi, K., and Fukushima, M., Antibacterial activity of flavonoids against *Staphylococcus epidermis*, a skin bacterium, *Agric. Biol. Chem.*, 51, 139, 1987.

106. Miski, M., Ulubelen, A., Johansson, C., and Mabry, T. J., Antibacterial activity studies of flavonoids from *Salvia palaestina*, *J. Nat. Prod.*, 46, 874, 1983.

107. Hedin, P. A. and Waage, S. K., Roles of flavonoids in plant resistance to insects, *Prog. Clin. Biol. Res.*, 213, 87, 1986.

108. Waage, S. K. and Hedin, P. A., Quercetin 3-O-galactosyl-(1→6)-glucoside, a compound from narrowleaf vetch with antibacterial activity, *Phytochemistry*, 24, 243, 1985.

109. Waage, S. K. and Hedin, P. A., Biologically-active flavonoids from *Gossypium arboreum*, *Phytochemistry*, 23, 2509, 1984.

110. Mori, A., Nishino, C., Enoki, N., and Tawata, S., Antibacterial activity and mode of action of plant flavonoids against *Proteus vulgaris* and *Staphylococcus aureus.*, *Phytochemistry*, 26, 2231, 1987.

111. Hartwig, U. E., Maxwell, C. A., Joseph, C. M., and Phillips D. A., Chrysoeriol and luteolin released from alfalfa seeds induce *nod* genes in *Rhizobium meliloti*, *Plant Physiol.*, 92, 116, 1990.

112. Vaughn, S. F., Phytotoxic and antimicrobial activity of 5,7-dihydroxychromone from peanut shells, *J. Chem. Ecol.*, 21, 107, 1995.

113. Harborne, J. B. and Grayer, R. J., Flavonoids and insects, in *The Flavonoids: Advances in Research Since 1986*, Harborne, J. B., Ed., Chapman and Hall, London, 1993, chap. 14.

114. Shaver, T. N. and Lukefahr, M. J., Effect of Flavonoid Pigments and Gossypol on Growth and Development of the Bollworm, Tobacco Budworm, and Pink Bollworm, *J. Econ. Entomol.* 62, 643, 1969.

115. Todd, G. W., Getahun, A., and Cress, D. C., Resistance in Barley to the Greenbug, *Schizaphis graminum*. 1. Toxicity of Phenolic and Flavonoid Compounds and Related Substances, *Ann. Entomol. Soc. Am.*, 64, 718, 1971.

116. Matsuda, K., Feeding stimulation of flavonoids for various leaf beetles, *Appl. Ent. Zool.*, 13, 228, 1978.

117. Echeverri, F., Cardona, G., Torres, F., Pelaez, C. Q., Quinones, W., and Renteria, E., Ermanin: an insect deterrent flavonoid from *Passiflora foetida* resin, *Phytochemistry*, 30, 153, 1991.

118. Mahoney, N. E., Roitman, J. N., and Chan, B. C., Structure-activity relationship of flavones as growth inhibitors of the navel orangeworm, *J. Chem. Ecol.*, 15, 285, 1989.

119. Waiss, A. C., Jr., Chan, B. G., Elliger, C. A., Wiseman, B. R., Mc Millian, W. W., Widstrom, N. W., Zuber, M. S., and Keaster, A. J., Maysin, a flavone glycoside from corn silks with antibiotic activity toward corn earworm, *J. Econ. Ent.*, 72, 256, 1979.

120. Dreyer, D. L. and Jones, K. C., Feeding deterrency of flavonoids and related phenolics towards *Schizaphis gramimum* and *Myzus persicae*: aphid feeding deterrents in wheat, *Phytochemistry*, 20, 2489, 1981.

121. Hedin, P. A., Jenkins, J. N., and Parrott, W. L., Evaluation of flavonoids in *Gossypium arboreum* (L.) cottons as potential source of resistance to tobacco budworm, *J. Chem. Ecol.*, 18, 105, 1992.

122. Isman, B. I. and Rodriguez, E., Larval growth inhibitors of *Parthenium* (*Asteraceae*), *Phytochemistry*, 22, 2709, 1983.

123. Abou-Zaid, M. M., Beninger, C. W., Arnason, J. T., and Nozzolillo, C., The effect of one flavanone, two catechins and four flavonols on mortality and growth of the European corn borer (*Ostrinia nubilalis* Hubner), *Biochem. Syst. Ecol.*, 21, 415, 1993.

124. Lindroth, R. L. and Peterson, S. S., Effects on plant phenols on performance of southern armyworm larvae, *Oecologia*, 75, 185, 1988.

125. Elliger, C. A., Chan, B. C., and Waiss, A. C., Jr., Flavonoids as larval growth inhibitors. Structural factors governing toxicity, *Naturwissenschaften*, 67, 358, 1980.

126. Griffiths, L. A. and Smith, G. E., Metabolism of myricetin and related compounds in rat metabolite formation in vivo and by the intestinal/microflora in vitro, *Biochem. J.*, 130, 379, 1972.

127. Lahaun, H. and Perucker, M. I., *Methoden der Organischen Chemie*, Muller, E., Ed., Theime, Stuttgart, 1975, 62.

128. Haresteen, B., Flavonoids, a class of natural products of high pharmacological potency, *Biochem. Pharmacol.*, 32, 1141, 1983.

129. Robbins, R. C., Stabilization of flow properties of blood with phenylbenzo-γ-pyrone derivatives (flavonoids), *Internat. J. Vit. Nutr. Res.*, 47, 373, 1977.

130. Huet, R., Constituants des agrumes à effet pharmacodynamique: les citroflavonoides, *Fruits*, 37, 267, 1982.

131. Horowitz, R. M., Flavonoids, Mutagens, and Citrus, *ACS Syp. Ser. 170 (Qual. Sel. Fruits Veg. North Am.)*, 1981, chap. 5.

132. MacGregor, J. T., Genetic toxicology of dietary flavonoids, in *Prog. Clin. Biol. Res. (Genetic toxicology of the diet)*, Alan R. Liss, Inc., 1986, 33.

133. Woerdenbag, H. J., Merfort, I., Passreiter, C. M., Schmidt, T. J., Willuhn, G., Van Uden, W., Pras, N., Kampinga, H. H., and Konings W. T., Cytotoxicity of flavonoids and sesquiterpene lactones from *Arnica* species against the GLC$_4$ and the COLO 320 cell lines, *Planta Med.*, 60, 434, 1994.

134. Nakayama, T., Yamada, M., Osawa, T., and Kawakishi, S., Suppression of active oxygen-induced cytotoxicity by flavonoids, *Biochem. Pharmacol.*, 45, 265, 1993.

135. Shi, Q., Chen, K., Li, L., Chang, J.-J., Autry, C., Kozuka, M., Konoshima, T., Estes, J. R., Lin, C. M., Hamel, E., McPhail, A. T., McPhail, D. R., and Lee, K.-H., Antitumor agents, 154. Cytotoxic and antimitotic flavonols from *Polanisia dodecandra, J. Nat. Prod.*, 58, 475, 1995.

136. Scambia, G., Ranelletti, F. O., Benedetti Panici, P., De Vincenzo, R., Bonanno, G., Ferrandina, G., Piantelli, M., Bussa, S., Rumi, C., Cianfriglia, M., and Mancuso, S., Quercetin potentiates the effect of Adriamycin in a multidrug-resistant MCF-7 human breast-cancer cell line: P-glycoprotein as possible target, *Cancer Chemother. Pharmacol.*, 34, 459, 1994.

137. Sahu, S. C. and Gray, G. C., Interactions of flavonoids, trace metals, and oxygen: nuclear DNA damage and lipid peroxidation induced by myricetin, *Cancer Lett.*, 70, 73, 1993.

APPENDIX 8.1
Trivial Names of Some Common Flavones Cited in Alphabetical Order

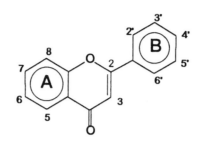

Trivial Name[a]	Substituents
Acacetin	5,7-diOH-4′-OMe
Almetin	5-OH-6,7,8-triOMe
Apigenin	5,7,4′-triOH
Baicalin	5,6-diOH-7-D-glucuronic
Candirone ≡ penduletin	5,4′-diOH-3,6,7-triOMe
Chrysin	5,7-diOH
Chrysoeriol	5,7,4′-triOH-3-OMe
Ermanin	3,5-diOH-7,4′-diOMe
Galangin	3,5,7-triOH
Gomphrenol	3,5,4′-triOH-6,7-methylenedioxy
Gossypetin	3,5,7,8,3′,4′-hexaOH
Isokaempferide	5,7,4′-triOH-3-OMe
Isoorientin	5,7,3′,4′-tetraOH-6-C-glucosyl
Isorhamnetin	3,5,7,4′-tetraOH-3′-OMe
Isoquercetin	5,7,3′,4′-tetraOH-3-O-glucosyl
Isoswertisin	5,4′-diOH-7-OMe-6-C-glucosyl
Isovitexin	5,7,4′-triOH-6-C-glucosyl
Kaempferide	3,5,7,-triOH-4′-OMe
Kaempferol	3,5,7,4′-tetraOH
Kumatakenin	5,4′-diOH-3,7-diOMe
Luteolin	5,7,3′,4′-tetraOH
Maysin	5,7,3′,4′-tetraOH-2″-O-α-L-rhamnosyl-6-C-(6-deoxy-*xylo*-hexos-4-ulosyl)
Molludistin	5,4′-diOH-7-OMe-8-C-arabinosyl
Morin	3,5,7,2′,4′-pentaOH
Myricetin	3,5,7,3′,4′,5′-hexaOH
Myricitrin	5,7,3′,4′,5′-pentaOH-3-O-rhamnosyl
Nobiletin	5,6,7,8,3′,4′-hexaOMe
Quercetin	3,5,7,3′,4′-pentaOH
Quercimeritrin	3,5,3′,4′-tetraOH-7-O-glucosyl
Quercitrin	5,7,3′,4′-tetraOH-3-O-rhamnosyl
Rutin	5,7,3′,4′-tetraOH-3-O-rutinosyl
Schaftoside	5,7,4′-triOH-6-C-glucosyl-8-C-arabinosyl
Sinensetin	5,6,7,3′,4′-pentaOMe
Tangeretin	5,6,7,8,4′-pentaOMe
Tectochrysin	5-OH-7-OMe
Topazolin	5,7,4′-triOH-3-OMe-6-(3,3-dimethylallyl)
Umuhengerin	5-OH-6,7, 3′,4′,5′-tetraOMe
Vitexin	5,7,4′-triOH-8-C-glucoside
Xanthomicrol	5,4′-diOH-6,7,8-triOMe

[a] For other cited structures, see Figure 8.2.

9.2.2 COUMESTAN

The biosynthesis of coumestrol as a coumestan has been proposed to occur by hydroxylation at C-2 of daidzein, and subsequent rearrangement of 2-hydroxy daidzein (Figure 9.2).[9] Coumestrol has been investigated by many researchers for its estrogenic effects in humans and animals.[10,11,12,13,14]

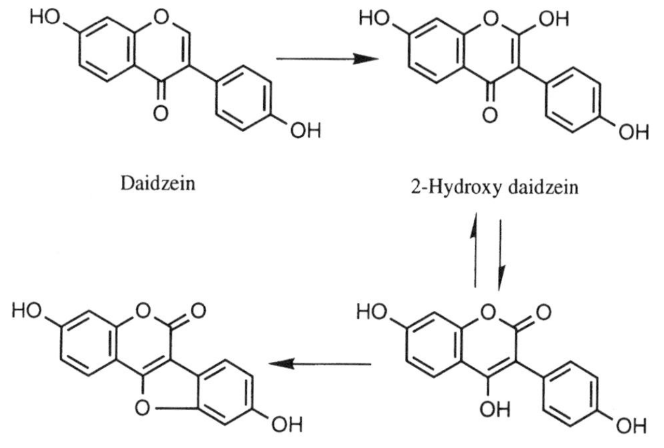

Daidzein   2-Hydroxy daidzein

Coumestrol

FIGURE 9.2 Derivation of coumestrol by rearrangement of 2-hydroxy daidzein corresponding to daidzein.

9.2.3 PTEROCARPAN

Oxygen heterocyclic phytoalexins having a 6a, 11a-dihydro-6*H*-benzo-furobenzopyran nucleus shown in Figure 9.1 are called pterocarpans and are produced mainly by plants belonging to Leguminosae. Accumulation of phytoalexins can be induced by pathogenic fungi such as *Phytophthora megasperma* or by specific glucan elicitors which are components of mycelia call walls of pathogen. Abiotic elicitors such as heavy metal ions or UV irradiation can also induce the pterocarpan components. These components are believed to play a role in disease resistance, and are detectable within 12 to 24 hr, with the highest concentrations at 2 to 4 days after induction.[15]

Banks and Dewick have been reported that glyceollins I, II, and III as phytoalexins in soybean (Figure 9.3) were synthesized as follows: daidzein is hydroxylated in 2'-position, then successively reduced in stereospecific manner via the corresponding isoflavone and isoflavanol to give 3, 9-dihydroxypterocarpan, and hydroxylated while keeping the configuration at C-6a of pterocarpans, before any prenylation and cyclization occur.[15]

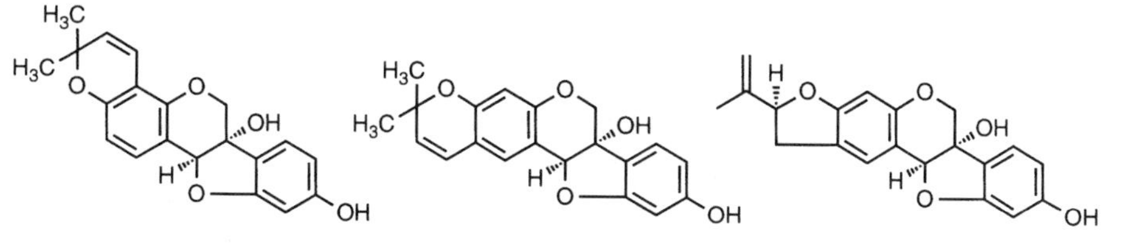

Glyceollin I   Glyceollin II   Glyceollin III

FIGURE 9.3 Chemical structures of glyceollins I, II and III as phytoalexins in soybean.

genistin and 2′-hydroxyl genistin-6-C-(-L-rhamnosyl (1 → 2) glucopyranoside from *Cassia nodos*.[32]

The nitrogen-containing isoflavones were identified as 4′-amino-5,7,3′-trihydroxy-5′-methoxy-2′,6-di-(3,3-dimethylallyl) isoflavone, 4′-amino-5,7,3′-trihydroxy-5′-methoxy-8,2′-di-(3,3-dimethylallyl) isoflavone, and 7-hydroxy-5′-methoxy-2′-(3,3-dimethylallyl)-oxazole-[4′,5′′′:4′,3′] isoflavone from *Piscidia erythrina* which have been reported to possess the fish poisoning and medicinal properties (Figure 9.5).[33]

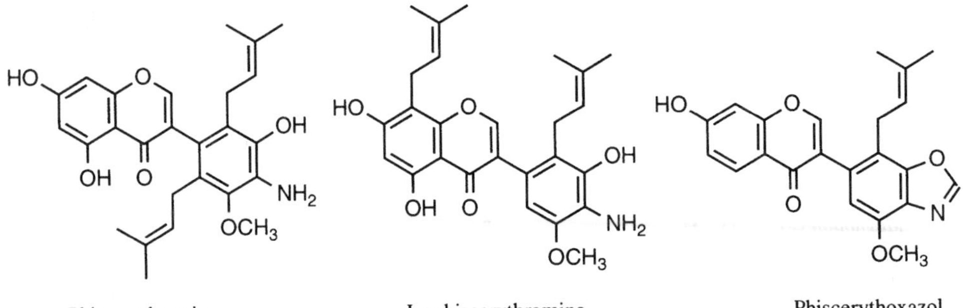

Phiscerythramine   Isophiscerythramine   Phiscerythoxazol

FIGURE 9.5 Nitrogen-containing isoflavones from *Piscidia eryhrina*.

Bryoflavones and heterobryoflavones as isoflavone-flavone dimers from *Bryum capillare* have been reported.[34] The other type of dimers also have been reported, that is, isoflavone-isoflavan dimer 2′, 7-dihydroxy-2-[(3S)-6′,7-dihydroxy-4′-methoxyisoflavane-3′-yl]-4′-methoxyisoflavone from *Dalbergia nitidul*,[35] isoflavane-flavanone and isoflavane-isoflavane dimers from *Dalbergia odorifera*,[36] and isoflavanone dimers hexaspermone A, B, and C from *Ouratea hexasperma* (Figure 9.6).[37]

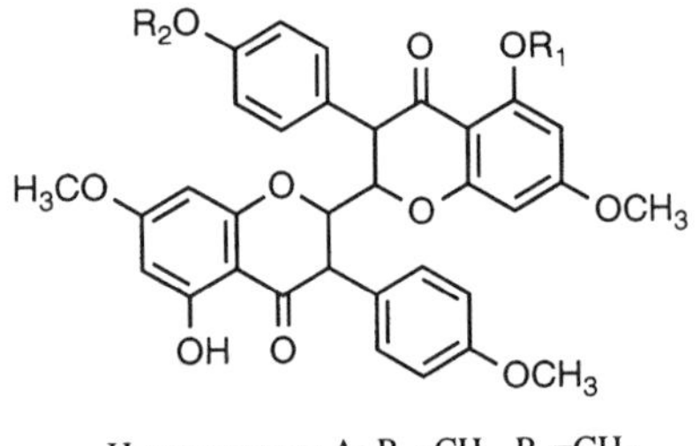

FIGURE 9.6 Isoflavanone dimers hexaspermone A, B and C from *Ouratea hexasperma*.

Hexaspermone A; R_1=CH$_3$, R_2=CH$_3$
Hexaspermone B; R_1=CH$_3$, R_2=H
Hexaspermone C; R_1=H, R_2=H

9.4 METHODS OF EXTRACTION, SEPARATION, AND STRUCTURAL ELUCIDATION

9.4.1 EXTRACTION AND SEPARATION

Isoflavonoids are generally isolated by extraction with organic solvents such as methanol, petroleum ether, chloroform, diethyl ether, aqueous alcohol, etc., based on properties of isoflavone components, after the plant materials are dried or freeze-dried, followed by partitioning between water and nonpolar organic solvent to remove lipids and pigments. Subsequently, the water layer fraction is partitioned by water : butanol mixture to remove sugars. In some cases, extractions with 70% ethanol or 80% methanol are efficient for isoflavone glycosides, since fats are not extracted and a defatted step is omitted.[27] The extraction temperature has to be paid attention to in order to obtain the intact isoflavonoids in plants. Although the extraction over 80°C results in higher recovery, it has been reported that malonylated isoflavone glucosides in soybean seeds were converted into the corresponding isoflavone glucosides and acetylated isoflavone glucosides which were artifacts derived by decarboxylation of malonic acid.[28]

The extract remaining after removal of lipids, pigments, and sugars is further fractionated by recrystallization and/or column chromatography with silica gel and C_{18}-solid phase. The simple step fractionation has been achieved by extraction with aqueous methanol and selective elution from C18-solid-phase column with a gradient of 10 to 60% methanol in the case of soybean isoflavones. After the methanol extract is applied on an ODS column, the column is washed with water to remove sugars and then carried out a gradient elution of 10 to 60% methanol. Isoflavone glycosides, malonyl isoflavone glycosides, and isoflavones can be eluted in 20 to 30% methanol, 30 to 40% methanol, and 60% methanol, respectively. Isoflavone compounds in each fraction are further isolated by preparative HPLC with a reversed phase column and an aqueous acetonitrile as a mobile phase. The separation by HPLC shows excellent resolution, especially on a gradient elution.

9.4.2 STRUCTURAL ELUCIDATION

Although UV spectrum analysis of isoflavones requires only a small amount of pure material, the UV spectrum provides considerable information about substitution patterns in the isoflavone skeleton. Specifically, the use of specific reagents which react with one or more functional groups on the isoflavonoid nucleus induces structurally significant shifts in the UV spectrum.[38] Sodium methoxide, sodium acetate, sodium acetate/boric acid, aluminum chloride, and aluminum chloride/hydrochloric acid as the specific reagents are used. The spectra of most flavonoids consists of two major absorption maxima, one of which occurs in the range 240 to 285 nm (band II) and the other in the range 300 to 400 nm (band I). In general terms, the band II absorption may be considered as having originated from the A-ring benzoyl system and band I from the B-ring cinnamoyl system. The UV spectra of isoflavones exhibit a low intensity band I absorption which often appears as a shoulder to the band II peak. The spectra of these compounds are largely unaffected by changes in the oxygenation and substitution patterns in the B-ring. However, increased oxygenation in the A-ring leads to a bathochromic shift in the band II absorption.

MS analysis provides much information about molecular weight, formula, and structures. Conventional EI-MS is used for volatile isoflavonoid aglycones; however, it usually requires permethylation or peracethylation of hydroxy groups in the molecular structures for poor volatile glycosides. The functional groups binding to isoflavone compounds can be readily estimated by EI-MS as a result of the detailed analyses of the produced fragment ions. The cleavage of the heterocyclic ring in isoflavone skeleton produces A- and B-ring fragments. These fragments are the most useful in terms of isoflavonoid identification; such fragment ions provide some information about the positions and numbers of substituent groups such as hydroxy, methoxy, and prenyl on the respective rings. This cleavage reaction occurs via one of two routes, designated as pathway-I corresponding to a retro-Diels-Alder (RDA) and pathway-II, depending on the kind of isoflavone aglycone. 1 FAB (fast atom bombardment) as the newer ionization technique is efficient for glycosides and biflavones possessing poor volatility, since it does not need to be derived to increase volatility. Although this technique provides less information than EI on the isoflavone skeletons, it provides valuable information about molecular ions, acylated organic acids, and sugars, and allows determination of the binding order of these groups. It is commonly observed that the fragment ion peaks at M+-162, M+-146, M+-86 when lacking hexose, methylpentose, and malonate, respectively.

^{1}H- and ^{13}C-NMR have been employed for analyses of isoflavonoid structures since the technique is useful for determination of isoflavonoid skeleton, the position of substituent group,

and the nature of the aglycone-sugar linkage without a requirement for derivatization.[2,39] The new developed techniques such as ^{13}C-^{1}H COSY, HMQC, HMBC, NOE, and NOESY permitted the elucidation of "complex" isoflavonoid structures with a slight amount of sample, including the complete assignments of proton and carbon, and also stereochemistry.[33,40,41,42]

Assignments from the ^{13}C-NMR spectra of daidzin, glycitin, genistin, and their malonylated compounds are summarized in Table 9.2. The ^{13}C-^{1}H COSY spectra of malonylglycitin is shown in Figure 9.7. The ^{13}C-NMR spectra of malonyldaidzin, malonylglycitin, and malonylgenistin show signals of δ 167.9, 166.8, and 41.4 (COOR, COOH, and -CH_2- in a malonyl group), neither of which are detected in the spectra of their corresponding isoflavone glycosides. The low field methylene protons Ha-6″ and Hb-6″, and the chemical shifts of C-6″ and C-5″ in malonylated compounds when compared with those of daidzin, glycitin, and genistin indicate acylation of the sugar moieties at C-6″.

Fukai et al. reported the effect of prenylation at C-6 or C-8, and hydroxylation at C-2′ on the chemical shift of the 5-OH signal of isoflavone.[41] The 5-OH signal of 6-prenylated isoflavone appears more downfield (0.28ppm) than that of the corresponding isoflavone. In contrast, the 5-OH signal of 8-prenylated isoflavone appears more upfield (0.06ppm) than

TABLE 9.2
^{13}C-NMR Data for Isoflavone Glycosides Isolated from the Hypocotyl of Soybean (100 MHz, in DMSO-d_6)

	Daidzin	Malonyl daidzin	Glycitin	Malonyl glycitin	Genistin	Malonyl genistin
A and B rings						
C-2	153.3	153.3	153.1	153.1	154.6	154.6
C-3	123.7	123.7	123.1	123.1	122.6	122.6
C-4	174.8	174.8	174.3	174.4	180.5	180.5
C-5	127.0	127.1	104.8	104.8	161.7	161.7
C-6	115.6	115.4	147.4	147.4	99.6	99.5
C-7	161.4	161.1	151.1	151.2	163.0	162.7
C-8	103.4	103.6	103.6	103.6	94.6	94.6
C-9	157.0	157.0	151.2	151.2	157.2	157.2
C-10	118.5	118.6	117.9	117.9	106.1	106.2
B ring						
C-1′	122.3	122.3	122.5	122.6	121.0	121.0
C-2′ and C-6′	130.1	130.1	130.0	130.0	130.2	130.2
C-3′ and C-5′	115.0	115.0	114.9	115.0	115.1	115.1
C-4′	157.3	157.3	157.2	157.2	157.5	157.5
OCH$_3$			55.8	55.9		
Glucose moiety						
C-1″	100.0	99.8	99.7	99.4	99.9	99.5
C-2″	73.2	73.0	73.0	72.9	73.1	73.0
C-3″	76.5	76.2	76.8	76.5	76.4	76.1
C-4″	69.7	69.7	69.6	73.8	69.6	69.6
C-5″	77.2	73.8	77.2	73.8	77.2	73.7
C-6″	60.7	64.1	60.6	64.0	60.6	64.0
Malonyl						
COOR		167.9		168.0		167.9
CH$_2$		41.4		41.5		41.5
COOH		166.8		167.0		166.9

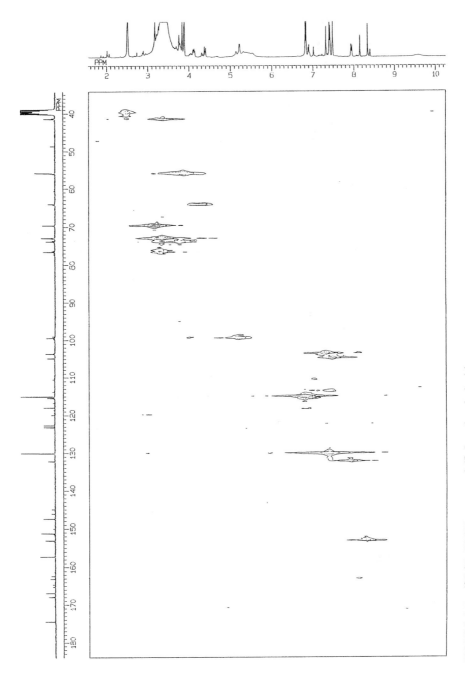

FIGURE 9.7 ^{13}C-^1H COSY spectra of 6″-O-malonyl glycitin.

that of the corresponding isoflavone. The 5-OH signal of 2′-hydroxyisoflavone appears more upfield (0.26ppm) than that of the corresponding isoflavone, which may be caused by the formation of a weak hydrogen-bond between the 2′-hydroxyl and 4-carbonyl group. These parameters are useful for the structural determination of prenylated isoflavones.

9.5   QUANTITATIVE ANALYSIS

The quantitative analyses of isoflavonoids in foods, plasma, and plants have been attempted from breeding,[43,44,45] phytoestrogenic,[46,47,48,49,50] chemotaxonomic, and biochemical aspects. Naim et al. developed a gas chromatographic method to determine daidzein and genistein as trimethyl derivatives after extraction of defatted soybean meal isoflavone glycosides by ether and methanol, subsequent chromatography on polyamide columns, and acid hydrolysis.[51] Recently, HPLC has been used for isoflavonoids with a gradient system of water- acetonitrile or water-methanol (sometimes modified with acetic acid, THA and some buffers) as mobile phase, C_{18}-reversed phase column and commonly UV detection. Chemical structures and HPLC Profile of soybean isoflavones are shown in Figure 9.8 and 9.9. The minimum detectable amount for daidzein and genistein was 14 and 24 ng, respectively.[46] Fluorescence,[52,53,54] and electrochemical detection[55,56] were shown to be very useful for increasing the sensitivity of commonly used UV detection. With amperometric detection, genistein was detectable at a level of 0.05 ng.[56] Photodiode array detection has been used to carry out quantification and identification simultaneously, with minimum detectable levels of 3.7 ng for daidzein and 2 ng for genistein.[47] It appears to provide a powerful tool for chemotaxonomic studies which need to assay and identity isoflavonoids in many plant materials. Capillary electrophoresis has been shown to be applicable for determination of five isoflavones and coumestrol in soybean seeds.[57] Radioimmunoassay for quantitative analysis of formononetin in blood plasma and rumen fluid has been developed by Wang et al., and their detectable amount was 0.7 pmol.[58]

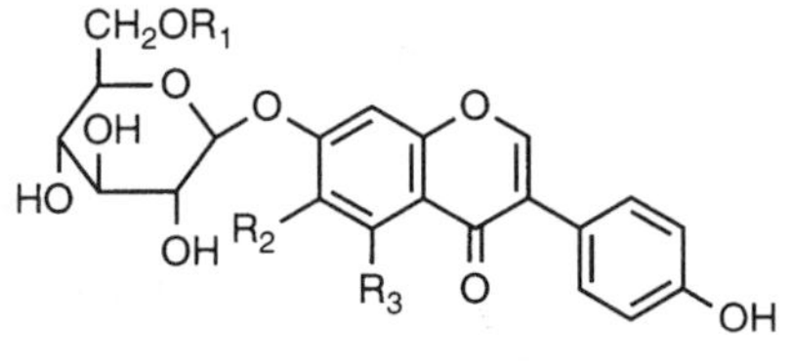

Compound	R_1	R_2	R_3
Daidzin	H	H	H
Glycitin	H	OCH_3	H
Genistin	H	H	OH
6″-O-Malonyldaidzin	$COCH_2COOH$	H	H
6″-O-Malonylglycitin	$COCH_2COOH$	OCH_3	H
6″-O-Malonylgenistin	$COCH_2COOH$	H	OH
6″-O-Acetyldaidzin	$COCH_3$	H	H
6″-O-Acetylglycitin	$COCH_3$	OCH_3	H
6″-O-Acetylgenistin	$COCH_3$	H	OH

FIGURE 9.8   Chemical structures of soybean isoflavones.

9.6   LEVELS

Soybeans are an important protein source, and many researchers have investigated isoflavone contents of soybeans from a nutritional viewpoint. Kudo et al. investigated components responsible for undesirable taste characteristics such as bitterness, astringency, and dry mouth feeling in soybeans and isolated 12 kinds of isoflavones.[28] The threshold values for daidzin and its related compounds were in the order of daidzin > daidzein = acetyl daidzin > malonyldaidzin. Those of the related compounds of glycitin and genistin were in the order of glycoside > aglycone > = 6″-O-acetyl glycoside (Table 9.3). The total isoflavone content of the hypocotyl

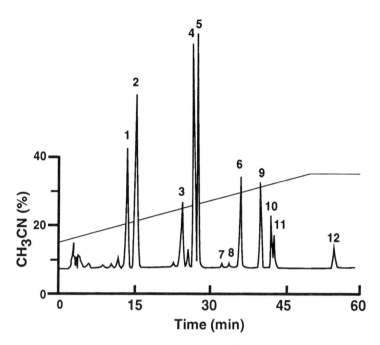

FIGURE 9.9 High-performance liquid chromatogram of 70% ethanol extract from hypocotyls of soybean seeds. 1, daidzin; 2, glycitin; 3, genistin; 4, 6″-O-malonyldaidzin; 5, 6″-O-malonylglycitin; 6, 6″-O-malonylgenistin; 7, 6-O-acetyldaidzin; 8, 6-O-acetylglycitin; 9, 6-O-acetylgenistin; 10, daidzein; 11, glycitein; 12, genistein. Conditions; HPLC were performed on a YMC-pack ODS-AM-303 column (250(4.6 mm), using a linear gradient of acetonitrile from 15 to 35% containing constant 0.1% acetic acid in 50 min. Flow rate was 1ml/min and the absorption was measured at 260 nm.

part was 5.5 or 6.0 times higher than that of the cotyledons, and glycitin and its derivative only occurred in the hypocotyl part of the soybean seed. Isoflavones were absent from the seed coat of soybean. While malonylated isoflavone glycosides were major constituents in the extract at room temperature, the contents of malonylated isoflavone glycosides in the extract at 80°C decreased significantly, and an increase in all isoflavone glycosides and acetyl isoflavone glycosides, with the exception of 6″-O-acetylgenistin, was observed. The acetyl derivatives may thus have arisen from corresponding malonyl derivatives as artifacts during the work-up procedures.

Isoflavonoids have been isolated from leaves, stems, roots, flowers, buds, and seeds. The amount of isoflavonoid in the respective parts might be influenced by environmental rather than genetic factors. In the case of soybean seed, the isoflavone contents decreased in the seeds harvested after growth at a high temperature for all soybean varieties tested. While the isoflavone contents of cotyledons exhibited large changes in response to high temperature during seed development, the isoflavone contents remained high in the hypocotyls.[42] Tsukamoto et al. studied the factors affecting the isoflavone contents of soybean seeds to provide a basis for attempts to improve seed quality by reduction of isoflavone content. The contents of isoflavone in soybean whole seed and hypocotyl were in the range of 94 to 1560 mg kg^{-1} and 4 to 15 g kg^{-1}, respectively. Within varieties, temperature resulted in 14–16-fold differences in isoflavone content in the whole seed, and 3–6-fold differences in the hypocotyl. The isoflavone contents were influenced by the temperatures during seed development and declined significantly (1% level) at elevated temperatures. Thus one of the factors affecting isoflavone content in soybean seeds is high temperature during seed development, and higher temperatures result in lower isoflavone content.

TABLE 9.3
Threshold Values and Contents of Isoflavones in Soybean Seeds.

Compound	Threshold Value (mM)	Room Temp (mg%)		80°C (mg%)	
		Hypocotyl	Cotyledon	Hypocotyl	Cotyledon
Daidzin	10^{-2}	320	45	838	145
Glycitin	10^{-3}	485	—	1004	—
Genistin	10^{-1}	118	80	246	210
6″-O-Malonyl daidzin	10^{-4}	423	70	8	3
6″-O-Malonyl glycitin	10^{-5}	445	—	11	—
6″-O-Malonyl genistin	10^{-2}	144	117	4	—
6″-O-Acetyl daidzin	10^{-3}	2	2	57	8
6″-O-Acetyl glycitin	10^{-5}	6	—	89	—
6″-O-Acetyl genistin	10^{-2}	105	1	39	1
Daidzein	10^{-3}	102	33	35	11
Glycitein	10^{-4}	—	—	15	—
Genistein	10^{-2}	35	48	16	14
Total		2185	396	2362	392

To evaluate phytoestrogens in legumes and legume foods, the isoflavonoid contents of soy ingredients, traditional soy foods, and second-generation soy foods have been measured.[47,49,59] Most Asian and American soy products, with the exception of soy sauce, alcohol-extracted soy protein concentrate, and soy protein isolate, have total isoflavone concentrations similar to those in the intact soybean. Asian fermented soy foods contain predominantly isoflavone aglycones, whereas in nonfermented soy foods of both American and Asian origin, isoflavones are present mainly as β-glycoside conjugates.[49] The second-generation soy foods contained isoflavone concentrations ranging from 34 to 289 μg g^{-1}. Their total isoflavone content was only 6 to 20% of the content of whole soybeans because most of food matrices in these foods were non-soybean constituents.[47] High levels of daidzein and genistein were found in soy products and black beans, whereas sprout items were found to be rich in coumestrol and formononetin.[45]

9.7   BIOLOGICAL ROLES IN PLANTS

The antifungal and antibacterial actions of isoflavonoids probably occur through multi-site actions that result in disruption of membrane systems and intracellular metabolism.[1] A variety of cellular effects have been described, such as cessation of cytoplasmic streaming, disorganization of cellular contents, and breakdown of the cell membrane.[1] Isoflavonoids also play a regulatory role in the unique symbiosis of legumes with nitrogen-fixing *Rhizobium* bacteria. The formation of root nodules that will house the bacteria is initiated by the secretion of flavones and flavonols that bind and activate NodD, a bacterial protein that induces transcription of other bacterial genes that stimulate growth of plant epidermal cells that will form the root nodule.[60,61] Isoflavonoids appear to regulate this process and the number of nodules formed by inhibiting the activation of NodD.[62] These actions parallel steroid-receptor interactions that regulate gene transcription in animals and structural and functional homologies between NodD and the steroid receptor superfamily suggest that these regulatory factors share a common ancestral origin, which may help to explain the biological actions of isoflavonoids and other flavonoids in animals.[63,64]

9.8 ACTIONS IN ANIMAL CELLS

9.8.1 PROLIFERATIVE ACTIONS

The isoflavonoids are notable for their ability to induce estrogenic responses in animals.[65] Although they are much less potent than endogenous steroidal estrogens, isoflavonoids can produce significant effects due to their high concentrations in plant material.[66] Moreover, they have the capacity to influence estrogen action via a variety of pathways[67] and may exert multi-site actions in animals as they do in microorganisms.

Several structural features shared with steroidal estrogens and other estrogenic substances appear to be responsible for the estrogenicity of many isoflavonoids. These features are the phenolic rings and two hydroxyl groups at either end of an essentially planar structure, features that appear to facilitate binding and activation of the estrogen receptor.[68,69] Competitive inhibition of the binding of estradiol to the estrogen receptor has been reported for isoflavones, coumestans, isoflavans, and isoflavanones (see Table 9.4).

TABLE 9.4
Estrogenic Potency of Isoflavonoids in Relation to Estradiol

Isoflavonoid	Binding[a]	Transcription[b] Protein Synthesis	Proliferation Min[c]	ED50[d]	Effect[e]	Uterine Growth[f]
Coumestan						
Coumestrol	1.4–19.7	0.06–0.07	100	0.03	93–98	0.2-0.8
Coumestrol diacetate	5					0.2
4-methoxycoumestrol	< 0.01					
Isoflavone						
Biochanin A	< 0.01–0.08	0.0005	0.01		64	0.004–0.01
Daidzein	0.1	0.003–0.01				0.005–0.01
Formononetin	< 0.01	0.0003–0.003				0.002–0.006
Genistein	0.6–3	0.01–0.04	0.01	0.005	72	0.01–0.03
Isoflavan						
Equol	0.4	0.01		0.01		
Isoflavanone						
O-desmethylangolensin	0.05					0.01

[a] Relative binding affinity for the estrogen receptor, expressed as a percentage of the affinity of estradiol (=100). Based on references 11, 71, 130, 131, 132, 133, 134.

[b] Transcriptional potency, calculated from concentration producing a half-maximal stimulation of transcriptional activity of the human estrogen receptor or protein synthesis, expressed as a percentage of the potency of estradiol (=100). Based on references 135, 136.

[c] Lowest concentration producing a stimulation of MCF-7 cell proliferation, compared to the minimal estradiol concentration (=100). Based on reference 72.

[d] Relative proliferative potency, calculated from concentration producing a half-maximal stimulation of MCF-7 cells. Based on references 71, 133.

[e] Relative proliferative effect, degree of maximal stimulation, compared to proliferation produced by estradiol (=100). Based on references 72, 73.

[f] Relative uterotrophic action, calculated from oral concentrations required to produce a 25 mg uterus in immature mice, compared to estimated concentrations required for estradiol (=100). Based on references 130, 137, 138.

Although binding affinities are low compared to estradiol, that does not preclude biological activity. Consistent with the actions of other estrogens, isoflavonoids increase estrogen receptor binding in the nuclear fraction,[12,70] transcription of estrogen-regulated genes,[69] induction of progestin receptors and other estrogen-regulated proteins,[13,69,71,72] and proliferation in estrogen-responsive cells and tissues.[71,72,73] Relative potencies vary with the endpoint examined. Whereas coumestrol and the isoflavones are weak estrogens with respect to the concentrations required for comparable transcription and proliferation rates, they are more similar to estradiol in their ability to initiate mitosis and can fully replicate estradiol actions if sufficient concentrations are achieved. Even though the concentrations (0.01–100 nM) required for isoflavonoid actions may be low relative to those required for estradiol action, they are well within the range of circulating concentrations reported in humans and other animals (see below).

9.8.2 ANTIPROLIFERATIVE ACTIONS

In addition to acting directly as estrogens, isoflavonoids might modulate the action of endogenous estrogens by competitive inhibition of estrogen receptors. There is evidence that isoflavonoids can limit the growth of cultured cancer cells and tumor growth *in vivo*.[74] However, because inhibition of proliferation has been observed in estrogen-receptor negative as well as estrogen-receptor positive cell lines, this action may not be mediated through the estrogen receptor.[75] Moreover, concentrations of more than 2 µM are required for isoflavone inhibition of cell proliferation, whereas cell growth is stimulated at concentrations of 10 nM to 2 µM, suggesting that isoflavones should be agonists at physiologic concentrations.[76,77]

Isoflavones inhibit a number of enzymes that regulate steroid biosynthesis or transduce steroid or growth factor action. These enzymes include aromatase, the enzyme that synthesizes estrogen from androgen,[78,79] and 17β-hydroxysteroid dehydrogenase 1, the enzyme that catalyzes the conversion of estrone to estradiol.[80] These actions may antagonize estrogen action indirectly by reducing the availability of endogenous estrogen in serum and target tissues. Genistein is also a specific inhibitor of protein tyrosine kinases.[81] Many of these kinases are responsible for autophosphorylation and activation of growth factor receptors and intracellular kinases involved in growth factor signal transduction. It has been proposed that genistein may inhibit cell proliferation by disrupting phosphorylation cascades that mediate cell growth and differentiation.[82] However, these actions also require concentrations that may not be physiologic.

9.9 METABOLISM AND ABSORPTION

The major isoflavonoids reported in mammalian urine and plasma are daidzein,[83] genistein,[84] equol,[85] and *O*-desmethylangolensin.[86] Daidzein and genistein are obtained from plant food as aglycones or through the action of intestinal bacteria resulting in deconjugation of their glycosides, daidzin and genistin, or metabolism of the plant precursors formononetin and Biochanin A.[87] Equol and *O*-desmethylangolensin are produced by intestinal bacteria from daidzein, by two separate pathways that appear to vary across individuals.[87,88,89] In ruminants, genistein and biochanin A are extensively metabolized to phenolic acids whereas daidzein and formononetin are metabolized to equol.[90] Equol is the major isoflavonoid in livestock plasma and appears to be the isoflavonoid responsible for the induction of clover disease.[66] In contrast, in humans the major isoflavones in plasma and urine are daidzein and genistein, and the ability to produce equol varies across individuals.[87,92] Although most individuals excrete some equol, only about a third increase equol excretion in response to increased isoflavone intake.[89] The consequences of differences in equol formation in humans are not clearly understood. Although equol's relative binding affinity is higher than that of daidzein, its potency in inducing protein synthesis and cell proliferation does not appear to differ markedly from that of genistein or daidzein (Table 9.4). However, *in vivo* studies in rodents showing negligible estrogenic activity for equol[93,94] suggest that it may represent a deactivated form in some species.

Isoflavones are present in plasma as the free aglycone or, primarily, as conjugates of glucuronic acid.[92] Circulating isoflavone concentrations can be quite high in livestock, with levels as high as 1 to 12 μM for total isoflavones and 0.04 to 0.3 μM for unconjugated isoflavones.[91] Circulating concentrations in humans are significantly lower. In Western human populations consuming an omnivorous diet, plasma levels of total isoflavones range from 1–67 nM with free levels from 1–5 nM.[92] Plasma concentrations are higher in Japanese populations and among vegetarians in Western populations, with total isoflavone levels ranging from 1 nM to 1 μM and free isoflavone levels of 2 to 55 nM.[50,92]

9.10　REPRODUCTIVE EFFECTS

A common characteristic of estrogenic isoflavonoids is their ability to stimulate proliferation of the female reproductive tract: isoflavones, isoflavans, and coumestans are uterotrophic in mice, rats, sheep, and pigs.[95] A variety of other disruptions of normal reproductive processes have been observed in livestock and laboratory animals following consumption or parenteral treatment with isoflavones or coumestans. These effects include premature sexual maturation,[13,96] reduced birth[96,97,98,99] and ovulation rates,[100,101] and altered steroid and gonadotropin secretion.[102,103,104,105] In human volunteers, daily consumption of approximately 50 mg of isoflavones, an intake similar to that of Japanese men and women, increased follicular phase estradiol and reduced periovulatory LH and FSH but did not prevent ovulation.[106] Isoflavonoids are also capable of influencing male reproductive function, reducing testicular growth,[107,108] spermatogenesis,[109] and seminal vesicle size[110] in prepubertal rodents and birds.

9.11　ORGANIZATIONAL EFFECTS

Permanent alterations of reproductive function have also been observed following chronic exposure to isoflavonoids in adulthood or briefer isoflavonoid exposure during sensitive periods in development. Prolonged consumption of high phytoestrogen levels induces permanent infertility in female sheep, cows, and pigs, known as "clover disease."[111,112] The symptoms of chronic exposure appear to be more severe in sheep than cattle, which may be due to differences in phytoestrogen metabolism.[91] In sheep, the primary cause of infertility is an alteration in cervical mucous that interferes with sperm transport, but a variety of other symptoms are evident in the reproductive tract and central nervous system,[111,113] including hyperplasia of the uterus, cervix, and adrenal cortex; polycystic ovaries; neuronal damage and gliosis in the hypothalamus;[114,115,116] and reduced responsiveness to estrogen.[117,118]

Perinatal isoflavonoid treatments also induce permanent infertility in rats and mice but in this case females are anovulatory[119,120,121] and gonadotropin secretion is altered or disordered,[122,123,124] resembling the defeminization that results from perinatal exposure to steroidal or environmental estrogens. Disorders of the reproductive tract[119,125,126] are also apparent although in rodents this is not the primary cause of infertility. Precocious breast development is apparent at high doses, which may have some beneficial effects by protecting against chemically induced carcinogenesis.[127] Male development can also be altered by perinatal coumestrol, resulting in significant reductions in mount and ejaculation frequencies in adulthood and prolongation of the latencies to mount and ejaculate.[128,129]

9.12 CONCLUSIONS

Isoflavonoids have potent effects on cell growth and estrogen-dependent processes. The prominence of isoflavonoids in legumes that are common food staples makes their actions of particular interest to humans. Plant concentrations result in biologically significant levels in human and animal circulation. Both beneficial and adverse effects have been reported in mammals, ranging from cancer protection to infertility, implicating isoflavonoids in both health and disease.

REFERENCES

1. Smith, D. A. and Banks, S. W., Biosynthesis, elicitation and biological activity of isoflavonoid phytoalexins, *Phytochemistry*, 25, 979, 1986.
2. Harborne, J. B., Mabry, T. J., and Mabry, H., *The Flavonoids*, Chapman and Hall, London, 1975.
3. Watanabe, K., Kinjo, J., and Nohara, R., Studies on leguminous plants. Part XXXIX. Three new isoflavonoid glycosides from *Lupinus luteus* and *L. polyphyllus* x *arboreus*, *Chem. Pharm. Bull.*, 41, 394, 1993.
4. Corsaro, M. M., Lanzetta, R., Mancino, A., and Parrilli, M., Homoisoflavanones from *Chionodoxa luciliae*, *Phytochemistry*, 31, 1395, 1992.
5. Barone, G., Corsaro, M. M., and Lanzetta, R. and Parrilli, M., Homoisoflavanones from *Muscari neglectum*, *Phytochemistry*, 27, 921, 1988
6. Zahringer, U., Schaller, E., and Grisebach, M. F., Induction of phytoalexin synthesis in soybean. Structure and reactions of naturally occurring and enzymically prepared prenylated pterocarpans from elicitor-treated cotyledons and cell cultures of soybean, *Z. Naturforsch.*, 36c, 234, 1981.
7. Adensaya, S. A., O'Neill, M. J. and Roberts, M. F., Structure-related fungitoxicity of isoflavonoids, *Physiol. Mol. Plant. Pathol.*, 29, 95, 1986.
8. Lane, G. A., Biggs, D. R., Russell, G. B., Sutherland, O. R. N., Williams, E. M., Maindonald, J. H., and Donnell, D. J., Isoflavonoid feeding deterrents for *Costelytra zealandica*. Structure-activity relationships, *J. Chem. Ecol.*, 11, 1713, 1985.
9. Sharma, R. D., Isoflavones and hypercholesterolemia in rats, *Lipids*, 14, 535, 1979.
10. Whitten, P. L., Russell, E., and Naftolin, F., Influence of phytoestrogen diets on estradiol action in the rat uterus, *Steroids*, 59, 443, 1994.
11. Markiewicz, L., Garey, J., Gurpide, E., and Adlercreutz, H., In vitro bioassays of non-steroidal phytoestrogens, *J. Steroid Biochem. Mol. Biol.*, 45, 399, 1993.
12. Whitten, P., Russell, E., and Naftolin, F., Effects of a normal, human-concentration, phytoestrogen diet on rat uterine growth, *Steroids*, 57, 98, 1992.
13. Whitten, P. L. and Naftolin, F., Effects of a phytoestrogen diet on estrogen-dependent reproductive processes in immature female rats, *Steroids*, 57, 56, 1992.
14. Pelissero, C., Bennetau, B., Babin, P., Le Menn, F., and Dunogues, J., The estrogenic activity of certain phytoestrogens in the Siberian Sturgeon *Acipenser baeri*, *J. Steroid Biochem. Mol. Biol.*, 38, 293, 1991.
15. Whitehead, I. M. and Threlfall, D. R., Production of phytoalexins by plant tissue cultures., *J. Biotechnol.*, 26, 63, 1992.
16. Stephen W. B. and Paul M. D., Biosynthesis of glyceollins I, II and III in soybean, *Phytochemistry*, 22, 27, 1983.
17. Li, L., Wang, H-K., Chang, J-J., Lee, K-H., McPhail, A. T., Mcphail, D. R., Terada, H., Kozuka, M., and Estes, J. R., Antitumor agents, 138. Rotenoids and isoflavones as cytotoxic constituents from *Amorpha fruticosa*, *J. Nat. Prod.*, 56, 690, 1993.
18. Tahara, S., Katagiri, Y., Mizutani, J., and Ingham, J. L., Prenylated flavonoids in the roots of yellow *Lupin*, *Phytochemistry*, 36, 1261, 1994.
19. Tahara, S., Moriyama, M., Orihara, S., Kawabata, J., Mizutani, J., and Ingham, J. L., Naturally Occurring Coumaronochroman-4-ones: A new class of isoflavonoids from Lupines and Jamaican Dogwood, *Z. Naturforsch. Sect. C.*, 46, 331, 1991.
20. Tahara, S., Shibaki, S., Mizutani, J., and Ingham, J. L., Further isoflavonoids from white *Lupin* roots, *Z. Naturforsch. Sec. C.*, 45, 147, 1990.

21. Ferrari, F., Messana, I., and Sant'ana, A. E. G., Two new isoflavonoids from *Boerhaavia coccinea, J. Nat. Prod.*, 54, 597, 1991.
22. Hanawa, F., Tahara, S., and Mizutani, J., Isoflavonoids produced by *Iris pseudacorus* leaves treated with cupric chloride, *Phytochemistry*, 30, 157, 1990.
23. Stein, W. and Zinsmeister, H. D., The occurrence of flavonoids in the moss family Bryaceae, *J. Hattori Bot. Lab.*, 69, 195, 1991.
24. Geiger, H., Stein, W., Mues, R., and Zinsmeister, H. D., Bryoflavone and heterobryoflavone, two new isoflavone-flavone dimers from *Bryum capillare, Z. Naturforsch. Sect.* C, 42, 863, 1987.
25. Anhut, S., Zinsmeister, H. D., Mues, R., Barz, W., Mackenbrock, K., Koester, J., and Markham, K. R., The first identification of isoflavones from a bryophyte, *Phytochemistry*, 23, 1073, 1984.
26. Sanduja, R., Weinheimer, A. J., and Alam, M., Albizoin: isolation and structure of a deoxybenzoin from the marine mollusc *Nerita albicilla, J. Chem. Res.* (Synopses), 2, 56(S), 1985.
27. Grayer-Barkmeijer, R. J. and Hegnauer, R., Relevance of seed polysaccharides and flavonoids for the classification of the Leguminosae: a chemotaxonomic approach, *Phytochemistry*, 34, 3, 1993.
28. Kudou, S., Uchida, T., Fleury, Y., Welti, D., Magnolato, D., Kitamura, K., and Okubo, K., Malonyl isoflavone glycosides in soybean seeds (*Glycine max* MERRILL)., *Agric. Biol. Chem.*, 55, 2227, 1991.
29. Saxena, V. K. and Bhadoria, B. K., 3'-Prenyl-4'-methoxy-isoflavone-7-*O*-β-D-(2''-*O*-*p*-coumaroyl) glucopyranoside, a novel phytoestrogen from *Sopubia delphinifolia, J. Nat. Prod.*, 53, 62, 1990.
30. Yadava R, N. and Syeda, Y., An isoflavone glycoside from the seeds of *Trichosanthes anguina, Phytochemistry*, 36, 1519, 1994.
31. He, Z-Q. and Findlay, J. A., Constituents of *Astragalus membranaceus, J. Nat. Prod.*, 54, 810, 1991.
32. Ilyas, M., Parveen, M., Khan, M. S., and Kamil, M., Nodosin, a novel *C*-glycosylisoflavone from *Cassia nodosa, J. Chem. Res.* (Synopses), 88, 1994.
33. Moriyama, M., Tahara, S., Mizutani, J., and Ingham, J. L., Isoflavonoid alkaloids from *Piscidia erythrina, Phytochemistry*, 32, 1317, 1993.
34. Geiger, H., Stein, W., Mues, R., and Zinsmeister, H. D., Bryoflavone and heterobryoflavone, two new isoflavone-flavone dimers from *Bryum capillare, Z. Naturforsch., Sect.* C, 42, 863, 1987.
35. Bezuidenhoudt, B. C. B., Brandt, E. V., Steenkamp, J. A., Roux, D. G., and Ferreira, D., Oligomeric isoflavonoids. Part 1. Structure and synthesis of the first (2,3')-isoflavone-isoflavan dimer, *J. Chem. Soc. Perkin. Trans.* 1, 5, 1227, 1988.
36. Ogata, T., Yahara, S., Nohara, T., Hisatsune, R., and Konishi, R., The studies on the constituents of leguminous plants. Part XVII. Isoflavan and related compounds from *Dalbergia odorifera*. II, *Chem. Pharm. Bull.*, 38, 2750, 1990.
37. Moreira, I. C., Sobrinho, D. C., DeCarvalho, M. G., and Braz-Filho, R., Isoflavanone dimers hexaspermone A, B and C from *Ouratea hexasperma, Phytochemistry*, 35, 1567, 1994.
38. Mabry, T. J., Markham, K. R., and Thomas, M. B., *The systematic identification of flavonoids*, Spring-Verlag, Berlin, 1970.
39. Murthy, M. S. R., Rao, E. V., and Ward, R. S., Carbon-13 nuclear magnetic resonance spectra of isoflavones, *Magn. Reson. Chem.*, 24, 225, 1986.
40. Lin, Y.-L. and Kuo, Y.-H., 6a, 12a-Dehydro-(-toxicarol and Derricarpin, two new isoflavonoids, from the roots of Derris oblonga BENTH., *Chem. Pharm. Bull.*, 41, 1456, 1993.
41. Fukai, T., Nishizawa, J., and Nomura, T., Variation in the chemical shift of the 5-hydroxyl proton of isoflavones; two isoflavones from *Licorice, Phytochemistry*, 36, 225, 1994.
42. Tsukamoto, C., Shimada, S., Igita, K, Kudou, S., Kokubun, M., Okubo, K., and Kitamura, K., Factors affecting isoflavone contents in soybean seeds: changes in isoflavones, saponins, and composition of fatty acids at different temperatures during seed development, *J. Agric. Food Chem.*, 43, 1184, 1995.
43. Tahara, S., Hanawa, F., Mizutani, J., and Ingham, J. L., ¹H-NMR chemical shift value of the isoflavone 5-hydroxyl proton as a convenient indicator of 6- substitution or 2'-hydroxylation, *Phytochemistry*, 30, 1683, 1991.

44. Wang, H-J. and Murphy, P. A., Isoflavone composition of American and Japanese soybeans in Iowa: effects of variety, crop year, and location, *J. Agric. Food Chem.*, 42, 1674, 1994.

45. Kitamura, K., Igita, K., Kikuchi, A., Kudou, S., and Okubo, K., Low isoflavone content in some early maturing cultivars, so-called summer-type soybeans (*Glycine max* (L) MERRILL), *Jpn. J. Breed.*, 41, 651, 1991.

46. Franke, A. A., Custer, L. J., Cerna, C. M., and Narala, K. K., Quantitation of phytoestrogens in legumes by HPLC, *J. Agric. Food Chem.*, 42, 1905, 1994.

47. Wang, H-J. and Murphy, P. A., Isoflavone content in commercial soybean foods, *J. Agric. Food Chem.*, 42, 1666, 1994.

48. Xu, X., Wang, H-J., Murphy, P. A., Cook, L., and Hendrich, S., Daidzein is a more bioavailable soymilk isoflavone than is genistein in adult women, *J. Nutr.*, 124, 825, 1994.

49. Coward, L., Barnes, S., Barnes, N. C., and Setchell, K. D. R., Genistein, daidzein, and their β-glycoside conjugates: antitumor isoflavones in soybean foods from American and Asian diets, *J. Agric. Food Chem.*, 41, 1961, 1993.

50. Adlercreutz, H., Markkanen, H., and Watanabe, S., Plasma concentrations of phyto-oestrogens in Japanese men, *Lancet*, 342, 1209, 1993.

51. Niam, M., Gestetner, B., Zilkah, S., Birk, Y., and Bondi, A., Soybean isoflavones. Characterization, and antifungal activity, *J. Agric. Food Chem.*, 22, 806, 1974.

52. Lundh, T. J.-O., Pettersson, H., and Kiessling, K.-H., Liquid chromatographic determination of the estrogens daidzein, formononetin, coumestrol and equol in bovine blood plasma and urine, *J. Assoc. Off. Anal. Chem.*, 71, 983, 1988.

53. Wang, G., Kuan, S. S., Francis, O. J., Ware, G. M., and Carman, A. S., A simplified HPLC method for the determination of phytoestrogen in soybean and its processed products, *J. Agric. Food Chem.* 38, 185, 1990.

54. Pettersson, H. and Kiessling, K. H., Liquid chromatographic determination of the plant estrogens coumestrol and isoflavones in animal feed, *J. Assoc. Off. Anal. Chem.*, 67, 503, 1984.

55. Setchell, K. D. R., Welch, M. B., and Lim, C. K., HPLC analysis of phytoesterogens in soy protein preparation with ultraviolet, electrochemical, and thermospray mass spectrometric detection, *J. Chromatogr.*, 386, 315, 1987.

56. Kitada, Y., Ueda, Y., Yamamoto, M., Ishikawa, M., Nakazawa, H., and Fujita, M., Determination of isoflavones in soybean by high-performance liquid chromatography with amperometric detection, *J. Chromatogr.*, 366, 403, 1986.

57. Shihabi, Z. K., Kute, T., Garcia, L. L., and Hinsdal, E. M., Analysis of isoflavones by capillary electrophoresis, *J. Chromatogr. A.*, 680, 181, 1994.

58. Wang, W., Tanaka, Y., Han, Z., and Cheng, J., Radioimmunoassay for quantitative analysis of formononetin in blood plasma and rumen fluid of wethers fed red clover, *J. Agric. Food Chem.*, 42, 1584, 1994.

59. Bec-Ferte, M-P., Prome, D., Savagnac, A., Prome, J-C., Krishnan, H. B., and Pueppke, S. G., Structures of nodulation factors from the nitrogen-fixing soybean symbiont *Rhizobium fredii* USDA257, *Biochemistry*, 33, 11782, 1994.

60. Redmond, J. W., Batley, M., Djordjevic, M.A., Innes, R. W. Kuempel, P. L., and Rolfe, B. G., Flavones induce expression of nodulation genes in *Rhizobium*, *Nature*, 323, 632, 1986.

61. Peters, K.N., Frost, J. W., and Long, S. R., A plant flavone, luteolin, induces expression of *Rhizobium meliloti* nodulation genes, *Nature*, 323, 977, 1986.

62. Firmin, J. L., Wilson, K. E., Rossen, L., and Johnston, A. W. B., Flavonoid activation of nodulation genes in *Rhizobium* reversed by other compounds present in plants, *Nature*, 324, 90, 1986.

63. Gyorgypal, Z. and Kondorosi, A., Homology of the ligand-binding regions of *Rhizobium* symbiotic regulatory protein NodD and vertebrate nuclear receptors, *Mol. Gen. Genet.*, 226, 337, 1991.

64. Baker, M. E., Origins of regulation of gene transcription by steroid, retinoid and thyroid hormones, in *The New Biology of Steroid Hormones*, Hochberg, R. B. and Naftolin, F., Eds., Raven Press, New York, 1991, p. 187.

65. Livingston, A. L., Forage plant estrogens, *J. Toxicol. Envir. Hlth.*, 4, 301, 1978.

66. Shutt, D. A., The effects of plant oestrogens on animal reproduction, *Endeavour*, 35, 110, 1976.

67. Whitten, P. L. and Naftolin, F., Dietary estrogens: A biologically active background for estrogen action, in *The New Biology of Steroid Hormones*, Hochberg, R. B. and Naftolin, F., Eds., Raven Press, New York, 1991, p. 155.

68. Duax, W. L. and Griffin, J. F., Structure-activity relationships of estrogenic chemicals, in *Estrogens in the Environment II. Influences on Development*, McLachlan, J. A., Ed., Elsevier, New York, 1995, p. 15.

69. Miksicek, R. J., Estrogenic flavonoids: structural requirements for biological activity, *Proc. Soc. Exp. Biol. Med.*, 208, 44, 1995.

70. Miksicek, R. J., In situ localization of the estrogen receptor in living cells with the fluorescent phytoestrogen coumestrol, *J. Histochem. Cytochem.*, 41, 801, 1993.

71. Welshons, W. V., Murphy, C. S., Koch, R., and Jordan, V. C., Stimulation of breast cancer cells in vitro by the environmental estrogen enterolactone and the phytoestrogen equol, *Breast Canc. Res.Treat*, 10, 169, 1987.

72. Mäkelä, S., Davis, V. L., Tally, W. C., Korkman, J., Salo, L., Vihko, R., Santti, R., and Korach, K.S., Dietary estrogens act through estrogen receptor mediated processes and show no antiestrogenicity in cultured breast cancer cells, *Env. Hlth. Persp.*, 102, 572, 1994.

73. Soto, A.M., Lin, T.-M., Justicia, H., Silvia, R. M., and Sonnenschien, C., An "in culture" bioassay to assess the estrogenicity of xenobiotics (E-Screen), in: *Chemically-Induced Alterations in Sexual and Functional Development: The Wildlife/Human Connection*, Colburn, T. and Clement, C., Eds., Princeton Scientific Publishing Co., Inc., Princeton, NJ, 1992, p. 295.

74. Barnes, S., Effect of genistein on *in vitro* and *in vivo* models of cancer, *J. Nutr.*, 125 Suppl. 3S, 777S, 1995.

75. Peterson, G. and Barnes, S., Genistein inhibition of the growth of human breast cancer cells: independence from estrogen receptors and the multi-drug resistance gene, *Biochem. Biophys. Res. Commun.*, 179, 661, 1991.

76. Martin, P. M., Horwitz, K. B., Ryan, D. S., and McGuire, W. L., Phytoestrogen interaction with estrogen receptors in human breast cancer cells, *Endocrinology*, 103, 1860, 1978.

77. Zava, D. T. and Duew, G., Estrogenic bioactivity of phytoestrogens in human breast cancer cells in monolayer culture, *J. Nutr.*, 125 Suppl. 3S, 807S, 1995.

78. Campbell, D. R. and Kurzer, M. S., Flavonoid inhibition of aromatase enzyme activity in human preadipocytes, *J. Steroid Biochem. Molec. Biol.*, 46, 381, 1993.

79. Wang, C., Makela, T. H., Hase, T. A., Adlercreutz, C. H. T., and Kurzer, M. S., Lignans and isoflavonoids inhibit aromatase enzyme in human preadipocytes, *J. Steroid Biochem. Molec. Biol.*, 50, 205, 1994.

80. Mäkelä, S., Poutanen, M., Lehtimäki, J., Kostian, M. L., Santti, R., and Vihko, R., Estrogen-specific 17β-hydroxysteroid oxidoreductase type 1 (E.C.1.1.1.62) as a possible target for the action of phytoestrogens, *Proc. Soc. Exp. Biol. Med.*, 208, 51, 1995.

81. Akiyama, T., Ishida, J., Nakagawa, S., Ogaward, H., Wantanabe, S., Itoh, N., Shibuya, M., and Fukami, Y., Genistein, a specific inhibitor of tyrosine-specific protein kinases, *J. Biol. Chem.*, 262, 5592, 1987.

82. Barnes, S. and Peterson, T. G., Biochemical targets of the isoflavone genistein in tumor cell lines, *Proc. Soc. Exp. Biol. Med.*, 208, 103, 1995.

83. Bannwart, C., Fotsis, T., Heikkinen, R., and Adlercreutz, H., Identification of the isoflavonic phytoestrogen daidzein in human urine, *Clin. Cheim. Acta*, 136, 165, 1984.

84. Adlercreutz, H., Fotsis, T., Bannwart, C., Wähalä, K., Brunow, G., and Hase, T., Isotope dilution gas chromatography-mass spectrometric method for the determination of lignans and isoflavonoids in human urine, including identification of genistein, *Clin. Chim. Acta*, 199, 263, 1991.

85. Axelson, M., Kirkk, D. N., Farrant, R. D., Cooley, G., Lawson, A. M., and Setchell, K. D. R., The identification of the weak oestrogen equol [7-hydroxy-3(4-hydroxyphenyl)chroman] in human urine, *Biochem. J.*, 201, 353, 1982.

86. Bannwart, C., Adlercreutz, H. , Fotsis, T., Wähalä, K., Hase, T., and Brunow, G., Identification of *O*-desmethylangolensin, a metabolite of daidzein, and of matairesinol, one likely plant precursor of the animal lignan enterolactone, in human urine, *Finn. Chem. Lett.*, 4-5, 120, 1984.

87. Setchell, K. D. R. and Adlercreutz, H., Mammalian lignans and phyto-estrogens: recent studies on their formation, metabolism and biological role in health and disease, in *Role of the Gut Flora in Toxicity and Cancer*, Rowland, I. R., Ed., Academic Press, London, 1988, p. 315.

88. Kelly, G. E., Nelson, C., Waring, M. A., Joannou, G. E., and Reeder, A. Y., Metabolites of dietary (soya) isoflavones in human urine. *Clin. Chim. Acta*, 223, 9, 1993.

89. Hutchins, A. M., Slavin, J. L., Lampe, J. W., Urinary isoflavonoid phytoestrogen and lignan excretion after consumption of fermented and unfermented soy products. *J. Am. Diet. Assoc.*, 95, 545, 1995.

90. Cox, R. I. and Davies, L. H., Modification of pasture oestrogens in the gastrointestinal tract of ruminants, *Proc. Nutr. Soc. Austr.* 13, 61, 1988.

91. Lundh, T., Metabolism of estrogenic isoflavones in domestic animals, *Proc. Soc. Exp. Biol. Med.*, 208,33, 1995.

92. Adlercreutz, H., Fotsis, T., Lampe, J., ., Wähälä, K., Mäkelä, T., Brunow, G., and Hase, T., Quantitative determination of lignans and isoflavonoids in plasma of omnivorous and vegetarian women by isotope dilution gas-chromatography mass-spectrometry, *Scand. J. Clin. Lab. Invest.*, 53 (Supp 215), 5, 1993.

93. Thompson, M. A., Lasley, B. L., Rideout, B. A., and Kasman, L. H., Characterization of the estrogenic properties of a nonsteroidal estrogen, equol, extracted from urine of pregnant macaques, *Biol. Reprod.*, 31, 705, 1984.

94. Medlock, K. L., Branham, W. S., and Sheehan, D. M., Effects of coumestrol and equol in the developing reproductive tract of the rat, *Proc. Soc. Exp. Biol. Med.*, 208, 67, 1995.

95. Stob, M., Naturally occurring food toxicants: oestrogens, in *Handbook of Naturally Occurring Food Toxicants*, Rechcigel, M., Jr., Ed., CRC Press, Boca Raton, FL., 1983, p. 81.

96. Carter, M. W., Matrone, G., and Smart, W. W. G., Jr., Effect of genistein on reproduction of the mouse, *J. Nutr.*, 55, 639, 1955.

97. Thain, R. I., Residual herd infertility in cattle, *Aust. Vet. J.*, 44, 218, 1968.

98. Chang, K., Kurtz, H. J., and Mirocha, C. J., The effect of the mycoestrogen zearalenone on swine reproduction, *Am. J. Vet. Res.*, 40, 1260, 1979.

99. Leopold, A. S., Erwin, M., and Browning, B., Phytoestrogens: Adverse effects on reproduction in California quail, *Science*, 191, 98, 1976.

100. Jagusch, K. T., Smith, J. F., and Kelly, R. W., Effect of feeding lucerne during mating on the fertility of ewes, *Proc. Nutr. Soc. N. Z.*, 2, 161, 1977.

101. Scales, G. H. and Moss, R. A., Mating ewes on lucerne, *N. Z. J. Agric.*, 132, 21, 1976.

102. Smith, J. F., Jagusch, K. T., Brunswick, L. F. L., and Kelly, R. W., Coumestans in lucerne and ovulation in ewes, *N. Z. J. Agric. Res.*, 22, 411, 1979.

103. Leavitt, W. W. and Wright, P. A., The plant estrogen, coumestrol, as an agent affecting hypophyseal gonadotropic function, *J. Exp. Zool.* 160, 319, 1965.

104. Obst, J. M. and Seamark, R. F., Plasma progesterone concentrations during the reproductive cycle of ewes grazing Yarloop clover, *J. Reprod. Fert.*, 21, 545, 1970.

105. Newsome, F. E. and Kitts, W. D., Effect of alfalfa consumption on estrogen levels in ewes, *Can. J. Anim. Sci.*, 57, 531, 1977.

106. Cassidy, A., Bingham, S., and Setchell, K. D. R., Biological effects of a diet of soy protein rich in isoflavones on the menstrual cycle of premenopausal women, *Am. J. Clin. Nutr.*, 60, 333, 1994.

107. Bornstein, S. and Adler, J. H., The oestrogenic effect of alfalfa meal on the growing chicken, *Refu. Vet.*, 20, 175, 1963.

108. Magee, A. C., Biological responses of young rats fed diets containing genistin and genistein, *J. Nutr.*, 80, 151,1963.

109. Matrone, G., Smart, W. G., Jr., Carter, M. W., and Smart, V. W., Effect of genistin on growth and development of the male mouse, *J. Nutr.*, 59, 235, 1955.

110. Chury, J., Uber einen die samenblaschen hemmenden Stoff aus Luzerne, *Experientia* 23, 285, 1967.

111. Adams, N. R., Pathological changes in the tissues of infertile ewes grazing on oestrogenic subterranean clover, *Res. Vet. Sci.* 22, 216, 1978.

112. Livingston, A. L., Forage plant estrogens, *J. Toxic. Envir. Hlth.* 3, 301, 1978.

113. Adams, N. R., Altered response of cervical and vaginal epithelia to oestradiol benzoate in ewes after prolonged exposure to oestrogenic pasture, *J. Reprod. Fert.* 53, 203, 1979.

114. Adams, N. R., Pathological changes in the tissues of infertile ewes with clover disease, *J. Comp. Pathol.*, 86, 29, 1976.

115. Adams, N. R., Morphological changes in the organs of ewes grazing on oestrogenic subterranean clover, *Res. Vet. Sci.*, 22, 216, 1977.

116. Adler, J. H. and Trainin, D., A hyperoestrogenic syndrome in cattle, *Refu. Vet.*, 17, 108, 1960.

117. Adams, N. R., Hearnshaw, H., and Oldham, C. M., Abnormal function of the corpus luteum in some ewes with phyto-oestrogenic infertility, *Aust. J. Biol. Sci.*, 34, 61, 1980.

118. Findlay, J. K., Buckmaster, J. M., Chamley, W. A., Cumming, I. A., Hearnshaw, H., and Goding, J. R., Release of luteinizing hormone by oestradiol-17β and a gonadotropin-releasing hormone in ewes affected with clover disease, *Neuroendocrinology* 11, 57, 1973.

119. Burroughs, C. D., Mills, K. T., and Bern, H. A., Long-term genital tract changes in female mice treated neonatally with coumestrol, *Reprod. Toxicol.* 4, 127, 1990.

120. Whitten, P. L., Lewis, C., and Naftolin, F., A phytoestrogen diet induces the premature anovulatory syndrome in lactationally exposed female rats, *Biol. Reprod.*, 49, 1117, 1993.

121. Leavitt, W. W. and Meismer, D. M., Sexual development altered by nonsteroidal oestrogens, *Nature*, 218, 181, 1968.

122. Register, B., Bethel, M. A., Thompson, N., Walmer, D., Blohm, P., Ayyash, L., and Hughes, C., Jr., The effect of neonatal exposure to diethylstilbestrol, coumestrol, and β-sitosterol on pituitary responsiveness and sexually dimorphic nucleus volume in the castrated adult rat, *Proc. Soc. Exp. Biol. Med.*, 208, 72, 1995.

123. Faber, K. A. and Hughes, C. L., Jr., The effect of neonatal exposure to diethylstilbestrol, genistein, and zearalenone on pituitary responsiveness to gonadotropin releasing hormone and volume of the sexually dimorphic nucleus of the preoptic area (SDN-POA) in postpubertal castrated female rats, *Reprod. Toxicol.*, 7, 35, 1991.

124. Faber, K. A. and Hughes, C. L., Jr., Dose-response characteristics of neonatal exposure to genistein on pituitary responsiveness to gonadotropin releasing hormone and volume of the sexually dimorphic nucleus of the preoptic area (SDN-POA), *Reprod. Toxicol.*, 7, 35, 1993.

125. Medlock, K. L., Branham, W. S., and Sheehan, D. M., Effects of coumestrol and equol on the developing reproductive tract of the rats, *Proc. Soc. Exp. Biol. Med.*, 208, 67, 1995.

126. Sheehan, D. M., Branham, W. S., Medlock, K. L., and Shamugasundaram, E. R. B., Estrogenic activity of zearalenone and zearalanol in the neonatal rat uterus, *Teratology*, 29, 383, 1984.

127. Lamartiniere, C., Neonatal genistein chemoprevents mammary cancer, *Proc. Soc. Exp. Biol. Med.*, 208, 120, 1995.

128. Whitten, P. L., Lewis, C., Russell, E., and Naftolin, F., Phytoestrogen influences on the development of behavior and gonadotropin function, *Proc. Soc. Exp. Biol. Med.*, 208, 82, 1995.

129. Whitten, P. L., Lewis, C., Russell, E., and Naftolin, F., Potential adverse effects of phytoestrogens, *J. Nutr.*, 125 Suppl 3S, 771S, 1995.

130. Shutt, D. A. and Cox, R. I., Steroid and phyto-oestrogen binding to sheep uterine receptors in vitro, *J. Endocrinol.*, 52, 299, 1972.

131. Shemesh, M., Lindner, H. R., and Ayalon, N., Affinity of rabbit uterine oestradiol receptor for phyto-oestrogens and its use in a competitive protein-binding radioassay for plasma coumestrol, *J. Reprod. Fertil.*, 29, 1, 1972.

132. Lee, Y. J., Notides, C. A., Tsay, UY., and Kende, A. A., coumestrol, NBD-norhexestrol, and dansyl-norhexestrol, fluorescent probes of estrogen-binding protein, *Biochemistry*, 16, 2896, 1977.

133. Martin, P. M., Horwitz, K. B., Ryan, D. S., and McGuire, W. L., Phytoestrogen interaction with estrogen receptors in human breast cancer cells, *Endocrinology*, 103, 1860, 1978.

134. Verdeal, K., Brown, R. R., Richardson, T., and Ryan, D. S., Affinity of phytoestrogens for estradiol-binding proteins and effect of coumestrol and growth of 7,12-dimethylbenz[a]anthracene-induced rat mammary tumors, *J. Nat. Canc. Inst.*, 64, 285, 1980.

135. Mayr, U., Butsch, A., and Schnieder, S., Validation of two in vitro test systems for estrogenic activities with zearalenone, phytoestrogens and cereal extracts, *Toxicology*, 74, 135, 1992.

136. Miksicek, R. J., Interaction of naturally occurring nonsteroidal estrogens with expressed recombinant human estrogen receptor, *J. Steroid Biochem. Molec. Biol.*, 49, 153, 1994.

137. Bickoff, E. M., Livingston, A. L., Hendrickson, A. P., and Booth, A. N., Relative potencies of several estrogen-like compounds found in forages, *Agric. Fd. Chem.*, 10, 410, 1962.

138. Tang, B. Y. and Adams, N. R., Effect of equol on oestrogen receptors and on synthesis of DNA and protein in the immature rat uterus, *J. Endocrinol.*, 85, 291, 1980.

10 Pyrimidine Glycosides

R. R. Marquardt, N. Wang, and M. S. Arbid

CONTENTS

10.1 INTRODUCTION

Pyrimidine glycosides of greatest concern are vicine and convicine. They are found primarily in faba beans (*Vicia faba* L) which are one of the most important pulse crops in the world, being consumed in large quantities in the Middle East, Far East, and North Africa, particularly Egypt. Except for the presence of vicine and convicine and relatively low concentrations of several other antimetabolites, faba beans are an excellent source of protein with the amino acid balance complementing that of cereals.[1]

Vicine and convicine are compounds which are hydrolyzed by intestinal microflora[2] to highly reactive free-radical generating compounds,[3,4] divicine and isouramil. Divicine and isouramil have been strongly implicated as the causative agents in favism,[5–17] a hemolytic disease in humans, particularly young males, that have a deficiency of erythrocytic glucose-6-phosphate dehydrogenase (G-6-PD) activity.[18] These free-radical generators may also cause other adverse effects including lipid peroxidation,[19] altered fat[20] and mitochondrial metabolism,[21] and possibly diabetes.[22] Some of these adverse effects may be neutralized by increasing the dietary concentration of free-radical scavenging compounds such as vitamins A, C, and E,[20,23,24] and through the use of chelating agents such as EDTA or deferoxamine.[25] The adverse effects of these compounds can also be reduced or eliminated by developing glycoside-free cultivars of faba beans [26] or by using processing techniques that selectively extract [27,28] or hydrolyze these compounds,[29,30] or by microbiological degradation.[31]

Vicine and convicine also appear to have certain beneficial properties including the prevention of cardiac arrhythmia[32] and under certain conditions are able to inhibit the growth of malaria parasite, *Plasmodium falciparum*.[33–37] The role of vicine and convicine in the seed of *V. faba* has not been established, but it may provide a storage source of nitrogen since it has a high content of nitrogen, or it may protect the seed against certain microorganisms, particularly fungi.[38] Other reviews on vicine and convicine have been published.[1,9]

10.2   PROPERTIES OF VICINE, CONVICINE, DIVICINE, AND ISOURAMIL

10.2.1   STRUCTURES AND PROPERTIES

Vicine [2,6-diamino-5-(ß-D-glucopyranosyloxy)-4(1H)-pyrimidinone; 2,4-diamino-6-oxypyrimidine-5-(ß-D-glucopyranoside); 2,6-diamino-4,5-dihydroxypyrimidine 5-(ß-D-glucopyranoside); divicine 5-glucoside; vicioside; divicine-ß-glucoside] has an empirical formula of $C_{10}H_{16}N_4O_7$, a molecular weight of 304.26, and has a structural formula as shown in Figure 10.1.[39] Vicine decomposes at 243 to 244°C, has an optical rotation $[\alpha]_D^{26}$ of −11.7° at a concentration of 39 g l^{-1} in 0.2N NaOH,[39] is white in color and has a needle-like structure.[40] Vicine can be stored in a neutral or basic (1N NaOH) solution at either 2 or 30°C for up to 7 days without decomposition but is unstable when stored at a low pH for a similar period of time. It does not, however, decompose to a significant degree in acid when kept at a low temperature for relatively short time periods.[40] It is rapidly hydrolyzed to its aglycone, divicine, at high temperature in the presence of concentrated acids.[5,39] Vicine is minimally soluble in water at pH levels between 4 and 9 with the values being 3.0 and 3.6 mg ml^{-1}, respectively. It is, however, highly soluble at pH values of greater than 10.5 and less than 1.0. The solubility of vicine in acetone, absolute ethanol, and methanol at room temperature is approximately 0, 0.4, and 10 mg ml^{-1}, respectively.[1] The ultraviolet absorption spectrum of vicine is influenced by pH[39] with the maximum molar extinction coefficients (ϵ) at different pH values being shown in Table 10.1.[39]

The structure of convicine [6-amino-5-(ß-D glucopyranosyloxy)-2,4(1H,3H)-pyrimidinedion; 2,4,5-trihydroxy-6-aminopyrimidine 5-(ß-D glucopyranoside)] was established by Bien et al. (Figure 10.1).[41] It has an empirical formula of $C_{10}H_{15}N_3O_8$ with a molecular weight of 305.24. The plate-like crystals are yellow in color and decompose without melting at 287°C.[40–41] Convicine is relatively stable in aqueous solutions at a neutral or basic pH but is hydrolyzed at a low pH. In general, convicine is more readily hydrolyzed than vicine. Convicine has a relatively low solubility at low pH levels in contrast to vicine which is highly soluble in aqueous solutions at low pH values. The solubility of convicine, which is high at high pH values, is influenced by temperature in a manner similar to that of vicine. The ultraviolet absorption spectrum is also affected by pH in a manner similar to that of vicine (Table 10.1).[41]

Divicine [2,6-diamino-5-hydroxy-4(3H)-pyrimidinone; 2,6-diamino-4,5-dihydroxypyrimidine; 2,6-diamino-1,6-dihydro-4,5-pyrimidinedione; 2,6-diamino-4,5-pyrimidinediol; 2,4-diamino-5,6-dihydroxypyrimidine] is the aglycone of the glycoside vicine and has an empirical formula of $C_4H_6N_4O_2$ with a molecular weight of 142.12. The pale yellow[42] needles decompose at 300°C, and 1g dissolves in 100 ml boiling water and in about 350 ml cold water. It is soluble in 10% KOH and has a structural formula as shown in Figure 10.1.[39] The molar extinction coefficient at two pH levels is presented in Table 10.1.[11,55]

TABLE 10.1
Ultraviolet Absorption Spectra of Vicine, Convicine, and their Aglycones

Compound	Oxidation State	pH	Absorption Maxima	Extinction Coefficient (e) (M^{-1} cm^{-1})	Ref.
Vicine		1	274	16,400	39
		6.8	275; 236	13,200; 4400	39
		13	269; 235	9500; 5300	39
Convicine		1	271; 220	17,400; 6000	41
		7	271	14,450	41
		13	273	14,450	41
Divicine	Reduced	1	282	12,500	55
	Reduced	7	285	9800	11
	Oxidized		245		11
Isouramil	Reduced	1	281	13,000	58
	Reduced	7	280	14,200	11
	Oxidized	7	255		11

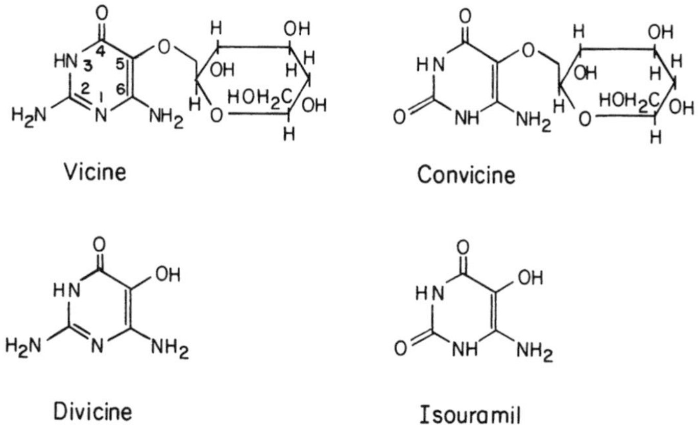

FIGURE 10.1 Structures of vicine, convicine, and their aglycones.

Isouramil [6-amino-5-hydroxy-2,4(1H,3H) pyrimidinedione; 2,4,5-trihydroxy-6-amino-pyrimidine; 6-amino-2,4,5-(1H,3H,6H) pyrinidinetrione] is the aglycone of the glycoside convicine and has an empirical formula of $C_4H_5N_3O$ with a molecular weight of 143.10 (Figure 10.1).[39,41] Pale yellow isouramil is insoluble in hot water, alcohol, or acetic acid. It dissolves in sodium hydroxide, ammonia, and dilute hydrochloric acid. Its molar extinction coefficient at two pH levels is presented in Table 10.1.[11,58]

10.2.2 CHEMICAL REACTIONS AND FREE RADICAL INTERMEDIATES

Divicine and isouramil, as well as the corresponding glycosides, react with Folin-Ciocalteu phenol reagent. Both divicine and isouramil vigorously reduce alkaline solutions of 2,6-dichlorophenolindophenol, phosphomolybdate, or phosphotungstate and produce an intense blue color reaction with an ammoniacal ferric chloride solution which is indicative of the presence of an enolic hydroxyl group.[9,41,43]

Divicine and isouramil in the absence of oxygen (i.e., under nitrogen)[11] or in the presence of reducing reagents[7,11,44] are stable at either a low or a neutral pH and exhibit single absorption peaks with absorbance maxima and extinction coefficients (Table 10.1). Exposure of an aqueous solution of the two compounds to oxygen in the absence of a reducing reagent results in a rapid decline in their absorbency peaks with the concomitant appearance of new absorption bands at 255nm (isouramil)[11] and 245nm (divicine).[11,39] This hypochromatic effect, which is readily reversed by the addition of reducing reagents,[7,11,44] can be attributable to formation of the oxidized species of the respective pyrimidine molecules.[7,11] The compounds are relatively stable at a low pH ($t_{1/2} > 20$h) and reach near maximal decomposition rates at pH 7.3 for divicine and 7.75 for isouramil.[10] The oxidative decomposition of the two aglycones is also markedly enhanced by the presence of traces of transition metals (Cu^{2+} and Fe^{3+})[39,45] and are protected by chelating agents such as EDTA. Prolonged exposure to oxygen in the absence of a reducing reagent results in total obliteration of the characteristic peaks. This latter change is no longer reversible by reduction and is suggestive of the rupture of the pyrimidine ring structure.[11] These spectral transitions are closely similar to that obtained when dialuric acid is oxidized to alloxan.[11,39,46] The spectral changes that occur when isouramil or divicine reacts with oxygen are summarized in Figure 10.2. Chevion et al.[11] demonstrated that the oxidation of GSH to GSSG by an oxygen-saturated solution was dramatically accelerate by the presence of isouramil or divicine in neutral pH at room temperature. Oxidized divicine and isouramil are converted nonenzymatically to their reduced forms (Figure 10.2) by 2 mol of GSH, 1 mol of NAD(P)H + H[+], or other reducing compounds with the concomitant formation of GSSG, NAD(P), or oxidized reductant. The oxidized forms of the pyrimidines are not only reduced by different reducing reagents, but also interact directly and nonenzymatically with them. GSH, for example, forms GSH-pyrimidine adducts that have absorption maxima at 305nm[7,11,46] and at 320nm.[44] Similar adducts are formed between alloxan and GSH but not between its reduced form (dialuric acid) and GSH.[46]

Bendich and Clements[39] pointed out in their comprehensive study that some of the distinctive features of divicine and its congeners — that is, their powerful reducing activity, spectral characteristics, and molecular instability, show a striking resemblance to those of ascorbic acid and reductic acid (2,3 dihydroxy-2-cyclopenten-1-one). These authors concluded that a common structural denomination underlying these properties is a carbonyl-conjugated enediol or aminoenol system. Compounds with these structural arrangements serve to characterize a broad class of compounds designated by the general name reductones.[1] Consequently, all the characteristic properties of the aglycone are abolished by substitution of the hydroxyl group at C-5, such as that represented by the glucosidic linkage present in vicine and convicine. Thus, unlike the free aglycones, the glycosides show no reducing ability, do not undergo oxidation in the presence of oxygen, resists destruction by boiling in aqueous solution, and their ultraviolet spectra differ significantly from those of their constituent pyrimidines.[11,39,41]

The production of free radicals by divicine as predicted[47] was confirmed using electron spin resonance (ESR) spectroscopy.[4] The intensity of the ESR signal was strongly dependent on pH, being maximal at pH 9.0 and low at pH 5.0. The signal was completely suppressed upon the addition of 1mM GSH. No ESR signal was generated by vicine or by hydrolyzed vicine in the complete absence of oxygen, indicating the release of aglycone and its reaction with oxygen are essential for radical formation. A summary of the one-electron oxidation-reduction of divicine or isouramil[4,16] is presented in Figure 10.3. The reaction of these compounds with oxygen is similar to the analogous formation of the semiquinone free radical and the quinone, alloxan, from dialuric acid.[46]

10.3   QUANTITATION, ISOLATION, AND SYNTHESIS OF VICINE, CONVICINE, AND THEIR AGLYCONES

Several spectrophotometric and chromatographic methods have been utilized for the quantitation of vicine and convicine in plant material and animal tissues, including thin-layer

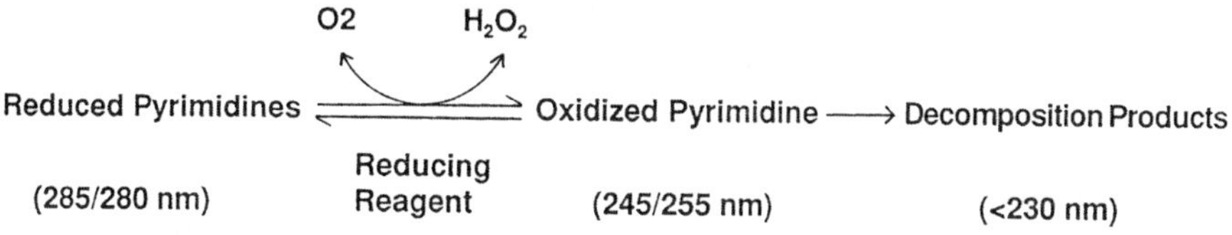

FIGURE 10.2 Interconversion of divicine and isouramil in the presence of oxidizing and reducing reagents. The numbers in parentheses refer to the wavelength absorption maxima of divicine/ isouramil.[11]

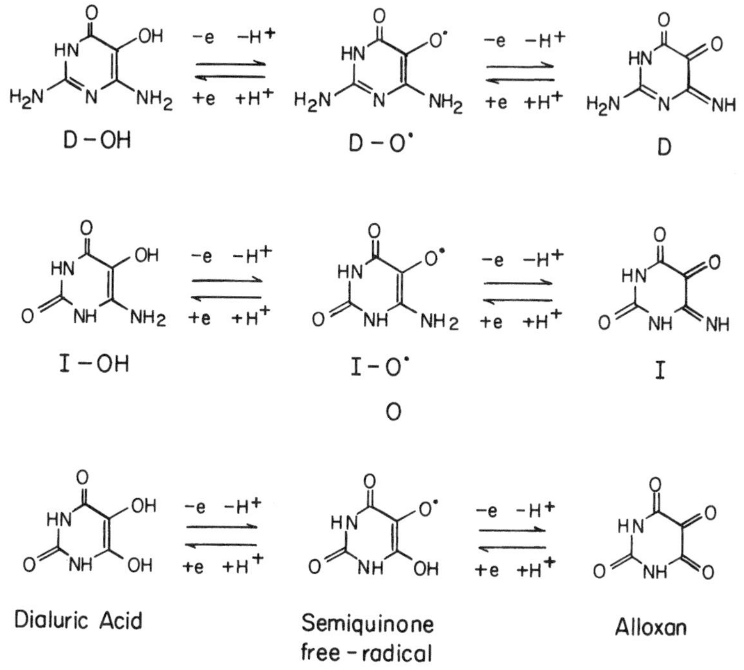

FIGURE 10.3 Stepwise one-electron oxidation of divicine, isouramil and dialuric acid. D-OH (hydroquinone, reduced divicine); D-O˙ (semiquinone free radical): D (quinone, oxidized form); I, isouramil. Dialuric acid and alloxan undergo a similar type oxidation-reduction.

chromatography (TLC), gas chromatography (GC), and HPLC.[1] HPLC procedures are superior to previously published methods since they are highly sensitive, rapid, and accurate, and are able to clearly resolve vicine and convicine with the procedure of Marquardt and Frohlich[48] appearing to be one of the simplest. Divicine and isouramil can also be resolved and quantitated using cation exchange or reverse-phase HPLC,[48] mass spectrometry, and NMR spectroscopy.[49] Special precautions must be used to exclude traces of oxygen to prevent rapid degradation of the latter compounds.

Vicine was discovered and first isolated by Ritthausen and Kreusler[50] from vetch seeds (*Vicia sativa*) with several modifications being published.[1] Convicine was also originally isolated from vetch seeds by Ritthausen[51] using procedures similar to those for vicine. Marquardt et al.[40] have published a simple procedure for the simultaneous isolation of pure vicine and convicine from an air-classified fraction of faba beans in aqueous solution according to their solubility at different pH values.

The glycoside nature of vicine was recognized by Ritthausen[52] who succeeded in isolating the aglycone, divicine from this compound.[53] Divicine can be prepared from vicine by either acid[5,54] or enzymatic hydrolysis,[6] or it can be synthesized chemically.[42,55,56] Isouramil can also be prepared by acid hydrolysis[57,58] or chemical synthesis.[43,56]

Hemoglobinuria appears within 5 to 30 h following exposure to the bean, and jaundice is observed a few hours later. Fever and marked leukocytosis may also occur. A typical attack lasts from 2 to 6 days. A more detailed review of favism and its relationship to G6PD deficiency have been presented.[9] As discussed elsewhere in this paper, divicine and isouramil have been strongly implicated as the causative agents of favism with a deficiency of G6PD being a predisposing factor.

Favism has been observed in many different population groups and has a seasonal incidence that coincides with the harvesting of the bean.[69] The incidence of favism varies considerably and depends on the distribution of the genetic defect, the presence of faba beans in the local diet, and the availability and utilization of medical facilities.[69] Favism has been associated with the Mediterranean type G6PD deficiency with wide distribution all over the Mediterranean, near East, and even Asia, including China.[18,70] The deficiency is an X-linked trait and it fully expressed in hemizygote male.[18,70] Patients with a history of favism have a diminished concentration of GSH and stability of red cells.[6,9,71] Carson et al.[71] revealed that the GSH instability of susceptible erythrocytes was attributable to a deficiency of NADP-linked G6PD. As a result, these cells are unable to maintain an adequate supply of NADPH required for the continuous reduction of GSSG by GSH reductase. It became apparent that all patients with favism were G6PD deficient, yet many patients with G6PD deficiency were able to ingest faba beans without experiencing hemolytic episodes or being subject to the disease.[1,9,69,70] Some factor(s) in addition to G6PD deficiency must, therefore, be required for the development of favism. One possibility, as discussed subsequently, is that the nutritional status, particularly that of vitamin E, may influence the development of the disease.

The aglycones of vicine and convicine also seem to produce other, perhaps subclinical, effects in humans, such as hemolytic crisis associated cataracts,[72] optical atrophy,[73] glucose metabolism disorder, and hyperlipidemia.[74] Divicine and isouramil which are free-radical producers may be expected to induce metabolic disturbances similar to those produced by other free radical generators.[3,4]

10.8 METABOLIC EFFECTS OF DIVICINE AND ISOURAMIL

10.8.1 INTERACTIONS OF DIVICINE AND ISOURAMIL

Arses et al.[47] demonstrated *in vitro* that there was a parallel and rapid induction of the synthesis of both GSH and NADPH in human normal RBCs in the presence of divicine or isouramil. There is a much slower rate of NADPH regeneration in the G6PD-deficient as compared to the normal RBC.[18] Similar results were obtained with GSH *in vivo* when rats were injected with divicine.[17] Although NADH is also subjected to oxidation by divicine/isouramil, it does not decrease dramatically in either the normal or the G6PD-deficient RBC.

As indicated previously, the reduced forms of divicine and isouramil are rapidly and nonenzymatically converted in the presence of oxygen of their oxidized form (Figures 10.2 and 10.3) with the concomitant formation of H_2O_2. The oxidized forms of these compounds, however, are able to react with reducing reagents such GSH,[7,11,44] NAD(P)H,[44] and ascorbate,[7] which results in the regeneration of the reduced aglycones. This shuttle system results in the net transfer of electrons from various reducing compounds to oxygen to form hydrogen peroxide with divicine and isouramil being the cofactors.[7,11] In the presence of glutathione peroxidase as would occur *in vivo*, H_2O_2 reacts with two molecules of GSH and is converted to H_2O and O_2. The shuttle mechanism as proposed by Chevion et al.[11] and modified to be consistent with the results of Albano et al.,[4] Benatti et al.,[44] and the review by Hebbel[75] is presented in Figure 10.4. In the overall reaction, four molecules of GSH are oxidized during a single oxidation-reduction cycle of isouramil or divicine. Glutathione reductase regenerates GSH from GSSG and NADPH + H⁺. *In vivo* this futile cycle would continue to operate until the aglycones either spontaneously decomposed or formed irreversible adducts with other

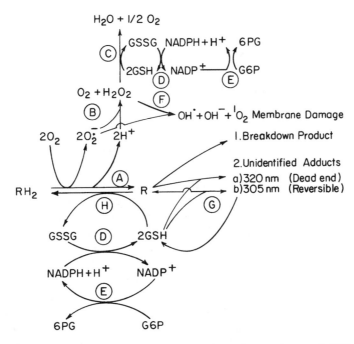

FIGURE 10.4 Oxidation-reduction shuttle mechanism for divicine or isouramil (RH_2, reduced pyrimidine; R, oxidized pyrimidine) and other side reactions. A, nonenzymatic autoxidation of RH_2; B, SOD (superoxide dismutase); C, GSH-peroxidase; D, GSH-reductase; E, G6PD (glucose-6-phosphate dehydrogenase); F, nonenzymatic formation of hydroxy and other radicals in the presence of metal ions; G, limited spontaneous formation of 320-nm dead-end product and 305-nm reversible product; and H, nonenzymatic reduction of oxidized pyrimidine (R); G6P, glucose-6-phosphate; 6PG, 6-phosphoglu-conate; O_2^-, superoxide radical; $OH^{\cdot}$, hydroxyl radical; OH^-, hydroxyl anion, $^1O_2$, singlet oxygen.

components in the cell, or until all of the reducing equivalents were exhausted, after which irreversible cell damage would occur. Of the two adducts, the 305-nm divicine (isouramil)-GSH adduct is a good substrate for glutathione reductase, whereas the 320-nm adduct does not interact with the enzyme, thereby possibly representing a dead-end species of divicine and isouramil.[44] Based on those observations, the 305-nm adduct may be considered to be involved in the red-ox cycling of the pyrimidines under *in vivo* conditions, while the 320-nm complex is virtually removed from this cycle and may act as an irreversible trap for divicine and isouramil. H_2O_2 that is not rapidly removed by the action of GSH reductase is converted in the presence of the superoxide radical (O_2^-) and the appropriate cofactors, such as the transition metals, into very reactive species including the hydroxyl radical ($OH^{\cdot}$) and singlet oxygen ($^1O_2$). These products can cause irreversible cell damage.

10.8.2   ROLE OF SUPEROXIDE, HYDROGEN PEROXIDE, AND TRANSITION METAL IONS

Winterbourn et al.,[16] demonstrated that autoxidation of divicine can occur by at least three different pathways. The normally predominate mechanism is O_2^--dependent and appears to be analogous to the mechanism of adrenaline autoxidation which has the following reaction sequence where RH_2, $RH^{\cdot}$ and R represent the hydroquinone, semiquinone and quinone forms of divicine, respectively (Figure 10.4).

$$RH_2 + O_2 \rightarrow RH^{\cdot} + O_2^- + H^+ \text{ (slow)} \qquad (10.1)$$

$$RH_2 + O_2^- + H^+ \rightarrow RH^{\cdot} + H_2O_2 \text{ (fast)} \qquad (10.2)$$

$$RH^{\cdot} + O_2 \rightarrow R + O_2^- + H^+ \text{ (fast)} \qquad (10.3)$$

After the initial step, steps 10.2 and 10.3 would constitute a chain and would be responsible for most of the reaction. In the normal biological system, superoxide dismutase (SOD) would rapidly dismutate all O_2^-, and the H_2O_2 that is formed would be rapidly converted in the presence of glutathione reductase to H_2O and O_2. A second pathway, which is prevented by catalase or metal chelators, depends on transition metal ions and H_2O_2. It may be exogenous or produced from the O_2^- formed by reaction 10.1. Normally this pathway is much slower than reactions 10.2 and 10.3 and is of minor significance. However, it predominates if SOD is present or if the concentrations of transition metal ions, e.g., Cu^{2+} and Fe^{2+}, are micromolar or more. The rate of autoxidation of divicine or isouramil are accelerated by hemoglobin and H_2O_2 due to the formation of peroxide complexes such as ferrylhemoglobin. When both of the above pathways are prevented by SOD, chelating agents, catalase or glutathione peroxidase, divicine autoxidation would still proceed slowly, presumably via reaction 10.1, up to a point where another mechanism takes over, which then results in rapid autoxidation. Winterbourn et al.[16] concluded that in the stressed erythrocyte, such as the G6PD-deficient erythrocyte, the impaired function of the GSH-peroxidase pathway due to a shortage of reducing equivalents would result in the accumulation of hydroxy peroxide by the autoxidation of divicine and isouramil. The H_2O_2 would react with mainly oxyhemoglobin ($HbFe^{2+}O_2$ or $HbFe^{3+}O_2^-$), but also methemoglobin ($HbFe^{3+}$) to form the predominantly ferryl species [$(Hb^{IV} OH^-)^{3+}$, a hemoglobin-H_2O_2 complex]. Methemoglobin is also produced by divicine. Ferrylhemoglobin and the even more-reactive intermediate formed during the reaction of H_2O_2 with methemoglobin are also capable of oxidizing many electron donors. These oxidized hemoglobin derivatives are, therefore, likely to be major contributors to cell damage in the oxidant-stressed erythrocyte. The particular oxidation-reduction pathway that occurs partially depends on the relative activities of SOD, GSH peroxidase, and catalase which as discussed above catalyze reactions that are relevant to the steady-state concentration of potentially toxic oxygen derivatives such as O_2^- and H_2O_2. These enzymes, in concert, keep the concentration of all active oxygen derivatives, including H_2O_2 and hydroxy radical ($OH^{\cdot}$) which arises from the secondary interactions of both O_2^- and H_2O_2, at a minimal value. In favic individuals the activity of G6PD is reduced and therefore, the ability of GSH reductase and GSH peroxidase to operate at maximal value is greatly impaired,[9,11,44,76] and undesirable secondary reactions occur in the oxidant-stressed cell.

Mavelli et al.[77] reported that RBCs of patients with an acute hemolytic anemia, in addition to being deficient in G6PD, had markedly higher SOD activity and less GSH-peroxidase activity than either normal controls or G6PD-deficient patients without hemolytic symptom. The rise of SOD and the drop of GSH peroxidase may assist in perpetuating a consistently increased H_2O_2 flux, thereby making the RBCs of favic subjects more susceptible to peroxidative injuries associated with the O_2/H_2O_2 sequence of reactions. The effect of divicine on the activities of the two enzymes coupled with the deficiency of GSH in G6PD-deficiency seem to act synergistically to precipitate a favic crisis and the associate peroxidative reactions. The concerted action of GSH and SOD in preventing redox cycling of divicine and isouramil has been reviewed by Winterbourn and Munday.[78]

Divicine and isouramil, instead of causing peroxidative reactions, can directly reduce methemoglobin to ferrous hemoglobin, which is the opposite of the above indicated mechanism. This particularly occurs in cells stabilized against autoxidation, i.e., red cells incubated in a nitrogen atmosphere.[16]

10.8.3 HEINZ BODY FORMATION, PROTEIN AGGREGATION, LIPID PEROXIDATION, AND RHEOLOGICAL EFFECTS IN RBCS

Oxidative stress in the red cell as induced by divicine and isouramil can cause changes in the cell membrane as well as hemoglobin. Methemoglobin which is formed from hemoglobin can be oxidized to reversible and irreversible hemichromes, where hemichromes represent oxidation states of altered iron-porphyrin complexes. Irreversible hemichromes are unstable and rapidly undergo denaturation and precipitation in the membrane as Heinz bodies.[9,13,75,79] Protein aggregates have also been shown to form in cells from patients with chronic hemolysis and in cells that have been subjected to oxidants stress[80] including divicine. Cross-linking of protein which occurs in GSH-depleted cells greatly reduces cell deformity.

Lipid peroxidation has been implicated in free radical reactions and membrane alterations associated with aging of cells and tissues.[79] Hochstein and Jain [81] suggested that increases in the rates of intrinsic membrane rigidity is accelerated in RBCs exposed to radical-generating hemolytic agents and that this effect may be attributed to the *in vivo* polymerization of membrane proteins consequent to radical-induced peroxidation of membrane lipids. Clear evidence on the role of lipid peroxidation during a favic crisis or as elicited by divicine or isouramil have not, however, been demonstrated.[14] Several researchers have reported that vitamin E was highly effective at protecting animals against the toxic effects of divicine.[1,20] Further studies are required to clarify the role of vitamin E and lipid peroxidation in favism.

Divicine and isouramil, as indicated in the previous section, cause rheological effects.[10,14,17,66,82] Studies with ^{51}Cr have demonstrated that the half-life of divine-treated rabbit erythrocytes was markedly reduced.[12] The kinetic study showed that in all treated cells there were two populations with different half-lives. The relative weight of the fraction of cells characterized by the short life span was dependent on the concentration of isouramil in the incubation system.

10.8.4 EFFECTS ON CALCIUM HOMEOSTASIS

Calcium is an important regulating agent in cell metabolism. Perturbation of this control system by oxidizing substances presumably could alter the cytoplasmic concentration of calcium and thereby result in a disturbance of cellular Ca^{2+} homeostasis, cell damage, and eventually cell death.[83] It has been shown that calcium homeostasis in the erythrocyte is severely impaired *in vitro* following divicine treatment [76,84] and *in vivo* during a hemolytic crisis. Several oxidant compounds, including divicine,[21] hydroperoxides,[85] and alloxan[86] can also induce Ca^{2+} release from rat liver mitochondria. Recently, it has been demonstrated that divicine leads to an accumulation of intracellular calcium causing a marked inactivation of calcium ATPase and a concomitant loss of spectrin. It was postulated that both enhanced calcium permeability and a calcium-stimulated degradation of the calcium pump are the mechanism responsible for the perturbation of erythrocyte calcium homeostasis in favism.[87]

10.9 REDUCTION OF THE TOXICITY OF FAVISM-CAUSING FACTORS

The oxidation-reduction shuttle that is autocatalyzed by divicine and isouramil is counteracted *in vivo* by enzymes and nonenzymatic scavengers and quenchers denoted by the term antioxidants and including compounds such as vitamin E, vitamin C, flavonoids, ß-carotene, urate, and plasma proteins.[88] Vitamin E, when incorporated into diets, provided protection against of the toxic effects of divicine in rats,[23] alleviated some of the vicine-induced depression in fertility and hatchability of eggs in laying hens,[20] and decreased chronic hemolysis by improving red-cell span and increasing red-cell hemoglobin concentration in G6PD-deficient patients.[24] In the rat, other free-radical scavengers, such as vitamin A and vitamin C, also prevented divicine-induced depletion of GSH, therefore protecting it against the toxic effects of divicine. Riboflavin, being a cofactor for GSH reductase, should also provide protection against reduced GSH-reductase activity.[1] The iron chelating agent deferoxamine has been shown to block the antimalarial and the hemolytic activity of divicine.[25,36,89,90] In addition, transition metal chelating agents have been shown to provide complete protection, *in vitro*, against enzyme inactivation caused by favism-inducing agents.[45] These observations suggest that certain compounds may be capable of greatly modifying individual susceptibility to favism and may be one of the important additional factors proposed by Mager et al.,[9] along with G6PD deficiency and exposure to faba beans, that influence the degree and severity of the hemolytic attack. A better understanding of the interactions that occur among these factors may assist in clarifying the bizarre and rather unpredictable mode of occurrence of favism in susceptible individuals.

10.10   MALARIA, FAVISM, AND GLUCOSE-6-PHOSPHATE DEHYDROGENASE

A possible link between G6PD deficiency and malaria was first proposed on the basis of epidemiological evidence.[91] Studies of Golenser et al.[35] and those of Clark et al.[33,37] demonstrated *in vitro* that both G6PD-deficient and normal erythrocytes supported similar growth of the parasite (*Plasmodium falciparum*) that causes malaria in humans. In contrast, G6PD-deficient erythrocytes, after treatment with the favism-inducing agent isouramil, were unable to support the parasites growth, whereas a similar treatment of normal erythrocytes had no effect on the parasite. Similar results have been observed in mice infected with *Plasmodium vinckei* after injection of divicine. Both death of the parasites and hemolysis of RBCs following divicine and isouramil administration were blocked by the iron chelator, deferoxamine.[36,92] These results suggest that hydroxy radicals are formed from H_2O_2 to initiate peroxidative reactions which not only cause hemolysis but also the killing of the parasite. Degenerated parasites were frequently observed inside intact, circulating erythrocytes, implying that the membrane of the parasites were more likely affected by the free radical-mediated oxidative damage than that of their host erythrocytes.

Several other activated oxygen generators, such as primaquine, alloxan, and t-butyl peroxide also have antimalarial activity *in vivo*.[37,93] In addition, vitamin E deficiency in animals reduces the survivability of plasmodia, presumably because more activated oxygen is present in these cells.[94] The concentration of other free radical scavenging compounds would probably have a similar effect to that of vitamin E. These results demonstrate that any factor that will increase the concentration of activated oxygen or free radical production in the host cell will have a negative effect on plasmodial survivability. Malaria parasites may be particularly susceptible to activated oxygen as they are not able to synthesize SOD.[95]

Faba bean consumption under control conditions may therefore provide sufficient protection against malaria for the host to acquire immunity. Selection of faba bean cultivars that are capable of producing high concentrations of vicine or convicine,[1,26,60] harvesting faba beans at a immature state,[61] or the use of air-classified faba bean fractions would provide a means of supplying malaria-infected individuals with a relatively inexpensive source of vicine and convicine.[1,40] These compounds may also be beneficial if administrated in pure form either orally or intraperitoneally. Research is required to establish the effectiveness of such procedures under different nutritional regimens in combination with the use of the drugs that are currently used to control malaria and means to rapidly counteract any possible deleterious side effects.

10.11 CONCLUSIONS

Divicine and isouramil participate in an oxidation-reduction shuttle system in which the net effect is the transfer of electrons from reducing compounds such as GSH, NADP(H), and ascorbic acid through the divicine/isouramil cofactor to O_2 with the resulting formation of O_2^-. These effects are usually neutralized by the free-radical scavenging system, but can lead to irreversible cell damage if the cell is unable to cope with the rapid depletion of reducing equivalents when challenged with an oxidant. This occurs upon exhaustion of GSH, particularly in G6PD-deficient erythrocytes. Under these conditions, there is formation of H_2O_2 which can be converted to the highly reactive hydroxyl radical in the presence of iron. This radical can then react with hemoglobin to form methemoglobin and ultimately Heinz bodies. Cross-bonding of proteins, and perhaps an increased degree of lipid peroxidation of the erythrocyte membrane are associated effects. This oxidant stress results in a loss of the structure and function of the properties of erythrocytes as evidenced by altered shape, deformability, and filterability, susceptibility to spontaneous hemolysis, shortened life span, increased degree of phagocytosis of damaged erythrocytes, transient electrolyte imbalances, and finally loss of the ability of the RBC to exchange respiratory gases. The net effect is that the animal or favic individual will die of asphyxiation if a sufficiently high portion of the RBCs are affected.

The adverse effects of divicine not only in favism susceptible individuals but by all individuals that consume faba beans can be reduced by the consumption of faba beans that have a low concentration or no vicine and convicine, by the use of faba beans subjected to certain processing methods, and by modification of the diet to enhance intake of the antioxidant vitamins and to reduce intake of pro-oxidants. Vicine and convicine, although harmful to humans under normal circumstances, may be beneficial as an antimalarial drug, particularly if used in combination with those drugs that are currently used to control malaria.

REFERENCES

1. Marquardt, R. R., Vicine, convicine, and their aglycones—divicine and isouramil, in *Toxicants of Plant Origin*, II., Cheeke, P., Ed., CRC Press, Boca Raton, FL, 1989, Chap. 6.
2. Hegazy, M. I. And Marquardt, R. R., Metabolism of vicine and convicine in rat tissue: absorption and excretion patterns and sites of hydrolysis, *J. Sci. Food. Agric.*, 35, 139, 1984.
3. Flohe, L., Niebch, G., and Reiber, H., Zur Wirkung von Divicine in menschlichen Erythrocyten, *Z. Klin. Chim. Klin. Biochem.*, 9, 431, 1971.
4. Albano, E., Tomasi, A., Mannuzzu, L., and Arese, P., Detection of a free radical intermediate from divicine of *Vicia faba, Biochem. Pharmacol.*, 33, 1701, 1984.
5. Lin, J.-Y. and Ling, K.-H., Studies on favism. I. Isolation of an active principle from faba beans (*Vicia faba*), *J. Formosan Med. Assoc.*, 61, 484, 1962.
6. Mager, J., Glaser, G., Razin, A., Izak, G., Bien, S., and Noam, M., Metabolic effects of pyrimidines derived from faba bean glucosides on human erythrocytes deficient in glucose-6-phosphate dehydrogenase, *Biochem. Biophys. Res. Commun.*, 20, 235, 1965.
7. Razin, A., Hershko, A., Glaser, G., and Mager, J., The oxidant effect of isouramil on red cell glutathione and its synergistic enhancement by ascorbic acid and 3,4-dihydroxyphenylalanine. Possible relation to the pathogenesis of favism, *Isr. J. Med. Sci.*, 4, 852, 1968.
8. Lin, J.-Y., Lee, S.-W., and Ling, K.-H., Divicine and favism, *J. Chin., Biochem. Soc.*, 6, 92, 1977.
9. Mager, J., Chevion, M., and Glaser, G., Favism, in *Toxic Constituents of Plant Foodstuffs*, 2nd ed., Liener, I. E., Ed., Academic Press, New York, 1980, 265.
10. Arese, P., Naitana, L., Mannuzzu, L., Turrini, F., Haest, C. W. M., Fisher, T. M., and Deuticke, B., Biochemical and micro-rheological modifications in normal and glucose-6-phosphate dehydrogenase-deficient red cells treated with divicine, in *Advances in Red Cell Biology*, Weatherall, D. J., Fiorelli, G., and Gorini, S., Eds., Raven Press, New York, 1982, 375.

11. Chevion, M., Navok, T., Glaser, G., and Mager, J., The chemistry of favism-inducing compounds: the properties of isouramil and divicine and their reaction with glutathione, *Eur. J. Biochem.*, 127, 405, 1982.

12. Chevion, M., Navok, T., and Glaser, G., Favism inducing agents: biochemical and mechanistic considerations, in Advances in Red Cell Biology, Weatherall, D. J., Fiorelli, G., and Gorini, S., Eds., Raven Press, New York, 1982, 381.

13. Baker, M. A., Bosia, A., Pescarmona, G., Turrini, F., and Arese, P., Mechanism of action of divicine in a cell-free system and in glucose-6-phosphate dehydrogenase-deficient red cells, *Toxicol. Pathol.*, 12, 331, 1984.

14. Chevion, M., Navok, T., Pfafferott, C., Meiselman, H. J., and Hochstein, P., The effects of isouramil on erythrocyte mechanics: implications for favism, *Microcirc. Endothel. Lymphat.*, 1, 295, 1984.

15. Yanni, S. and Marquardt, R. R., Induction of favism-like symptoms in the rat: effects of vicine and divicine in normal and buthionine sulfoximine-treated rats, *J. Sci. Food Agric.*, 36, 1161, 1985.

16. Winterbourn, C., Benatti, U., and DeFlora, A., Contributions of superoxide, hydrogen peroxide and transition metal ions to autoxidation of the favism-inducing pyrimidine aglycone, divicine and its reaction with hemoglobin, *Biochem. Pharmacol.* 35, 2009, 1985.

17. Arbid, M. S. S. and Marquardt, R., R., Effect of intraperitoneally injected vicine and convicine on the rat: induction of favism-like signs, *J. Sci. Food Agric.*, 37, 539, 1986.

18. Yoshida, A., Hemolytic anemia and G6PD deficiency, *Science*, 179, 532, 1973.

19. D'Aquino, M., Gaetani, S., and Spadoni, M. A., Effect of factors of favism on the protein and lipid components of rats erythrocyte membrane, *Biochim. Biophys. Acta*, 731, 161, 1983.

20. Muduuli, D. S., Marquardt, R. R., and Guenter, W., Effect of dietary vicine and vitamin E supplementation on productive performance of growing and laying chickens, *Br. J. Nutr.*, 47, 53, 1982.

21. Graf, M., Frei, B., Winterhalter, K. H., and Richter, C., Divicine induces calcium release from rat liver mitochondria, *Biochem. Biophys. Res. Commun.*, 129, 18, 1985.

22. Rocic, B., Rocic, S., Ashcroft, J. H., Harrison, D. E., and Poje, M., Diabetogenic action of alloxan-like compounds: the cytotoxic effect of dehydro-isouramil hydrochloride on the rat pancreatic ß-cells, *Diabetes Croat.*, 14, 143, 1985.

23. Marquardt, R. R. and Arbid, M. S. S., Protection against the toxic effects of the favism factor (divicine) in rats by vitamin E, A, and C and iron chelating agents, *J. Sci. Food Agric.*, 43, 155, 1988.

24. Corash, L. M., Sheetz, M., Bieri, J. G., Bartsocas, C., Moses, S., Bashan, N., and Schulman, J. P., Chronic hemolytic anemia due to glucose-6-phosphate dehydrogenase deficiency or glutathione synthetase deficiency: the role of vitamin E in its treatment, *Ann. N.Y. Acad. Sci.*, 393, 348, 1982.

25. Vanella, A., Campisi, A., Castroina, C., Sorrenti, V., Attaguile, G., Samperi, P., Azzia, N., Di-Giacomo, C., and Schiliro, G., Antioxidant enzymatic system and oxidative stress in erythrocytes with G6PD deficiency: effect of deferoxamine, *Pharmacol. Res.,* 24, 25, 1991.

26. Duc, G., Sixdenier, G., Lila, M., and Furstoss, V., Search of genetic variability for vicine and convicine content in *Vicia faba* L. A first report of a gene which codes for nearly zero-vicine and zero-convicine content, in *Recent Advances of Research in Antinutritional Factors in Legume Seeds*, Huisman, J., van der Poel, T. F. B., and Liener I. E., Eds., Pudoc Wagenungen, Netherlands, 305, 1988.

27. Hegazy, M. I. and Marquardt, R. R., Development of a simple procedure for complete extraction of vicine and convicine from fababeans (*Vicia faba* L.), *J. Sci. Food Agric.*, 34, 100, 1983.

28. Arntfield, S. D., Ismond, M. A. H., and Murray, D. E., The fate of antinutritional factors during the preparation of a faba bean protein isolate using a micellization technique, *Can. Inst. Food Sci. Technol. J.,* 18, 137, 1985.

29. Arbid, M. S. S. and Marquardt, R. R., Hydrolysis of the toxic constituents (vicine and convicine) in faba bean (*Vicia faba* L.) food preparations following treatment with ß-glucosidase, *J. Sci. Food Agric.,* 36, 839, 1985.

30. Hussein, L., Motawei, H., Nassib, A., Khalil, S., and Marquardt R., R., The complete elimination of vicine and convicine from the faba bean by combinations of genetic selection and processing techniques, *Qual. Plant Foods Hum. Nutr.*, 36, 231, 1986.

31. Mckay, A. M., Hydrolysis of vicine and convicine from faba beans by microbial beta-glucosidase enzymes, *J. Appl. Bacteriol.*, 72, 475, 1992.

32. Chaumeil, J. C., Extraction of vicine from vegetable material and its use in the treatment of cardiac arrythmia, *Chem. Abstr.*, 86, 1111 68f, 1977.

33. Clark, I. A., Cowden, W. B., Hunt, N. H., Maxwell, L. E., and Mackie, E. J., Activity of divicine in *Plasmodium vinckei*-infected mice has implications for treatment of favism and epidemiology of G6PD deficiency, *Br. J. Haematol.*, 57, 479, 1984.

34. Roth, E. F., Raventos-Suarez, C., Rinaldi, A., and Nagel, R. L., Glucose-6-phosphate dehydrogenase deficiency inhibits *in vitro* growth of *Plasmodium falciparum*, *Proc. Natl. Acad. Sci.*, 80, 298, 1983.

35. Golenser, J., Miller, J., Spira, D. T., Navok, T., and Chevion, M., The effect of a favism inducing agent on the *in vitro* development of *Plasmodium falciparum* in normal and glucose-6-phosphate dehydrogenase deficient erythrocytes, *Blood*, 61, 507, 1983.

36. Clark, I. A. and Cowden, W. B., Antimalarials, in *Oxidative Stress*, Sies, H., Ed., Academic Press, New York, 1985, Chap.7.

37. Clark, I. A., Mackie, E. J., and Cowden, W. B., Injection of free radical generators causes premature onset of tissue damage in malaria-infected mice, *J. Pathol.*, 148, 301, 1986.

38. Bjerg, B., Heide, M., Knudsen, J. C. N., and Sorensen, H., Inhibitory effects of convicine, vicine and dopa from *Vicia faba* on the *in vitro* growth rates of fungal pathogens, *Z. Pflanzenkr. Pflanzenschutz*, 91, 483, 1984.

39. Bendich, A. and Clements, G. C., A revision of the structural formulation of vicine and its pyrimidine aglycone divicine, *Biochim. Biophys. Acta*, 12, 462, 1953.

40. Marquardt, R. R., Muduuli, D. S., and Frohlich, A. A., Purification and some properties of vicine and convicine isolated from faba bean (*Vicia faba* L.) protein concentrate, *J. Agric. Food Sci.*, 31, 839, 1983.

41. Bien, S., Salemnik, G., Zamir, L., and Rosenblum, M., The structure of convicine, J. Chem. Soc. C., 496, 1968.

42. Chesterfield, J. H., Hurst, D. T., McOmie, J. F. W., and Tute, M. S., Pyrimidines, XIII. Electrophilic substitution at position 6 and a synthesis of divicine (2,4-diamino-5,6-dihydroxypyrimidine), *J. Chem. Soc.*, 1001, 1964.

43. Davoll, J. and Laney, D. H., Synthesis of divicine (2,4-diamono-5,6-dihydroxypyrimidine) and other derivatives of 4,5(5,6)-dihydrixypyrimidine, *J. Chem. Soc.*, p.2124, 1956.

44. Benatti, U., Guida, L., and DeFlora, A., The interaction of divicine with glutathione and pyridine nucleotides, *Biochem. Biophys. Res. Commun.*, 120, 747, 1984.

45. Navok, T. and Chevion, M., Transition metals mediate enzymatic inactivation caused by favism-inducing agents, *Biochem. Biophys. Res. Commun.*, 122, 297, 1984.

46. Malaisse, W. J., Alloxan toxicity to the pancreatic ß-cell; a new hypothesis, *Biochem. Pharmacol.*, 31, 3527, 1982.

47. Arese, P., Bosia, A., Naitana, A., Gaetani, S., D'Aquino, M., and Gaetani, G. F., Effect of divicine and isouramil on red cell metabolism in normal and G6PD-deficient (Mediterranean variant) subjects. Possible role in the genesis of favism, in *The Red Cell: Fifth Annu. Arbor. Conf.*, Brewer, G., Ed., Alan R. Liss, New York, 1981, 725.

48. Marquardt, R. R. and Frohlich, A. A., Rapid reversed-phase high-performance liquid chromatographic method for the quantitation of vicine and convicine and related compounds, *J. Chromatogr*, 208, 373, 1981.

49. McMillan, D. C., Schey, K. L., Meier, G. P., and Jollow, D. J., Chemical analysis and hemolytic activity of the faba bean aglycone divicine, *Chem. Res. Toxicol.*, 6, 439, 1993.

50. Ritthausen, H. and Kreusler, U., Ueber vorkommen von amygdalin und eine neue dem asparagin achnliche substanz in wickensamen (Vicia sativa), *J. Prakt. Chem.*, 2, 233, 1870.

51. Ritthausen, H., Ueber vicin und eine zweite stickstoffreiche substanz der wickensamen, convicin, *J. Prakt Chem.*, 24, 202, 1881.

52. Ritthausen, H., Vicin ein glucosid, *Ber. Dtsch. Chem. Ges.*, 29, 2108, 1896.

53. Ritthausen, H., Über die divicin, *J. Prakt Chem.*, 59, 482, 1899.

54. Levene, P. A. and Senior, J. K., Vicine and divicine, *J. Biol. Chem.*, 25, 607, 1916.
55. Marquardt, R. R., Frohlich, A. A., and Arbid, M. S. S., Isolation of crystalline divicine from an acid hydrolysate of vicine and a crystalline decomposition product of divicine and some properties of these compounds, *J. Agric. Food Chem.*, 37, 455, 1989.
56. McOmie, J. F. W. and Chesterfiels, J. H., Synthesis of 5-hydroxypyrimidine and a new synthesis of divicine, *Chem. Ind. (London)*, 2, 1453, 1956.
57. Ritthausen, H., Über alloxantin als spaltungsproduct des convins aus saubohnen (*Vicia faba* minor) und wicken (*Vicia sativa*), *Ber. Dtsch. Chem. Ges.*, 29, 894, 1896.
58. Marquardt, R. R., Frohlich, A. A., and Arbid, M. S. S., Preparation of crystalline isouramil by acid hydrolysis and isolation of two decomposition products of isouramil, *J. Agric. Food Chem.*, 37, 448, 1989.
59. Pitz, W. J., Sosulski, F. W., and Hogge, L. R., Occurrence of vicine and convicine in seeds of some *Vicia* species and other pulses, *Can. Inst. Food Sci. Technol. J.*, 13, 35 1980.
60. Pitz, W. J., Sosulski, F. W., and Rowland, G. G., Effect of genotype and environment of vicine and convicine levels in fababeans (*Vicia faba* minor), *J. Sci. Food Agric.*, 32, 1, 1981.
61. Jamalian, J. and Bassiri, A., Variation in vicine concentration during pod development in broad beans (*Vicia faba* L.), *J. Agric. Food Chem.*, 26, 1454, 1978.
62. Brown, E. G. and Roberts, F. M., Formation of vicine and convicine by *Vicia faba, Phytochemistry*, 11, 3203, 1972.
63. Olaboro, G., Marquardt, R. R., Campbell, L. D., and Frohlich, A. A., Purification, identification and quantification of an egg-weight-depressing factor (vicine) in faba beans (*Vicia faba* L.), *J. Sci. Food Agric.*, 32, 1163, 1981.
64. Olaboro, G., Campbell, L. D., and Marquardt, R. R., Influence of faba bean fractions on egg weight among laying hens fed test diets for a short time period, *Can. J. Anim. Sci.*, 61, 751, 1981.
65. Muduuli, D. S., Marquardt, R. R., and Guenter, W., Effect of dietary vicine on the productive performance of laying chickens, *Can. J. Anim. Sci.*, 61, 757, 1981.
66. Yannai, S. and Marquardt, R. R., Effect of divicine, one of its degradation products and hydrogen peroxide on normal and pre-treated erythrocytes, *Vet. Human. Toxicol.*, 29, 393, 1987.
67. Arbid, M. S. S. and Marquardt, R. R., Favism-like effects of divicine and isouramil in the rat: acute and chronic effects on animal health, mortalities, blood parameters and ability to exchange respiratory gases, *J. Sci. Food Agric.*, 43, 75, 1988.
68. Aurichio, L., Sul favismo, *Rass. Clin. Sci.*, 13, 20, 1935.
69. Belsey, M. A., The epidemiology of favism, *Bull. W.H.O.*, 48, 1, 1973.
70. Motulsky, A., Hemolysis in glucose-6-phosphate dehydrogenase deficiency, *Fed. Proc. Fed. Am. Soc. Exp. Biol.*, 31, 1286, 1972.
71. Carson, P. E., Flanagan, C. L., Ickes, C. E., and Alving, A. S., Enzymatic deficiency in primaquine-sensitive erythrocytes, *Science*, 124, 484, 1956.
72. Harley, J. D., Agar, N. S., and Gruca, M. A., Cataracts with a glucose-6-phosphate dehydrogenase variant, *Br. Med. J.*, 2, 86, 1975.
73. Escobar, M. A., Heller, P., and Trobaugh, F. E., Jr., "Complete" erythrocyte glucose-6-phosphate dehydrogenase deficiency, *Arch. Intern. Med.,* 113, 428, 1964.
74. Schmitz, G., Hohage, H., and Ullrich, K., Glucose-6-phosphate: a key compound in glycogenesis and favism leading hyper- or hypolipidermia, *Eur. J. Pediatr.*, 152 suppl.1, S77, 1993.
75. Hebbel, R. P., Autoxidation and the sickle erythrocyte membrane: a possible model of iron decompartmentalization, in *Free Radicals, Aging and Degenerative Diseases,* Johnson, J. E., Walford, R., Harmon, D., and Miguel, J., Eds., Alan R. Liss, New York, 8, 395, 1986.
76. Benatti, U., Guida, L., Forteleoni, G., Meloni, T., and De Flora, A., Impairment of the calcium pump of human erythrocytes by divicine, *Arch. Biochem. Biophys.*, 239, 334, 1985.
77. Marvelli, I., Ciriolo, M. R., Rossi, L., Meloni, T., Forteleoni, G., DeFlora, A., Benatti, U., Morelli, A., and Rotilio, G., Favism: a hemolytic disease associated with increased superoxide dismutase and decreased glutathione peroxidase activity in red blood cells, *Eur. J. Biochem.*, 139, 13, 1984.
78. Winterbourn, C. and Munday, R., Concerted action of reduced glutathione and superoxide dismutase in preventing redox cycling of dihydroxypyrimidines and their role in antioxidant defence, *Free Radic. Res. Commun.*, 8, 287, 1990.

79. Stern, A., Red cell oxidative damage, in *Oxidative Stress*, Sies, H., Ed., Academic Press, New York, 1985, Chap. 14.

80. Johnson, G. J., Allen, D. W., Cadman, S., Fairbank, V. F., White, J. G., Lampkin, B. C., and Kaplan, M. E., Red-cell-membrane polypeptide aggregates in glucose-6-phosphate dehydrogenase mutants with chronic hemolytic disease. A clue to the mechanism of hemolysis, *N. Engl. J. Med.*, 301, 522, 1979.

81. Hochstein, P. and Jain, S. K., Association of lipid peroxidation and polymerization of membrane proteins with erythrocyte aging, *Fed. Proc. Fed. Am. Soc. Exp. Biol.*, 40, 183, 1981.

82. Fischer, T. M., Meloni, T., Pescarmona, G. P., and Arese, P., Sequestration of red cells in favic crises, *Clin. Hemorheol.*, 3, 228, 1983.

83. Nicotera, P. and Orrenius, S., Molecular mechanism of toxic cell death: an overview, in *In Vitro Toxicity Indicators*, Tyson, C. A. and Frazier. J. M., Eds., Academic Press, New York, 1994, 23.

84. Bosia, A., Passow, H., Arese, P., Lepke, S., and Mannuzzu, L., Effect of divicine on Ca^{++}-stimulated K^+ efflux in normal and G6PD-deficient erythrocytes, *Ital. J. Biochem.*, 29, 393, 1980.

85. Bellomo, G., Matino, A., Richelmi, R., Moore, G. A., Jewell, S. A., and Orrenius, S., Pyridine-nucleotide oxidation, Ca^{++} cycling and membrane damage during tert-butyl hydroperoxide metabolism by rat liver mitochondria, *Eur. J. Biochem.*, 140, 1, 1984.

86. Frei, B., Winterhalter, K. H., and Richter, C., Mechanisms of alloxan induced calcium release from rat liver mitochondria, *J. Biol. Chem.*, 260, 7394, 1985.

87. Damonte, G., Guida, L., Sdraffa, A., Benatti, U., Melloni, E., Forteleoni, G., Meloni, T., Carafoli, E., and De-Flora, A., Mechanisms of perturbation of erythrocyte calcium homeostasis in favism, *Cell Calcium*, 13, 649, 1992.

88. Sies, H., Oxidative stress: introductory remarks, in *Oxidative Stress*, Sies, H., Ed., Academic Press, New York, 1985, 1.

89. Meloni, T., Forteleoni, G., and Gartani, G. F., Desferrioxamine and favism, *Br. J. Haematol.*, 63, 394, 1986.

90. Ferrali, M., Signorini, C., Ciccoli, D., and Comporti, M., Iron release and membrane damage in erythrocytes exposed to oxidizing agents, phenylhydrazine, divicine and isouramil, *Biochem. J.*, 285, 295, 1992.

91. Motulsky, A. G., Metabolic polymorphism and the role of infectious disease in human evolution, *Human Biol.*, 32, 28, 1960.

92. Gutteridge, J. M. C., Richmond, R., and Halliwell, B., Inhibition of the iron-catalysed formation of hydroxyl radicals from superoxide and of lipid peroxidation by desferrioxamine, *Biochem. J.*, 184, 469, 1979.

93. Clark, I. A., Butcher, G. A., Buffinton, G. D., Hunt, N. H., and Cowden, W. B., Toxicity of certain products of lipid peroxidation to the human malaria parasite *Plasmodium falciparum*, *Biochemical. Pharmacology*, 36, 543, 1987.

94. Eaton, J. W., Eckman, J. R., Berger, E., and Jacob, H. S., Suppression of malaria infection by oxidant-sensitive host erythrocytes, *Nature*, 264, 758, 1976.

95. Fairfield, A. S., Meshnick, S. R., and Eaton, J. W., Host superoxide dismutase incorporation by intra-erythrocytic plasmodia, in *Malaria and the Red Cell*, Eaton, J. W. and Brewer, G. J., Eds., Alan R. Liss, New York, 1984, 13.

11 Allergens in Fruits and Vegetables

S. Vieths

CONTENTS

11.1   INTRODUCTION

There is no doubt that fruits and vegetables belong to the most important elicitors of food allergy in adults and adolescents.[1–5] The same is true for other plant-derived foods like nuts and some legumes which are not within the scope of this chapter. However, contrary to media perception, adverse reactions to food occur less often than estimated by the patients. The prevalences of food allergies (FA) and/or food intolerances (FI) reported by patients were between 12 and 19%, whereas the confined prevalences varied from 0.8 to 2.4%.[3] Food additives, residues, and contaminants are frequently suspected to be the reasons for an increasing number of adverse reactions to food. However, all food allergens that have been identified and characterized so far are natural molecules, i.e., proteins or glycoproteins.[3–8]

The overwhelming majority of hypersensitivity reactions against fruits and vegetables is highly associated with several pollen allergies.[9–12] At the molecular level, this phenomenon is based on cross-reactions of human antibodies which are directed against pollen allergens. Therefore, these clusters of "cross-allergenicity" are a main subject of this contribution. Approximately 15% of the population of developed countries are allergic to pollen[13] and between 50 to 93% of birch pollen allergic patients suffer from allergies against fruits and vegetables.[5] From these data, the prevalence of fruit and vegetable hypersensitivity can be estimated to be higher than 1%. Generally, the allergenic activity of most fruits and vegetables is very sensitive to heating and processing.[5,6,12,14,15] Some vegetables, however, have a partly thermostable allergenicity, for example celery[12] or potato.[4] In addition, a relatively high number of cases of severe anaphylatic reactions due to ingestion of celery has been reported from Switzerland.[2,12]

Allergic diseases are highly individual. Moreover, they tend to change when progressing. Consequently, general features are difficult to describe and looking at case reports can give rise to complete confusion. Furthermore, the terms FA and FI are widely misused even in the scientific literature. Almost all sorts of symptoms and diseases including psychological syndromes have been misinterpreted as FI.[3] To clarify this puzzling situation, the next sub-chapter deals with terms and basic mechanisms in the context of food allergy.

A chapter like this cannot present a complete evaluation of the literature. Focusing the attention on important review articles and the recently published literature about allergens in fruits and vegetables may simplify the reader's familiarization with the special literature.

11.2   ADVERSE REACTIONS TO FOOD

The following definitions have been recommended by the European Academy of Allergy and Clinical Immunology.[3,6] Adverse reactions occurring after ingestion of food are divided into toxic reactions and nontoxic reactions. Nontoxic reactions depend on individual susceptibility to a certain food. They are either immune mediated (FA) or nonimmune mediated (FI). The mechanism of FI can be enzymatic, pharmacological, or undefined. Examples for the first and the second are lactase deficiency and abnormal reactivity to vasoactive amines present in some foods, respectively (Figure 11.1).

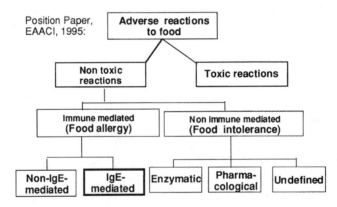

FIGURE 11.1   Classification of adverse reactions to food. (From Wüthrich, B., *Proc. XVI. European Congress of Allergology and Clinical Immunology,*[3] Madrid, 1995, 859. With permission.)

In this chapter, the term FA refers to a type I allergic reaction caused by antigen-specific immunoglobulin E (IgE) antibodies.[7] "Atopy" is used for a genetic predisposition to develop type I allergic reactions. "Food allergens" are defined as the antigenic molecules giving rise to the immunologic response.[6] "Sensitization" to allergens occurs during an asymptomatic period and the "class switch" from a nonpathologic IgG- to an IgE-response is due to a misfunction at the T cell level.[16–18] Once sensitized to food, atopic persons can experience symptoms that normally occur immediately within the first 60 minutes after ingestion of the offending food.

The essential event of the type I allergic reaction is allergen specific degranulation of mast cells and blood basophils leading to spontaneous mediator release (e.g., histamine, tryptase, acid hydrolases) and *de novo* synthesis of mediators[19] (e.g., leukotrienes, prostaglandins)

(Figure 11.2). High affinity receptors on the surface of these cells can specifically bind IgE via its Fc region. Bridging of membrane bound IgE by an at least divalent allergen initiates mediator release and the following cascade of the type I reaction.[19] The high diversity of symptoms that can be caused by FAs depends on the locations and intensity of these immunoreactions. Unfortunately, all these symptoms are nonspecific.[3,6] However, if based on an immunological reaction, reproducibility by allergen challenge is an essential criterium for allergy.

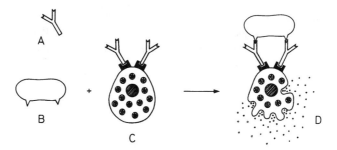

FIGURE 11.2 Schematic representation of allergen induced effector cell degranulation. IgE antibodies specific for allergen (A) can bind to high affinity receptors on the surface of mast cells and basophils (C). Specific binding of allergen (B) to membrane bound IgE results in antibody crosslinking followed by mast cell degranulation (D), mediator release and de novo synthesis of mediators. (From Baltes W., *Food Chemistry*, Springer, Berlin, 1994. With permission.)

Most patients who are sensitive to fruits and vegetables suffer from local symptoms at the site of the primary allergen contact, i.e., the oropharynx. Typical phenomena are swelling of the lips, tongue, and throat. Other food allergic patients more frequently exhibit symptoms in varying organs, for example the gastrointestinal tract, the skin, the respiratory tract, and the cardiovascular system (systemic anaphylaxis).[3,6]

11.3 FOOD ALLERGY AND CROSS-REACTIVITY

Cross-reactivity of allergen-specific antibodies is a well known phenomenon in food allergy. These cross-reactions often include phylogenetically unrelated species, for example birch and kiwi.[15] In many cases, respiratory uptake of inhalant allergens is thought to be the causative factor for development of food hypersensitivities. Important examples are the so-called "bird-egg syndrome" which designates egg yolk allergy induced by inhalation of serum proteins from birds,[20] food allergies to flour components which occur in persons with bakers asthma,[20–22] and sunflower seed allergy due to inhalation of food for domestic birds.[23] Moreover, IgE cross-reactivities frequently occur between proteins from phylogenetically related plants. A relatively high number of these seems to be without clinical significance, for example cross-reactions among the legume botanical family[24] and between grass pollen and cereal proteins.[6,25,26] However, the clinical phenomenon of pollen related food allergy which is discussed in the next subchapter is undoubtedly due to cross-reactivity.[1,27] The latter belongs to the rare cases where molecular data of the allergens are available.

11.4 CLINICAL SYNDROMES

11.4.1 POLLEN RELATED FOOD ALLERGIES

Due to the existence of immunologically related structures in particular parts of the plant kingdom, only a minor proportion of allergies to fruits and vegetables is independent of pollen allergy.[1,5,10,11,28,29] The foods that are most frequently mentioned in this context are listed and referenced in Table 11.1. In areas with birch trees, for example northern and central Europe, birch pollen related food allergies are most important. Patients sensitized to birch pollen are very often hypersensitive to other pollen of trees belonging to the Fagales order and to apple, many stone fruit, kiwi, celery, carrot, and potato. The latter, however, predominantly cause contact urticaria of the hands or allergic rhinitis during peeling. Allergic reactions are usually due to the ingestion of fresh fruits but not to consumption of their heated and processed products.[1,3,6,15] The thermostable allergenic component of celery allergy seems to be associated with a cosensitization to mugwort pollen[12] (*Artemisia vulgaris*). Although they do not belong to fruits in the narrower sense, it has to be noted that additional allergies to nuts and almonds are very common in birch pollen allergic patients.[5,9]

TABLE 11.1
Pollen Related Food Allergies against Fruits and Vegetables

Pollen	Food	Ref.
Birch	Apple, pear, cherry, peach, nectarine, apricot, plum, kiwi, celery, carrot, potato	Eriksson et al., 1982;[9] Calkhoven et al., 1987;[30] Dreborg, 1988;[5] Wüthrich et al., 1990;[12] Helbling et al., 1993;[31] Gall et al., 1994[15]
Birch / mugwort	Celery, carrot, spices	Wüthrich and Dietschi, 1985;[10] Thiel, 1988;[11] Bauer et al., 1996;[32] Jankiewicz et al., 1995[33]
Grass	Melon, watermelon, orange, tomato, potato, swiss chard	Calkhoven et al., 1987;[30] De Martino et al., 1988;[29] Ortolani et al., 1993;[1] Boccafogli et al., 1994;[34] Varela et al., 1994;[35] Garcia Ortiz et al., 1995[36]
Ragweed	Watermelon and other melons, banana, zucchini, cucumber	Anderson et al., 1970;[37] Enberg et al., 1987;[38] Ross et al., 1991[40]

Allergy to celery, carrot, and umbelliferous spices has originally been described as being mainly associated with mugwort pollen allergy.[10,28] However, the more recent studies listed in Table 11.1 revealed a cosensitization to birch pollen in most of the cases. At the molecular level, no birch pollen independent food allergens have been identified in celery (see below). Reports about typical food allergy clusters in grass pollen allergic patients are mainly from the mediterranean area (Table 11.1). Tomatos, potatoes, melons, and oranges have been particularly named in this context.[1,29] In addition to these plant foods, peanut allergy has also been found to be associated with grass pollen hypersensitivity.[29] In contrast to birch pollen allergy, only a minor proportion of grass pollen allergics is affected by food hypersensitivities.[5]

Allergenic pollen of ragweed (Ambrosia) frequently cause pollinosis in North America, some parts of France, Northern Italy, and Austria. In 1970, Anderson et al.[37] investigated 1447 pollen-sensitized patients and diagnosed a co-occurring allergy against ragweed pollen, melon, and banana in 90 cases. These results have been confirmed by Enberg et al.[38,39] who also showed an extended reactivity to other cucurbitaceae such as cucumber and zucchini and who demonstrated partial cross-reactivity between the food and pollen allergens involved (Table 11.1). This cross-reactivity was confirmed later by other authors' results.[40]

Although parietaria (Urticaceae) and trees from the oleacae family are common elicitors of pollen allergies in the mediterranean area,[41] there are no reports about specific food allergies associated with these diseases.

11.4.2   FRUIT AND VEGETABLE ALLERGY INDEPENDENT OF POLLEN SENSITIZATION

As stated above, allergy to particular fruits and vegetables without concomitant pollen sensitization is much less important than the previously described phenomena.[4] Apart from monosensitization to a particular plant derived food, a specific cluster of sensitivities has recently been described which is also due to cross-reactivity. Allergy to natural rubber latex from the rubber tree (*Hevea brasiliensis*) is an important medical problem among health care workers and patients who have undergone multiple operations and patients with spina bifida. The wide use of latex products such as household gloves, condoms, and balloons has made this allergy also common in the general population. There is an increased number of reports about fruit and vegetable allergies in these patients. A partial cross-reactivity has been demonstrated between these allergen sources.[42–48] The offending foods were banana, avocado, chestnut, and kiwi fruit. Furthermore, an important proportion of the allergenicity of stone fruit belonging to the prunoideae has been found to be independent of pollen allergy in patients from Italy.[49]

Almost all other kinds of fruits and vegetables have been found to be allergenic in some patients. The number and the content of scientific reports on this subject is scarce and is frequently restricted to single case reports (for review see Kreft et al. 1995[4]). Molecular characteristics are not available in most of the cases. Some of these hypersensitivities may be due to IgE reactivity with ubiquitous plant allergens like profilins, the characteristics of which will be described in the next subchapter.

11.5 MOLECULAR CHARACTERISTICS OF IMPORTANT ALLERGENS IN FRUITS AND VEGETABLES

11.5.1 General Properties

As mentioned, all known food allergens are proteins or glycoproteins. Most of them have relatively low molecular weights ranging between 10 kDa and 70 kDa and are highly soluble in water or physiologically buffered solutions.[7,50–53] The sizes of the most intensively characterized fruit and vegetable allergens Mal d 1 (apple), Api g 1 (celery), and profilin (all food plants) are below 20 kDa. However, our knowledge of the structure of food allergens is strikingly poor.[6] This is particularly true for allergens from plant derived food and in contrast to many major pollen allergens that have been intensively studied by modern immunochemical techniques as well as at the nucleic acid level. Moreover, many major pollen allergens have been expressed as recombinant proteins[54,55] and have been used for research purposes. Recombinant allergens represent the purified allergenic reference material of the future that can, theoretically, be produced in unlimited amounts.

Application of these advanced methods to certain food allergens have provided us with chemical data about important cross-reactive allergens and cross-reactive structures from pollen, fruit, and vegetables. As a result, we are now able to describe the molecular basis of some important clusters of cross-reactivity, which make the majority of allergies against fruits and vegetables explainable. However, despite the enormous progress in allergen research, one basic question in the context of allergy remains open: what are the specific structural features which make a molecule allergenic?

11.5.2 Birch Pollen Related Allergens in Fruits and Vegetables

Due to its predominance as a clinical phenomenon, many efforts have been made to characterize birch pollen related plant allergens. Bet v 1, the major birch pollen allergen, is a 159 residue protein with a pI of 5.5 that has been cloned and sequenced.[55] Bet v 1 consists of up to 20 isoallergens with slightly different molecular weights, pIs and immunologic properties.[56,57] More than 95% of birch pollen allergic patients are sensitive to Bet v 1.[58] Ebner et al.[59] were the first who could demonstrate common epitopes on Bet v 1 and a 17 to 18 kDa protein of apple. Moreover, the authors were able to show homology between the two proteins at the nucleic acid level by Northern blotting. One-dimensional and two-dimensional immunoblotting and immunoblot inhibition led to the identification of major fruit and vegetable allergens in certain birch pollen related fruits and vegetables, for example apple, pear, cherry, celery, and carrot. All these allergens clearly cross-reacted with Bet v 1-specific IgE antibodies.[60,61,62] The molecular data of these allergens are summarized in Table 11.2. Microsequencing of 15 N-terminal amino acid residues revealed significant sequence identities of four major fruit allergens with Bet v 1, that ranged from 28% (carrot) to 67% (cherry).[60] These results clearly demonstrate the molecular relationship of the food allergens with the major pollen allergen. The major allergens of apple, Mal d 1 (allergen 1 from Malus domestica), and celery, Api g 1 (allergen 1 from Apium graveolens), have also been investigated at the nucleic acid level.[63,64] Both cDNAs have been sequenced and the allergens have been cloned into *Escherichia coli*. The deduced amino acid sequence of Mal d 1 showed a 158 residue protein with a molecular weight of 17.5 kDa. The whole protein sequence is 56% identical with Bet v 1.[63] In the case of Api g 1, the major celery allergen, cDNA analysis revealed a 153 residue protein with 40% amino acid sequence identity to Bet v 1.[64] Comparison of Mal d 1 with Api g 1 showed 39% sequence identity. Moreover, the whole sequences of Bet v 1, Api g 1, Mal d 1, and the partial sequences of the cherry and carrot allergens are approximately 50% identical with some intracellular "pathogenesis related proteins" (PRP) from food plants, for example parsley[60] (Table 11.2), pea, or soybean.[63] The exact physiological functions of these proteins are unknown. Since they can be elicited by "stress factors," it has been speculated that these PRP are involved in defence reactions of food plants and allergenic trees of the Fagales order.[59] Finally, it has to be noted that mugwort pollen do not contain allergens possessing high homology with Api g 1 and Mal d 1. Therefore, Api g 1, for example, cannot be the causative agent of celery allergy in mugwort pollen-sensitive patients.

Some additional studies on Mal d 1 revealed very interesting results. Isolation of Mal d 1 and comparison of its allergenic potency with purified Bet v 1 by basophil histamine release and specific IgE inhibition showed an at least 15-fold higher allergenic potency of the birch

---

TABLE 11.2
Molecular Characteristics of Bet v 1[a] Related Major Allergens in Fruits and Vegetables

Food	MW [kDa]	pI	N-Terminal Sequence Identity[b] with Bet v 1 [%]	N-Terminal Sequence Identity[b] with PcPR 1-1[c] [%]	Ref.
Apple (Mal d 1)	17–18	5.5	53	40	Schöning et al., 1995[60]
Pear	16–18	5.1	n.d.	—	Schöning, 1995 (unpublished)
Cherry	18	5.8	67	40	Schöning et al., 1995[60]
Celery (Api g 1)	15–16	4.4-4.6	40	66	Schöning et al., 1995;[60] Ebner et al., 1995[61]
Carrot	16	4.4	28	57	Schöning et al., 1995;[60] Ebner et al., 1995[61]

Note: Acronyms of named food allergens are given in parenthesis. n.d.: not done

[a] Bet v 1: major birch pollen allergen.

[b] Calculated from 15 N-terminal amino acid residues.

[c] PcPR 1-1: pathogenesis related protein from parsley, sequence data from Somssich et al., *Mol. Gen. Genet.*, 213, 93, 1988.

pollen allergen in birch pollen allergic patients affected by a cross-allergy to apple.[65] These data support the view that Bet v 1 acts as the primary immunogen and that apple hypersensitivity is due to low-avidity cross-reactions of IgE. This interpretation is also in accordance with the clinical history of the patients who usually become sensitized to pollen before developing fruit allergy.[5] Furthermore, the Mal d 1 content and allergenicity differs drastically between apple strains,[66] and the major allergen content of Golden Delicious apples increases during maturation and storage.[67] When looking at the sequence identity to pathogenesis related proteins and the distribution of Mal d 1 in apple varieties one might speculate that the high allergen content of many apple strains (e.g., Granny Smith, Golden Delicious, Braeburn) may correlate with economical importance. One may speculate that Mal d 1 mediates properties which make these strains interesting for cultivation, for example, resistance against microbial attacks.

Finally, apple allergy represents an excellent model to gain information about the nature of the lability of a fruit allergen. With commonly used extraction buffers (for example PBS or carbonate buffers) the allergenicity of apples gets completely lost within a very short time resulting in inactivity of diagnostic skin test solutions.[14,65,68] This also happens after grinding or heating of apple fruit. It has been shown that allergen degradation in apple fruit during extraction is most likely due to activities of polyphenol oxidases.[14,65,68,69] After crushing of apple tissue, cell compartimentation gets destroyed and plant phenolics are enzymatically oxidized. The resulting quinones are able to bind covalently to, e.g., free amino groups of proteins and rapid denaturation takes place during this process. In contrast, active apple extracts can be obtained when enzyme inhibitors are used or by means of a low temperature acetone powder method which has been established in our laboratory.[14,66] Heat inactivation of the allergen is again most likely due to interactions with constituents of the fruit matrix. This can be concluded indirectly from our isolation procedure. We have purified Mal d 1 from apple extract by preparative SDS-PAGE under reducing and denaturing conditions and were able to obtain a highly allergenic preparation by electro-elution.[65] It is therefore unlikely that the obvious lability of Mal d 1 is due to the destruction of conformational epitopes. As an example, the high lability of apple proteins during extraction is illustrated by the SDS-PAGE analysis shown in Figure 11.3.[14]

Summarizing, it has to be concluded that the majority of allergies against fruits and vegetables in birch pollen allergic patients is due to analogues of the Bet v 1 allergen which occur in many food plants.

11.5.3 PROFILINS AS UBIQUITOUS CROSS-REACTING PLANT PAN ALLERGENS

Profilins constitute a family of conserved proteins with molecular weights ranging from 12 kDa to 15 kDa corresponding to 124 to 153 amino acid residues. They are present in almost all eukaryotic cells. They probably function as important mediators of membrane-cytoskeleton communication. Profilins specifically bind to several ligands, i.e., monomeric actin, phosphatidylinositol 4,5 bisphosphate (PIP 2), and poly-L-proline. These characteristics enable them to participate in the regulation of actin polymerization and to interact with the PIP2 pathway of signal transduction.[70–75]

Resulting from its ubiquitous distribution, the question must be raised as to why profilin has only been described as a pan allergen from plants and why the literature lacks reports about allergenic profilins in food derived from animals or autoimmune diseases caused by human profilins. Indeed, a weak immunoreactivity of IgE from a birch pollen allergic patient with human profilin has been described by Valenta and colleagues.[70] Human profilin was also able to trigger histamine release from sensitized basophils of allergic patients.[70,71]

For a conclusive interpretation of the complex data regarding profilin allergy, some additional molecular information about profilins has to be taken into consideration. Despite

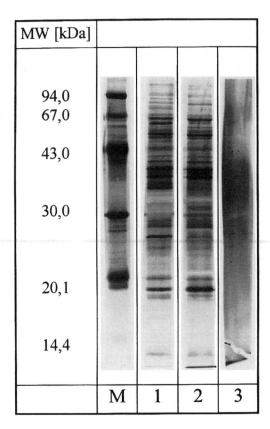

FIGURE 11.3 SDS-PAGE analysis of apple extracts. 1: Granny Smith (low temperature extract), 2: Golden Delicious (low temperature extract); 3: Allergen extract "apple" prepared by a conventional buffer extraction technique without application of enzyme inhibitors or temperature below 0°C, M: Low molecular weight marker proteins. The gel was silver stained. (From Vieths, S., Aulepp, H., Becker, W.-M., Buschmann L, in: *Deutsche Forschungsgemeinschaft, Food Allergies and Intolerances, Symposium*,[14] Eisenbrand, G., Aulepp, H., Dayan, A. D. Elias, P. S. Grunow, W., Ring J., and Schlatter J., Eds., VCH, Weinheim 1995. With permission.)

their identical functionality, the molecular data of profilins from distantly related organisms are very different. The pI-values of known profilins range between 4.6 and 9.2.[72] The sequences of many profilins show a relatively low degree of identity (Table 11.3).[72–75] Profilins from organisms of six different orders had only four common amino acid residues in identical positions.[74] Mammalian profilins had sequence identities of approximately 90% (Table 11.3). The identities of some known plant profilins are between 70 and 80% and these profilins share an identical sequence motif of 9 amino acid residues (VVERLGDYL) near their C-termini.[75] Due to the common functions and ligand-binding properties, the tertiary structures have been found to be highly conserved even between bovine and amoeba profilin.[76]

Returning to allergenicity, the close relationship between plant profilins and common partial sequences as mentioned above may be responsible for the observed differences in allergenic potency between profilins from mammals and plants. Moreover, a high uptake of native plant profilin via the respiratory route (pollen) and by oral ingestion (fresh fruits and vegetables) can be assumed, while animal profilins are usually consumed in a heat denatured state. Weak IgE reactivity of pollen allergic patients with animal and human profilin may be below the "symptom threshold." However, it has been speculated that autoimmune reactivity of birch pollen allergic patients to human profilin is responsible for maintaining a high IgE level outside the pollen season, whereas IgE against Bet v 1 decreases in winter.[70,71]

TABLE 11.3
Degree of Amino Acid Sequence Identity between Some Known Profilins

Profilin	Sequence identity with birch pollen profilin [%]	Sequence identity with timothy grass pollen profilin [%]
Saccharomyces	31	33
Acanthamoeba I	38	40
Acanthamoeba II	40	40
Physarum	41	44
Tetrahymena	29	31
Dictyostelium I	45	49
Dictyostelium II	39	44
Sea urchin	28	29
Mouse	32	26
Bovine	34	30
Human	33	27
Wheat (isoform TaPro 1)	79	80
Timothy grass pollen	79	100
Birch pollen	100	79

Note: Identities were calculated from aligned protein sequences.

Source: Data from Valenta, R., Ferreira, F., Grote, M., Swoboda, I., Vrtala, S., Duchene, M., Deviller, P., Meagher, R.B., Mckinney, R.B., Heberle-Bors, E., Kraft, D., and Scheiner, O., *J. Biol. Chem.*, 268: 22777-22781, 1993;[73] and Rihs, H.-P., Royznek, P., May-Taube, K., Welticke, B., and Baur, X., *Int. Arch. Allergy Immunol.*, 105, 190–194, 1994.[75]

Examples for food sensitizations mediated by profilins are celery,[77,78] tomato,[79] lychee fruit,[80] potato, apple, and peach.[27,81] Furthermore, immunoreactive profilins have been detected in more than 20 different fruits, vegetables, nuts, legumes, and cereals.[14,27,82]

The role of profilins as allergens can be summarized as follows:

1. Allergies against almost every particular kind of fruits and vegetables can be due to profilin sensitization.
2. In addition to Bet v 1 related allergens, profilin can also cause all typical birch pollen related food hypersensitivities.[27] Moreover, unusual cross-reactions to birch pollen exceeding the cluster of typical birch pollen associated foods can be due to profilin.
3. A high number of mugwort pollen associated food hypersensitivities seems to be caused by profilins.[32,33,77,78] These are frequently concomitted by cross-reactivity and co-sensitization to birch pollen profilin.
4. Food allergies in patients suffering from grass pollinosis may also be mediated by profilins. This has been confirmed in Spanish patients with peach and apple hypersensitivity[81] and for other fruits in Dutch patients.[27] The particular nature of the "grass pollen food cluster" may be influenced by specific nutritional habits of mediterranean countries.
5. The occasionally observed lack of cross-reactions between some profilin containing source materials may be due to structural differences of the involved profilins which can result in different epitope specificities of IgE.

However, intensive research is required to answer many open questions in the context of profilin allergy.

11.5.4   CARBOHYDRATE EPITOPES INVOLVED IN CROSS-REACTIVITY BETWEEN POLLEN AND FOOD

This paragraph deals with a third distinct structure of cross-reactivity which has been the subject of controversial discussions over many years. Allergen extracts are complex mixtures of proteins of which many are glycosylated. With some sera, the IgE binding capacity of many allergens can be significantly reduced by periodate treatment but is resistant to proteolytic cleavage of the peptide chain.[27,30,83] This points to the involvement of carbohydrate structures in IgE binding. Immunoblotting of electrophoretically separated allergen extracts has shown that these carbohydrate structures can be present on many different glycoproteins from one allergen source.[78,79,83] Sera with anti-carbohydrate IgE show a very broad spectrum of cross-reactivity including pollen, many fruits and vegetables, and also allergens from invertebrates like molluscs and arthropods.[27] Example for allergens with IgE carbohydrate moieties are phospholipase A 2 (PLA 2) from bee venom[84] or a β-fructofuranosidase from tomato.[79] The antigenic structure of PLA 2 and of the tomato allergen was identified as an α-1,3 linked fucose at the core of an N-linked oligosaccharide.[79,84] In addition, β-1,2 fucose xylose residues linked to the proximal mammose are thought to be IgE binding constituents N-glycans from food plants.[27] Fucose and xylose linked in this way are unknown in mammalian glycoproteins and are therefore highly immunogenic.[27] As an example for the detection of the immunoreactive glycans, Figure 11.4 shows IgE-reactivity and the binding of the lectin Con A to electrophoretically separated extract of gum arabic by immunoblotting.[83] Periodate treatment completely abolished the reactivity of both IgE and lectin. Despite the additional reactivity to invertebrate glycans, the cross-reactivity patterns of patients sensitized to these structures are comparable to profilin. In our studies, for example, we could demonstrate cross-reactivity of celery extract with apple, strawberry, orange, lychee fruit, onion, zucchini, brussels sprouts, aniseed, and basil. However, different reaction patterns observed with sera from individual patients may imply that IgE can distinguish carbohydrate groups with small structural differences.[27,83]

However, the clinical significance of these serological results is in doubt. Van Ree and Alberse[27] concluded from their data that IgE against carbohydrate epitopes generally occurs in asymptomatic patients. Our group of 12 carbohydrate sensitized patients included symptomatic as well as asymptomatic individuals.[83] Unfortunately, we could demonstrate a cosensitization to protein determinants in many cases. However, the carbohydrate moiety of PLA 2 may be involved in the induction of symptoms of bee venom allergy.[84] Ongoing investigations may clarify this crucial point.

11.5.5   OTHER CROSS-REACTIVE ALLERGENS IN POLLEN, FRUITS, AND VEGETABLES

Recently, Wellhausen and her colleagues[85] isolated a 35-kDa protein from birch pollen that bound IgE of 9/67 (12%) sera from birch pollen allergic patients. Despite the molecular weight estimated from SDS-PAGE, a pI of 7.4, and the existence of several isoallergens in the pI-range between 6.2 and 7.5, no additional molecular data of this protein is available. However, the allergen cross-reacted with IgE-binding protein bands of comparable size from apple, pear, orange, banana, mango, lychee fruit, and carrot. These 35-kDa proteins did not cross-react with Bet v 1. Although the clinical relevance of this observation needs further investigation, these minor allergens may be responsible for some typical birch pollen related food hypersensitivities (apple, pear, carrot) and allergies to exotic fruit which occasionally affect birch pollen allergic patients.[9] Like Bet v 1 and in contrast to profilin, the 35-kDa proteins effect a limited cluster of cross-reactivity.

Pastorello and her co-authors[49] recently studied 21 Italian patients with oral allergy syndrome due to the ingestion of peach, cherry, apricot, and plum. Peach was the example that

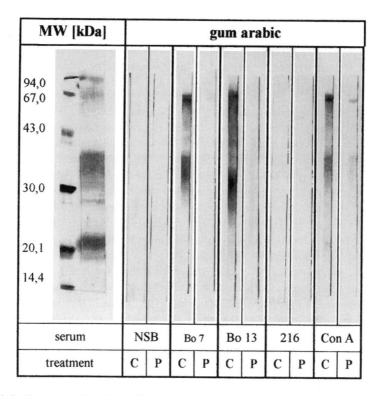

MW [kDa]	gum arabic									
serum	NSB		Bo 7		Bo 13		216		Con A	
treatment	C	P	C	P	C	P	C	P	C	P

FIGURE 11.4 Demonstration of specific binding of human IgE and the lectin Con A to plant glycans. Extract from gum arabic was separated by SDS-PAGE and transferred to a nitrocellulose (NC) membrane by electroblotting. Allergic human sera as indicated. NSB: Buffer control, C: control treated NC strips, P: NC strips treated with sodium periodate (0.1 M). The negative control serum 216 contained IgE against the major allergen Bet v 1 from birch pollen, but no evidence for IgE binding to carbohydrate epitopes was found. (From Vieths, S., Mayer, M., and Baumgart, M., *Food Agric. Immunol.*,[83] 6, 453, 1994. With permission.)

was most intensively studied. The major IgE-binding components of this fruit had molecular weights between 13 kDa and 20 kDa. A 14-kDa band of peach cross-reacted with grass pollen extract, birch pollen extract, and all fruit extracts and most likely represents profilin. One 20-kDa protein in peach extract that cross-reacted with birch pollen extract but not with grass pollen extract probably belonged to the "Bet v 1 allergen family." Furthermore, a 13-kDa major fruit allergen could be identified that was immunologically unrelated to pollen allergens but caused intensive cross-reactivity among all these four members of the prunoideae family (see Section 11.5.6).

No data are, however, available about specific allergens exclusively shared by grass pollen and associated plant food like orange, melon, or tomato except the finding that profilin can be the cross-reactive allergen.[27,49,79,80,81] As in mugwort pollen allergy, there is no evidence that major grass pollen allergens which occur in the molecular weight range between 30 kDa and 40 kDa have a significant degree of sequence identity nor that they cross-react with allergens in foods that have been described as allergenic for grass pollen hypersensitive patients.

The data about the structural background of fruit allergies in ragweed-sensitive patients are very poor. However, the clinical and immunological correlation is well documented. Out of 120 patient sera with anti-ragweed IgE, 48% reacted with watermelon, 46% with zucchini, 38% with cantaloupe, 35% with cucumber, and 18% with banana.[38,39] Cross-reactivity between the allergens has been documented by IgE-inhibition experiments.[38,39,40] Watermelon and ragweed extract were investigated by isoelectric focusing and immunoprinting. Both extracts contained 6 IgE binding bands with comparable isoelectric points. However, common migration characteristics in IEF and IgE binding do not necessarily indicate cross-reactivity. Besides these initial studies, no further data of the cross-reactive material is available.

11.5.6   FRUIT AND VEGETABLE ALLERGENS IMMUNOLOGICALLY INDEPENDENT OF POLLEN ALLERGENS

In the study of 23 patients with oral allergy syndrome caused by peach and prunoideae,[49] a common 13-kDa major allergen has been identified in peach, apricot, plum, and cherry. This allergen was cross-reactive between all four fruits. In contrast, the protein was not cross-reactive with whole protein extracts from grass pollen and birch pollen. The authors suspected this protein to be the main reason for the patients' clinical syndrome. Additional information about the allergen, for example pI or sequence data, has not been published.

Some interesting data have been reported about the participation thiol proteases in fruit allergy against kiwi, papaya fruit, and pineapple. Despite the findings that will be described below, it has to be mentioned that many kiwi fruit allergies seem to be due to the existence of a birch pollen related kiwi fruit allergen. This view is strengthened by a recent study of Gall et al.[15] who investigated 10 patients with severe and 12 subjects with mild kiwi hypersensitivity. Cross-allergenicity has been concluded from the co-occurrence of elevated specific IgE in the whole patient group and IgE inhibition experiments in three patients.[15]

Kiwi fruit and other tropical fruit contain proteolytic enzymes that are categorized as plant thiol proteases: actinidine in kiwi fruit, bromelain in pineapple, and papain in papaya. These are low molecular weight glycoproteins. Actinidine, which has also been sequenced, has a molecular weight of 12.8 kDa. Approximately one third of the patient group of Gall et al. was weakly reactive to either bromelain or papain in RAST and/or skin test. In a previous study, the same authors described a patient who was allergic against kiwi, pineapple, and papaya and whose IgE antibodies cross-reacted with all three proteases.[86] Furthermore, enzyme preparations of bromelain and papain are used for commercial purposes, for example in food technology, medicine, and as additives for detergent formulae. These applications have led to occupational sensitizations via the respiratory route.[87] Like in pollen allergy, inhalative sensitization in some cases results in allergic reactions after the ingestion of fruit containing these proteases. This observation has been described for bromelain/pineapple, for example.[20,88] The carbohydrate moiety of bromelain is known to contain an α1,3-linked fucose residue. Tretter et al.[84] could demonstrate cross-reactivity between PLA 2 from bee venom and the enzyme when using sera from carbohydrate sensitized patients with bee-venom allergy. Therefore, it cannot be excluded that protease allergy again is due to a well known allergenic structure, i.e., cross-reactive carbohydrate determinants. However, due to the low number of cases described thiol proteases are only minor fruit allergens with a relatively weak sensitization potency.[15,85]

As mentioned in Section 11.4.1, native potatoes are allergenic for patients sensitized to birch pollen. Two cross reactive allergens have been characterized by Calkhoven et al.[30] However, ingestion of cooked potatoes and even inhalation of steam released during cooking are known to cause severe symptoms in some allergic individuals.[4,89,90] The offending allergens seem to be immunologically independent of pollen allergens. Serological investigations have been performed with sera of 12 children and adolescents who were allergic to boiled potatoes.[90] The major IgE binding proteins had molecular weights between 16 kDa and 30 kDa and pIs in the range of 4.5 to 5.2. Further molecular data of the allergens are not available.

Finally, the structural basis of the latex fruit cluster has to be discussed. As mentioned, banana and avocado are the fruit involved. Immunoblot and immunoblot inhibition experiments have been performed with sera of 6 patients sensitized to latex and banana.[46] Latex extract contained 8 IgE reactive bands with molecular weights between 40 kDa and 75 kDa. The 13 IgE binding proteins of banana ranged between 14 kDa and 150 kDa. All allergenic bands from latex cross-reacted with banana extract and 10/13 banana allergens cross-reacted with latex extract. The authors did not investigate whether the allergens carried carbohydrate epitopes or not. A comparable study has been performed with avocado extract.[42] 14/18 sera had IgE against a total number of 17 avocado proteins with molecular weights between 16 and 91 kDa. 100 µg of latex protein inhibited IgE-binding to 15 of these 17 avocado proteins. Due to its increasing importance as a clinical phenomenon, the cross-reactive material eliciting symptoms needs further characterization.

Although many other kinds of fruits and vegetables have been described as being allergenic for individual patients[4] the literature lacks characteristics of the causative allergens in almost all cases.

In addition to the low number of the fruit and vegetable allergens with known sequence data (Mal d 1, Api g 1, and profilin) there are some other plant food allergens that have been studied in more detail. The most important are Ara h 1 and Ara h 2 from peanut which are highly potent elicitors of severe and sometimes lethal allergic reactions.[91,92] Other examples are Sin a 1, a 14-kDa allergen from mustard[93] and a 16-kDa rice allergen.[94]

11.6   CONCLUSIONS

Those plant derived foods with a high water content are classified as fruits and vegetables. This chapter represents a review of the current knowledge about allergy against these foods and about the allergenic molecules. From the clinical and epidemiological point of view, food allergies in pollen sensitized patients are of particular interest. These allergies are due to a serological cross-reactivity. Plausible reasons for the phenomena might not only be the higher sensitization potency of pollen allergens. Allergen uptake via the respiratory route appears to be an important mechanism for sensitization. Evidence for this view may also be taken from other important food allergies caused by allergen inhalation. Examples are oral allergy to bread ingestion in persons with bakers asthma, the bird egg-syndrome or milk allergies in farmers caused by inhalation of dander proteins from cattle.[20]

Detailed molecular data are only available for two families of fruit and vegetable allergens. Both belong to the pollen-related allergens: first, proteins of the Bet v 1-family which are homologous to pathogenesis related proteins and may have specific defense functions in the food plants, and second, profilins which have several known physiological functions and are therefore highly abundant in eukaryotic cells. Both are very small highly soluble proteins with molecular weights below 20 kDa. Specific carbohydrate structures being constituents of sugar moieties of many different plant glycoproteins may also contribute to allergies against fruits and vegetables. Despite these phenomena, our knowledge about the offending allergens is strikingly low. This is particularly so in the case of fruit and vegetable allergens that are not related to pollen allergens. One important allergenic molecule, again very small (13 kDa), that did not cross-react with pollen proteins has recently been described as a major allergenic protein of the prunoideae family. In addition, some groups have started to investigate the allergens of the "latex fruit cluster" at the molecular level and we can expect interesting results on both subjects in the near future.

REFERENCES

1. Ortolani, C., Pastorello, E. A., Farioli, L., Ispano, M., Prevettoni, V., Berti, C., Incorvaia, C., and Zanussi, C., IgE-mediated allergy from vegetable allergens, *Ann. Allergy,* 71, 470, 1993.
2. Wüthrich, B., Zur Nahrungsmittelallergie: Häufigkeit der Symptome und der allergieauslösenden Nahrungsmittel bei 402 Patienten, *Allergologie 1993,* 16, 280.
3. Wüthrich, B., Food Allergy: definition, prevalence, social impact, in *Proc. XVI. European Congress of Allergology and Clinical Immunology,* Madrid, 1995, 859.

4. Kreft, D., Bauer, R., and Goerlich, R., *Nahrungsmittelallergene—Charakteristika und Wirkungs-weisen,* de Gruyter, Berlin, 1995.

5. Dreborg, S., Food allergy in pollen-sensitive patients, *Ann. Allergy,* 61, 41, 1988.

6. Bruijnzeel-Koomen, C., Ortolani, C., Aas, K., Bindslev-Jensen, C., Björksten, B., Moneret-Vautrin, D., and Wüthrich, B., Position Paper: Adverse reactions to food, *Allergy* 50, 623, 1995.

7. Matsuda, T. and Nakamura, R., Molecular structure and immunologic properties of food allergens, *Trends Food Sci. Technol.* 41, 289, 1993.

8. Liebers, V., Kampen, V. v., Sander, I., Raulf-Heimsoth, M., Royznek, P., and Baur, X., Aktuelle Aspekte der Allergieforschung, *Allergo J.,* 4, 280, 1995.

9. Eriksson, N. E., Formgren, H., and Svenonius, E., Food hypersensitivity in patients with birch pollen allergy, *Allergy,* 37, 437, 1982.

10. Wüthrich, B. and Dietschi, R., Das "Sellerie-Karotten-Beifuß-Gewürz-Syndrom:" Hauttest und RAST-Ergebnisse, *Schweiz. Med. Wochenschr.,* 115, 358, 1985.

11. Thiel, C., Nahrungsmittelallergien bei Pollenallergikern (sogenannte pollenassoziierte Nahrungsmittelallergien), *Allergologie,* 11, 397, 1988.

12. Wüthrich, B., Stäger, J., and Johansson, S. G. O., Celery allergy associated with birch and mugwort pollinosis, *Allergy,* 45, 566, 1990.

13. Wüthrich, B., Schindler C., and Zemp, E., Prevalence of atopy and pollinosis in the adult population of Switzerland, *Int. Arch. Allergy Immunol.,* 106, 149, 1995.

14. Vieths, S., Aulepp, H., Becker, W.-M., and Buschmann, L., Characterization of labile and stable allergens in foods of plant origin, in: *Deutsche Forschungsgemeinschaft, Food Allergies and Intolerances, Symposium,* Eisenbrand, G., Aulepp, H., Dayan, A. D., Elias, P. S., Grunow, W., Ring, J., and Schlatter, J., Eds., VCH, Weinheim 1995.

15. Gall, H., Kalveram, K.-J., Forck, G., and Sterry, W., Kiwi fruit allergy: a new birch pollen associated food allergy, *J. Allergy Clin. Immunol.,* 94, 70, 1994.

16. Del Prete, G., Human Th1 and Th2 lymphocytes: their role in the pathophysiology of atopy, *Allergy,* 47, 450, 1992.

17. Prinz, J., Immunologische Grundlagen allergischer Erkrankungen, *Allergologie,* 16, 126-133, 1993.

18. Barnaba, V., Immunopathologic effect of antigen recognition, *Allergy,* 48, 137, 1993.

19. Roitt, I. M., Brostoff J., and Male, D. K., *Immunology,* Gower Medical, London, 1985.

20. De Blay, F., Pauli, G., and Bessot, J. C., Cross-reactions between respiratory and food allergens, *Allergy Proc.,* 12, 313, 1991.

21. Leonhardt, L. and Molitor, S. J., Nahrungsmittelallergien bei Bäckern, *Allergologie,* 16, 91, 1993.

22. Dummer, R., Bircher, A., and Wüthrich, B., Chronische Urticaria, berufsbedingte Rhinokonjunktivits und Asthma bronchiale bei Typ 1-Sensibilisierungen auf Johannisbrotkernmehl (E 410), *Allergologie,* 17, 217, 1994.

23. Axelsson, I. G. K., Ihre, E., and Zetterström, O., Anaphylactic reactions to sunflower seed, *Allergy,* 49, 517., 1994

24. Bernhisel-Broadbent, J. and Sampson, H. A., Cross allergenicity in the legume botanical family in children with food hypersensitivity, *J. Allergy Clin. Immunol.,* 83, 435, 1989.

25. Vestergard, H., Bindslev-Jensen, C., and Poulsen, L. K., Specific IgE to wheat in patients with grass pollen allergy, *Allergy Clin. Immunol. News, Suppl. 2,* Abstr. No. 1618, p. 445 (EAACI Congress, Stockholm, Sweden, 1994).

26. Jones, S. M., Magnolfi, C. F., Cooke, S. K., and Sampson, H. A., Immunologic cross-reactivity among cereal grains and grasses in children with food hypersensitivity, *J. Allergy Clin. Immunol.,* 96, 341, 1995.

27. Van Ree, R. and Aalberse, R. C., Pollen-vegetable food cross-reactivity: serological and clinical relevance of cross-reactive IgE, *J. Clin. Immunoassay,* 16, 124, 1993.

28. Pauli, G., Bessot, J. C., Dietemann-Molard, A., Braun, P. A., and Thierry, R., Celery sensitivity: clinical and immunological correlations with pollen allergy, *Clinical Allergy,* 15, 273, 1985.

29. De Martino, M., Novembre, E., Cozza, G., De Marco, A., Bonazza, P., and Vierucci, A., Sensitivity to tomato and peanut allergens in children monosensitized to grass pollen, *Allergy,* 43, 206, 1988.

30. Calkhoven, P. G., Aalbers, M., Koshte, V. L., Pos, O., Oei, H. D., and Aalberse, R. C., Cross-reactivity among birch pollen, vegetables and fruits as detected by IgE antibodies is due to at least three distinct cross-reactive structures, *Allergy,* 42, 382, 1987.

31. Helbling, A., Lopez, M., Schwartz, H. J., and Lehrer, S. B., Reactivity of carrot-specific IgE antibodies with celery, apiaceous spices, and birch pollen, *Ann. Allergy,* 70, 495, 1993.

32. Bauer, L., Hirschwehr, R., Wüthrich, B., Pichler, W., Fritsch, R., Scheiner, O., Ebner, C., and Kraft, D., IgE-cross-reactivity between birch pollen, mugwort pollen and celery is due to at least three distinct cross-reacting allergens. Immunoblot investigation of the birch-mugwort-celery syndrome. (submitted).

33. Jankiewicz, A., Aulepp, H., Baltes, W., Bögl, K. W., Dehne, L. I., Zuberbier, T., and Vieths, S., Allergic sensitization to native and heated celery root in pollen-sensitive patients by skin test and IgE binding, *Int. Arch. Allergy Immunol.* (in press).

34. Boccafogli, A., Vicentini, L., Camerani, A., Cogliati, P., D'Ambrosi, A., and Scolozzi, R., Adverse food reactions in patients with grass pollen allergic respiratory disease, *Ann. Allergy,* 73, 301, 1994.

35. Varela, S., Martinez-Cocera, C., Subiza, J. L., Marco, F., Jiminez, S., and Robledo, T., Anaphylaxis caused by swiss chard, *Allergy Clin. Immunol. News, Suppl. 2,* Abstr. No. 1622, p. 446 (EAACI Congress, Stockholm, Sweden, 1994).

36. Garcia Ortiz, J. C., Cosmes Martin, P., and Lopez-Ansulo, A., Melon sensitivity shares allergens with plantago and grass pollens, *Allergy,* 50, 269, 1995.

37. Anderson, L. B., Dreyfuss, E. M., Logan, J., Johnston, D. E., and Glaser, J., Melon and banana sensitivity coincident with ragweed pollinosis, *J. Allergy Clin. Immunol.,* 45, 310, 1970.

38. Enberg, R. N., Leickly, F. E., McCullough, J., Baily, J., and Ownby, D. R., Watermelon and ragweed share allergens, *J. Allergy Clin. Immunol.,* 79, 867, 1987.

39. Enberg, R. N., McCullough, J., and Ownby, D. R., Antibody responses in watermelon sensitivity, *J. Allergy Clin. Immunol.,* 82, 795, 1988.

40. Ross, B. D., McCullough, J., and Ownby, D. R., Evidence for allergenic cross-reactivity between banana and watermelon, *J. Allergy Clin. Immunol.,* 87, 274, 1991.

41. D'Amato, G. D. and Liccardi, G., Pollen-related allergy in the european mediterranean area, *Clin. Exp. Allergy,* 24, 210, 1994.

42. Alroth, M., Alenius, H., Turjanmaa, T., Mäkinen-Kiljunen, S., Reunala, T., and Palosuo, T., Cross-reacting allergens in natural rubber latex and avocado, *J. Allergy Clin. Immunol.,* 96, 167, 1995.

43. Blanco, C., Carillo, T., Castillo, R., Quiralte, J., and Cuevas, M., Latex allergy: clinical features and cross-reactivity with fruits, *Ann. Allergy,* 73, 309, 1994.

44. Rodriguez, M., Vega, F., Garcia, M. T., Panizo, C., Laffond, E., Montalvo, A., and Cuevas, M., Hypersensitivity to latex, chestnut and banana, *Ann. Allergy,* 69, 31, 1993.

45. De Corres, L. F., Moneo, I., Munoz, D., Bernaloa, G., Fernandez, E., Audicana, M., and Urrutiam I., Sensitization from chestnuts and bananas in patients with urticaria and anaphylaxis from contact with latex, *Ann. Allergy,* 70, 35, 1993.

46. Mäkinen-Kiljunen, S., Banana allergy in patients with immediate type hypersensitivity to natural rubber latex: characterization of cross-reacting antibodies and allergens, *J. Allergy Clin. Immunol.,* 93, 990, 1994.

47. M'Raihi, L., Charpin, D., Pons, A., Bongrand, P., and Vervloet, D., Cross-reactivity between latex and banana, *J. Allergy Clin. Immunol.,* 87, 129, 1991.

48. Andre, L., Guilloux, L., and Andre, C., Frequency of cross-sensitivities between latex, banana, chestnut, avocado and kiwi in 50 carriers of food allergy, *Allergy,* 50, 125, 1995.

49. Pastorello, E. A., Ortolani, C., Farioli, L., Pravettoni, V., Ispano, M., Borga, A., Bengtsson, A., Incorvaia, C., Berti, C., and Zanussi, C., Allergenic cross-reactivity among peach, apricot, plum and cherry in patients with oral allergy syndrome: an in vivo and in vitro study, *J. Allergy Clin. Immunol.,* 94, 699, 1994.

50. Metcalfe, D. D., Food allergens, *Clin. Rev. Allergy,* 3, 331, 1985.

51. Taylor, S. L., Lemanske, R. F., Bush, R. K., and Busse, W. W., Food allergens: structure and immunologic properties, *Ann. Allergy,* 59, 93-99, 1987.

52. Anderson, J. A., Food allergy, *Immunology and Allergy Clinics of North America,* Vol. 11, No. 4, Saunders, Philadelphia, 1991.

53. Yunginger, J. W., Classical food allergens, *Allergy Proc.*, 11, 7, 1990.

54. Scheiner, O. and Kraft, D., Basic and practical aspects of recombinant allergens, *Allergy*, 50, 384, 1995.

55. Breiteneder, H., Pettenburger, K., Bito, A., Valenta, R., Kraft, D., Rumpold, H., and Breitenbach, M., The gene coding for the major birch pollen allergen Bet v I is highly homologous to a pea disease resistance response gene, *EMBO J.*, 8, 1935, 1989.

56. Swoboda, I., Jilek, A., Ferreira, F., Engel, E., Hoffmann-Sommergruber, K., Scheiner, O., Kraft, D., Breiteneder, H., Pittenauer, E., Schmidt, E., Vicente, O., Heberle-Bors, E., Ahorn, A., and Breitenbach, M., Isoforms of Bet v 1, the major birch pollen allergen, analyzed by liquid chromatography, mass spectrometry and cDNA cloning, *J. Biol. Chem.*, 270, 2607, 1995.

57. Ebner, C., Ferreira, F., Breiteneder, H., Schenk, S., Scheiner, O., and Kraft, D., Recombinant isoforms of allergenic molecules. Possible role for therapeutical purposes in type I allergy, in: *Proc. XVI. European Congress of Allergology and Clinical Immunology*, Madrid, 1995, 523.

58. Jarolim, J., Rumpold, E., Endler, A. T., Ebner, H., Breitenbach, M., Scheiner, O., and Kraft, D., IgE and IgG antibodies as tools to define the allergen profile of Betula verrucosa, *Allergy*, 44, 385, 1989.

59. Ebner, C., Birkner, T., Valenta, R., Rumpold, H., Breitenbach, M., Scheiner, O., and Kraft, D., Common epitopes of birch pollen and apples—studies by western and northern blot, *J. Allergy Clin. Immunol.*, 88, 588, 1991.

60. Schöning, B., Vieths, S., Petersen, A., and Baltes, W., Identification and characterization of allergens related to Bet v I, the major birch pollen allergen, in apple, cherry, celery and carrot by two-dimensional immunoblotting and microsequencing, *J. Sci. Food Agric.*, 67, 431, 1995.

61. Ebner, C., Hirschwehr, R., Bauer, L., Breitender, H., Valenta, R., Ebner, H., Kraft, D., and Scheiner, O., Identification of allergens in fruits and vegetables: IgE cross-reactivities with the important birch pollen allergens Bet v 1 and Bet v 2 (birch profilin), *J. Allergy Clin. Immunol.*, 95, 962, 1995.

62. Vieths, S., Schöning, B., Aulepp, H., and Baltes, W., Identifizierung kreuzreagierender Allergene in Pollen und pflanzlichen Lebensmitteln, *Lebensmittelchemie*, 47, 49, 1993.

63. Schöning, B., Ziegler, W. H., Vieths, S., and Baltes, W., Apple allergy: The cDNA sequence determined by performing PCR with a primer based on the N-terminal amino acid sequence of the major allergen of apple is highly homologous to the sequence of the major birch pollen allergen, *J. Sci. Food Agric*, (in press).

64. Breiteneder, H., Hoffmann-Sommergruber, K., O'Riordain, G., Susani, M., Ahorn, H., Ebner, C., Kraft, D., and Scheiner, O., Molecular characterization of Api g 1, the major allergen of celery (Apium graveolens), and its immunological and structural relationships to a group of 17 kDa tree pollen allergens, *Eur. J. Biochem.*, 233, 484, 1995.

65. Vieths, S., Janek, K., Aulepp, H., and Petersen, A., Isolation and characterization of the 18-kDa major apple allergen and comparison with the major birch pollen allergen (Bet v I), *Allergy*, 50, 421, 1995.

66. Vieths, S., Jankiewicz, A., Schöning, B., and Aulepp, H., The IgE binding potency of apple strains is related to the occurrence of the 18-kDa apple allergen, *Allergy*, 49, 262, 1994.

67. Vieths, S., Schöning, B., and Jankiewicz, A., Occurrence of IgE binding allergens during ripening of apple fruits, *Food Agric. Immunol.*, 5, 93, 1993.

68. Vieths, S., Aulepp, H., Schöning, B., and Tschirnich, R., Untersuchungen zur Apfelallergie bei Birkenpollenallergikern: Hauttestergebnisse und Immunoblot-Untersuchungen mit Tieftemperaturextrakten unterschiedlicher Apfelsorten, *Allergologie*, 18, 89, 1995

69. Björksten, F., Halmepuro, L., Hannuksela, M., and Lahti, A., Extraction and properties of apple allergens, *Allergy*, 35, 671, 1980.

70. Valenta, R., Duchene, M., Pettenburger, K., Sillaber, S., Valent, P., Bettelheim, P., Breitenbach, M., Rumpold, H., Kraft, D., and Scheiner, O., Identification of profilin as a novel pollen allergen. IgE autoreactivity in sensitized individuals, *Science*, 253, 557, 1991.

71. Valenta, R., Duchene, M., Ebner, C., Valent, P., Sillaber, C., Deviller, P., Ferreira, F., Teijkl, M., Edelmann, H., Kraft, D., and Scheiner, O., Profilins constitute a novel family of functional plant pan-allergens, *J. Exp. Med.*, 175, 377, 1992.

72. Giehl, K., Valenta, R., Rothkegel, M., Ronsiek, M., Mannherz, H. G., and Jockusch, B. M., Interaction of plant profilins with mammalian actin, *Eur. J. Biochem.*, 226, 681, 1994.

73. Valenta, R., Ferreira, F., Grote, M., Swoboda, I., Vrtala, S., Duchene, M., Deviller, P., Meagher, R. B., Mckinney, E., Heberle-Bors, E., Kraft, D., and Scheiner, O., Identification of profilin as an actin-binding protein in higher plants, *J. Biol. Chem.*, 268, 22777, 1993.

74. Pollard, T. D. and Rimm, D. L., Analysis of cDNA clones for Acanthamoeba profilin-I and profilin-II shows end to end homology with vertebrate profilins and a small family of profilin genes, *Cell Motil. Cytoskeleton*, 20, 169, 1991.

75. Rihs, H.-P., Royznek, P., May-Taube, K., Welticke, B., and Baur, X., Polymerase chain reaction based cloning of wheat profilin: a potential plant allergen, *Int. Arch. Allergy Immunol.*, 105, 190, 1994.

76. Cedergren-Zeppezauer, E. S., Gooneskere, N. C. W., Rozycki, M. D., Myslik, J. C., Dauter, Z., Lindberg, U., and Schutt, C. E., Crystallization and structure determination of bovine profilin at 2.0 A resolution, *J. Mol. Biol.*, 240, 459, 1994.

77. Vallier, P., Dechamp, C., Valenta, R., Vial, O., and Deviller, P., Purification and characterization of an allergen from celery immunochemically related to an allergen present in several other plant species. Identification as a profilin, *Clin. Exp. Allergy*, 22, 774, 1992.

78. Vieths, S., Jankiewicz, A., Wüthrich, B., and Baltes, W., Immunoblot study of IgE binding allergens in celery roots, *Ann. Allergy Asthma Immunol.*, 1995; 74, 48, 1995.

79. Petersen, A., Vieths, S., Aulepp, H., Schlaak, M., and Becker, W.-M., Ubiquitous structures responsible for IgE cross-reactivity between tomato fruit and grass pollen allergens, *J. Allergy Clin. Immunol.* (in press).

80. Fäh, J., Wüthrich, B., and Vieths, S., Anaphylactic reaction to lychee fruit - evidence for sensitization to profilin, *Clin. Exp. Allergy*, 25, 1018, 1995.

81. van Ree, R., Fernandez-Rivas, M., Cuevas, M., van Wijngaarden, M., and Aalberse, R. C., Pollen-related allergy to peach and apple: an important role for profilin, *J. Allergy Clin. Immunol.*, 95, 726, 1995.

82. van Ree, R., Voitenko, V., van Leeuwen, W. A., and Aalberse, R. C., Profilin is a cross-reactive allergen in pollen and vegetable food, *Int. Arch. Allergy Immunol.*, 98, 97, 1992.

83. Vieths, S., Mayer, M., and Baumgart, M., Food Allergy: specific binding of IgE antibodies from plant food sensitized individuals to carbohydrate epitopes, *Food Agric. Immunol.*, 6, 453, 1994.

84. Tretter, V., Altmann, F., Kubelka, V., März, L., and Becker, W.-M., Fucose (α1,3-linked to the core region of glycoprotein N-glycans creates an important epitope for IgE from honeybee venom allergic individuals, *Int. Arch. Allergy Immunol.*, 102, 259, 1993.

85. Wellhausen, A., Schöning, B., Petersen, A., and Vieths, S., IgE binding to a new cross-reactive structure: a 35 kDa protein in birch pollen, exotic fruit and other plant foods, *Z. Ernährungswiss.* (in press).

86. Gall, H., Kalveram, K.-J., Forck, G., and Tümmers, U., Kreuzallergie zwischen Kiwi, Thiol-proteinasen, Pollen, und Nahrungsmitteln, *Allergologie*, 13, 447, 1990.

87. Merget, R., Stollfuss, J., Wiewrodt, R., Frühauf, H., Koch, U., Bolm-Audorff, U., Bienfait, H.-G., Hiltl, G., and Schulze-Werninghaus, G., Respiratory pathophysiologic responses. Diagnostic tests in enzyme allergy, *J. Allergy Clin. Immunol.*, 92, 264, 1993.

88. Baur, X. and Fruhmann, G., Allergic reactions, including asthma, to pineapple protease bromelain following occupational exposure, *Clin. Allergy*, 9, 443, 1979.

89. Kästner, H., Kalveram, K. F., and Forck, G., Soforttypallergie gegen Kartoffeln, *Allergologie*, 7, 354, 1984.

90. Wahl, R., Lau, S., Maasch, H.-J., and Wahn, U., IgE mediated allergic reactions to potatoes, *Int. Arch. Allergy Appl. Immunol.*, 92, 168, 1990.

91. Burks, A. W., Williams, L. W., Helm, R. M., Connaugthon, C., Cockrell, G., and O'Brian, T., Identification of a major peanut allergen, Ara h I, in patients with atopic dermatitis and positive peanut challenges, *J. Allergy Clin. Immunol.*, 88, 172, 1991.

92. Burks, A. W., Williams, L. W., Connaugthon, C., Cockrell, G., O'Brian, T., and Helm, R. M., Identification and characterization of a second major peanut allergen, Ara h II, with use of sera of patients with atopic dermatitis and positive peanut challenge, *J. Allergy Clin. Immunol.*, 90, 962, 1992.

93. Mendenez-Arias, L., Moneo, I., Dominguez, J., and Rodriguez, R., Primary structure of the major allergen of yellow mustard (Sinapis alba L.,) seeds, Sina a I, *Eur. J. Biochem.,* 177, 159, 1988.
94. Matsuda, T., Suiyama, M., Nakamura, R., and Torii, S., Purification and properties of an allergenic protein in rice grain, *Agric. Biol. Chem.,* 52, 1465, 1988.

12 Linear Furanocoumarins

M. M. Diawara and J. T. Trumble

CONTENTS

12.1   INTRODUCTION

The linear furanocoumarins are plant metabolites that have been used since ancient times to treat skin disorders such as psoriasis, conditions of skin depigmentation (such as leprosy, vitiligio, and leukoderma), mycosis fungoides, polymorphous dermatitis, and eczema.[1,2,3,4,5,6] Use of furanocoumarin-containing plants for medicinal purposes dates as far back as 2000 B.C.[4] The legume *Psorelea coryfolia* and the umbelliferous plant *Ammi majus*, for example, have been used since ancient times in North African civilizations, in the Hindu culture, and by the Chinese.[1,4,7] The increased use of the linear furanocoumarins in medicine has occasionally been linked, however, to higher incidence of skin cancer.[1,2,3,8] and other disorders such as sister chromatid exchanges, gene mutation, and chromosomal aberrations in humans.[9,10] Because these biosynthetic compounds are active against herbivores (including humans) and distributed among both wild and domesticated plant species,[8,11] they have garnered substantial scientific attention and thus have been the subject of a great deal of research in the last several decades. To date, the linear furanocoumarins have been characterized and identified in at least 15 plant families: Amaranthaceae, Compositae, Cyperaceae, Dipsacaceae, Fabaceae, Goodeniaceae, Guttiferae, Leguminosae, Moraceae, Pittosporaceae, Rosaceae, Rutaceae, Samydaceae, Solanaceae, and Umbelliferae (Apiaceae).[8,11,12]

From an evolutionary standpoint, the driving force(s) leading to the production of furanocoumarins has generated considerable speculation. In Apiaceae, the presence of furanocoumarins is suspected to have evolved in response to several physical and biological stress factors. Beier and Oertli,[13] Zangerl and Berenbaum,[14] and Zobel and Brown,[15] found that furanocoumarins were induced by UV light, suggesting that these chemicals may provide protection against mutagenic UV radiation. Zobel and associates[16] recently reported that a significant portion of these compounds can be exuded on the plant surface, where they may act as "sunscreens" (UV blockers).

The furanocoumarins have been reported to be active against a wide variety of organisms. Inhibition of bacterial, fungal, as well as viral infections have been associated with increased concentrations of linear furanocoumarins in plants.[17,18,19,20,21,22,23] Recent studies also suggest the potential allelopathic role of furanocoumarins against nearby plants through retardation of germination and growth.[16,24] The furanocoumarins have also been shown to provide the chemical basis for feeding deterrency in *Apium graveolens* (celery) cultivars selected for, or found to have, resistance to insects[13,25,26,27] or, in other cases, they prevented feeding adaptation by specialist herbivores.[28] In addition, furanocoumarins have proven to be toxic to a broad spectrum of insects, suggesting that they may have evolved, at least in parts, in response to herbivory.[8,29,30] The allocation of a high proportion (>60%),[31] of the available complement of linear furanocoumarins to the outer leaves of celery (as opposed to the interior leaves, petioles, or roots), appears to support all of the previously mentioned rationales for the development and/or maintenance of these compounds (e.g., UV radiation, fungi, bacteria, herbivory), at least in celery.

Due to the diverse array of documented biological activities (from medicinal to agricultural to ecological) of the linear furanocoumarins, reporting in detail on all of the available literature is not feasible. Therefore, this chapter has been organized into a selection of what we believe are key topic areas including the biosynthesis of linear furanocoumarins, their toxicity to a wide range of organisms, the factors affecting production and toxicity, metabolic detoxification, and the outlook for research on linear furanocoumarins.

12.2 BIOSYNTHESIS OF LINEAR FURANOCOUMARINS

Like many of the more than 800 coumarin (**1**) derivatives that have been identified and characterized primarily from green plants,[8] the linear furanocoumarins are structurally derived from shikimic and chorismic acids via phenylalanine (**2**).[8,11] Their biosynthesis "begins with the transformation, catalyzed by phenylalanine ammonia lyase, of phenylalanine to *trans*-cinnamic acid (**3**)."[8] *Trans*-cinnamic acid is first ortho-hydroxylated to 2-hydroxycinnamic acid. Umbelliferone (or 7-hydroxycoumarin) (**4**), which is considered to be the mother compound of all linear furanocoumarins,[8,32] is derived when cinnamic acid is hydroxylated at the 4' position to p-coumaric acid (**5**) and then at the 2' position.

Several workers have reviewed the biosynthesis of coumarins.[8,11,32,33,34] The specific biosynthetic pathway varies among plant taxa.[11,32,33,34,35,36,37,38,39] In Rutaceae and Apiaceae, demethylsuberosin (**6**) is the coumarin widely reported to be the intermediate in the conversion of umbelliferone to marmesin (**7**) via prenylation.[8,32,39,40,41] This reaction is reportedly catalyzed by an enzyme referred to as dimethylallylpyrophosphate: umbelliferone dimethylallyl-transferase.[41] Ebel[42] recently proposed that marmesin loses its hydroxypropyl group via oxidation to yield psoralen (**8**); it is believed that this conversion is catalyzed by a P450 monooxygenase. Psoralen is hydroxymethylated to 5-methoxypsoralen (bergapten) (**9**), 8-methoxypsoralen (xanthotoxin) (**10**), or 5,8-methoxypsoralen (isopimpinellin) (**11**) in the presence of site-specific methylases.[8,32,37,38,40] Psoralen, bergapten, xanthotoxin, and isopimpinellin are the four linear furanocoumarins that have been widely characterized and identified in green plants. Except for some reported fungal toxicity,[23,43] isopimpinellin has generally not proved to have the photosensitizing properties of the other three linear furanocoumarins;[44] consequently, this compound has received less attention. In addition to the linear furanocoumarins, some plant species also have angular furanocoumarins such as angelicin (**12**), but the linear furanocoumarins usually constitute a higher proportion of their total furanocoumarins.[8]

Murray and colleagues[11] reported that coumarin synthesis occurred primarily in younger leaves of legumes. In contrast, the buds and seeds of *Pastinaca sativa* (Apiaceae) had the highest concentrations of these compounds.[8] Leaves of celery have been shown to have much higher concentrations of furanocoumarins than petioles.[25,26,27,31,45] Similar results were found

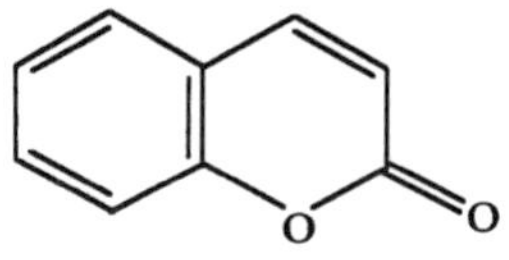

1. Coumarin

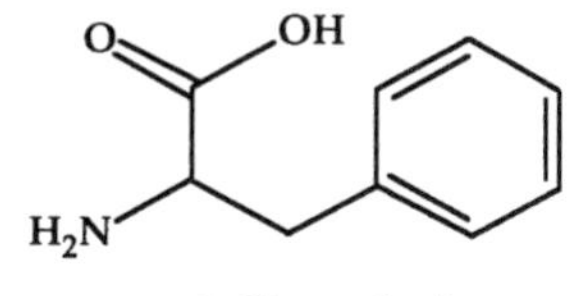

2. Phenyalanine

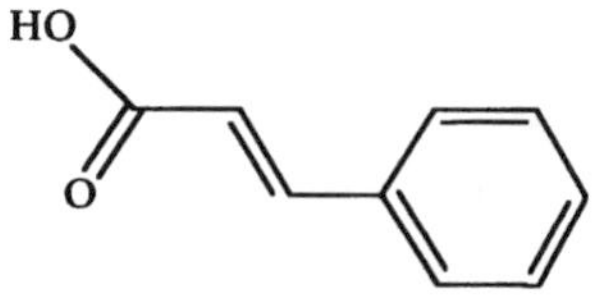

3. *trans*-Cinnamic acid

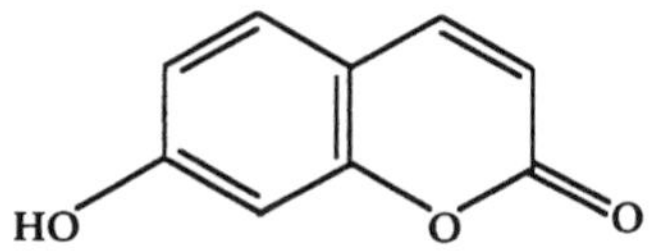

4. Umbelliferone

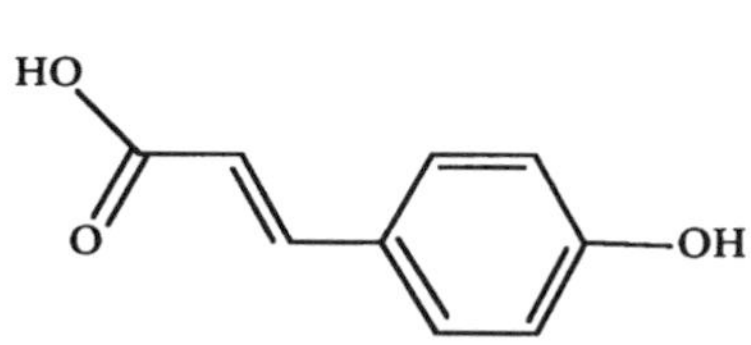

5. p-Coumaric acid

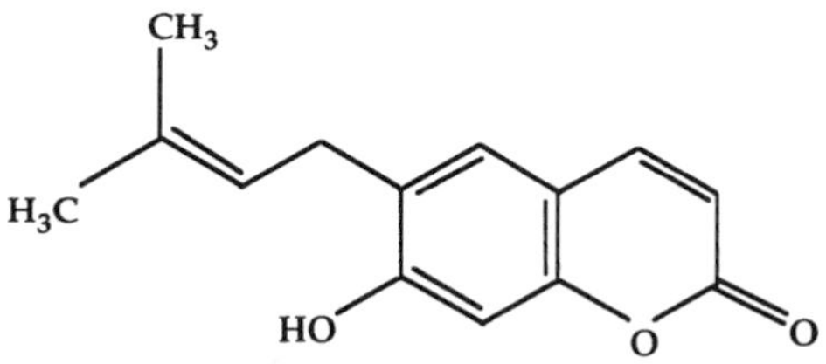

6. Demethylsuberosin

for other species of Apiacae and among Rutaceae; upper green leaves of *Ruta graveolens* contained more furanocoumarins than lower green leaves and green leaves of the plant had more furanocoumarins than yellow leaves, which also contained more than dry leaves.[46] Lime pulp was shown to have much less furanocoumarins than the peel.[47] Coumarins are, however, generally found in all plants parts.[8,31] As for their biosynthesis, the composition[21,25,31,45,46] as well as the localization[8,31,48] of linear furanocoumarins within specific plant structures vary among and within plant taxa. To date, it remains unclear whether the furanocoumarins are translocated between different parts of the plant either during or after synthesis; this has been previously discussed (see Diawara et al.[31]).

12.3 TOXICITY OF LINEAR FURANOCOUMARINS

12.3.1 MECHANISM OF TOXICITY

Like most other coumarins, the linear furanocoumarins are photoactivated plant biosynthetic compounds.[8,11,14,29,46,49,50,51,52] The effective ultraviolet A (UVA) wavelength range for this photoreactivity is between 320 and 400 nm;[11,52,53,54,55,56] the addition of UVB radiation does

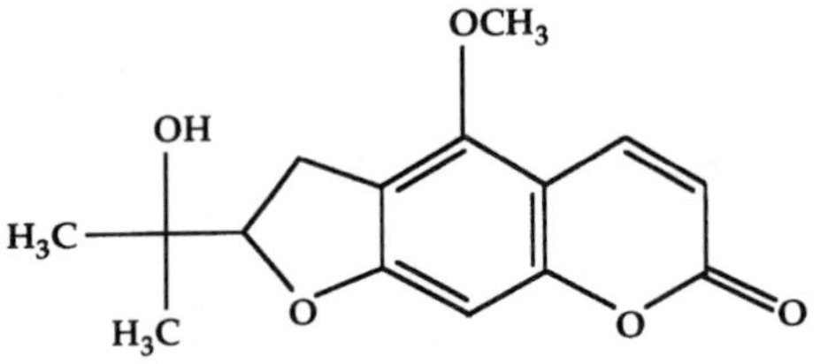

7. (+)-Marmesin

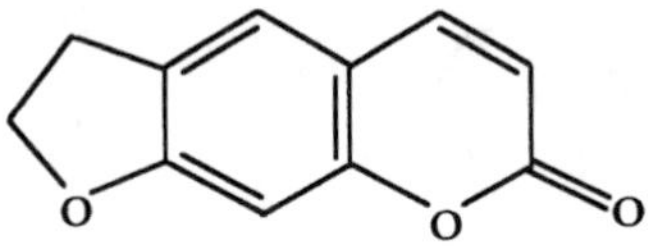

8. Psoralen

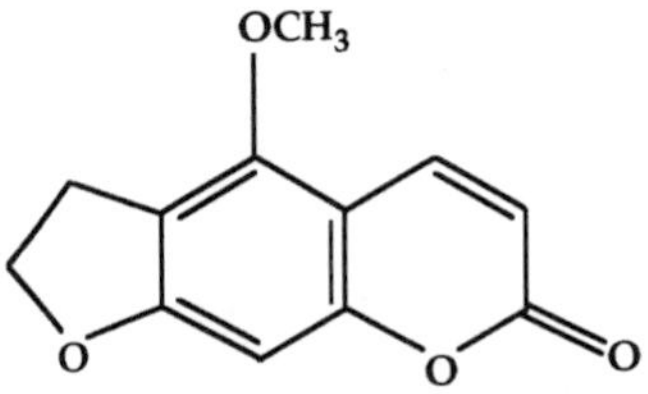

9. Bergapten

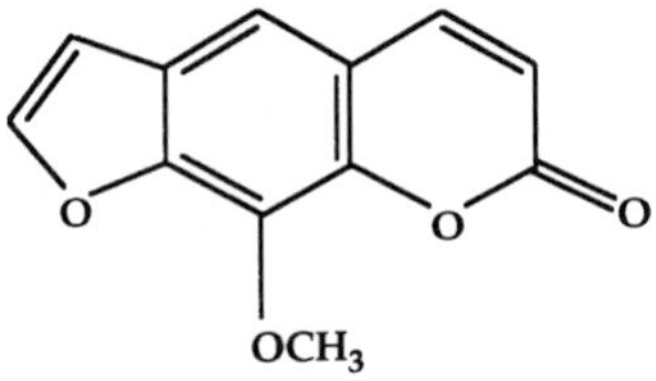

10. Xanthotoxin

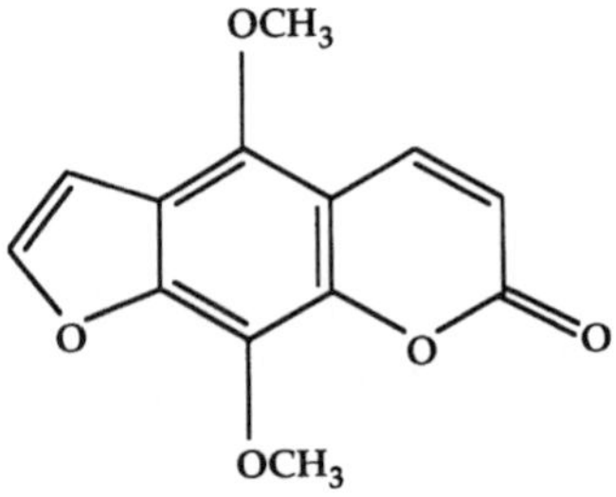

11. Isopimpinellin

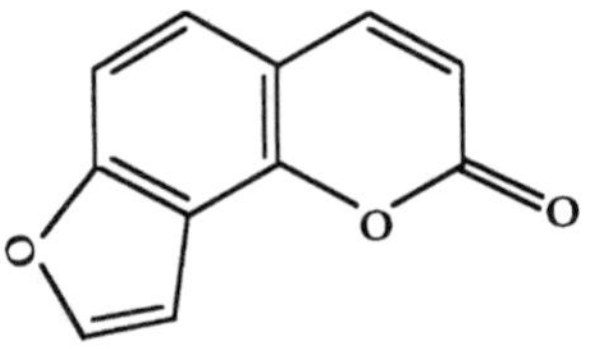

12. Angelicin

not seem to significantly affect activity.[57] Following absorption of a photon, the furanocoumarins form an excited triplet state which can react with molecules such as pyrimidine bases or with ground state oxygen, resulting in the formation of singlet oxygen or toxic oxyradicals such as superoxide anion radicals or hydroxyradicals.[8] All of these molecules can react with DNA, RNA, proteins, and lipids. The furanocoumarins have been shown to bind to the pyrimidine base of DNA.[53,58] This binding can result in formation of monoadducts, where furanocoumarins bind to a single pyrimidine base, and thus cause cytoplasmic mutations.[59] Recent studies by Laquerbe and colleagues[60] comparing data obtained from normal human lymphoblasts, rodents, and yeast cells suggest that the mutagenic potential of the monoadducts vs. diadducts (also called cross-links) may be species-specific. Diadducts, which cross-link complementary strands of DNA and prevent their separation,[52,61] are formed when UV-activated monoadducts react with additional pyrimidine bases in opposing strands of DNA.[50,58] The furanocoumarins have also been shown to inhibit enzymes by degrading protein constituents due to production of singlet oxygens and photobinding.[62,63] They can be photoactive with proteins and lipids in both oxygen-dependent and oxygen-independent reactions.[8] The DNA binding ability of furanocoumarins and their reactivity with and ability to damage lipids, proteins, RNA, as well as DNA, constitute the basis for their toxicity to a wide range of organisms including mammals, insects and other arthropods, nematodes, viruses, and bacteria, and even plants and fungi.

Despite the vast volume of literature on toxicity of furanocoumarins, the actual mechanism(s) of their action at the molecular level is not well understood; for instance, it is unclear how pigment deposition in photodermatitis (see toxicity to mammals) is affected by cross-linking of DNA. However, it is known that activity of tyrosinase in pigment cells is stimulated by the compound trimethylpsoralen.[8] Xanthotoxin and other furanocoumarins inhibit both mammalian[64,65,66,67] and insect cytochrome P450s,[68,69] which are among the most important insect enzymes involved in metabolism of allelochemicals.[70] During a recent study designed to test coumarin, benzofuran, and 16 furanocoumarins for inhibitory effects on the insect *Manduca sexta* midgut cytochrome P450-catalyzed *O*-demethylation of *p*-nitroanisole, Neal and Wu[69] found that "all of the inhibitory furanocoumarins tested were mechanism-based irreversible inhibitors" and proposed that "the furanocoumarin is oxidized by cytochrome P450 at the double bond of the furan ring forming an unstable epoxide that can react with cytochrome P450."

12.3.2 TOXICITY TO MAMMALS

The most commonly reported manifestation of linear furanocoumarin toxicity to higher animals is phytophotodermatitis, an epidermal reaction symptomized by bullous eruptions, pigmentation, erythema, and potential vesicle formation.[8,32,51,71,72] These manifestations may be seen simply at the point of contact with high-furanocoumarin content material or over the entire body of an individual, depending on whether there was dermal contact only or an oral ingestion, respectively. Thus, furanocoumarins can reach the skin by direct contact or by blood-borne transfer to the skin following ingestion. Most humans show little reaction, or at least symptoms are not visible in the absence of UVA light exposure; thus the terms "photosensitization" and "photoactivation" are sometimes used to describe the reaction in the medical literature.

Scientific interest in phytophotodermatitis started in Europe in the 17th century. According to Brown,[32] a major step toward understanding the role of furanocoumarins in the causation of dermatitis in humans was a study reported in 1938 which demonstrated that synthetic pure bergapten and xanthotoxin induced the effects of furanocoumarin-containing plants when directly applied to the skin; this was later confirmed by several workers in Italy and the U.S.

Crop plants that have been reportedly associated with human health hazards as a result of high contents of linear furanocoumarins belong to four plant families: Apiaceae (anise, caraway, carrot, celery, chervil, dill, fennel, lovage, parsley, and parsnip), Moraceae (figs), Rutaceae (grapefruit, lemon, lime, and orange), and Solanaceae (potato).[11,12,8,45,47,73,74,75,76] Of these, celery has been among the most extensively studied because of the occasionally high concentrations of linear furanocoumarins in the plant and risks of phytophotodermatitis associated with harvesting, handling, or ingestion.[45,75,76,77,78,79] These hazards apparently are even more serious when plants are infected with disease-causing pathogens.[17,20]

The chemicals responsible for crop plant-induced phytophotodermatitis have been known since the mid-1970s to be the linear furanocoumarins psoralen, 5-methoxypsoralen (bergapten), and 8-methoxypsoralen (xanthotoxin).[13,18,77] The threshold level for toxicity to humans was determined to be 18 μg g^{-1} fresh weight for development of acute dermatitis,[77] and 7–9 μg g^{-1} for repeated or chronic exposures.[78] This may vary for different body regions.[80]

The furanocoumarins have been shown to be both mutagenic and carcinogenic.[45,55,60,81,82,83,84,85,86,87,88] *In vitro* bioassays with bacterial and mammalian cells demonstrated that these chemicals are lethal and carcinogenic.[81] Beier[47] recently reported the death of a 45-year-old woman due to complications from severe burns that the patient received in a tanning salon while under psoralen medication. The World Health Organization recognizes these psoralens as causal agents of skin cancer in humans.[89] Further, these compounds can interact with other medications: furanocoumarins have been demonstrated to induce hypothermic activity and anticonvulsive activity in combination with several drugs when injected into rats.[90] These compounds have also been shown to deter feeding by grazing animals.[91]

Like other DNA-damaging agents, the furanocoumarins hypersensitize several rare hereditary and tumor-prone disorders in humans, including Fanconi anemia (see Bredberg et al.[92] and references therein). It is also well established that they can cause conjunctival hyperemia and decreased lacrimation[10] and increase plasma melatonin levels by inhibiting metabolism of this compound (see Garde et al.[93] and Rosselli et al.[94] and references therein). Thus, potential effects of these plant compounds on mammals are not only diverse, but occasionally debilitating or even lethal.

12.3.3 TOXICITY TO INSECTS, PLANTS, FUNGI, BACTERIA, AND VIRUSES

Berenbaum[8] recently reviewed the toxicity of the furanocoumarins to insects, plants, fungi, bacteria, and viruses; the reader is referred to her article for details on earlier studies. In reference to furanocoumarin toxicity to plants, Berenbaum did suggest the involvement of furanocoumarins in regulation of seed germination because of their localization in seed coats in many species, and their absence from the endosperm. More recent reports have confirmed that the furanocoumarins can exert allelopathic effects against nearby plants through retardation of germination and growth.[16,95] Kupidlowska and co-workers[24] suggested that this allelopathy may be due to the furanocoumarins' ability to retard mitosis, to decrease oxygen uptake by meristematic cells, and to cause structural and physiological alterations in the mitochondrial matrix, as observed in *Allium cepa*. New advances in activity of furanocoumarins against fungi, bacteria, and viruses that attack plants include reports on their toxicity to several species of fungus,[20,21,22,23] to the yeast *Saccharomyces cerevisiae*,[96,97] and the green alga *Chlamydomonas reinhardtii*.[98]

Studies on furanocoumarin-containing plants and their relationships to herbivores have provided excellent model systems of plant-insect interactions. A prospective case of coevolution based on these interactions has been described.[8] Since Berenbaum's[8] report on toxicity of furanocoumarins to arthropods, other studies have confirmed that these chemicals cause loss of fitness through delayed developmental times or growth reductions,[30,99,100] and in some cases have demonstrated activity as feeding deterrents.[30]

Some recent studies suggest that some chemical precursors of linear furanocoumarins may be more toxic to fungi than the linear furanocoumarins. Afek et al.[23] observed that marmesin, an immediate precursor to the linear furanocoumarins, played a more important role than psoralen, bergapten, xanthotoxin, and isopimpinellin in celery resistance to pathogenic agents during storage. If this pattern is repeated for other plant pathogens, it would provide substantial evidence regarding the driving forces responsible for the evolution of the linear furanocoumarins. Also, such a pattern would suggest that the linear furanocoumarins provide additional advantages that outweigh those of the precursors such as stability, storability, lack of autotoxicity, etc.

One such possible advantage of linear furanocoumarins over the precursors is an increased activity against herbivores. In tests with the coumarin derivatives ostruthin and osthol, no effects on insect survival or growth were found.[99] However, there is ample evidence that the linear furanocoumarins can effectively limit insect herbivory. Although psoralen has been reported to be the most photodynamically active compound among furanocoumarins according to earlier studies,[1,54,102,103] a growing number of recent studies[30,66,67,69] suggest that xanthotoxin is more toxic.

12.4   FACTORS AFFECTING LINEAR FURANOCOUMARIN
PRODUCTION AND TOXICITY

Ultraviolet A radiation is certainly a major factor determining the toxicity of linear furano-coumarins to most organisms.[8,11,14,29,49,50] In addition to UV light, a number of other factors can increase the toxic effects of furanocoumarins. Exposure to fungal, bacterial, and viral agents has been associated with increased plant furanocoumarin-content, and consequently increased human and animal health hazards. Infection of celery with the disease-causing pathogens such as *Sclerotinia sclerotiorum* or *Fusarium oxysporum* resulted in induction of a new furanocoumarin and/or an increased production of existing ones compared with healthy plants.[13,17,20,22] This initially led scientists to suggest that only diseased celery contained linear furanocoumarins (perhaps produced by the pathogen, rather than the plant), but it has now been established that healthy celery contains furanocoumarins and can cause photodermatitis.[45,75,76,78] Higher production of linear furanocoumarins in plants infected with pathogenic agents compared with healthy ones has also been documented in parsnip,[19] parsley,[104] citrus, and fig leaves.[74] Because of the increased concentrations of linear furanocoumarins in plants following exposure to pathogens, these compounds have also been referred to as phytoalexins.[105] Increased production of furanocoumarins in plants as a response to environmental stress has been confirmed at the molecular level (see Berenbaum[8] and references therein).

A variety of anthropogenic stresses also have the potential to induce furanocoumarin production in plants. Experiments by Dercks and colleagues,[106] showed significant increases (more than 500%) in linear furanocoumarin production in celery as a result of a 4 h exposure to acidic fogs. Beier and Oertli[13] and Beier and Nigg[12] demonstrated that application of copper sulfate served as a general elicitor in celery. Mechanical damage occurring during harvesting and storage has also been shown to increase concentrations from about 2 $\mu g\ g^{-1}$ to 95 $\mu g\ g^{-1}$.[56] Other factors such as temperature,[13,23] cold storage practices,[56] and growing conditions such as light and nutrient regime,[8,14] have all been implicated in increased furanocoumarin production in plants. Purohit and colleagues[107] recently demonstrated that tissue culturing induced higher production of xanthotoxin in *Ammi majus* than any other technique previously described. Thus, human activities related to production, storage, and pollutant generation can substantially increase the hazards associated with furanocoumarins.

In oral medicinal use, the potential carcinogenicity and toxicity of furanocoumarins to humans can be influenced by the environmental factors related to the ingestion of the compounds. When used in skin phototherapy, the eating habits of patients under psoralen treatment have been shown to impact treatment efficacy. During a study designed to determine the impact of food consumption vs. fasting conditions on the pharmacokinetics of bergapten, Ehrsson et al.[6] observed that administration of bergapten tablets with food greatly increased the bioavailability of the medication. Dietary omega-3 and omega-6 fatty acid sources decreased inflammatory responses and allowed relatively rapid repair of psoralen-induced cutaneous toxicity, but these lipids did not affect psoralen-induced tumorigenesis.[108]

12.5   METABOLIC DETOXIFICATION OF LINEAR
FURANOCOUMARINS

The ability of certain insect species to specialize on furanocoumarin-containing plants[8] sparked a series of studies designed to document the metabolic detoxification mechanisms of these compounds by arthropods. The chief metabolic detoxification pathway appears to be through mixed function oxidases (MFOs), a series of enzymatic reactions which allow organisms to break down complex molecules into smaller ones that are more easily degraded or excreted.[109]

In one early study, Brattsten et al.[110] reported that plant secondary compounds increased MFOs in larvae of *Spodoptera eridania*. Both the xanthotoxin-tolerant *Papilio polyxenes* and the xanthotoxin-susceptible *Spodoptera frugiperda* were found to metabolize this chemical by oxidative cleavage of the furan ring, but the rate of the metabolism was much higher in *P. polyxenes*.[111] Ingestion of xanthotoxin has also been shown to increase enzymatic activity in larvae of *Trichoplusia ni*[112] and *Depressaria pastinacella*.[113] Several studies[113,114,115,116] indicate involvement of cytochrome P450 in excretion and metabolism of furanocoumarins. The observation that the furanocoumarins deter feeding by herbivores might be linked to the fact that a slow feeding rate would allow herbivores to better metabolize toxic chemicals by either detoxifying them in the midgut prior to absorption[109,111,115,117] or by efficiently excreting them.[117]

However, such metabolic detoxification of plant defensive compounds by herbivores reportedly is not without costs. For example, nicotine can be detoxified by P450s in *Spodoptera eridania* larvae, but concentrations of 0.05% dietary nicotine have proven to significantly reduce relative growth rates and the efficiency of food conversion.[118] In a series of experiments conducted with increasing concentrations of proteins, Berenbaum and Zangerl[115] recently evaluated the cost of cytochrome P450-mediated detoxification of xanthotoxin by *Depressaria pastinacella*, an insect restricted to feeding on plants in two genera of Apiaceae. They noted a progressive decline in growth rates with decreasing protein levels, but silk spinning and detoxification rates were only affected with 0% protein in the artificial diet. Much higher (almost threefold) metabolism of xanthotoxin was also induced when there was no protein in the diet; this, however, resulted in nearly 80% reduction in growth rates. As reported by the authors, these results suggest that "xanthotoxin detoxification capacity is maintained at the expense of growth."

Using natural concentrations co-occurring in fruits of *Pastinaca sativa*, Berenbaum and co-workers[119] observed a synergism between six different furanocoumarins in their toxicity to the insect *Helicoverpa zea*. This synergistic interaction was confirmed by subsequent studies designed to test co-occurring natural concentrations of *Pastinaca sativa* furanocoumarins for toxicity against the insect *Papilio polyxenes*. Rates of cytochrome P450-mediated metabolism were significantly reduced when equimolar concentrations of the linear furanocoumarins xanthotoxin and bergapten and the angular furanocoumarin angelicin were combined.[120] Conversely, using much higher artificial dietary concentrations of these chemicals based rather on LC$_{50}$ values, Diawara and colleagues[30] found that the combination of bergapten and xanthotoxin produced an additive effect on *Spodoptera exigua* mortality, but combining psoralen with either bergapten, xanthotoxin, or both resulted in significant antagonistic effects.

In mammals, the toxicity of furanocoumarins is primarily reduced through quick excretion after ingestion[3] and metabolic breakdown into nonphototoxic compounds.[121] For instance, the phototoxic 4,8-dimethylpsoralen is degraded into the nonphototoxic 4,8-dimethyl-5' carboxypsoralen in both humans and mice.[121] The skin photosensitizing property of furanocoumarins is believed to be due to psoralen ring system;[54] therefore, the furanocoumarins lacking methyl substituents also lack potency to photosensitize.[8] Ma and colleagues,[116] recently observed that the nature of substituents on the benzene ring determines the efficiency of metabolism of furanocoumarins by larvae of *Papilio polyxenes*. Presence of nonmetabolizable angular furanocoumarins also proved to inhibit metabolism of linear furanocoumarins,[116] and this may increase the latter's toxicity.

Plant extracts have also been shown to detoxify linear furanocoumarins. Rizzi and colleagues[122] reported *in vitro* antimutagenetic effect against xanthotoxin-induced photomutagenesis to *Salmonella typhimurium* from unspecified compounds in extracts and chromatographic fractions of the bark of the plant *Uncaria tomentosa*.

12.6 OUTLOOK FOR LINEAR FURANOCOUMARIN RESEARCH

The furanocoumarins will certainly continue to receive attention in skin therapy due to their proven medicinal value. In addition to the use of furanocoumarins in the treatment of psoriasis, xanthotoxin likely will continue to be a successful skin photochemotherapy agent for use against several other skin disorders including T-cell lymphoma. Because psoralen derivatives are used in 1) nucleic acid research, 2) in human immunodeficiency syndrome (AIDS) research as possible treatments of this condition or its related complications (see Danheiser and Trova[123] and references therein) and 3) in cancer research, more studies can be expected in the near future. In addition, new psoralen analogs[124,125] and derivatives[87,126,127] are being synthesized and tested for potential use in skin phototherapy in an attempt to reduce the phototoxic side effects associated with the use of psoralens. An increasing number of studies are reporting on the repair of psoralen plus UV-induced DNA damage.[97,128,129]

The concerns of potential human and animal health hazards associated with furanocoumarins can be significantly reduced by continued research efforts to better understand the mechanisms of their toxicity and localization of these compounds in specific plant parts and structures. For instance, it has recently been observed that over 60% of the linear furanocoumarins in celery occurs in leaves on outer plant petioles.[31] These highly localized concentrations of furanocoumarins are of considerable importance given a trend in marketing intact organically-grown celery (leaves not trimmed) rather than the more common "topped" celery (outer petioles and most leaves removed) marketed by most commercial producers.[106] Consequently, potential hazards to consumers could be greatly minimized by avoiding contact with these plant parts. Finally, we expect more studies on photochemotherapy techniques, such as the newly described method where xanthotoxin is administered in a "relaxing" bath, which may be more effective and have fewer side effects than standard ingestion therapy.[130]

Currently, little is known about the biological activity of the immediate chemical precursors of the furanocoumarins. Elucidation of the biological effects of the precursors seems likely to provide insight into the evolution of these compounds. To date, much of the research on toxicity of furanocoumarins has focused on testing of single compounds, particularly 8-methoxypsoralen (xanthotoxin). Continued efforts to study chemical combinations that occur naturally in plants under natural conditions would also further our understanding of their activity. Some of these studies will require investigation at the molecular level.

ACKNOWLEDGMENTS

The authors thank Drs. S. Bonetti and L. A. Martínez of the University of Southern Colorado's Department of Chemistry and Department of Biology, respectively, for their technical review. We are also grateful to Joey Irby of the University of Southern Colorado and Greg Hund of the University of California Riverside for their assistance in the literature search. Research supported in part by NIEHS/NIH Grant No. ES00288.

REFERENCES

1. Musajo, L. and Rodighiero G., The skin-photosensitizing furocoumarins, *Experientia*, 18, 153, 1962.
2. Van Scott, E. J., Therapy of psoriasis. *J. Am. Med. Assoc.*, 235, 197, 1976.
3. Scott, B. R., Pathak, M. A., and Mohn, G. R., Molecular and genetic basis of furanocoumarin reactions, *Mutat. Res.*, 39, 29, 1976.
4. Pathak, M. A. and Fitzpatrick, T. B., The evolution of photochemotherapy with psoralens and UVA (PUVA): 2000 BC to 1992 AD, *Photochem. Photobiol. B*, 14, 3, 1992.

5. Chadwick, C. A., Potten, C. S., Cohen, A. J., and Young, A. R., The time of onset and duration of 5-methoxypsoralen photochomoprotection from UVR-induced DNA damage in human skin, *Br. J. Dermatol.*, 131, 483, 1994.

6. Ehrsson, H., Wallin, I., Ros, AM., Eksborg, S., and Berg, M., Food-induced increase in bio-availability of 5-methoxypsoralen, *Eur. J. Clin. Pharrnacol.*, 46, 375, 1994.

7. Pathak, M. A. and Fitzpatrick, T. B., Relationship of molecular configuration to the activity of furocoumarins which increase the cutaneous responses following long wave ultraviolet radiation, *J. Invest. Dermatol.*, 32, 255, 1959a.

8. Berenbaum, M. R., Coumarins, in *Herbivores Their Interactions with Secondary Plant Metabolites, The Chemical Participants, Vol. I,* Rosenthal, G. A. and Berenbaum, M. R., Eds., Academic Press, New York, 1991, chap. 6.

9. Bredberg, A. and Lambert, B., Induction of SCE by DNA cross-links in human fibroblasts exposed to 8-MOP and UVA irradiation, *Mutat. Res.*, 118, 191, 1983.

10. Calzavara-Pinton, P., Carlino, A., Manfedi, E., Semeraro, F., Zane, C., and Panfilis, G., Ocular side effects of PLTVA-treated patients refusing eye sun protection, *Acta. Derm. Venereol. Suppl. Stockh.*, 186, 164, 1994.

11. Murray, R. D. H., Mendez, J., and Brown, S. A., *The Natural Coumarins: Occurrence, Chemistry and Biochemistry.* J. Wiley and Sons, Chichester, UK., 1982.

12. Beier, R. C. and Nigg, H. N., Toxicity of Naturally occurring chemicals in food, in *Foodborne Disease Handbook*, 3, Hui, Y. H., Richard Gorham, J., Murrell, K. D., and Cliver, D. O., Eds., Marcel Decker, New York, 1994.

13. Beier, R. C. and Oertli, E. H., Psoralen and other linear furanocoumarins as phytoalexins in celery, *Phytochemistry*, 22, 2595, 1983.

14. Zangerl, A. R. and Berenbaum, M. R., Furanocoumarins in wild parsnip: effects of photosynthetically active radiation, ultraviolet light, and nutrients, *Ecology*, 68, 516, 1987.

15. Zobel, A. and Brown, S., Influence of low-intensity ultraviolet radiation on extrusion of furanocoumarins to the leaf surface, *J. Chem. Ecol.*, 19, 939, 1993.

16. Zobel, A. M., Crellin, J., Brown, S. A., and Glowniak, K., Concentration of furanocoumarins under stress conditions and their histological localization, in *International Symposium on Natural Phenols in Plants, Vol. II Acta Hort.* 381, Giebel, M., Treutter, D., and Feucht, W., Eds., Weihenstephan, 1994, pages 510-516.

17. Scheel, L. D., Perone, V. B., Larkin, R. L., and Kupel, R. E., The isolation and characterization of two phototoxic furanocoumarins (psoralens) from diseased celery, *Biochemistry*, 2, 1127, 1963.

18. Ashwood-Smith, M. J., Ceska, O., and Chaudhary, S. K., Mechanism of photosensitivity reactions to diseased celery, *Br. Med. J.*, 290, 1249, 1985.

19. Desjardins, A. E., Spencer, G. F., Plattner, R. D., and Beremand, M. N., Furanocoumarin phytoalexins, trichothecene toxins, and infection of *Pastinaca sativa* by *Fusarium sporotrichioides*, *Phytopathol.*, 79, 170, 1989.

20. Heath-Pagliuso, S., Matlin, S. A., Fang, N., Thompson, R. H., and Rappaport, L., Stimulation of furanocoumarin accumulation in celery and celeriac tissues by *Fusarium oxysporum* F sp. *apii.*, *Phytochemistry*, 31, 2683, 1992.

21. McCloud, E., Berenbaum, M., and Tuveson, R. W., Furanocoumarin content and phototoxicity of rough lemon (*Citrus jambhiri*) foliage exposed to enhanced UVB irradiation, *J. Chem. Ecol.*, 18, 1125, 1992.

22. Afek, U., Aharoni, N., Carmeli, S., and Roiser, L., A suggestion for new mechanism of celery resistance to pathogens, *Acta Hort.*, 342, 357, 1993.

23. Afek, U., Aharoni, N., and Carmeli, S., The involvement of marmesin in celery resistance to pathogens during storage and the effect of temperature on its concentration, *Phytopathology*, 85, 711, 1995.

24. Kupidlowska, E., Kowalec, M., Sulkowski, G., and Zobel, A., The effect of coumarins on root elongation and ultrastructure of meristematic cell protoplast, *Annals Botany*, 73, 525, 1994a.

25. Trumble, J. T., Dercks, W., Quiros, C. F., and Beier, R. C., Host plant resistance and linear furanocoumarin contents of *Apium* accessions, *J. Econ. Entomol.*, 83, 519, 1990.

26. Diawara, M. M., Trumble, J. T., Quiros, C. F., and Millar, J. G., Resistance to *Spodoptera exigua* in *Apium prostratum*, *Entomol. Exp. Appl.*, 64, 125, 1992.

27. Diawara, M. M., Trumble, J. T., and Quiros, C. F., Linear furanocoumarins of three celery breeding lines: implications for integrated pest management, *J. Agric. Food Chem.*, 41, 819, 1993.

28. Berenbaum, M. R., Zangerl, A. R., and Lee, K., Chemical barriers to adaptation by a specialist herbivore, *Oecologia*, 80, 501, 1989.

29. Trumble, J. T., Moar, W. T. Brewer, M. J., and Carson, W. G., Impact of UV radiation on activity of linear furanocoumarins and *Bacillus thuringiensis* var *kurstaki* against *Spodoptera exigua*: implications for tritrophic interactions, *J. Chem. Ecol.*, 17, 973, 1991.

30. Diawara, M., Trumble, J., White, K., Carson, W., and Martinez, L., Toxicity of linear furano-coumarins to *Spodoptera exigua*: evidence for antagonistic interactions, *J. Chem. Ecol.* 19, 2473, 1993.

31. Diawara, M. M., Trumble, J. T., Quiros, C. F., and Hansen, R., Implications of distribution of linear furanocoumarins within celery, *J. Agric. Food Chem.*, 43, 723, 1995.

32. Brown, S, A., Biochemistry of the coumarins, in *Recent Advances in Phytochemistry Vol. 12*, Swain, T., Harborne, J. B., and Sumere, C. H., Eds., Plenum Press, New York, 1978. pages 249–286.

33. Edwards, K. G. and Stoker, J. R., Biosynthesis of coumarin: the isomerization stage, *Phytochemistry*, 6, 655, 1967.

34. Floss, H. G., Biosynthesis of furanocoumarins, in *Recent Advances in Phytochemistry Vol. 4*, Runeckles, V. C. and Watkin, J. E., Eds., Appleton-Century-Crofts, New York, 1972, pages 143–164.

35. Brown, S. A., Towers, G. H. N., and Wright, D., Biosynthesis of the coumarins. Tracer studies on coumarin formation in *Hierochloe odorata and Melilotus officinalis, Can. J. Biochem. Physiol.*, 38, 143, 1960.

36. Floss, H. G. and Mothes, U., Zur Biosynthese von Furanocumarinen in *Pimpinella magna, Z. Naturforsch.* 19b: 770, 1964.

37. Caporale, G., Dall'Acqua, F., Capozzi, A., Marciani, S., and Crocco, R., Studies on the bio-synthesis of some furocoumarins present in *Ruta graveolens, Z. Naturforsch.*, 26b, 1256, 1971.

38. Caporale, G., Dall'Acqua, F., Capozzi, A. and Marciani, S., Studies on the biosynthesis of furocoumarins in the leaves of *Ficus carica* L., *Ann. Chim. (Rome)*, 62, 536, 1972.

39. Games, D. E. and James, D. H., The biosynthesis of the coumarins of *Angelica archangelica, Phytochemistry*, 11, 868, 1972.

40. Brown, S. A. and Steck, W., 7-Demethylsuberosin and osthenol as intermediates in furanocou-marin biosynthe*sis, Phytochemistry, 12*, 1315, 1973.

41. Ellis, B. E. and Brown, S. A., Isolation of dimethylallylpyrophosphate: umbelliferone dimeth-ylallyltransferase from *Ruta graveolens, Can. J. Biochem.*, 52, 734, 1974.

42. Ebel, J., Phytoalexin synthesis: the biochemical analysis of the induction process, *Ann. Rev. Phytopathol.*, 24, 235, 1986.

43. Martin, J. T., Baker, E. A., and Byrde, R. J. W., The fungitoxicities of cuticular and cellular components of citrus lime leaves, *Ann. Appl. Biol.*, 57, 491, 1966.

44. Ashwood-Smith, M., Poulton, G. A., and Liu, M., Photobiological activity of 5,7-dimethoxy-coumarin, *Experientia*, 39, 262, 1983.

45. Berkley, S. F., Hightower, A. W., Beier, R. C., Fleming, D. W., Brokopp, C. D., Ivie, G. W., and Broome, C. V., Dermatitis in grocery workers associated with high natural concentrations of furanocoumarins in celery, *Ann. Intern. Med.*, 105, 351, 1986.

46. Zobel, A. M. and Brown, S. A., Psoralens in senescing leaves of *Ruta graveolens, J. Chem Ecol.*, 17, 1801, 1991.

47. Beier, R. C., Natural pesticides and bioactive components in foods, *Rev. Environ. Contam. Toxicol.*, 113, 47, 1990.

48. Zobel, A. M. and March R., Autofluorescence reveals differential histological localizations of furanocoumarins in fruits of some Umbelliferae and Leguminosae, *Annalas Botany*, 71, 251, 1993.

49. Dall'Acqua, F., Marciani, S., and Rodighiero, G., The action spectrum of xanthotoxin and bergapten for the photoreaction with native DNA, *Z. Naturforsch.*, 24b, 667, 1969.

50. Cole, R. S., Light-induced cross-linking of DNA in the presence of a furocoumarin (psoralen). Studies with phase 1, *Escherichia coli* and mouse leukemia cells, *Biochim. Biophys. Acta*, 217, 30, 1970.

51. Parrish, J. A., Fitzpatrick, T. B., Tanenbaum, L., and Pathak, M. A., Photochemotherapy of psoriasis with oral methoxsalen and longwave ultraviolet light, *New Engl. J. Med.*, 291, 1207, 1974.

52. Ashwood-Smith, M. J. and Grant, E., Conversion of psoralen DNA monoadducts in *E. coli* to interstrand DNA cross links by near UV light (320-360 nm): Inability of angelicin to form cross links, in vivo, *Experientia*, 33, 384, 1977.

53. Musajo, L., Rodighiero, G., and Dall'Acqua, F., Evidence of a photoreaction of the photosensitizing furocoumarins with DNA and with pyrimidine nucleosides and nucleotides, *Experientia*, 21, 24, 1965.

54. Pathak, M. A., Worden, L. R., and Kaufman, K. D., Effect of structural alterations on the photosensitizing potency of furocoumarins (psoralens) and related compounds, *J. Invest. Dermatol.*, 48, 103, 1967.

55. Igali, S., Bridges, B. A., Ashwood-Smith, M. J., and Scott, B. R., Mutagenesis in *Escherichia coli* IV. Photosensitization to near ultraviolet light by 8-methoxypsoralen, *Mutat. Res.*, 9, 21, 1970.

56. Chaudhary, S. K., Ceska, O., Warrington, P. J., and Ashwood-Smith, M. J., Increased furanocoumarin content of celery during storage, *J. Agric. Food Chem.*, 33, 1153, 1985.

57. Granstein, R., Morison, W., and Kripke, L., Carcinogenicity of combined ultraviolet B radiation and psoralen plus ultraviolet A irradiation treatment of mice, *Photoderm. Photoimm. Photomed.*, 9, 198, 1993.

58. Dall'Acqua, F., Marciani, S., Ciavatta, L., and Rodighiero, G., Formation of interstrand crosslinkings in the photoreactions between furocoumarins and DNA, *Z. Naturforsch.*, 26b, 561, 1971.

59. Rodighiero, G. and Dall'Acqua, F., Biochemical and medical aspects of psoralens, *Photochem. Photobiol.*, 24, 647, 1976.

60. Laquerbe, A., Moustacchi, E., and Papadopoulo, D., Genotoxic potential of psoralen cross-links versus monoadducts in normal human lymphoblasts, *Mutat. Res.*, 346, 173, 1994.

61. Cole, R. S. and Zusman, D., Sedimentation properties of phage DNA molecules containing light-induced psoralen cross-links, *Biochim. Biophys. Acta*, 224, 660, 1970.

62. Veronese, F. M., Schiavon, O., Bevilacqua, R., Bordin, F., and Rodighiero, G., Photoinactivation of enzymes by linear and angular furocoumarins, *Photochem. Photobiol.*, 36, 25, 1982.

63. Wagner, S., White, R., Wolf, L., Capman, J., Robinette, D., Lawlor, T., and Dodd, R., Determination of residual 4'-aminomethyl-4,5',8-trimethylpsoralen and mutagenicity testing following psoralen plus UVA treatment of platelet suspensions, *Photochem. Photobiol.*, 57, 819, 1993.

64. Letteron, P., Descator, V., Larrey, D., Tinel, M., Geneve, J., and Pessayre, D., Inactivation and induction of cytochrome P450 by various psoralen derivatives in rats, *J. Pharmacol. Exp. Ther.*, 238, 685, 1986.

65. Mays, D. C., Hilliard, J. B., Wong, D. D., and Gerber, D. E., Activation of 8-methoxypsoralen by cytochrome P450, *Biochem. Pharmacol.*, 38, 1647, 1989.

66. Maenpaa, J., Sigusch, H., Raunio, H., Syngelma, T., Vuorela, P., Vuorela, H., and Pelkonen, O., Differential inhibition of coumarin 7-hydroxylase activity in mouse and human liver microsomes, *Biochem. Pharmacol.*, 45, 1035, 1993.

67. Maenpaa, J., Juvonen, R., Raunio, H., Rautio, A., and Pelkonen, O., Metabolic interactions of methoxsalen and coumarin in humans and mice, *Biochem. Pharma.*, 48, 1363, 1994.

68. Zumwalt, J. G. and Neal, J. J., Cytochrome P450 from Papilio polyxenes: adaptations to host plant allelochemicals, *Comp. Biochem. Physiol.*, 106C, 111, 1993.

69. Neal, J. and Wu, D., Inhibition of insect cytochromes P450 by furanocoumarins, *Pest. Biochem. Phys.*, 50, 43, 1994.

70. Brattsten, L. B., Metabolic defenses against plant allelochemicals, in *Herbivores: Their Interactions with Secondary Plant Metabolites*, 1, Rosenthal, G. A. and Berenbaum, M. R., Eds., Academic Press, New York, 1991, 175.

71. Giese, A. C., Photosensitization by natural pigments, *Photophysiology*, 6, 77, 1971.

72. Musajo, L. and Rodighiero, G., Mode of photosensitizing action of furocoumarins, *Photophysiology*, 7, 115, 1972.

73. Legrain, P. M. and Barthe, R., Dermite des mains et des avant-bras chez un ramasseur de celeris, *Bull. Fr. Dermatol. Syphiligr.*, 33, 662, 1926.

74. Zaynoun, S. T., Alfimos, B., G., Ali, L. A., Tenekjian, K. K., Khalidi, U., and Kurban, A. K., *Ficus carica*: isolation and quantification of the photoactive components, *Contact Dermatitis*, 11, 21, 1984.

75. Ljunggren, B., Severe phototoxic burn following celery ingestion, *Arch. Dermat.*, 126, 1334, 1990.

76. Finkelstein, E., Afek, U., Gross, E., Aharoni, N., Rosenberg, L., and Havely, S., An outbreak of phytophotodermatitis due to celery, *Internat. J. Dermat.*, 33, 116, 1994.

77. Austad, J. and Kavli, G., Phototoxic dermatitis caused by celery infected by *Sclerotinia sclerotiorum*, *Contact Dermatitis*, 9, 448, 1983.

78. Seligman, P. J., Mathias, C. G., O'Malley, M. A., Beier, R. C., Fehrs, L. J., Serril, W. S., and Halperin, W. E., Photodermatitis from celery among grocery store workers, *Arch Dermatol.*, 123, 1478, 1987.

79. Fleming, D., Dermatitis among grocery workers associated with high natural concentrations of furanocoumarins in celery, *Allergy Proc.*, 11, 125, 1990.

80. Kuusilehto, A., Lehtinen, R., and Jansen, C., Comparison of the minimal phototoxic dose in topical 4,5' 8-trimethylpsoralen PUVA treatment of Caucasian skin and of oral mucous membrane, *Acta. Derm. Venereol.*, 70, 508, 1990.

81. Ashwood-Smith, M. J., Poulton, G. A., Barker, M., and Mildenberger, M., 5-methoxypsoralen, an ingredient in several suntan preparations, has lethal, mutagenic and cleistogenic properties, *Nature*, 285, 407, 1980.

82. Bauluz, C., Paramio, J. M., and de-Vidania, R., Further studies on the lethal and mutagenic effects of 8-methoxypsoralen-induced lesions on plasmid DNA, *Cell. Mol. Biol.*, 37, 481, 1991

83. Boesen, J. J., Stuivenberg, S., Thyssens, C. H., Panneman, H., Darroudi, F., Lohman, P. H., and Simons, J. W., Stress response induced by DNA damage leads to specific, delayed and untargeted mutations, *Mol. Gen. Genet.*, 234, 217, 1992.

84. Mathews, M. M., Comparative studies of lethal photosensitization of *Sarcina lutea* by 8-methoxypsoralen and by toluidine blue, *J. Bacteriol.*, 85, 322, 1963.

85. Roelandts, R., Mutagenicity and carcinogenicity of methoxypsoralen plus UV-A, *Arch. Dermatol.*, 120, 662, 1984.

86. Papadopoulo, D., and Moustacchi, E., Mutagenic effects photoinduced in normal human lymphoblasts by a monofunctional pyridopsoralen in comparison to 8-methoxypsoralen, *Mutat. Res.*, 245, 259, 1990.

87. Yang, S. C., Lin, J. G., Chion, C. C., Chen, L. Y., and Yang, J. L., Mutation specificity of 8-methoxypsoralen plus two doses of UVA radiation in the hprt gene in diploid human fibroblasts, *Carcinogenesis*, 15, 201, 1994.

88. Young, A. R., Photocarcinogenicity of psoralens used in PUVA treatment-present status in mouse and man, *J. Photochem. Photobiol. B-Biol.*, 6, 237, 1990.

89. International Agency for Research on Skin Cancer, *Evaluation of the Carcinogenic Risk of Chemicals to Humans, Supplement 4: Chemicals, Industrial Process and Industries Associated with Cancer in Humans*, International Agency for Research on Cancer, Lyon, 1983.

90. Chandhoke, N. and Ghatak, G. J., Pharmacological investigations of angelicin: a tranquillosedative and anticonvulsant agent, *Indian J. Med. Res.*, 63, 833, 1975.

91. Ashkenazy, D., Kashman, Y., Nyksa, A. and Friedman, J., Furanocoumarins in shoots of *Pituranthos triadiatus* (Umbelliferae) as protectants against grazing by hyrax (Procaviidae: *Procavia capensis syriaca*), *J. Chem. Ecol.*, 11, 231, 1985.

92. Bredberg, A., Sandor, Z., and Brant, M., Mutational response of Fanconi anemia cells to shuttle vector site-specific psoralen cross-links, *Carcinogenesis*, 16, 555, 1995.

93. Garde, E., Micic, S., Knudsen, K., Angelo, H., and Wulf, H., 8-Methoxypsoralen increases daytime plasma melatonin levels in humans through inhibition of metabolism, *Photochem. Photobiol.*, 60, 475, 1994.

94. Rosselli, F., Duchaud, E., Averbeck, D., and Moustacchi, E., Comparison of the effects of DNA topoisomerase inhibitors on lymphoblasts from normal and Fanconi anemia donors, *Mutat. Res.*, 325, 137, 1994.

95. Kupidlowska, E., Dobrzynska, K., Parys, E., and Zobel, A., Effect of coumarin and xanthotoxin on mitochondrial structure, oxygen uptake, an succinate dehydrogenase activity in onion root cells, *J. Chem. Ecol.*, 20, 2471, 1994b.

96. Saffran, W., Greenberg, R., Thaler-Scheer, M., and Jones, M., Single strand and double strand DNA damage-induced reciprocal recombination in yeast. Dependence on nucleotide excision repair and RADI recombination, *Nuc. Acids Res.*, 22, 2823, 1994.

97. Dardalhon, M. and Averbeck, D., Pulsed-field gel electrophoresis analysis of the repair of psoralen plus LJVA induced DNA photoadducts in *Saccharomyces cerevisiae, Mutat. Res.*, 336, 49, 1995.

98. Schimmer, O. and Kuhne, I., Mutagenic compounds in an extract from *Rutae Herba* (*Ruta graveolens* L.), *Mutat. Res.* 243, 57, 1990.

99. Hadacek, F., Muller, C., Wemer, A., Greger, H., and Proksch, P., Analysis, isolation and insecticidal activity of linear furanocoumarins and other coumarin derivatives from *Peucedanum* (Apiaceae: Apioideae), *J. Chem. Ecol.*, 20, 2035, 1994.

100. Brewer, M., Meade, T., and Trumble, J., Development of insecticide-resistant nd susceptible *Spodoptera exigua* (Lepidoptera: Noctuidae) exposed to furanocoumarins found in celery, *Environ. Entomol.*, 24, 392, 1995.

101. Berenbaum, M., Toxicity of a furanocoumarin to armyworms: a case of biosynthetic escape from insect herbivores, *Science*, 201, 532, 1978.

102. Musajo, L., Rodighiero, G., and Caporale, G., L'activité photodynamique des coumarines naturelles, *Bull. Soc. Chim. Biol.*, 36, 1213, 1954.

103. Pathak, M. A. and Fitzpatrick, T. B., Bioassay of natural and synthetic furocoumarins (psoralens), *J. Invest. Dermatol.*, 32: 509, 1959b.

104. Knogge, W., Kombrink, E., Schmelzer, E., Hahlbrock, K., Occurrence of phytoalexins and other putative defense-related substances in uninfected parsley plants, *Planta*, 171, 279, 1987.

105. Paxton, J., A new working definition of the term "phytoalexin," *Plant Dis.*, 64, 734, 1980.

106. Dercks, W., Trumble, J. T., and Winter, C., Impact of atmospheric pollution on linear furanocoumarin content in celery, *J. Chem. Ecol.*, 16, 443, 1990.

107. Purohit, M., Pande, D., Datta, A., and Srivastava, P., Enhanced xanthotoxin content in regenerating cultures of *Ammi majus* and micropropagation, *Planta. Med.*, 61, 481, 1995.

108. Yen, A., Black, H., and Tschen, J., Effect of dietary omega-3 and omega-6 fatty acid sources on PUVA-induced cutaneous toxicity and tumorigenesis in the hairless mouse, *Arch. Dermatol. Res.*, 286, 331, 1994.

109. Ivie, G. W., Bull, D. L., Beier, R. C., Pryor, N. W., and Oertli, E. H., Metabolic detoxification: mechanism of insect resistance to plant psoralens, *Science*, 221, 374, 1983.

110. Brattsten, L. B., Wilkinson, C. F., and Eisner, T., Herbivore-plant interactions: mixed function oxidases and secondary plant substances, *Science*, 196, 1349, 1977.

111. Bull, D. L., Ivie, G. W., Beier, R. C., Pryor, N. W., and Oertli, E. H., Fate of photosensitizing furanocoumarins in tolerant and sensitive insects, *J. Chem. Ecol.*, 10, 893, 1984.

112. Lee, K. and Berenbaum, M. R., Action of antioxidant enzymes and cytochrome P450 monooxygenases in the cabbage looper in response to plant phototoxins, *Arch. Insect Biochem. Physiol.*, 10, 151, 1989.

113. Nitao, J. K., Enzymatic adaptation in a specialist herbivore for feeding on furanocoumarin-containing plants, *Ecology*, 70, 629, 1989.

114. Cohen, M. B., Berenbaum, M. R., and Schuler, M. A., Induction of Cytochrome P450-mediated detoxification of xanthotoxin in the black swallowtail, *J. Chem. Ecol.*, 15, 404, 1989.

115. Berenbaum, M. and Zangerl, A., Costs of inducible defense: protein limitation, growth, and detoxification in parsnip webworm, *Ecology*, 75, 2311, 1994.

116. Ma, R., Cohen, M., Berenbaum, M., and Schuler, M., Black swallowtail (Papilio polyxenes) alleles encode cytochrome P450s that selectively metabolize linear furanocoumarins, *Arch. Biochem. Biophys.*, 310, 332, 1994.

117. Nitao, J. K., Metabolism and excretion of the furanocoumarin xanthotoxin by parsnip webworm, *Depressaria pastinacella, J. Chem. Ecol.*, 16, 417, 1990.

118. Cresswell, J. E., Merritt, S. Z., and Martin, M. M., The effect of dietary nicotine on the allocation of assimilated food to energy metabolism and growth of fourth instar larvae of the southern armyworm, *Spodoptera eridania* (Lepidoptera: Noctuidae), *Oecologia (Berlin)*, 89, 449, 1992.

119. Berenbaum, M. R., Nitao, J. K., and Zangerl, A. R., Adaptive significance of furanocoumarin diversity in *Pastinaca sativa* (Apiaceae), *J. Chem. Ecol.*, 17, 207, 1991.

120. Berenbaum, M. and Zangerl, A., Furanocoumarin metabolism in *Papilio polyxenes*: biochemistry, genetic variability, and ecological significance, *Oecologia*, 95, 370, 1993.

121. Mandula, B. B., Pathak, M. A., and Dudek, G., Photochemotherapy: identification of a metabolite of 4,5',8-trimethylpsoralen, *Science*, 193, 1131, 1976.

122. Rizzi, R., Re, F., Bianchi, A., De-Feo, V., de-Simone, F., Bianchi, L., and Stivala, L. A., Mutagenic and antimutagenic activities of *Uncaria tomentosa* and its extracts, *J. Ethnopharmacol.*, 38, 63, 1993.

123. Danheiser, R. and Trova, M., Synthesis of linear furocoumarins via photochemical aromatic annulation strategy. An efficient total synthesis of bergapten, *Synlett.*, 573, 1995.

124. Vedaldi, D., Caffieri, S., Miolo, G., Dall'Acqua, F., Baccichetti, F., Guiotto, A., Benetollo, F., Bombieri, G., Recchia, G., and Cristofolini, M., Azapsoralens: new potential photochemotherapeutic agents for psoriasis, *Farrnaco.*, 46, 1407, 1991.

125. Caffieri, S., Moor, A., Beijersbergen van Henegouwen, M., Acgua, F., Guiotto, A., Chilin, A., and Rodighlero, P., Difurocoumarins, psoralen analogs: synthesis and DNA photobinding, *Chem. Sci.*, 50, 1257, 1995.

126. Baccichetti, F., Bordin, F., Simonato, M., Toniolo, L., Marzano, C., Rodighiero, P., Chillin, A., and Carlassare, F., Photobiological activity of certain new methylazapsoralens, *Farrnaco.*, 47, 1529, 1992.

127. Zagotto, G., Gia, O., Baccicbetti, F., Uriarte, E., and Palumbo, M., Synthesis and photobiological properties of 4-hydroxymethyl-4'-methylpsoralen derivatives, *Photochem. Photobiol.*, 58, 485, 1993.

128. Spielmann, H., Dwyer, T., Sastry, S., Hearst, J., and Wemmer, D., DNA Structural reorganization upon conversion of a psoralen furan-side monoadduct to an interstrand cross-link: Implications for DNA repair, *Proc. Natl. Acad. Sci. U.S.A.*, 92, 2345, 1995.

129. Wang, G. and Galzer, P., Altered repair of targeted psoralen photoadducts in the context of an oligonucleopeptide-mediated triple helix, *J. Biol. Chem.*, 270, 22595, 1995.

130. Kerscher, M., Lehmann, P., and Plewig, G., PUVA bath therapy. Indications and practical implementation, *Hautarzt*, 45, 526, 1994.

13 Photosensitization Disorders

Arne Flåøyen and Arne Frøslie

CONTENTS

13.1 INTRODUCTION

Photosensitization in livestock has been a problem of economic importance in various parts of the world for several hundred years. For example, the Arabs in Tunis painted their horses with henna or tobacco juice as a protection against hypericism and Cirillo reported in 1787 that black sheep were used on the Tarantine fields because of the susceptibility of white animals to poisoning by *Hypericum crispum*.[1] More than 500,000 head of small-ruminant livestock were reported to have been photosensitized in severe outbreaks of geeldikkop in South Africa.[2] Production losses due to severe outbreaks of facial eczema in livestock from New Zealand have been calculated to be in the order of $100 million.[3]

13.1.1 MANIFESTATION AND CLINICAL SIGNS

Photosensitization disorders are seen in a wide range of grazing farm animals as well as other animals that are kept outdoors. In general, only unpigmented animals or animals with unpigmented areas of skin become photosensitized. The clinical signs are similar in all diseases regardless of type of photosensitizing agent. The most salient clinical signs in sheep are increasing restlessness, head shaking, scratching of the face and ears with the hind feet, and rubbing of the irritated skin against the ground. The skin changes develop rapidly, and include edema and reddening. The eyelids, muzzle, and lips become swollen and turgid. However, the most obvious signs in serious cases are the thickened edematous, heavily drooping ears (Figure 13.1). Signs progress to include seepage of sticky, honey-colored serum from the thickened skin which then forms extensive scabs that mat the covering hair after one or two days. In addition, jaundice is often visible in animals suffering from hepatogenous photosensitization.

FIGURE 13.1   Ten-week-old lamb photosensitized after ingesting *Narthecium ossifragum.*

Similar skin lesions can be seen in cattle and horses, although only on unpigmented areas of skin. Cattle also have lesions where hair is thin or absent such as on the udder, teats, and escutcheon. After some days or weeks necrotic skin may peel off in large, dry, and leathery pieces.

Depression, severe keratoconjunctivitis, photophobia, dermatitis in the beak, and inflammation in areas that are not covered with feathers are the most common clinical signs in photosensitized chickens, turkeys, ducks, and geese.

13.1.2   MODE OF ACTION

Information about the photodynamic action of photosensitizing agents is still not complete, and only the main principles will be outlined in this review. Most of the natural photosensitizing agents generate oxidative reactions that require molecular oxygen to be available for the reaction to proceed. Sunlight can excite most photosensitizing molecules. These excited or photo activated molecules can interact and transfer their energy either directly to biomolecules or to oxygen. If the interaction is with oxygen, singlet oxygen or oxygen radicals are generated. These chemicals cause pathological changes in the cells in various manners. The pathological changes include abnormal cell division, changes in the permeability of membranes and active transport processes, interference with glycolysis and cellular respiration, disruption of protein and DNA synthesis, mitochondrial damage, lysosomal damage, and cell death.[4] A given photosensitizing agent can be activated only by light in a given wavelength range corresponding to its absorption spectrum.

13.1.3   PATHOGENESIS OF PHOTOSENSITIZING DISORDERS IN ANIMALS

Clare divided the photosensitizing disorders into three main categories of photosensitization and his classification is generally still considered to be valid.[5]

Type I. Primary photosensitization
Type II. Photosensitization due to aberrant pigment synthesis
Type III. Hepatogenous photosensitization

Type I photosensitization results when the photosensitizing agent is absorbed directly in the digestive tract and is transported by the blood to the skin where it causes the characteristic lesions when exposed to sunlight.

Type II photosensitization results when the photosensitizing substance is a pigment, not normally found in animals, that is produced endogenously by an aberrant metabolic process. Alternatively, the aberration may be the excessive formation of a pigment that is ordinarily produced only in small, harmless amounts.

Type III photosensitization (hepatogenous photosensitization) results when a toxin, normally produced by a plant, fungus, or alga, causes liver damage or liver dysfunction resulting in retention of the photosensitizing agent phylloerythrin. Phylloerythrin is a metabolic product of chlorophyll produced by rumen microorganisms.[6,7] Unlike chlorophyll, which has a long phytyl hydrocarbon side chain preventing it from being absorbed from the digestive tract, phylloerythrin has lost the side chain and can be absorbed from the gut (Figure 13.2). Normally, the absorbed phylloerythrin is removed from circulation by the liver.

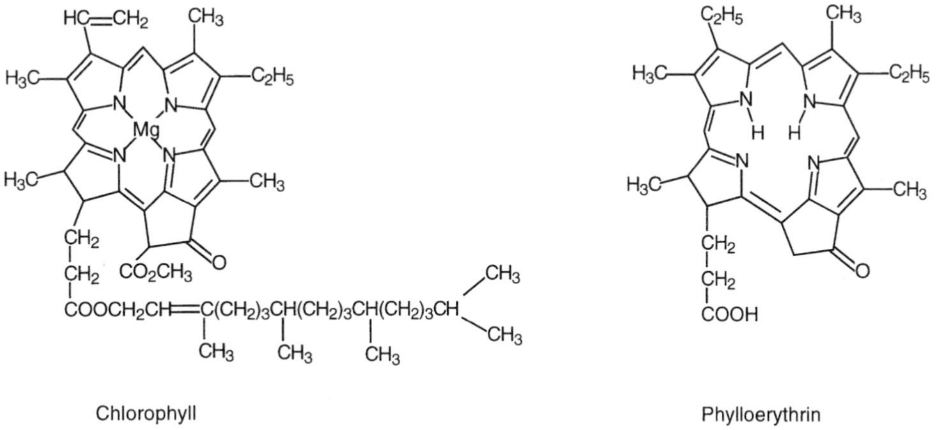

Chlorophyll

Phylloerythrin

FIGURE 13.2 Chlorophyll and phylloerythrin molecules.

Some types of liver damage or liver dysfunction in animals lead to depression or cessation of hepatic elimination of phylloerythrin. Therefore, phylloerythrin can reach the skin where reaction with sunlight can take place.

This chapter describes the major diseases caused by either Type I photosensitization or Type III photosensitization.

13.2 PRIMARY PHOTOSENSITIZATION DISORDERS

13.2.1 HYPERICISM (*HYPERICUM* INTOXICATION)

Hypericism is a disease of animals ingesting plants of the genus *Hypericum* and it is probably the oldest known cause of photosensitization in domestic animals. Plants of this genus are widely distributed. They are herbaceous or low shrubs, usually deciduous, but sometimes evergreen.[8] The flowers are usually yellow, but occasionally may be pink or purplish. Hypericin (7,14-dione-1,3,4,6,8,13-hexahydroxy-10,11-dimethyl-phenathrol[1,10,9,8-*opgra*]perylene) (Figure 13.3), the photosensitizing pigment of *Hypericum*, is present in minute glands located on different plant organs in various species. The definitive photodynamic action of hypericin

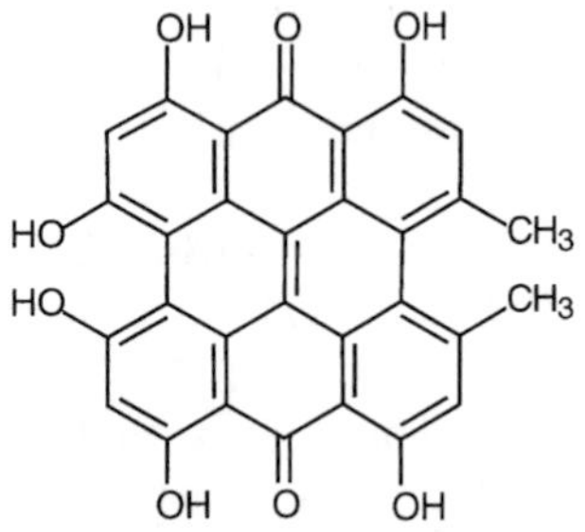

FIGURE 13.3 Hypericin molecule.

causing photosensitization is not known, but it is well known that oxygen and visible light are required.[9,10]

Hypericum perforatum (St. Johns wort) is the most common cause of hypericism. The plant was spread with European settlement and represents a problem in livestock on range lands in the western U.S. and Australia.[5,8] Cases have also occurred in Europe, New Zealand, and Iraq.[11] *Hypericum crispum* causes photosensitization in the Mediterranean regions.[8] *Hypericum triquetrifolium* has been reported to cause photosensitization in goats and sheep in Israel.[12] Other species of the genus are reported to contain hypericin, but none of those others have been reported to cause clinical problems. Although hypericin is apparently present at all times, the plants do not seem to be particularly palatable to animals and are eaten only when good forage is not available. Photosensitization occurs in sheep, cattle, horses, and goats, but the problems in sheep represent the greatest losses. Animals ingesting hypericin remain sensitive to sunlight for a week or more after exposure.[10]

13.2.2 FAGOPYRISM (BUCKWHEAT INTOXICATION)

The term fagopyrism has long been used to designate the photosensitization of sheep, cattle, goats, horses, and pigs eating *Fagopyrum esculentum* (buckwheat). *F. esculentum* grows readily on poor soil and it is common in parts of Asia, Europe, North America, and South Africa.[5,8,13]

Fagopyrin, a pigment found mainly in the flowers and seeds, normally causes the disease. However, all parts of the plant (fresh or dry) are capable of causing photosensitization. The chemical structure of fagopyrin is similar to the structure of hypericin.[13]

13.2.3 *AMMI MAJUS* INTOXICATION

A. majus (bishop's weed) grows on the coastal regions of the southern U.S. and it is widely distributed, and locally very common, in the Mediterranean area.[14,15] The plant has been used for centuries in folk medicine in the Middle East and India in the treatment of vitiligo in man.[16] Furocoumarins (psoralens), the toxic compounds in *A. majus*, are concentrated in the seeds. *A. majus* has caused photosensitization in cattle, sheep, geese, ducks, chickens, and turkeys.[14,15,17–19] Localized photosensitization due to dermal contact occurs in man.

13.3 HEPATOGENOUS PHOTOSENSITIZATION DISORDERS

13.3.1 STEROIDAL SAPONINS

Many hepatogenous photosensitization disorders are associated with ingestion of plants containing steroidal saponins.[20] Examples of such plants include: *Agave lecheguilla*, *Tribulus terrestris*, *Narthecium ossifragum*, *Brachiaria decumbens* (or *Avena sativa*), *Panicum dichotomiflorum*, *Panicum schinzii*, *Panicum miliaceum*, *Panicum coloratum*, and *Panicum virgatum*. The diseases caused by these plants have in common a sporadic occurrence, difficult experimental reproduction of symptoms when dosing with plant material, and inconsistent toxicity.

For example, the saponin-containing plants are not always toxic to grazing animals. Another common finding for the diseases caused by this group of plants is the accumulation of birefringent crystals in hepatocytes, and within and around bile ductuli and ducts. For the diseases that have been well studied, the crystals have been reported to be insoluble Ca^{2+} salts of sapogenin glucuronides originating from the saponins present in the corresponding plants.[20] The saponins have been suggested to be the liver toxin in these plants and photosensitization has been reproduced by dosing crude saponins to sheep.[21–23] Despite that, the doses needed to cause the typical symptoms have been large. Results from other experiments indicate that the saponins alone are unlikely to cause sufficient liver damage to result in photosensitization.[24,25]

The sapogenins exist as glycosides in the plants (Figure 13.4). Upon ingestion, the sugars from the saponins are hydrolyzed by the ruminal microorganisms into free sapogenins.[26] Some of the free sapogenins are then transported unchanged via the omasum and abomasum to the duodenum and jejunum. The rest of the free sapogenins are converted to epi-forms of the corresponding sapogenin before being transported further down the digestive tract. The conversion into epi-forms is due, at least in part, to metabolism by rumen microorganisms. The fate of the sapogenins after they leave the rumen is not well studied, but unpublished results from work on *N. ossifragum* [27] indicate that the hydrolysis and conversion into epi-forms that occur in the rumen continues in the omasum, and ceases in the abomasum.

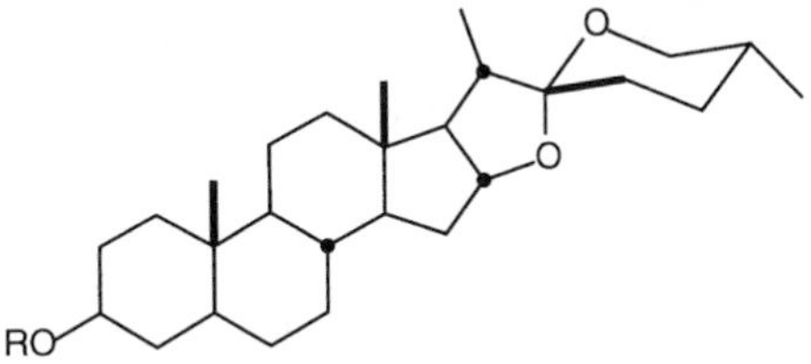

FIGURE 13.4 Typical structure of a steroidal saponin (R = sugars) and a steroidal sapogenin (R = H).

The major site for absorption of free sapogenins is probably the jejunum. After absorption, they are transported via the portal vein to the liver where further metabolism takes place.[27] A proportion of the free sapogenins will not be absorbed, remain unchanged in the intestines, and are excreted in the feces.

Studies that have been conducted to date indicate that the biliary crystals, from sheep photosensitized after ingestion of plants containing steroidal saponins, are insoluble salts of β-D-glucuronides of either episarsasapogenin or epismilagenin. Only those sapogenins capable of being metabolized to epismilagenin or episarsasapogenin can form biliary crystals. For example, conjugated forms of other sapogenins (e.g., tigogenin) are often abundant in the bile of affected animals, but are not found in the biliary crystals isolated from these animals.

Only trace amounts of sapogenin conjugates are apparently excreted into the urine, indicating that the metabolism and excretion of sapogenin conjugates in the liver is the most important metabolic pathway for these compounds.[25,26]

The glucuronides of episarsasapogenin and epismilagenin emptied into the duodenum (from the ductus choledochus) are probably not reabsorbed from the jejunum and they are likely to be transported unchanged to the colon.[27] In the colon and caecum, most of the episarsasapogenin and epismilagenin glucuronides are deconjugated into free episarsasapogenin and epismilagenin before being excreted in the feces.[27] The deconjugation that is taking place in the colon and caecum is probably due to microbial metabolism, but action by intestinal cells cannot be excluded as a possible role in this metabolism.

Characteristic liver lesions in these diseases include a varying degree of necrosis of hepatocytes, concentric lamellar periductal fibrosis, proliferation of the bile duct epithelium, and accumulation of sapogenin crystals in the bile ducts and the bile duct epithelium.[13,28–30] Work on *T. terrestris* intoxication indicates that the retention of phylloerythrin causing photosensitization is due to occlusion and/or obstruction of the biliary system.[13] The *T. terrestris* findings are contradictory to results from work on *N. ossifragum* intoxication which suggest that the damage to the hepatocytes is more severe than the damage to the biliary system.[30] Assuming that both observations are correct, it can be suggested that the sapogenins can cause or enhance liver damage, resulting in retention of phylloerythrin in two different ways. Perhaps (1) the sapogenins can cause damage to the hepatocytes resulting in reduced or ceased excretion of phylloerythrin into the biliary system. It is also possible that (2) the sapogenin crystals accumulate in the biliary system, causing fibroplasia, proliferation of the biliary epithelium, and/or obstruction of the bile canaliculi and bile ducts.

13.3.2   THE DIFFERENT DISORDERS

In addition to the conditions caused by the aforementioned plants, other photosensitization syndromes such as facial eczema, *Lantana* poisoning, and *Microcystis aeruginosa* intoxication are also reviewed.

13.3.2.1   *Agave lecheguilla* Intoxication

A. lecheguilla is a conspicuous, long-lived perennial with a thick, fibrous, toothed crown bearing a cluster of thick, fleshy basal leaves, and a tall flower stalk.[31] The plant grows thickly over a wide area of southwestern Texas, where it is reported to cause photosensitization of sheep.[31,32] The leaves contain, on average, 10 g kg^{-1} smilagenin, (dry basis) as the sole sapogenin constituent.[32] Analysis of crystals from the bile of sheep fed *A. lecheguilla* suggested the presence of either smilagenin or sarsasapogenin.[32]

13.3.2.2   Geeldikkop (*Tribulus terrestris* Intoxication)

T. terrestris is a prostrate, creeping plant with a semi-perennial underground stem and root system. Each year a mat of aerial branches emerge, bearing compound leaves with five to eight pairs of hairy leaflets. The small flowers are yellow.[13] The plant grows in different parts of the world and is reported to cause photosensitization of sheep in South Africa, the U.S., Argentina, Australia, South Africa, and Iran.[13,34–37] The greatest losses occur in South Africa, where the disease is called geeldikkop (yellow thick head).[13] *T. terrestris* has been reported to cause photosensitization in both sheep and goats.[13,38]

   T. terrestris is a highly nutritious plant that is often grazed by sheep in the Karoo area of South Africa, where it only sporadically causes geeldikkop.[13] Typically, outbreaks of geeldikkop occur when *T. terrestris* is wilted during the hot dry spells that follow summer rains. It has been suggested that the presence of the toxin, sporidesmin, produced by the fungus *Pithomyces chartarum* (see facial eczema), could explain why *T. terrestris* is sporadically toxic under special weather conditions.[13] Experiments have suggested that more sheep grazing wilted *T. terrestris* became photosensitized after being dosed subclinical doses of sporidesmin. However, results from more recent experiments indicate that the saponins alone cause the disease.[23]

   T. terrestris contains several steroidal sapogenins.[39] These include diosgenin, yamogenin, epismilagenin, tigogenin, neotigogenin, gitogenin, neogitogenin, ruscogenin, and neoruscogenin.[39] Biliary crystals from sheep photosensitized after ingesting *T. terrestris* have been found to consist of the insoluble calcium salts of epismilagenin β-D-glucuronide and episarsasapogenin β-D-glucuronide.[39]

13.3.2.3   Alveld (*Narthecium ossifragum* Intoxication)

N. ossifragum is a loosely to densely clonal, perennial herb that is up to 40 cm tall and has a creeping rhizome. The plant occurs on oligotrophic, mesotrophic, and eutrophic peat deposits in Scandinavia to 69°42′N, in the British Isles, the Netherlands, Belgium, Northwestern Germany, Western and Central France, Northern Spain, and Eastern Portugal.[40] Photosensitization of sheep grazing this plant has been reported from Norway, the Faroe Islands, and the British Isles.[21,41–43] The greatest problems occur in Norway where the disease is called alveld (literally "elf-fire").

Photosensitization normally occurs only in lambs that are 2–6 months of age and is rarely seen in adult sheep.[42] Up to 50% of the lambs in certain flocks are affected in some years, but in other years almost no lambs in the same flocks are affected. Special weather conditions have been connected to the incidence of alveld. More cases are seen in cold and rainy summers than in warm and dry summers. Some *Narthecium* pastures are known to be nontoxic to sheep whereas others are toxic almost every year. Hepatotoxic fungi have been suggested as the etiology of this disease. The fungus *Cladosporium magnusianum*, which is unique to *N. ossifragum*,[44] is apparently not toxic to sheep.[45] However, the possibility exists that, in some years as a response to *C. magnusianum* infection, the plant may produce phytoalexins that might be hepatotoxic to sheep.[42]

Differences in susceptibility to the disease between different breeds and age groups have also been demonstrated.[46] The differences in susceptibility to the disease may be due to differences in activities of the detoxifying microsomal or cytosolic enzymes in the livers of the affected animals.[46,47]

In addition to photosensitization of sheep, *N. ossifragum* will also cause nephrotoxicity in cattle.[48,49] Dosing experiments using *N. ossifragum* have also resulted in liver dysfunction in calves and nephrotoxicity in sheep and goats.[50–52] Results from ongoing dosing experiments indicate that the hepatotoxin and the nephrotoxin are two different toxins.[52]

N. ossifragum has been found to contain 4 different sapogenins, of which sarsasapogenin and smilagenin are the most important.[53–54] Whether these saponins act alone or in combination with other factors to cause the characteristic liver lesions is still debated.[42] Dosing with unrealistically large doses of crude saponins from *N. ossifragum* caused photosensitization in lambs.[21,22] But dosing with realistic doses daily for three weeks was found not to cause liver dysfunction.[24]

It is sometimes very difficult to find sapogenin crystals in sections of the liver from lambs that have been photosensitized after ingesting *N. ossifragum*.[30] Insoluble salts of episarsasapogenin β-D-glucuronide and epismilagenin β-D-glucuronide have been found in bile from sheep with aveld.[54]

13.3.2.4   *Brachiaria decumbens* Intoxication

B. decumbens belongs to the grass family *Panicoideae* and is the major species used to improve pasture in many places in the tropics due to its aggressive growth habit, efficient nitrogen utilization, ability to withstand heavy grazing, drought resistance, and relative freedom from pests and diseases. *B. decumbens* intoxication of sheep, cattle, and goats has been reported to occur in Australia, Malaysia, Indonesia, Nigeria, Papua New Guinea, and Brazil.[55–58] As well as causing photosensitization of sheep and cattle, the plant has been reported to cause neurological disorders in sheep.[59] The saponins reported to be present in *B. decumbens* are yamogenin and diosgenin.[60] Episarsasapogenin and epismilagenin have been isolated from the ruminal contents of sheep suffering from *B. decumbens* intoxication.[61]

13.3.2.5   *Panicum* spp. Intoxication

The *Panicum* grasses are annual or perennial grasses, either tufted or rhizomatous. The leaves are broad or narrow, and the ligule may be a membrane, a ciliate membrane, a rim of hairs, or completely absent. The silica bodies are cross- or dumb bell-shaped and micro-hairs are always present. *Panicum* plants are widely distributed from the tropics to the warm temperate regions. Many of the *Panicum* species are important pasture grasses or cereals, and some are aggressive weeds.[62]

Panicum dichotomiflorum intoxication: Photosensitization of sheep, cattle and goats due to grazing of *P. dichotomiflorum* is reported from New Zealand.[63,64] The only sapogenin reported to be found in *P. dichotomiflorum* is diosgenin,[64–65] and the biliary crystals from photosensitized sheep are composed of the calcium salt of epismilagenin β-D-glucuronide.

Panicum schinzii intoxication: Photosensitization of sheep grazing *P. schinzii* has been reported to occur in Australia.[29,65–67] The only sapogenin that has been found in *P. schinzii* is diosgenin, and as in *P. dichotomiflorum* intoxicated sheep, the calcium salt of epismilagenin β-D-glucuronide was the only component in the biliary crystals of sheep affected by *P. schinzii* toxicosis.

Panicum miliaceum intoxication: *P. miliaceum*-associated photosensitization of sheep is known to occur in New Zealand.[5,63,64] The plant contains a mixture of two sapogenins, diosgenin and yamogenin.[68]

Panicum coloratum intoxication: *P. coloratum* is known to cause photosensitization of sheep in the U.S. and in South Africa.[13,28] Two steroidal sapogenins have been found in *P. coloratum*. These are diosgenin and yamogenin.[69]

Panicum virgatum intoxication: *P. virgatum* has been reported to be implicated in photosensitization of sheep in the U.S.[70] Grass from this outbreak contained diosgenin saponins.[71]

13.3.2.6   Facial Eczema (Pithomycotoxicosis)

Sporidesmin (Figure 13.5) produced by the saprophytic fungus *Pithomyces chartarum* causes severe liver lesions resulting in retention of phylloerythrin and photosensitization in sheep, cattle, goats, and fallow deer.[3,72] The disease occurs in animals grazing on improved pastures, and it causes severe problems in New Zealand. It is also diagnosed in Australia, South Africa, the U.S., Argentina, France, Spain, Uruguay, and Paraguay.[3] The disease occurs only when specific weather conditions prevail. *P. chartarum* grows rapidly and sporulates freely in periods when high humidity is combined with grass-minimum temperatures of 12°C or greater on two consecutive nights. The substrate for growth of *P. chartarum* in the field is dead vegetative material.[3]

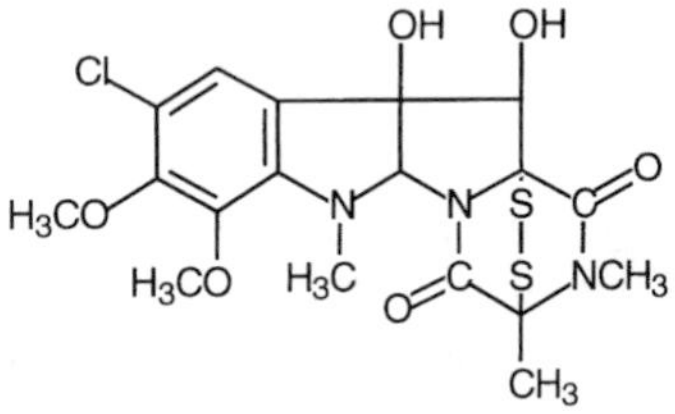

FIGURE 13.5   Sporidesmin molecule.

The incidence and severity of the disease is related to the number of spores ingested, their sporidesmin content, and the susceptibility of the animal.[3] Close grazing of leafy, well-controlled pastures increases the intake of toxic spores. Not all strains of *P. chartarum* produce spores containing sporidesmin. Nontoxic strains of *P. chartarum* have been introduced into pastures known to be toxic in attempts to reduce the concentration of sporidesmin in those pastures.[73]

P. chartarum produces several sporidesmins. The major toxin in the group, sporidesmin is often referred to as sporidesmin A. Results from *in vitro* studies indicate that the biological activity of the molecule is associated with the sulphur-containing ring.[3] In calves, a total oral dose of 3.0 mg kg⁻¹ of sporidesmin has been reported to cause 100% mortality within 3–5 days. A total dose of 0.8 mg kg⁻¹ caused photosensitization in 50% of calves but low mortality. In general, sheep and fallow deer are more susceptible to the toxin than are cattle.[3]

Ingested sporidesmin may be reduced in the rumen and oxidized before it is absorbed or it may be absorbed unchanged from the intestines. Sporidesmin in both changed and unchanged form is excreted from the liver into the bile. In the liver, the sporidesmin causes inflammation in the bile ducts, progressive obliterative cholangitis, obstructive jaundice, and phylloerythrin retention.[3] Damage to the hepatocytes has also been reported.[74] Enterohepatic circulation of sporidesmin adds little to the severity of the hepatic lesions.[3] Lesions in the urinary system do occur.[3] Edematous inflammatory changes can be seen in the mucosa and muscular wall of the bladder.

Prediction of dangerous periods (meteorological observations) and counting of spores are important tools for controlling the disease. Spraying of pastures with fungicides to reduce the number of fungi and keeping the animals on pastures that are not overgrazed both reduce the toxic hazard. In sheep, breeding for resistance to sporidesmin has been practiced for many years and has reduced the problem significantly. Oral dosing with zinc has been shown to cause effective protection in sheep and cattle.[75,76]

13.3.2.7 *Lantana camara* Intoxication

L. camara is native to central America and Africa and it has become distributed throughout many tropical and subtropical areas of the world. Intoxication of ruminants has been reported from Australia, India, South Africa, and the Americas.[77] Under natural conditions, *L. camara* poisoning occurs almost exclusively in cattle, but sheep and goats are similarly susceptible to the toxins.[13,77] Horses are not susceptible to the toxins.[77]

The toxic compounds of *L. camara* are pentacyclic triterpene acids, of which lantadene A and B are the most important.[13,77] These toxins are found in the leaves of the plant. Toxicity is not cumulative but occurs when a sufficient amount of toxic plant is consumed in one feeding.[77] Normally, toxicity occurs only during periods when feed is short.

Only a small proportion of the ingested lantadenes are absorbed. Toxicity is enhanced by the ruminal stasis that occurs in lantana poisoned animals due to inappetence in combination with reflex inhibition from the injured liver.[77] Lantadene A (Figure 13.6) is absorbed unchanged from the rumen and the intestines and is transferred to the liver via the portal vein. Results from studies of the hepatic metabolism of lantadenes indicate that the species difference in susceptibility to the toxins could be due to the action of metabolites rather than of the parent compound.[77] The toxic metabolites cause damage to the bile canalicular membranes, causing inhibition of biliary secretion. Thus, secretion of phylloerythrin is impaired.

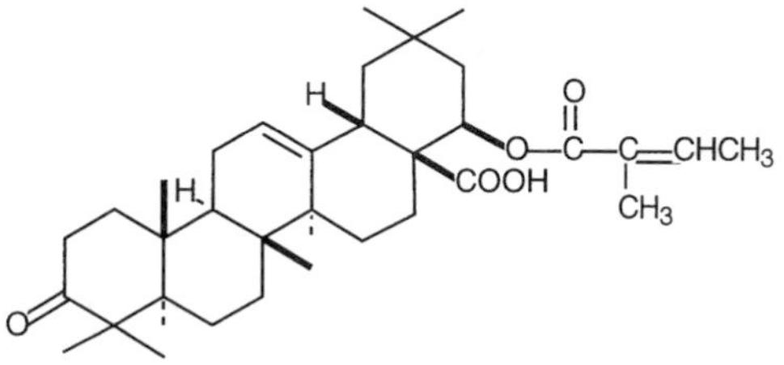

FIGURE 13.6 Lantadene A molecule.

The macroscopic lesions commonly seen in *L. camara* intoxicated animals are icterus, yellow to orange-brown discoloration and swelling of the liver, impaction of the caecum and colon, and nephrosis. Skin lesions characteristic for all photosensitization diseases are also present.[13,77] Characteristic histological findings in the liver include swelling and hydropic degeneration of the hepatocytes, single cell necrosis of hepatocytes, bile duct proliferation, fibrosis, and mononuclear cell infiltration in the portal area. Vacuolar degeneration of renal tubular cells is common.[13,77] The retention of phylloerythrin seen in *Lantana* poisoned animals is probably due to inhibited hepatocytic bile secretion.[77] Renal failure due to tubular degeneration and necrosis has been seen in chronic cases of *Lantana* poisoning.

TABLE 13.1
Photosensitization Disorders of Minor Importance

Agent	Plant/ Fungus	Toxin	Affected Species	Geographical Location
Cymopterus watsoni	Plant	Furocoumarins	Sheep	Western United States[80]
Thamnosma texana	Plant	Furocoumarins	Sheep	South-western United States[81]
Hepatogenous photosensitization disorders				
Nolina texana	Plant	Unknown	Sheep and goats	Texas, United States[82]
Tetradymia glabrata	Plan	Unknown	Sheep	Western United States[83,84]
Tetradymia canescens	Plant	Unknown	Sheep	Western United States[83,84]
Myoporum aff. *insulare*	Plant	Unknown	Cattle	Southern Australia[85]
Phomopsis leptostromiformis	Fungus	Phomopsin A	Sheep and cattle	Australia, Germany, New Zealand and South Africa[13]
Asaemia axillaris	Plant	Unknown	Sheep	South Africa[13]
Lasiospermum bipinnatum	Plant	Furanosesquiterpenoids	Sheep and cattle	South Africa[13]
Athanasia trifurcata L.	Plant	Unknown	Sheep	South Africa[13]
Nidorella foetida	Plant	Unknown	Sheep	South Africa[86]
Kochia scoparia	Plant	Unknown	Cattle	Central United States[87]
Mouldy hay	—	Unknown	Cattle	United States[88,89]
Photosensitization disorders where mechanisms are not known				
Brassica spp.	Plant	Unknown	Sheep	New Zealand, Norway[90]
Trifolium hybridum	Plant	Unknown	Horses	Australia, Canada, Denmark, England and United States[91–94]

(Primary photosensitization disorders heading appears above Cymopterus watsoni)

Lantana poisoning can be treated by oral dosing with relatively large quantities of powdered activated charcoal to adsorb the toxin in the rumen and the intestines, in an effort to minimize systemic absorption.[77] The effective dose of charcoal is about 500 g for a sheep and 2 to 2.5 kg for an adult cow.

13.3.2.8   *Microcystis aeruginosa* Intoxication

M. aeruginosa is a cosmopolitan fresh water, blue-green alga causing sporadic outbreaks of intoxication in horses, cattle, sheep, dogs, turkeys, ducks, and fish. Ruminants are the species most commonly affected.[13] Factors promoting the growth of the alga and thereby increasing the risk for the occurrence of intoxications are high water temperature, high electrolyte concentration, concentrating wind, and high concentration of other nutrients such as nitrogen and phosphorus (eutrophication).[78] The hepatotoxin causing photosensitization is a cyclic heptapeptide causing disruption of the hepatic tubular cytoskeleton which is released into the water when the algae die or disintegrate.[13,79] High doses of the toxin cause death in the animal before signs of photosensitization appear. Massive hepatic necrosis occurs in peracutely intoxicated sheep. In less severe cases, moderate to severe fatty degeneration of hepatocytes and single cell necrosis or necrosis of small foci can be present. Mild centrilobular fibrosis and mild bile duct proliferation are present in more chronically affected sheep.[13]

13.4 PHOTOSENSITIZATION DISORDERS OF MINOR IMPORTANCE

A number of photosensitization disorders of minor importance are recognized. These are summarized in Table 13.1. In many instances, the identity of the toxic agent remains elusive.

REFERENCES

1. Giese, A. C., Hypericism, *Photochem. Photobiol. Rev.*, 5, 229, 1980.
2. Steyn, D. G., *Vergiftiging van mens en dier*, Van Schaik, Pretoria, 1949.
3. Mortimer, P. H. and Ronaldson, J. W., Fungal-toxin-induced photosensitization, in *Handbook of Natural Toxins. Plant and Fungal Toxins*, Keeler, R. F. and Tu, A. T., Eds. Marcel Dekker, New York, 1983, Chap. 11.
4. Johnson, A. E., Photosensitizing toxins from plants and their biological effects, in *Handbook of Natural Toxins. Plant and Fungal Toxins*, Keeler, R. F. and Tu, A. T., Eds. Marcel Dekker, New York, 1983, Chap. 10.
5. Clare, N. T., *Photosensitization in Diseases of Domestic Animals*, CAB, Farnham Royal, 1952.
6. Rimington, C. and Quin, J. I., Photosensitizing agent in "geel-dikkop" phylloerythrin, *Nature*, 132, 178, 1933.
7. Quin, J. I., Rimington, C., and Roets, G. C. S., Studies on the photosensitization of animals in South Africa. VIII. The biological formation of phylloerythrin in the digestive tracts of various domesticated animals, *Onderstepoort J. Vet. Sci. Anim. Ind.*, 4, 463, 1935.
8. Blum, H. F., *Photodynamic Action and Diseases Caused by Light*, Reinhold Publishing Corporation, New York, 1941.
9. Durán, N. and Song, P.-S., Hypericin and its photodynamic action, *Photochem. Photobiol.*, 43, 677, 1986.
10. Knox, J. P., Samuels, R. L., and Dodge, A. D., Photodynamic action of hypericin, *Am. Chem. Soc. Symp. Ser.*, 339, 265, 1987.
11. Muslih, N. J., Al Kassim, N. A. H., Arslam, S. H., and Al Latif, A. R., Aran poisoning and its treatment in sheep, *Indian J. Vet. Med.*, 7, 52, 1987.
12. Bale, S., Poisoning of sheep, goats and cows by the weed *Hypericum triquetrifolium*, *Refuah Veterinarith*, 35, 36, 1978.
13. Kellerman, T. S., Coetzer, J. A. W., and Naudé, T. W., *Plant Poisonings and Mycotoxicoses of Livestock in Southern Africa*, Oxford University Press, Cape Town, 1988.
14. Egyed, M. N., Shlosberg, A., Eilat, A., and Malkinson, M., Photosensitization in domestic fowl caused by *Ammi majus*, in *Proceedings*. 20th World Veterinary Congress, Thessaloniki, 2353, 1975.
15. Witzel, D. A., Dollahite, J. W., and Jones. L. P., Photosensitization in sheep fed *Ammi majus* (Bishop's Weed) seed, *Am. J. Vet. Res.*, 39, 319, 1978.
16. Towers, G. H. N., Photosensitizers from plants and their photodynamic action, *Prog. Phytochem.*, 6, 183, 1980.
17. Eyged, M. N., Shlosberg, A., Eilat, A., Cohen, U., and Beemer, A., Photosensitization in dairy cattle associated with ingestion of *Ammi majus*, *Refuah Veterinarith*, 31, 128, 1974.
18. Shlosberg, A. and Egyed, M. N., *Ammi majus* induced photosensitization in chickens and turkey poults, *Refuah Veterinarith*, 35, 159, 1978.
19. Yeruham, I., Lemberg, D., Natan, A., and Egyed, M. N., Photosensitization in sheep due to ingestion of *Vicia* hay contaminated by *Ammi* majus, *Israel J. Vet. Med.*, 44, 147, 1988.
20. Flåøyen, A., Do steroidal saponins have a role in hepatogenous photosensitization diseases of sheep?, in *Saponins Used in Food and Agriculture*, Waller, G. R. and Yamaski, K., Eds., Plenum Press, 395, 1996.
21. Ender, F., Undersøkelser over alveldsykens etiologi. [Eng.: Etiological studies on "alveld"- a disease involving photosensitization and icterus in lambs], *Nordisk Veterinærmedicin*, 7, 329, 1955.
22. Abdelkader, S. V., Ceh, L., Dishington, I. W., and Hauge, J. G., Alveld-producing saponins. II. Toxicological studies, *Acta Vet. Scand.*, 25, 76, 1984.

23. Kellerman, T. S., Erasmus, G. L., Coetzer, J. A. W., Brown, J. M. M., and Maartens, B. P., Photosensitivity in South Africa. VI. The experimental induction of geeldikkop in sheep with crude steroidal saponins from *Tribulus terrestris, Onderstepoort J. Vet. Res.*, 58, 47, 1991.

24. Flåøyen, A., Hjorth Tønnesen, H., Grønstøl, H., and Karlsen, J., Failure to induce toxicity in lambs by administering saponins from *Narthecium ossifragum, Vet. Res. Comm.*, 15, 483, 1991.

25. Flåøyen, A., Smith, B. L., and Miles, C. O., An attempt to reproduce crystal-associated cholangitis in lambs by the experimental dosing of sarsasapogenin or diosgenin alone and in combination with sporidesmin, *N. Z. Vet. J.*, 41, 171, 1993.

26. Miles, C. O., Wilkins, A. L., Erasmus, G. L., and Kellerman, T. S., Photosensitivity in South Africa. VII. Ovine metabolism of *Tribulus terrestris* saponins during experimentally induced geeldikkop, *Onderstepoort J. Vet. Med.*, 61, 351, 1994.

27. Flåøyen, A., Wilkins A. L., and Miles, C.O., unpublished data, 1996.

28. Bridges, C. H., Camp, B. J., Livingston, C. W., and Bailey, E. M., Kleingrass (*Panicum coloratum L.*) poisoning in sheep, *Vet. Pathol.*, 24, 525, 1987.

29. Button, C., Paynter, D. I., Shiel, M. J., Colson, A. R., Paterson, P. J., and Lyford, R. L., Crystal-associated cholangiohepatopathy and photosensitisation in lambs, *Austr. Vet. J.*, 64, 176, 1987.

30. Flåøyen, A., Borrebæk, B., and Nordstoga, K., Glycogen accumulation and histological changes in the livers of lambs with alveld and experimental sporidesmin intoxication, *Vet. Res. Comm.*, 15, 443, 1991.

31. Mathews, F. P., Lechuguilla (*Agave lecheguilla*) poisoning in sheep, goats, and laboratory animals, in *Texas Agricultural Experiment Station Bulletin No. 554*, 1937.

32. Wall, M. E., Warnock, B. H., and Willaman, J. J., Steroidal sapogenins. LXVIII. Their occurrence in *Agave lecheguilla, Econ. Bot.*, 16, 266, 1962.

33. Camp, B. J., Bridges, C. H., Hill, D. W., Patamalai, B., and Wilson, S., Isolation of a steroidal sapogenin from the bile of a sheep fed *Agave lecheguilla, Vet. Human Toxicol.*, 30, 533, 1988.

34. Amjadi, A. R., Ahourai, P., and Baharsefat, M., First report of geeldikkop in sheep in Iran, *Arch. Inst. Razi*, 29, 71, 1977.

35. Coetzer, J. A. W., Kellerman, T. S., Sadler, W., and Bath, G. F., Photosensitivity in South Africa. V. A comparative study of the pathology of the ovine hepatogenous photosensitivity diseases, facial eczema and geeldikkop (*Tribulosis ovis*), with special reference to their pathogenesis, *Onderstepoort J. Vet. Res.*, 50, 59, 1983.

36. Glastonbury, J. R. W., Doughty, F. R., Whitaker, S. J., and Sergeant, E., A syndrome of hepatogenous photosensitisation, resembling geeldikkop, in sheep grazing *Tribulus terrestris, Austr. Vet. J.*, 61, 314, 1984.

37. Tapia, M. O., Giordano, M. A., and Gueper, H.G., An outbreak of hepatogenous photosensitization in sheep grazing *Tribulus terrestris* in Argentina, *Vet. Human Toxicol.*, 36, 311, 1994.

38. Glastonbury, J. R. W. and Boal, G. K., Geeldikkop in goats, *Austr. Vet. J.*, 62, 62, 1985.

39. Miles, C. O., Wilkins, A. L., Erasmus, G. L., Kellerman, T.S., and Coetzer, J. A. W., Photosensitivity in South Africa. VII. Chemical composition of biliary crystals from a sheep with experimental induced geeldikkop, *Onderstepoort J. Vet. Med.*, 61, 215, 1994.

40. Summerfield, R. J., Biological flora of British Isles, *J. Ecol.*, 162, 325, 1974.

41. Ford, E. J. H., A preliminary investigation of photosensitization in Scottish sheep, *J. Comp. Pathol.*, 74, 37, 1964.

42. Flåøyen, A., *Studies on the Aetiology and Pathology of Alveld with some Comparisons to Sporidesmin Intoxication*, Dr.med.vet.-thesis, Norwegian College of Veterinary Medicine, Oslo, 1993.

43. Flåøyen, A., Jóhansen, J., and Olsen, J., *Narthecium ossifragum* associated photosensitization in sheep in the Faroe Islands, *Acta Vet. Scand.*, 36, 277, 1995.

44. di Menna, M. E., Flåøyen, A., and Ulvund, M. J., Fungi on *Narthecium ossifragum* leaves and their possible involvement in alveld disease of Norwegian lambs, *Vet. Res. Comm.*, 16, 117, 1992.

45. Flåøyen, A., di Menna, M. E., Collin, R. G., and Smith, B. L., *Cladosporium magnusianum* (Jaap) M.B. Ellis is probably not involved in alveld, *Vet. Res. Comm.*, 17, 241, 1993.

46. Flåøyen, A., A difference in susceptibility of two breeds of sheep to the "alveld toxin," *Vet. Res. Comm.*, 15, 455, 1991.

47. Flåøyen, A. and Jensen, E. G., Microsomal enzymes in lambs and adult sheep, and their possible relationship to alveld, *Vet. Res. Comm.*, 15, 271, 1991.

48. Malone, F. E., Kennedy, S., Reilly, G. A. C., and Woods, F. M., Bog asphodel (*Narthecium ossifragum*) poisoning in cattle, *Vet. Rec.*, 131, 100, 1992.

49. Flåøyen, A., Binde, M., Bratberg, B., Djønne, B., Fjølstad, M., Grønstøl, H., Hassan, H., Mantle, P. G., Landsverk, T., Schönheit, J., and Tønnesen, M. H., Nephrotoxicity of *Narthecium ossifragum* in cattle in Norway, *Vet. Rec.*, 137, 259, 1995.

50. Flåøyen, A., Bratberg, B., Frøslie, H., and Grønstøl, H., Nephrotoxicity and hepatotoxicity in calves apparently caused by experimental feeding with *Narthecium ossifragum*, *Vet. Res. Comm.*, 19, 63, 1995.

51. Flåøyen, A., Bratberg, B., and Grønstøl, H., Nephrotoxicity in lambs apparently caused by experimental feeding with *Narthecium ossifragum*, *Vet. Res. Comm.*, 19, 75, 1995.

52. Flåøyen, A., Bratberg, B., Frøslie, H., Grønstøl, H., Langseth, W., Mantle, P. G., and von Krogh, A., unpublished data, 1996.

53. Ceh, L. and Hauge, J. G., Alveld-producing saponins. I. Chemical studies, *Acta Vet. Scand.*, 22, 391, 1981.

54. Miles, C. O., Wilkins, A. L., Munday, S. C., Flåøyen, A., Holland, P. T., and Smith, B. L., Identification of insoluble salts of the β-D-glucuronides of episarsasapogenin and epismilagenin in the bile of lambs with alveld and examination of *Narthecium ossifragum*, *Tribulus terrestris*, and *Panicum miliaceum* for sapogenins, *J. Agric. Food Chem.*, 41, 914, 1993.

55. Abas Mazni, O., Mohd Khusahry, Y., and Sheikh Omar, A. R., Jaundice and photosensitization in indigenous sheep of Malaysia grazing on *Brachiaria decumbens*, *Malaysian Vet. J.*, 7, 254, 1983.

56. Opasina, B. A., Photosensitization jaundice syndrome in West African Dwarf Sheep and goats grazed on *Brachiaria decumbens*, *Trop. Grass.*, 19, 120, 1985.

57. Graydon, R. J., Hamid, H., Zahari, P., and Gardiner, C., Photosensitisation and crystal-associated cholangiohepatopathy in sheep grazing *Brachiaria decumbens*, *Austr. Vet. J.*, 68, 234, 1991.

58. Low, S. G., Bryden, W. L., Jephcott, S. B., and Grant, I. McL., Photosensitisation of cattle grazing Signal grass (*Brachiaria decumbens*) in Papua New Guinea, *N. Z. Vet. J.*, 41, 220, 1993.

59. Abdullah, A. S., Noordin, M. M., and Rajion, M. A., Neurological disorders in sheep during Signal grass (Brachiaria decumbens) toxicity, *Vet. Human Toxicol.*, 31, 128, 1989.

60. Smith, B. L. and Miles, C. O., A role for *Brachiaria decumbens* in hepatogenous photosensitization of ruminants, *Vet. Human Toxicol.*, 35: 256, 1993.

61. Lajis, N. H., Abdullah, A. S., Salim, S. J. S., Bremner, J. B., and Khan, M. N., *Epi*-sarsasapogenin and *epi*-smilagenin: two sapogenins isolated from the rumen content of sheep intoxicated by *Brachiaria decumbens*, *Steroids*, 58, 387, 1993.

62. Dahlgren, R. M. T., Clifford, H. T., and Yeo, P. F., *The Families of the Monocotyledons. Structure, Evolution, and Taxonomy*, Springer-Verlag, Berlin, 1985, 442.

63. Holland, P. T., Miles, C. O., Mortimer, P.H., Wilkins, A. L., Hawkes, A. D., and Smith, B. L., Isolation of the steroidal sapogenin epismilagenin from the bile of sheep affected by *Panicum dichotomiflorum* toxicosis, *J. Agric. Food Chem.*, 1991, 39, 1963, 1993.

64. Miles, C. O., Munday, S. C., Holland, P. T., Smith, B. L., Embling, P. P., and Wilkins, A. L., Identification of a sapogenin glucuronide in the bile of sheep affected by *Panicum dichotomiflorum* toxicosis, *N. Z. Vet. J.*, 39, 150, 1991.

65. Miles, C. O., Wilkins, A. L., Munday, S. C., Holland, P. T., Smith, B. L., Lancaster, M. J., and Embling, P. P., Identification of the calcium salts of epismilagenin β-D-glucuronides in the bile crystals of sheep affected by *Panicum dichotomiflorum* and *Panicum schinzii* toxicoses, *J. Agric. Food Chem.*, 40, 1606, 1992.

66. Lancaster, M. J., Vit, I., and Lyford, R. L., Analysis of bile crystals from sheep grazing *Panicum schinzii* (sweet grass), *Austr. Vet. J.*, 68, 281, 1991.

67. Miles, C. O., Munday, S. C., Holland, P. T., Lancaster, M. J., and Wilkins, A. L., Further analysis of bile crystals from sheep grazing *Panicum schinzii* (sweet grass), *Austr. Vet. J.*, 69, 34, 1992.

68. Miles, C. O., A role for steroidal saponins in hepatogenous photosensitisation, *N. Z. Vet. J.*, 41, 221, 1993.

69. Patamalai, B., Hejtmancik, E., Bridges, C. H., Hill, D. W., and Camp, B. J., The isolation and identification of steroidal sapogenins in Kleingrass, *Vet. Human Toxicol.*, 32, 314, 1990.

70. Puoli, J. R., Reid, R. L., and Belesky, D. P., Photosensitization in lambs grazing switchgrass, *Agron. J.*, 84, 1077, 1992.

71. Smith, B. L., personal communication, 1996.

72. Smith, B. L. and Embling, P. P., Facial eczema in goats: The toxicity of sporidesmin in goats and its pathology, *N. Z. Vet. J.*, 39, 18, 1991.

73. Collin, R. G. and Towers, N. R., First reported isolation from New Zealand pastures of *Pithomyces chartarum* unable to produce sporidesmin, *Mycopathologia*, 130, 37, 1995.

74. Flåøyen, A. and Smith, B. L., Parenchymal injury and biliary obstruction in relation to photosensitization in sporidesmin-intoxicated lambs, *Vet. Res. Comm.*, 16, 337, 1992.

75. Smith, B. L., Embling, P. P., Towers, N. R., Wright, D. E., and Payne, E., The protective effect of zinc sulphate in experimental sporidesmin poisoning of sheep, *N. Z. Vet. J.*, 25, 124, 1977.

76. Towers, N. R. and Smith, B. L., The protective effect of zinc sulphate in experimental sporidesmin intoxication of lactating dairy cows, *N. Z. Vet. J.*, 26, 199, 1978.

77. Pass, M. A., Poisoning of livestock by *Lantana* plants, in *Handbook of Natural Toxins. Toxicology of Plant and Fungal Compounds*, Keeler, R. F. and Tu, A. T., Eds., Marcel Dekker, New York, 1991, Chap. 14.

78. Radostits, O. M., Blood, D. C., and Gay, C. C., *Veterinary Medicine. A Textbook of the Diseases of Cattle, Sheep, Pigs, Goats and Horses*, Eight edition, Baillière Tindall, London, 1994.

79. Galey, F. D., personal communication, 1996.

80. Binns, W., James, L. F., and Brooksby, W., *Cymopterus watsoni*: A photosensitizing plant for sheep, *Vet. Med./Small Anim. Clin.*, 59, 375, 1964.

81. Oertli, E. H., Rowe, L. D., Lovering, S. L., Ivie, G. W., and Bailey, E. M., Phototoxic effect of *Thamnosma texana* (Dutchman's breeches) in sheep, *Am. J. Vet. Res.*, 44, 1126, 1983.

82. Mathews, F.P., Poisoning in sheep and goats by sacahuiste (*Nolina texana*) bus and blooms, in *Texas Agricultural Experiment Station Bulletin No. 585*, 1940.

83. Johnson, A. E. Predisposing influence of range plants on *Tetradymia*-related photosensitization in sheep: Work of Drs. A. B. Clawson and W. T. Huffman, *Am. J. Vet. Res.*, 35, 1583, 1974.

84. Johnson, A. E., Tetradymia toxicity - a new look at an old problem, in *Effects of Poisonous Plants on Livestock*, Keeler, R. F., Van Kampen, K. R. and James, L. F., Eds., Academic Press, New York, 1978, 209.

85. Jerrett, I. V. and Chinnock, R. J., Outbreaks of photosensitisation and deaths in cattle due to *Myoporum* aff. *Insulare* R. Br. toxicity, *Austr. Vet. J.*, 60, 183, 1983.

86. Schneider, D. J., Green, J. R., and Collett, M. G., Ovine hepatogenous photosensitivity caused by the plant *Nidorella foetida* (Thunb.) dc. (asteraceae), *Onderstepoort J. Vet. Sci.*, 54, 53, 1987.

87. Dickie, C. W. and James, L. F., *Kochia scoparia* poisoning in cattle, *J. Am. Vet. Med. Assoc.*, 183, 765, 1983.

88. Bagley, C. V., McKinnon, J. B., and Asay, C. S., Photosensitization associated with exposure of cattle to moldy straw, *J. Am. Vet. Med. Assoc.*, 183, 802, 1983.

89. Scruggs, D. W. and Blue, G. K., Toxic hepatopathy and photosensitization in cattle fed moldy alfalfa hay, *J. Am. Vet. Med. Assoc.*, 204, 264, 1994.

90. Vermunt, J. J., West, D. M., and Cooke, M. M., Rape poisoning in sheep, *N. Z. Vet. J.*, 41, 151, 1993.

91. Fincher, M. G. and Fuller, H. K., Photosensitization-trifoliosis-light sensitization. *Cornell Vet.*, 32, 95, 1942.

92. Nation, P. N., Alsike clover poisoning: A review, *Can. Vet. J.*, 30, 410, 1989.

93. Traub, J. L., Potter, K. A., Bayly, W. M., and Reed, S. M., Alsike clover poisoning, *Mod. Vet. Pract.*, 63, 307, 1982.

94. Østergaard, H., Et tilfælde af fotosensibilisering hos hest, *Dansk Veterinærtidsskrift*, 75, 49, 1992.

14 Nitrate and Oxalates

J.P. Marais

CONTENTS

14.1 INTRODUCTION

Nitrate and oxalates are often regarded as anti-quality factors in food products and animal feeds. Nitrate is a key component of the global nitrogen cycle. Its uptake and assimilation in plants are of fundamental importance in the synthesis of plant proteins and other nitrogenous compounds, and therefore indirectly also of nitrogenous substances in animals. On decay, these organic nitrogenous substances again give rise to nitrate in the soil, which could be reutilized by the plant or could leach out of the soil and enter rivers or aquifers. Nitrate therefore occurs in most foodstuffs and water supplies and can have an adverse effect on human and animal health.

Oxalic acid (ethanedioic acid), in turn, is a product of plant, animal, and fungal metabolism. It accumulates in relatively large concentrations in the mycelium and growth media of many fungi and in the cell sap of members of certain plant families, notably the Amaranthaceae, Araceae, Chenopodiaceae, Mesembryanthemaceae, Oxalidaceae, Poaceae, and Polygonaceae. Due to its acidic properties, its extremely strong chelating ability, and its enzyme inhibitory action, oxalic acid could have far-reaching effects on metabolic processes. This chapter is a critical appraisal of current knowledge of the occurrence and toxicity of nitrate and oxalates.

14.2   NITRATE

14.2.1   ACCUMULATION IN PLANTS AND WATER

In higher plants, nitrate is assimilated in the cytoplasm, but some is also channeled into vacuoles (reserve pools) where, under conditions of excess nitrogen, large amounts can accumulate. Two important steps in the assimilation process involve the conversion of nitrate to nitrite, catalyzed by the enzyme nitrate reductase, and the reduction of nitrite to ammonia, catalyzed by nitrite reductase.

The main factors leading to nitrate accumulation are reduced photosynthesis and excess nitrogen fertilization. Vegetables are often the main source of nitrate in the human diet and could contribute 75 to 80% of the daily intake of nitrate.[1] The most common high-nitrate vegetables are listed in Table 14.1. Some forages are high in nitrate, especially when heavily fertilized. Forage species reported to have poisoned ruminants are listed in Table 14.2.

TABLE 14.1
The Nitrate Content of the Major High-Nitrate Vegetables

Vegetable	Botanical name	Nitrate content (mg kg^{-1} fresh)
Beetroot[2]	*Beta vulgaris*	900
Brinjal[2]	*Solanum melongena*	1300
Cabbage[3]	*Brassica oleracea*	810
Cauliflower[2]	*Brassica oleracea*	1310
Celery[4]	*Apium graveolens*	1200
Fennel[4]	*Foeniculum vulgare*	2000
Kohlrabi[3]	*Brassica oleracea*	1060
Lettuce (butterhead)[5]	*Lactuca sativa*	5360
Chinese mustard[5]	*Brassica* sp	5670
Radish[4]	*Raphanus sativus*	1100
Spinach[5]	*Spinacea oleracea*	3560
Turnip[4]	*Brassica rapa*	970
Watercress[4]	*Nasturtium officinale*	1300

TABLE 14.2
Forage Crops Responsible for Poisoning Ruminants

Crop	Nitrate content (g kg^{-1} dry weight)
Barley straw[6]	34
Green oats[7]	28
Hay[8]	39
Oat hay[9]	20–44
Perennial ryegrass[10]	36
Pigweed (*Amaranthus retroflexus*)[7]	36
Turnips[7]	51–56

A large amount of nitrogen, most of which is directly or indirectly derived from plant nitrogen, is tied up in organic form in the topsoil. Under suitable conditions, this nitrogen is mineralized to ammonia, which is then oxidized to nitrite and eventually to nitrate by soil organisms such as *Nitrosomonas*, and *Nitrobacter* spp., respectively. The nitrate is available for reabsorption by the plant, but can be lost from the surface horizons by leaching into the groundwater, especially from well-drained soils under warm, moist conditions in the absence of plant growth.

Considerable concern has been expressed regarding the gradual increase in the nitrate content of water supplies. The World Health Organization recommends a nitrate content of less than 50 mg l^{-1}.[11] In rural areas, however, wells are often the main source of water for human and animal consumption, in extreme cases containing nitrate levels in excess of 1000 mg l^{-1}.

14.2.2 TOXICITY OF NITRATE, NITRITE, AND NITROSAMINES

t is readily reduced to ammonia by microbial action
uation 14.1),

$$NO_2 \xrightarrow{\text{Nitrite reductase}} NH_3 \tag{14.1}$$

ian the conversion of nitrite to ammonia, nitrite readily
a and hypotension when absorbed into the blood. In
ry amino groups, it forms mutagenic N-nitroso com-
ertain microbial populations.

emoglobin (MetHb) by nitrate is its prior microbial
re iron-containing components of the blood, trans-
ral tissue. In the presence of nitrite, the iron (Fe II)
forming MetHb, which is unable to bind oxygen.
is) of the tissue which, in severe cases, could lead
s a large proportion of the Hb of the fetus and the
tHb than other forms of Hb, making infants more

ethemoglobinemia, including 21 deaths, have been
stances, the affected were babies less than 90 days
ll water containing high concentrations of nitrate.
important public health problem in rural areas and
er supplies.

by the chocolate-brown appearance of the blood,
a muddy appearance. Cyanosis becomes apparent
of MetHb [13] and death could occur at 44% MetHb.[14]
le, since commercial formula foods encourage the
lation in the infant gut, actively reducing nitrate to
s developing methemoglobinemia, the consumption
eetroot should be avoided.

2g kg^{-1} body weight resulted in death with 70%
kg^{-1} caused profuse perspiration and rendered the
symptoms include frequent urination and restless-
lly, animals are unable to rise and die after severe

The nitrate-reducing capacity of the rumen makes ruminants particularly vulnerable to poisoning by nitrate. Acute poisoning occurs when the nitrate content of forages exceeds 5g kg^{-1} on a dry weight basis and that of drinking water exceeds 500mg l^{-1}.[17] The rumen microbial population is remarkably adaptable. On ingesting nitrate, the proliferation of nitrate-reducing microbes is stimulated, but it could take several days before nitrite production in the rumen is optimized. Methemoglobinemia and toxic symptoms therefore often only manifest a week after exposure to high levels of nitrate. On prolonged nitrate intake (over several weeks), MetHb levels gradually decline and animals can tolerate higher concentrations than before, indicating the adaptation of the microbial population to a more efficient reduction of nitrite.

Lowland abortion syndrome, which affects cattle on lowland pastures in Wisconsin, has been attributed to the intake of nitrate.[18] Reproductive problems and abortions in ruminants have also been reported following nitrate administration.[19] Abortion has been attributed to hypoxia. Umbilical venous blood from treated pregnant animals shows low oxygen saturation, suggesting low oxygen transfer to the fetus, while high concentrations of MetHb were reported in still-born calves.[20]

Severely intoxicated animals can be treated with methylene blue as a 1% aqueous solution administered intravenously at a rate of 4.4 mg kg^{-1} body weight.[21]

14.2.2.2   Hypotension

Nitrite has a vasodilatory action on animals, causing circulatory disturbances such as a reduction in blood pressure.[22] Although a drop in blood pressure is followed by an increased cardiac output, which tends to normalize the blood pressure of the animal, vasodilation seriously affects the fetus, since the pregnant uterus is unable to autoregulate its blood flow.[23]

14.2.2.3   Antimicrobial Action

Due to their curing properties, nitrate and nitrite are at present extensively used as preservatives in foods such as meat, fish, and cheese. Nitrite derived from high-nitrate pastures markedly reduces digestibility *in vitro*[24] by reducing the cellulolytic and xylanolytic microbial population of rumen digests, with a concomitant reduction in cellulase and xylanase activity.[25] Furthermore, the growth in pure culture of three of the four major cellulolytic bacteria commonly found in the rumen, viz. *Ruminococcus flavefaciens*, *Butyrivibrio fibrisolvens*, and *Bacteroides succinogenes*, is reduced in the presence of nitrite.

Nitrite exerts its antimicrobial effect either by forming a Perigo type factor in heat-treated meat products,[26] or by forming undissociated nitrous acid at pH 4.5 to 5.5, which is capable of passing bacterial cell wall barriers. Although the site of inhibitory action of nitrite on aerobic microbes is not known, evidence strongly suggests that bacterial cell membranes and processes associated with these membranes are involved.[27] In anaerobes, electron transport mechanisms appear to be affected.[25,27]

14.2.2.4   Cancer

Gastric cancer is the most common form of cancer worldwide,[28] and is linked to a diet rich in nitrate and nitrite. According to the model of gastric carcinogenesis advanced by Correa (1988),[29] the intake of nitrate might lead to human cancer, since nitrate, under suitable conditions, gives rise to N-nitroso compounds, most of which are mutagens or potent carcinogens. The chemistry of N-nitroso compounds has recently been reviewed by Shuker (1988),[30] and their carcinogenicity by Rowland (1988).[31] Nitrite as such is not a nitrosating agent, but is readily oxidized to a series of nitrosating substances, viz. N_2O_3, N_2O_4, H_2ONO^+ and NO^+, depending on the acidity of the medium.[30] N-nitroso compounds form when a nitrosating agent reacts with secondary nitrogen groups under acidic conditions (pH < 4). If the nitrosatable compound is a secondary amine, the product is a N-nitrosamine (Equation 14.2).

$$\begin{array}{c} R1 \\ \diagdown \\ NH \\ \diagup \\ R2 \end{array} + \;\; X - NO \;\; \longrightarrow \;\; \begin{array}{c} R1 \\ \diagdown \\ N - N = O \\ \diagup \\ R2 \end{array}$$

Secondary amine   Nitrosating agent   N-nitrosamine   (14.2)

R1 and R2 could be any of a large number of chemical groups.[32] If the nitrosatable compound is an amide, the product is a N-nitrosamide, while ureas give rise to N-nitrosoureas. N-nitrosamines cause tumors at sites distant to the introduction site after activation by mammalian enzymes such as cytochromes P450.[32] In contrast, the N-nitrosamides and N-nitrosoureas are reactive, direct-acting mutagens which do not require activation. They form tumors only at the site of their introduction. Mutagenicity of N-nitroso compounds results from their affinity to alkylate oxygen atoms of nucleic acids in DNA and RNA, thus bringing about alterations in the expression of the genome. However, reactions other than alkylation of DNA may also be important in tumor formation.[34]

Since a low pH is required, the acidic stomach is the only organ in which N-nitroso compounds are likely to be formed in large quantities by acid-catalysis. The nitrosation reaction can also be catalyzed at neutral pH by certain bacterial populations, using both nitrate and nitrite as substrate.[35] N-nitroso compounds are likely to be formed whenever nitrate or nitrite, nitrosatable nitrogen compounds, and bacteria are present together.[12]

Gastric cancer develops in stages, usually over a period of many years.[12] It has been hypothesized that the first stage is the atrophy of the normal gastric mucosa and a concomitant rise in pH to above 4, which is normally associated with advancing age. At this stage a bacterial flora establishes in the stomach. These bacteria are capable of reducing nitrate to nitrite and catalyzing the N-nitrosation reaction, forming mutagenic N-nitroso compounds. These substances change the nature of the tissue, eventually leading to gastric carcinoma. Bartsch (1991)[36] pointed out that exposure to N-nitroso compounds formed by acid-catalysed nitrosation in the normal stomach early in life could be as important as bacterial nitrosation in the etiology of gastric cancer.

Ingested nitrate absorbed from the proximal small intestine is actively secreted into the saliva,[37] and to a lesser extent into gastric secretions, milk, and possibly vaginal secretions.[38] Some of the salivary nitrate is converted to nitrite by oral bacteria,[39] and enhances the synthesis of N-nitroso compounds on entering the acidic stomach. The major portion (55–65%) of the absorbed nitrate is excreted in the urine within 24h.[40] Urine also contains nitrosatable amines, while about 90% of the bacterial strains from urinary tract infections produces nitrosating enzymes.[41] Under these conditions, high levels of nitrite and N-nitroso compounds are formed. People with chronic urinary tract infection are therefore at a higher risk of developing bladder cancer than the general population.[42] *Lactobacillus* and *Streptococcus* spp in the normal vagina maintain a low pH and do not reduce nitrate to nitrite, or catalyze the N-nitrosation reaction. However, it has been suggested that the incidence of cervical cancer associated with *Trichomonas vaginalis* infection in black South Africans in the eastern Transvaal is due to the synthesis and activation of N-nitroso compounds by prolonged secondary bacterial infections.[45] Conditions for nitrosation reactions in the rumen and abomasum of dairy cattle appear unfavorable.[43,44]

Nitroso compounds acting at sites distant from the site of synthesis appear to be responsible for incidents of congenital tumors of the central nervous system of the newborn, after ingestion of high levels of nitrate by pregnant women.[46] Preformed N-nitroso compounds have been detected in sun-dried vegetables in the Kashmir region of India, a high-risk area for esophageal cancer.[3] N-nitroso compounds in products of the tobacco plant appear to be involved in causing cancer of the lungs, oral cavity, esophagus, and pancreas,[47] while high endogenous nitrosation was indicated in patients with liver cancer associated with liver fluke infection in Thailand.[48]

Many studies have shown that fresh fruit and raw vegetables have a protective effect against certain types of cancer.[29] The role of vitamin A, C, and E in inhibiting the formation of N-nitroso compounds has recently been reviewed.[49] Due to the importance of nitrite in the etiology of gastric and other cancers, vegetables and water excessively high in nitrate should be limited in the diet. The intake of foods high in vitamins A, C, and E should be increased, especially during pregnancy, and in subjects predisposed to infections of the mouth, stomach, and urinary bladder.

14.3   OXALATES

14.3.1   OCCURRENCE

Oxalic acid is the strongest dicarboxylic acid commonly present in living organisms, forming highly soluble neutral and acidic salts with monovalent metals such as potassium and sodium, and poorly soluble chelates with divalent cations such as calcium (Equation 14.3).

$$O = C - OH \atop O = C - OH \quad + M^{2+} \longrightarrow \quad O = C - O \atop O = C - O \Large{\diagdown} M + 2H^+$$

Oxalic acid   Chelate ring

$$(14.3)$$

Many fungi excrete oxalic acid, which disperses in the growth medium.[50] Intracellular crystals are also formed. An extensive list of oxalate-producing lichens and pathogenic and rhizobial fungi has been published recently.[51]

Higher plants vary in their ability to accumulate oxalate. Some important oxalate-accumulating plants are listed in Table 14.3. More detailed lists have appeared in the literature.[52–54]

Oxalic acid is bound as insoluble oxalates in large specialized cells called crystal idioblasts, often associated with specific tissue, such as the xylem, phloem, cambium, epidermis, or mesophyll. Developing idioblasts appear to be highly active cells. Prior to crystal formation

TABLE 14.3
Oxalate Content of Some Agriculturally Important Plants

Family	Species	Oxalate Content (g kg⁻¹)	
		Water Soluble	Total
Chenopodiaceae	*Halogeton glomeratus*[55]	172–346	218–387
	Spinacia oleracea[56]		97–112
Mesembryanthemaceae	*Mesembryanthemum nodiflorum*[57]	180	
Poaceae	*Brachiaria mutica*[58]	4	8
	Cenchrus ciliaris[59]		27–41
	Cenchrus ciliaris (hay)[60]	35–43	
	Digitaria decumbens[58]	4	9
	Panicum maximum (hay)[61]		10–23
	Pennisetum clandestinum[62]	6–12	11–18
	Pennisetum purpureum (hay)[61]		25–26
	Setaria italica[63]		7
	Setaria sphacelata (hay)[64]		34–91
Polygonaceae	*Rheum* sp.[65]		34–95
	Rumex crispus[66]		66–111

membranous complexes are formed in the vacuole, giving rise to membrane chambers in which the crystals eventually develop. A detailed account of oxalate crystal formation in plants has been given by Franceschi and Horner (1980).[53]

14.3.2   Toxicity

14.3.2.1   Host Invasion by Pathogenic Fungi

Fungal diseases of plants and wood-rotting molds have a worldwide distribution and cause severe crop losses and the destruction of timber. Oxalic acid has been identified as an important factor in the mechanism of infection of the host tissue.[67] Due to its chelating properties, oxalic acid sequesters calcium from calcium pectate in the cell walls of the host plants to form calcium oxalate, thus rendering them vulnerable to enzymic attack. Furthermore, oxalic acid lowers the tissue pH to near the optimum for fungal endopolygalacturonase, pectin methylesterase, and cellulase activity, which appear to act synergistically with the oxalic acid to depolymerize the host cell walls.[68] Sugars and uronic acids released during depolymerization stimulate the further production of oxalic acid by the fungus.[69]

Oxalic acid is rapidly oxidized by the destructive brown-rot fungus *Gloeophyllum trabeum*. The decrease in oxalic acid concentration parallels cellulose depolymerization.[70] Since oxalate produces hydrogen peroxide on oxidation, and wood contains considerable amounts of iron ions as Fe(III), it has been postulated that depolymerization of wood cellulose is caused by hydroxyl radicals generated by means of the Fenton reaction [70] (Equation 14.4).

$$\mathbf{Fe\ (II) + H_2O_2 \longrightarrow Fe\ (III) + \ ^{\cdot}OH + OH^-} \tag{14.4}$$

The oxidation of oxalate appears to be iron dependent, converting Fe(III) back to Fe(II).

Oxalate has also been implicated in the oxidative degradation of lignin by pathogenic fungi. The oxidation occurs through a series of reactions involving oxalate as a manganese chelator and the oxidation of Mn(II) to Mn(III).[71] The trivalent manganese is an oxidant capable of oxidizing many phenolic substrates in the host tissue.

In addition to its function as a sequestering agent, oxalic acid is regarded as a causal agent in wilt disease by acting as an inhibitor of polyphenoloxidase at low pH.[72] Furthermore, fungal oxalate suppresses the host's defence system by inhibiting o-diphenol oxidase, an enzyme producing o-quinones, essential in plant defence mechanisms.[73]

Infection of man with oxalate-producing fungi of the *Aspergillus* group is characterized by renal impairment due to oxalosis and the massive deposition of calcium oxalate at the site of infection, namely the lungs. In the lungs, iron cations complex onto calcium oxalate crystal surfaces, generating oxidants at the solid-solution interface, which are capable of causing tissue damage.[74] Moldy feed, due to contamination with *Aspergillus niger* and *A. flavus*, often contains large amounts of oxalate, which could cause severe intoxication of animals.[75]

14.3.2.2   Protection Against Invasion and Predation

Plants exposed to pathogenic fungi induce the synthesis of biosynthetic enzymes capable of synthesizing defence compounds (phytoalexins), with antifungal activity. Davis et al. (1992)[76] showed that oxalate could enhance the synthesis of these defence compounds in suspension cultures of cotton (*Gossypium hirsutum*). Oxalic acid is an elicitor of the phytoalexin, glyceollin I in soybean (*Glycine max*),[77] which decreases the activity of polygalacturonase produced by the fungus.[78]

Oxalate appears to play an important role in the development of plant defence responses through the production of hydrogen peroxide. Oxalate oxidase, a germin-like protein catalyzing the oxidation of oxalic acid to hydrogen peroxide, is induced in barley leaves during infection by the fungus *Erysiphe graminis*.[79] Increased hydrogen peroxide concentrations in plants lead to the induction of defence genes.[80] Although domestic animals are sometimes severely intoxicated by oxalate present in forage plants, the palatability of the plant does not appear to be reduced. Oxalate might, however, protect plants by discouraging more selective foragers. Genes in the rice plant (*Oryza sativa*), coding for resistance against the brown planthopper, appear to be associated with elevated oxalate levels in the leaf sheath.[81] Of greater survival value to the plant is the production by certain families, notably the Araceae, of calcium oxalate crystal raphides, often in conjunction with a protein toxin, which cause severe irritation when it comes in contact with the skin or mucous membranes.[82] Crystals are often abundant in the surface layers of these plants,[82] sometimes occurring in specialized stinging emergences.[84]

14.3.2.3   Oxalate Poisoning of Mammals

Due to its wide occurrence throughout the plant kingdom, oxalate is often ingested by man and domestic animals, sometimes with dire consequences. Only a small number of foods contribute significantly to the oxalate intake in a western diet. These are spinach, rhubarb, strawberries, tea, chocolate, and nuts.[85] Grazing animals are more often exposed to high levels of oxalate than human beings. Oxalate accumulation is more common in the panicoid than the festucoid grasses.[61] Of the tropical panicoid grasses indigenous to Africa, those belonging to the genera *Cenchrus, Setaria, Pennisetum, Digitaria*, and *Brachiaria* accumulate oxalate and form a major part of the diet of many wild and domesticated herbivores.[86] The introduction of these grasses into Australia has led to severe outbreaks of nutritional secondary hyperparathyroidism in horses.[58] In New Zealand, sporadic cases of oxalate poisoning have been reported on sorrel (*Rumex* sp.), fathen (*Chenopodium* sp.), *Oxalis* spp., and pigweed (*Amaranthus retroflexus*).[87] In North America, oxalate poisoning has been associated mainly with the intake of two introduced members of the Chenopodiaceae, viz. *Halogeton glomeratus* and *Saracobatus vermiculatus* (greasewood).[88] These palatable plants are readily grazed by sheep and cattle. The weeds *Oxalis pes-caprae* and *Rumex angiocarpus* are often responsible for animal deaths in South Africa. In semi-arid regions, *Mesembryanthemum* sp. (succulent vygies) are also associated with oxalate poisoning of sheep and goats.[89]

Calcium in the digestive tract, reacting with dietary oxalate, is one of the most important factors reducing oxalate absorption from the gut. The oxalate in the gut is partly destroyed by intestinal bacteria and the remainder is excreted in the feces.[52] Once absorbed, oxalate is poorly metabolized but has a marked effect on mammalian tissue, leading to acute or chronic poisoning.

Ruminants are generally less susceptible to oxalate poisoning than nonruminants. By gradually increasing the oxalate content of the ruminant diet over a period of 3 to 4 days, the animal becomes adapted to concentrations which would otherwise be lethal.[90] Adaptation is attributed to the proliferation of an obligate anaerobe, *Oxalobacter formigenes*, which uses oxalate as sole energy-yielding substrate. Similar microbial populations occur in the cecae and colons of a wide range of nonruminants such as rodents,[91] pigs,[92] horses[93] and man.[94]

The local corrosive action following the intake of high concentrations (2 to 30g) of oxalic acid by man could lead to acute hemorrhagic gastroenteritis and death.[52] The severity of poisoning depends on the amount of food, and in particular the amount of calcium present in the digestive tract. Further symptoms include renal insufficiency, cardiovascular and neuromuscular impairment and depression of the central nervous system. Extensive deposition of calcium oxalate crystals occurs in both the cells and lumen of tubules of the renal system, often causing irreparable damage to the tissue. Therapy is often of little avail once kidney damage has occurred.

Young and James (1988)[88] suggested that the principal cause of death from oxalate is its interference with energy metabolism. Since oxalate is an inhibitor of pyruvate kinase,[95] pyruvate carboxylase,[96] and malic enzyme,[97] excessive levels of tissue oxalate, due to absorption from the gut, could seriously affect glycolysis, NADPH formation and gluconeogenesis.

A single dose of 454g of oxalate is considered lethal to horses.[98] Acute toxicity has been recorded in cattle consuming pasture containing oxalate at a concentration of 69g kg^{-1}.[99] Sheep are more likely to be poisoned than cattle, since they are more inclined to graze high-oxalate weeds, such as *Rumex, Chenopodium, Amaranthus*, and *Oxalis* species, than cattle. The lethal oxalic acid dose for sheep is about 1.1g kg^{-1} body weight.[90] Unadapted sheep died over a period of 5 days after exposure to young buffalo grass (*Cenchrus ciliaris*) with a soluble oxalate content of about 25g kg^{-1}.[59] Affected sheep had dyspnoea, were stiff-gaited, collapsed, and died without struggling.

The continuous ingestion of small amounts of oxalate is characterized by chronic renal failure due to the precipitation of oxalate crystals, the formation of renal calculi, or diseases associated with a calcium deficiency or mineral imbalance.

An increasing incidence in Western countries of upper urinary tract stones in man, consisting mainly of calcium oxalate, has been reported.[52,100] Calcium oxalate often forms a minor component of uroliths, but oxalic acid can cause phosphate urinary calculi in sheep, by lowering the calcium concentration relative to phosphate in the diet.[101]

Most of the calcium in the body is present in the form of bone, which acts as a reserve to maintain a relatively uniform blood calcium concentration. Only 0.33% of total body calcium occurs in the blood,[102] and this calcium is involved in vital body functions. A prolonged calcium deficiency causes a softening and weakening of the bone structure due to the remobilization of calcium reserves and reduces voluntary feed intake,[103] growth,[104] and milk production.[105] Subtropical grasses are naturally low in calcium, compared to temperate grasses and legumes. Since many of these tropical grasses are also oxalate producers, they are responsible for many of the reported incidences of hypocalcemia in herbivores.[86] The oxalate in many of these grasses could, theoretically, bind the entire calcium content of the grasses,[106,62] considerably reducing calcium availability to herbivores.[107] The availability to the animal of the calcium in calcium oxalate depends on several factors. In certain plants, calcium oxalate idioblasts are protected from breakdown by associated poorly digestible vascular tissue.[108] The efficiency of utilizing oxalate depends on the type of animal involved. Horses appear to be completely unable to utilize calcium oxalate in grasses,[108] while pigs have a much better utilization.[109] The presence of oxalate-degrading microbes in the digestive tract of animals plays an important role in maintaining a normal calcium balance.[91]

Nutritional secondary hyperparathyroidism is a well-known calcium deficiency disease of horses fed high-oxalate pastures,[86] or diets composed mainly of low-calcium cereal grains. The time between introduction to a high-oxalate pasture and the onset of symptoms of hyperparathyroidism in horses may vary from 2 to 8 months.[106,86] Animals on high-oxalate pastures should be supplemented with calcium.[93]

14.4   CONCLUSION

Although nitrate as such is relatively nontoxic, its presence in food and drinking water can be a serious health hazard to both man and domestic animals due to the short- and long-term effects of its highly reactive reduction products. Methemoglobinemia is still a serious debilitating condition in rural communities exposed to high-nitrate water supplies and where health standards are poor. Due to the wide use of highly fertilized pastures and high-nitrate forage crops for animal consumption, the adverse effect on rumen digestion, reproduction, and animal health may be considerable. Although causality in man has not yet been established, the fact that N-nitroso compounds produce tumors in many animal species underlines the potential role of nitrate in human cancer etiology, particularly since exposure usually persists over a lifetime.

Wood-decaying fungi and plant pathogenic fungi which cause severe worldwide crop losses utilize oxalic acid as part of their host-invasive mechanisms, while some higher plants have evolved defence responses and antipredatory mechanisms involving oxalic acid or oxalate crystals. The intake of excessive amounts of oxalate by mammals lead to clinical conditions such as chronic renal failure, calcium oxalate urolithiasis, hypocalcemia, nutritional secondary hyperparathyroidism, and death. Chronic oxalate poisoning in herbivores has a depressing effect on animal production. Acute toxicity usually affects small numbers of animals, but could occasionally cause the death of hundreds of head of livestock in a single incidence,[88] thereby causing a severe set-back to agricultural enterprises.

REFERENCES

1. Walker, R., Nitrates, nitrites and N-nitroso compounds: a review of the occurrence in food and diet and the toxicological implications, *Food Add. Contam.*, 7, 717, 1990.
2. Gundimeda, U., Naidu, A. N., and Krishnaswamy, K., Dietary intake of nitrate in India, *J. Food Comp. Anal.* 6, 242, 1993.
3. Siddiqi, M. A., Tricker, A. R., Kumar, R., Fazili, Z., and Preussmann, R., Dietary sources of N-nitrosamines in a high-risk area for oesophageal cancer — Kashmir, India, in *Relevance to Human Cancer of N-nitroso Compounds, Tobacco Smoke and Mycotoxins,* O'Neill, I. K., Chen, J., and Bartsch, H., Eds., IARC, Lyon, 1991, 210.
4. Meah, M. N., Harrison, N., and Davies, A., Nitrate and nitrite in foods and the diet, *Food Add. Contam.*, 11, 519, 1994.
5. Tsuji, S., Kohsaka, M., Yukihiro, M., Shibata, T., Kaneta, N., Wakabayashi, K., Uchibori-Hase, S., Ide, S., Fujiwara, K., Suzuki, H., and Ito, Y., Naturally occurring nitrite and nitrate existing in various raw and processed food, *J. Food Hyg. Soc. Jpn.*, 34, 294, 1993.
6. Jones, T. O. and Jones, D. R., Nitrate/nitrite poisoning of cattle from forage crops, *Vet. Rec.*, 101, 266, 1977.
7. Dodd, D. C. and Coup, M. R., Poisoning of cattle by certain nitrate-containing plants, *N.Z. Vet. J.*, 5, 51, 1957.
9. Bradley, W. B., Eppson, H. F., and Beath, O. A., Nitrate as the cause of oat hay poisoning, *J. Am. Vet. Med. Assoc.*, 94, 541, 1939.
10. Low, I. C. S., Nitrite poisoning of calves grazing "Grasslands Tama" ryegrass, *N.Z. Vet. J.*, 22, 60, 1974.
11. Bruning-Fann, C. S. and Kaneene, J. B., The effects of nitrate, nitrite and N-nitroso compounds on human health: a review, *Vet. Human Toxicol.*, 35, 521, 1993.
12. Hill, M. J., Nitrates and nitrites from food and water in relation to human disease, in *Nitrates and Nitrites in Food and Water*, Hill, M. J., Ed., Ellis Horwood Limited, Chichester, England, 1991, 163.
13. Knotek, Z. and Schmidt, P., Pathogenesis, incidence and possibilities of preventing alimentary nitrate methemoglobinemia in infants, *Pediatrics*, 34, 78, 1964.
14. Robertson, H. E. and Riddell, W. A., Cyanosis of infants produced by high nitrate concentrations in rural waters of Saskatchewan, *Can. J. Public Health*, 40, 72, 1949.
15. Bradley, W. B., Eppson, H. F., and Beath, O. A., Livestock poisoning by oat hay and other plants containing nitrate, *Univ. Wyoming Agric. Exp. Stn. Bull.*, 241, 3, 1940.
16. London, W. T., Henderson, W., and Cross, R. F., An attempt to produce chronic nitrite toxicosis in swine, *J. Am. Vet. Med. Assoc.*, 150, 398, 1967.
17. Buck, W. B., Diagnosis of feed-related toxicoses, *J. Am. Vet. Med. Assoc.*, 156, 1434, 1970.
18. Simon, J., Sund, J. M., Wright, M. J., and Douglas, F. D., Prevention of noninfectious abortion in cattle by weed control and fertilization practices on lowland pastures, *J. Am. Vet. Med. Assoc.*, 135, 315, 1959.
19. Vermunt, J. and Visser, R., Nitrate toxicity in cattle, *N.Z. Vet. J.*, 35, 136, 1987.
20. Case, A. A., Some aspects of nitrate intoxication in livestock, *J. Am. Vet. Med. Assoc.*, 130, 323, 1957.

21. Bhikane, A. U., Bhoop Singh, and Ali, M. S., Therapeutic efficacy of methylene blue against experimental nitrite poisoning in crossbred calves, *Ind. Vet. J.*, 67, 459, 1990.

22. Ashbury, A. C. and Rhode, E. A., Nitrite intoxication in cattle: the effects of lethal doses of nitrite on blood pressure, *Am. J. Vet. Res.*, 25, 1010, 1964.

23. Malestein, A., Geurink, J. H., Schuyt, G., Schotman, A. J. H., Kemp, A., and Van't Klooster, A. Th., Nitrate poisoning in cattle: the effect of nitrate dosing during parturition on the oxygen capacity of maternal blood and the oxygen supply to the unborn calf, *Vet. Quart.*, 2, 149, 1980.

24. Marais, J. P., Effect of nitrate and non-protein organic nitrogen levels in kikuyu (*Pennisetum clandestinum*) pastures on digestibility *in vitro*, *Agroanimalia*, 12, 7, 1980.

25. Marais, J. P., Therion, J. J., Mackie, R. I., Kistner, A., and Dennison, C., Effect of nitrate and its reduction products on the growth and activity of the rumen microbial population, *Br. J. Nutr.*, 59, 301, 1988.

26. Chang, P. C., Aktilar, S. M., Burke, T., and Pivnick, H., Effect of sodium nitrite on *Clostridium botulinum* in canned luncheon meat: evidence for a Perigo-Type Factor in the absence of nitrite, *Can. Inst. Food Sci. Technol. J.*, 7, 209, 1974.

27. Roberts, T.A. and Dainty, R. H., Nitrite and nitrate as food additives: rationale and mode of action, in *Nitrates and Nitrites in Food and Water*, Hill, M. J., Ed., Ellis Horwood Limited, Chichester, England, 1991, 113.

28. Parkin, D. M., Läärä, E., and Muir, C. S., Estimates of the worldwide frequency of sixteen major cancers in 1980, *Int. J. Cancer*, 41, 184, 1988.

29. Correa, P., Cuello, C., Fajardo, L. F., Haenszel, W., Bolanos, O., and De Ramirez, B., Diet and gastric cancer: nutrition survey in a high-risk area, *J. Natl. Cancer Inst.*, 70, 673, 1983.

30. Shuker, D. E. G., The chemistry of N-nitrosation, in *Nitrosamines: Toxicology and Microbiology*, Hill, M. J., Ed., VCH Publishers, Cambridge, 1988, 48.

31. Rowland, I. R., The toxicology of N-nitroso compounds, in *Nitrosamines: Toxicology and Microbiology*, Hill, M. J., Ed., VCH Publishers, Cambridge, 1988, 117.

32. Sen, N. P., Nitrosamines, in *Toxic Constituents of Animal Foodstuffs*, Liener, I. E., Ed., Academic Press, New York, 1974, 131.

33. Yang, C. S., Smith, T., Ishizaki, H., and Hong, J. Y., Enzyme mechanisms in the metabolism of nitrosamines, in *Relevance to Human Cancer of N-nitroso Compounds, Tobacco Smoke and Mycotoxins*, O'Neill, I. K., Chen, J., and Bartsch, H., IARC, Lyon, 1991, 265.

34. Lijinsky, W., Alkylation of DNA related to organ-specific carcinogenesis by N-nitroso compounds, in *Relevance to Human Cancer of N-nitroso Compounds, Tobacco Smoke and Mycotoxins*, O'Neill, I. K., Chen, J., and Bartsch, H., Eds., IARC, Lyon, 1991, 305.

35. Leach, S., Thompson, M., and Hill, M., Bacterially catalysed N-nitrosation in the human stomach, *Carcinogenesis*, 8, 1907, 1987.

36. Bartsch, H., N-nitroso compounds and human cancer: where do we stand?, in *Relevance to Human Cancer of N-nitroso Compounds, Tobacco Smoke and Mycotoxins*, O'Neill, I. K., Chen, J., and Bartsch, H., Eds., IARC, Lyon, 1991, 1.

37. Bartholomew, B. A. and Hill, M. J., The pharmacology of dietary nitrate and the origin of urinary nitrate, *Food Chem. Technol.*, 22, 789, 1984.

38. Alsobrook, A. J., DuPlessis, L. S., Harrington, J. C., Nunn, A.J., and Nunn, J. R., Nitrosamines in the human vaginal vault, in *N-nitroso Compounds in the Environment*, Bogovski, P. and Walker, E., Eds., IARC, Lyon, 1974, 197.

39. Boss, P. M. J., Van Den Brandt, P. A., Wedel, M., and Ockhuizen, T. H., The reproducibility of the conversion of nitrate to nitrite in human saliva after a nitrate load, *Food Chem. Toxicol.*, 26, 93, 1988.

40. Packer, P. J., Leach, S. A., Duncan, S. N., Thompson, M. H., and Hill, M. J., The effect of different sources of nitrate exposure on urinary nitrate recovery in humans and its relevance to the methods of estimating nitrate exposure in epidemiological studies, *Carcinogenesis*, 10, 1989, 1989.

41. Calmels, S., Ohshima, H., Crespi, M., Leclerc, H., Cattoen, C., and Bartsch, H., N- nitrosamine formation by micro-organisms isolated from human gastric juice and urine: biochemical studies on bacteria-catalysed nitrosation, in *The Relevance of N-nitroso Compounds to Human Cancer: Exposure and Mechanisms*, Bartsch, H., O'Neill, I. K., and Schutte-Hermann, R., Eds., IARC, Lyon, 1987, 391.

42. Rodamski, J. L., Greenwald, D., Hearn, W. L., Block, N. L., and Wood, F. M., Nitrosamine formation in bladder infections and its role in the etiology of bladder cancer, *J. Urol.*, 120, 48, 1978.

43. Van Broekhoven, L. W. and Davies, J. A. R., The analysis of volatile N-nitrosamines in the rumen fluid of cows, *Neth. J. Agric. Sci.*, 29, 173, 1981.

44. Van Broekhoven, L. W., Davies, J. A. R., and Geurink, J. H., The metabolism of nitrate and proline in the rumen fluid of a cow and its effect on *in vivo* formation of N-nitrosoproline, *Neth. J. Agric. Sci.*, 37, 157, 1989.

45. Harrington, J. S., Epidemiology and etiology of cancer of the uterine cervix, *S. Afr. Med. J.*, 49, 1975.

46. Scragg, R. K., Dorsch, M. M., McMichael, A. J., and Baghurst, P.A., Birth defects and household water supply, *Med. J. Aust.*, 2, 577, 1982.

47. Hecht, S. S. and Hoffmann, D., N-nitroso compounds and tobacco-induced cancers in man, in *Relevance to Human Cancer of N-nitroso Compounds, Tobacco Smoke and Mycotoxins,* O'Neill, I. K., Chen, J., and Bartsch, H., Eds., IARC, Lyon, 1991, 54.

48. Srianujata, S., Tonbuth, S., Bunyaatvej, S., Valyseva, A., Promvanit, N., and Chiavatsagul, W., High urinary excretion of nitrate and N-nitrosoproline in opisthorchiasis subjects, in *The Relevance of N-nitroso Compounds to Human Cancer: Exposure and Mechanisms*, Bartsch, H., O'Neill, I. K., and Schulte- Hermann, R., Eds., IARC, Lyon 1987, 544.

49. Birt, D. F., Update on the effects of vitamins A, C, and E and selenium on carcinogenesis, *Proc. Soc. Exp. Biol. Med.*, 183, 311, 1986.

50. Lapeyrie, F., Chilvers, G. A., and Bhem, C. A., Oxalic acid synthesis by the mycorrhizal fungus *Paxillus involutus* (Batsch ex Fr) Fr., *New Phytol.*, 106, 139, 1987.

51. Pinna, D., Fungal physiology and the formation of calcium oxalate films on stone monuments, *Aerobiologia*, 9, 157, 1993.

52. Hodgkinson, A., *Oxalic Acid in Biology and Medicine*, Academic Press, New York, 1977, 325.

53. Franceschi, V. R. and Horner, H. T., Calcium oxalate crystals in plants, *Bot. Rev.*, 46, 361, 1980.

54. Libert, B. and Franceschi, V. R., Oxalate in crop plants, *J. Agric. Food Chem.*, 35, 926, 1987.

55. Williams, M. C., Effect of sodium and potassium salts on growth and oxalate content of *Halogeton, Plant Physiol.*, 35, 500, 1960.

56. Eheart, J. F. and Massey, P. H., Factors affecting the oxalate content of spinach, *J. Food Chem.*, 10, 325, 1962.

57. Jacob, R. H. and Peet, R. L., Acute oxalate toxicity of sheep associated with slender iceplant (*Mesembryanthemum nodiflorum*), *Aust. Vet. J.*, 66, 91, 1989.

58. Blaney, B. J., Gartner, R. J. W., and McKenzie, R. A., The effects of oxalate in some tropical grasses on the availability to horses of calcium, phosphorus and magnesium, *J. Agric. Sci. Camb.*, 97, 507, 1981.

59. McKenzie, R. A., Bell, A. M., Storie, G. J., Keenan, F. J., Cornack, K. M., and Grant, S. G., Acute oxalate poisoning of sheep by buffel grass (*Cenchrus ciliaris*), *Aust. Vet. J.*, 65, 26, 1988.

60. Silcock, R. G. and Smith, F. T.,Soluble oxalates in summer pastures on a Mulga soil, *Trop. Grassl.*, 17, 179, 1983.

61. Garcia-Rivera, J. and Morris, H. P., Oxalate content of tropical forage grasses, *Science*, 122, 1089, 1955.

62. Marais, J. P., Effect of nitrogen on the oxalate and calcium content of Kikuyu grass (*Pennisetum clandestinum* Hochst), *J. Grassl. Soc. S. Afr.* 7, 106, 1990.

63. Hintz, H. F. and Soderholm, L. V., Oxalic acid and millet, *Equine Practice*, 13, 5, 1991.

64. Roughan, P. G. and Warrington, I. J., Effect of nitrogen source on oxalate accumulation in *Setaria sphacelata* (cv. Kazungula), *J. Sci. Food Agric.*, 27, 281, 1976.

65. Libert, B. and Creed, C., Oxalate content of seventy-eight rhubarb cultivars and its relation to some other characters, *J. Hort. Sci.*, 60, 257, 1985.

66. Panciera, R. J., Martin, T., Burrows, G. E., Taylor, D. S., and Rice, L. E., Acute oxalate poisoning attributable to ingestion of curly dock (*Rumex crispus*) in sheep, *J. Am. Vet. Med. Assoc.*, 196, 1981, 1990.

67. Rowe, D. E., Oxalic acid effects in exudates of *Sclerotinia trifoliorum* and *S. sclerotiorum* and potential use in selection, *Crop Sci.*, 33, 1146, 1993.

68. Green, F., Larsen, M. J., Winandy, J. E., and Highley, T. L., Role of oxalic acid in incipient brown-rot decay, *Material und Organismen*, 26, 191, 1991.

69. Micales, J. A., Induction of oxalic acid by carbohydrate and nitrogen sources in the brown-rot fungus *Postia placenta*, *Material und Organismen*, 28, 197, 1993.

70. Espejo, E. and Agosin, E., Production and degradation of oxalic acid by brown rot fungi, *Appl. Environ. Microbiol.*, 57, 1980, 1991.

71. Kuan, I. C. and Tien, M., Stimulation of Mn peroxidase activity: a possible role for oxalate in lignin biodegradation, *Proc. Nat. Acad. Sci. U.S.A.*, 90, 1242, 1993.

72. Marciano, P., Dilenna, P., and Magro, P., Oxalic acid, cell wall degrading enzymes and pH in pathogenesis and their significance in the virulence of two *Sclerotinia sclerotiorum* isolates on sunflower, *Physiol. Plant Pathol.*, 22, 339, 1983.

73. Ferrar, P. H. and Walker, J. R. L., o-Diphenol oxidase inhibition — an additional role for oxalic acid in the phytopathogenic arsenal of *Sclerotinia sclerotiorum* and *Sclerotium rolfsii*, *Physiol. Mol. Plant Pathol.*, 43, 415, 1993.

74. Ghio, A. J. Peterseim, D. S., Roggli, V. L., and Piantadosi, C. A., Pulmonary oxalate deposition association with *Aspergillus niger* infection. An oxidant hypothesis of toxicity, *Am. Rev. Resp. Dis.*, 145, 1499, 1992.

75. Wilson, B. J. and Wilson, C. H., Oxalate formation in moldy feedstuff, as a possible factor in livestock toxic disease, *Am. J. Vet. Res.*, 22, 961, 1961.

76. Davis, D. A., Tsao, D., Seo, J-H, Emery, A., Low, P. S., and Heinstein, P., Enhancement of phytoalexin accumulation in cultured plant cells by oxalate, *Phytochemistry*, 31, 1603, 1992.

77. Favaron, F., Alghisi, P., Marciano, P., and Magro, P., Polygalacturonase iso-enzyme and oxalic acid produced by *Sclerotinia sclerotiorum* in soybean hypocotyls as elicitors of glyceollin, *Physiol. Mol. Plant Pathol.*, 33, 385, 1988.

78. Marciano, P., *In vitro* effect of the phytoalexin glyceollin I on polygalacturonase and oxalic acid production in *Sclerotinia sclerotiorum*, *Phytoparasitica*, 22, 101, 1994.

79. Dumas, B., Freyssinet, G., and Pallett, K. E., Tissue-specific expression of germin-like oxalate oxidase during development and fungal infection of barley seedlings, *Plant Physiol.*, 107, 1091, 1995.

80. Chen, Z., Silva, H., and Klessig, D. F., Active oxygen species in the induction of plant systemic acquired resistance by salicylic acid, *Science*, 262, 1883, 1993.

81. Yoshihara, T., Sogawa, K., Pathak, M. D., Juliano, B. O., and Sakamura, S., Oxalic acid as a sucking inhibitor of the brown plant hopper in rice (Delphacidae, Homoptera), *Entomol. Exp. Appl.*, 27, 149, 1980.

82. Sakai, W. S., Sheronia, S. S., and Nagao, M. A., A study of raphide microstructure in relation to irritation, *Scanning Electron Microsc.*, 2, 979, 1984.

83. Berg, R. H., A calcium oxalate-secreting tissue in branchlets of the Casuarinaceae, *Protoplasma*, 183, 29, 1994.

84. Thurston, E. L., Morphology, fine structure and ontogeny of the stinging emergence of *Tragia ramosa* and *T. saxicola* (Euphorbiaceae), *Am. J. Bot.*, 63, 710, 1976.

85. Massey, L. K. and Sutton, R. A. L., Modification of dietary oxalate and calcium reduces urinary oxalate in hyperoxaluric patients with kidney stones, *J. Am. Diet. Assoc.*, 93, 1305, 1993.

86. McKenzie, R. A., Poisoning of horses by oxalate in grasses, in *Plant Toxicology, Proceedings of the Australian-U.S. Poisonous Plant Symposium*, Brisbane, Australia, May 14-18, 1984, 1985.

87. Ellison, R. S., Poisonings in ruminants grazing pasture and fodder crops, *Surveillance*, 21, 23, 1994.

88. Young, J. A. and James, L. F., *The Ecology and Economic Impact of Poisonous Plants on Livestock Production*, James, L. F., Ralphs, M. H., and Nielsen, D. B., Eds., Westview Press, U.S.A., 1988.

89. Kellerman, T. S., Coetzer, J. A. W., and Naudé, T. W., *Plant Poisonings and Mycotoxicoses of Livestock in Southern Africa*, Oxford Univ. Press, Cape Town, 1988.

90. James, L. F. and Butcher, J. E., Halogeton poisoning of sheep: effect of high level oxalate intake, *J. Anim. Sci.*, 35, 1233, 1972.

91. Shirley, E. K. and Schmidt-Nielsen, K., Oxalate metabolism in the pack rat, sand rat, hamster, and white rat, *J. Nutr.*, 91, 496, 1967.

92. Wilson, G. D. A. and Harvey, D. G., Studies on experimental oxaluria in pigs, *Br. Vet. J.*, 133, 418, 1977.

93. McKenzie, R. A., Blaney, B. J., and Gartner, R. J. W., The effect of dietary oxalate on calcium, phosphorus and magnesium balance in horses, *J. Agric. Sci. Camb.*, 97, 69, 1981.

94. Allison, M. J., Cook, H. M., Milne, D. B., Gallagher, S., and Clayman, R. V., Oxalate degradation by gastrointestinal bacteria from humans, *J. Nutr.*, 116, 455, 1986.

95. Reed, G. H. and Morgan, S. D., Kinetic and magnetic resonance studies of the interaction of oxalate with pyruvate kinase, *Biochemistry*, 13, 3537, 1974.

96. Mildvan, A. S., Scutton, M. C., and Utter, M. F., Pyruvate carboxylase. VII. A possible role for tightly bound manganese, *J. Biol. Chem.*, 241, 3488, 1966.

97. Hsu, R. Y., Mildvan, A. S., Chang, G. G., and Fung, C. H., Mechanism of malic enzyme from pigeon liver, *J. Biol. Chem.*, 251, 6574, 1976.

98. Stewart, J. and McCallum, J. W., The anhydraemia of oxalate poisoning in horses, *Vet. Res.*, 56, 77, 1944.

99. Seawright, A. A., Groenendyk, S., and Silva, K. I., An outbreak of oxalate poisoning in cattle grazing *Setaria sphacelata, Aust. Vet. J.*, 46, 293, 1970.

100. Hesse, A., Siener, R., Heynck, H., and Jahnen, A., The influence of dietary factors on the risk of urinary stone formation, *Scan. Microsc.*, 7, 1119, 1993.

101. Emerick, R. J. and Embry, L. B., Calcium and phosphorus levels related to the development of phosphate urinary calculi in sheep, *J. Anim. Sci.*, 22, 510, 1963.

102. Grace, N. D., Amounts and distribution of mineral elements associated with fleece-free empty body weight gains in the grazing sheep, *N.Z. J. Agric. Res.*, 26, 59, 1983.

103. Field, A. C., Suttle, N. F., and Nisbet, D. I., Effect of diets low in calcium and phosphorus on the development of growing lambs, *J. Agric. Sci. Camb.*, 85, 435, 1975.

104. Benzie, D., Boyne, A. W., Dalgarno, A. C., Duckworth, J., Hill, R., and Walker, D. M., Studies on the skeleton of the sheep. IV. The effects and interactions of dietary supplements of calcium, phosphorus, cod-liver oil and energy, as starch, on the skeleton of growing blackface wethers, *J. Agric. Sci. Camb.*, 54, 202, 1960.

105. McDowell, L. R., *Minerals in Animal and Human Nutrition*, Acad. Press, New York, 1992.

106. Walthall, J. C. and McKenzie, R. A., Osteodystrophia fibrosa in horses at pasture in Queensland: field and laboratory observations, *Aust. Vet. J.*, 52, 11, 1976.

107. Blaney, B. J., Gartner, R. J. W., and Head, T. A., The effects of oxalate in tropical grasses on calcium, phosphorus and magnesium availability to cattle, *J. Agric. Sci. Camb.*, 99, 533, 1982.

108. McKenzie, R. A. and Schultz, K., Confirmation of the presence of calcium oxalate crystals in some tropical grasses, *J. Agric. Sci. Camb.*, 100, 249, 1983.

109. Brune, N. and Bredehorn, H., On the physiology of bacterial degradation of calcium oxalate and the ability to utilize calcium from calcium oxalate in the pig, *Z. Tierphysiol. Tierernahr. Futtermittelk.*, 16, 214, 1961.

15 Pinus Ponderosa Needle-Induced Toxicity

S. P. Ford, J. P. N. Rosazza, and R. E. Short

CONTENTS

15.1 INTRODUCTION

Pinus ponderosa or western yellow pine (Pinaceae) is widely distributed in the U.S., but is most abundant in the western U.S. and western Canada.[1,2] Ranchers and stockman as early as the 1920's suspected that ingestion of needles and buds from *Pinus ponderosa* caused a late term abortion in beef cattle.[3] These original claims were not taken seriously, however, due to the many other confirmed causes of bovine abortion including infectious diseases such as brucellosis or leptospirosis, as well as phosphorous and Vitamin A deficiency.[4] In 1952, MacDonald[5] provided the first definitive evidence for a direct effect of *Pinus ponderosa* needles in the induction of abortion. This researcher fed a control diet or a *Pinus ponderosa* needle supplemented diet to brucellosis free heifers during midgestation. In that study, 75% of the heifers fed the *Pinus ponderosa* needle supplemented diet calved prematurely, while heifers fed the control diet were unaffected. This experiment was followed by other studies by Tucker[6] and Faulkner[7] using the same protocol as MacDonald[5] and providing evidence eliminating leptospirosis and Vitamin A as causes for the abortions. Further, *Pinus ponderosa* is the only species of Pinus in the western U.S. and Canada known to cause abortion in cattle.[8,9] Both green and dry needles appear to cause abortion,[10] and bark and branch tips also appear to contain abortifacient principles.[11]

0-8493-8551-2/97/$0.00+$.50
© 1997 by CRC Press, Inc.

15.2   POSSIBLE ROLE OF *PINUS PONDEROSA* IN ELICITING ANTI-ESTROGENIC ACTIVITY

In search of the possible toxin in *Pinus ponderosa*, Allen and Kitts[12] speculated that *Pinus ponderosa* might contain estrogenic activity. Bradbury and White[13] had previously reported that several legumes and grasses contained estrogenic activity which was a suspected cause of reproductive disorders in a variety of herbivores. Using mice, Allen and Kitts[12] reported that the injection of *Pinus ponderosa* needle extracts reduced uterine weight and suggested an anti-estrogenic effect. Further studies by Allison and Kitts[9] and Cook and Kitts[14] also confirmed an anti-estrogenic effect of *Pinus ponderosa* needles in the rat. In these studies, it was reported that the injection of an extract of *Pinus ponderosa* needles inhibited the estradiol-17β-induced increase in uterine weight which occurs approximately 6 h after estradiol-17β injection. In contrast, Wagner and Jackson[15] reported that *Pinus ponderosa* needle extracts failed to inhibit the estrogen-induced increase in uterine dry weight which occurs 24 h after injection. This apparent disparity may stem from the time-related effects of estrogen on uterine tissues.[16]

Within 30-60 min of an estrogen administration, there is a progressive increase in uterine wet weight which is maximal by 4-6 h resulting from an increased uterine blood flow and vascular permeability.[17] These short-term effects of estrogen on the uterine vasculature may not function through activation of the estrogen receptor cascade.[18] The long-term effects of estradiol-17β are known to be receptor-mediated,[19] and result in increased uterine protein synthesis and cellular growth. These cellular effects are not evident until approximately 12–24 h after estrogen administration. When taken together, these observations suggest that a component in the *Pinus ponderosa* needle extract was capable of inhibiting the short-term effects of estrogen (i.e., vasodilation) but not estrogen-stimulated protein synthesis and cellular growth.

15.3   FACTORS PREDISPOSING COWS TO *PINUS PONDEROSA*-INDUCED ABORTION

It was suggested by James et al.[8] that in addition to the possible toxins in *Pinus ponderosa* needles, several other factors may be associated with and/or predispose animals to abort following their consumption. These include: stage of gestation when agent(s) are ingested, environmental stresses, condition of the animal, and the general physiology of the animal. The feeding of *Pinus ponderosa* needles to late pregnant cattle results in an abortion in from 3 days to 3 weeks, with most occurring 5–15 days after the initiation of daily feeding.[8] This abortifacient effect of *Pinus ponderosa* needles appears to be species specific, since the feeding of *Pinus ponderosa* needles to sheep and goats had no effect on gestation length in these species, while feeding them to cattle and bison induced early parturition.[20] Further, no deleterious reproductive effects are observed when *Pinus ponderosa* needles were fed to nonpregnant sheep or cows, suggesting a specific effect on the gravid uterus. Short et al.[21] examined if stage of gestation altered the susceptibility of the beef cow to the abortifacient effects of *Pinus ponderosa* needles. In that study, groups of cows at 116, 167, 215, and 254 days of pregnancy were fed *Pinus ponderosa* needles for 3 weeks or until abortion. Three weeks of *Pinus ponderosa* needle feeding to the 116 day group had no effect on gestation length, while 38, 50, and 100% of the cows in the 167, 215, and 254 day groups, respectively, exhibited abortion. Furthermore, the interval from the initiation of *Pinus ponderosa* needle feeding to abortion was shorter in the 254 day group (5.4 days) as compared to the 215 day (8 days) and the 167 day group (21 days). These results suggest increasing susceptibility of *Pinus ponderosa* needle toxicity with advancing gestational age.

15.4 EVIDENCE FOR A PREMATURE FETAL-INDUCED PARTURITION RATHER THAN A NONSPECIFIC ABORTION

Although *Pinus ponderosa* needle abortion is a common name for the syndrome afflicting cattle, it is in some ways misleading, since abortion is normally taken to mean expulsion of a nonviable fetus. While this is true if feeding occurs during midgestation, late pregnant cows (250 days) consuming *Pinus ponderosa* needles give birth to weak calves, which in most cases can be saved if given immediate attention.[22] Further, evidence supporting an early induction of parturition rather than a toxin inducing an abortion is also confirmed by several other studies. Visual observations of *Pinus ponderosa* needle fed to late pregnant cows suggest that the normal maternal sequence of events occurs prior to delivery, including edema of the udder, swelling of the genitalia, and relaxation of the pelvic ligaments.[22,23,24,25]

15.4.1 ENDOCRINE PROFILES DURING NORMAL AND *PINUS PONDEROSA*-INDUCED PARTURITION

Comparison of the steroid hormone profiles between parturition-induced *Pinus ponderosa* needle-fed and normally parturiating cows also indicate a similar sequence of periparturient endocrine changes.[22,26] Both groups of cows exhibited a gradual rise in estrogen concentrations which reached peak levels on the day of parturition, followed by a decline after delivery. Late pregnant cows induced to parturiate with prostaglandins or corticoid also show this same estrogen profile.[27,28] Although similar profiles of circulating estrogen were observed for *Pinus ponderosa* needle and control-fed cows, Christenson et al.[22] demonstrated that prepartum estrogen concentrations in systemic blood of *Pinus ponderosa* needle-fed cows were reduced when compared to those of control-fed cows (Figure 15.1). This resulted primarily from a decreased secretion of estrogen by the gravid uterus, as evidenced by the markedly reduced uterine venous concentrations of this steroid in *Pinus ponderosa* needle-fed cows.

Results of Christenson et al.[22] contrast those of Short et al.[26] in that they demonstrated no effect of *Pinus ponderosa* needle feeding on altering the prepartum progesterone profiles or concentrations when compared to control-fed cows (Figures 15.2 and 15.3). The pattern of progesterone concentrations in systemic blood observed by Christenson et al.[22] are again consistent with those reported by others to precede normal parturition[29] or a parturition induced by prostaglandins[28] or corticoids.[27] It was speculated that the acute increase in circulating progesterone observed by Short et al.[26] after the initiation of *Pinus ponderosa* needle feeding may have resulted from adrenal release due to the stress of handling untrained range beef cows. Both Short et al.[26] and Christenson et al.[22] observed a high incidence of retained placentae in the *Pinus ponderosa* needle-fed cows at parturition which may have resulted from their decreased estrogen:progesterone ratio during the prepartum period. This hypothesis is consistent with previous research demonstrating that placental retention was associated with a reduced estrogen:progesterone ratio in blood, whether the cows were induced to calve prematurely or not.[27,30,31]

The one major prepartum difference observed between *Pinus ponderosa* needle-fed and control-fed cows reported by Christenson et al.[22] was a marked and progressive decline in blood flow to the gravid uterine horn exhibited by *Pinus ponderosa* needle-fed cows from the initiation of feeding to fetal expulsion ($\approx$ 8 days) with no associated change in heart rate or blood pressure. In contrast, blood flow to the gravid horn of control-fed cows remained relatively constant through the day of parturition (Figure 15.4).

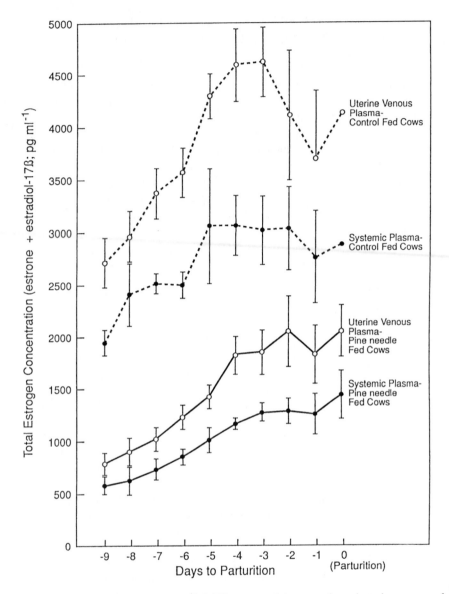

FIGURE 15.1 Total free estrogens (estradiol-17β + estrone) in systemic and uterine venous plasma in both control-fed cows and those fed *Pinus ponderosa* needles during the last 9 days of gestation. Means ± SEM represents systemic and uterine venous plasma obtained from five and seven control-fed cows, respectively, or five pine needle-fed cows. (From Christenson et al., *J. Anim. Sci.*, 70, 531, 1992. With permission.)

15.5 SPECIFIC EFFECTS OF *PINUS PONDEROSA* ON THE CARUNCULAR ARTERIAL BED

A specific effect of consumption of *Pinus ponderosa* needles by cattle on the uterine vasculature was first noted by Stuart et al.[25] These researchers reported a consistent and marked occlusion of the caruncular arterial bed, with an associated necrosis of the central portions of the caruncles. Upon examination of the associated cotyledonary tissue, however, no similar changes were evident. These data are consistent with the hypothesis that *Pinus ponderosa* needles contain a compound which preferentially reduces blood flow and thus nutrients and O_2 to the fetal-maternal interface (i.e., the placentome).

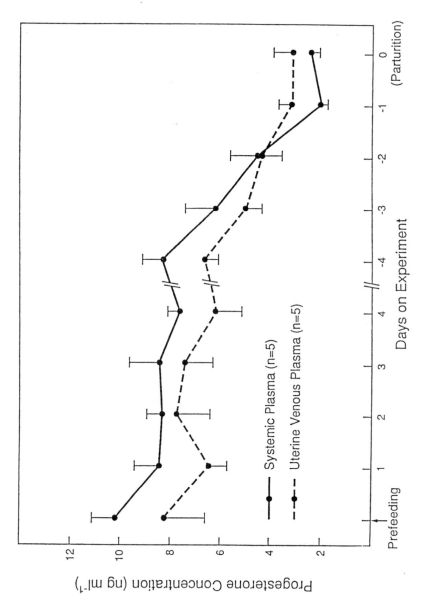

FIGURE 15.2 Progesterone concentrations in systemic and uterine venous plasma from day 250 of gestation through parturition in *Pinus ponderosa*-fed cows. Means ± SEM represent systemic and uterine venous plasma obtained from five pine needle-fed cows. (From Christenson et al., *J. Anim. Sci.*, 70, 531, 1992. With permission.)

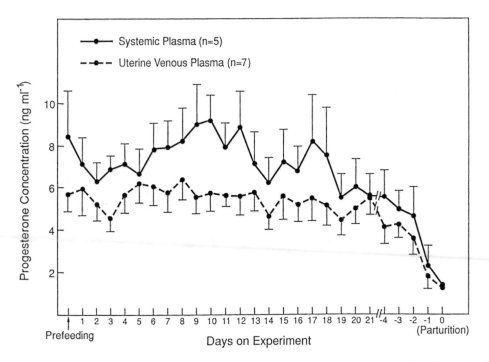

FIGURE 15.3 Progesterone concentrations in systemic and uterine venous plasma from day 250 of gestation through parturition in control-fed cows. Means ± SEM represent systemic plasma from five cows and uterine venous blood from seven cows. (From Christenson et al., *J. Anim. Sci.*, 70, 531, 1992. With permission.)

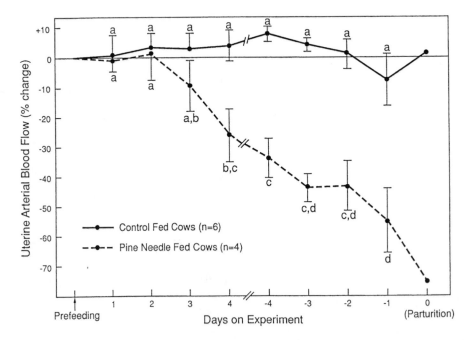

FIGURE 15.4 Blood flow to the gravid uterine horn (percentage change from the pre-feeding value) for control-fed cows and *Pinus ponderosa* needle-fed cows that calved early during the 4 days after initiation of the diets (day 250 of gestation) and the 4 days preceding parturition. [a,b,c,d]Means within treatments with different superscripts differ (P<0.05). Each mean ± SEM represents values from six control-fed cows or four pine needle-fed cows.(From Christenson et al., *J. Anim. Sci.*, 70, 531, 1992. With permission.)

15.5.1 Circulating Factor(s) in *Pinus ponderosa*-Fed Cows Which Increase Caruncular Arterial Tone

Christenson et al.[32] reported the presence of a factor in the plasma of *Pinus ponderosa* needle-fed cows, but not control-fed cows, which induced a prolonged and sustained contraction (i.e., increased caruncular arterial smooth muscle tone) in an isolated perfused bovine placentome model. These data would explain the results of Christenson et al.[22] who observed that late pregnant cows consuming *Pinus ponderosa* needles experienced a progressive and sustained decline in uterine blood flow culminating in the birth of weak premature calves and exhibiting retained placentae. Cows induced to parturiate early invariably exhibit retained placentae regardless of the method of induction.[8,33,34] Placental retention also predisposes the cow to secondary infections such as peritonitis, septicemia, and if not treated promptly, death, conditions which are frequently reported after *Pinus ponderosa* needle-induced abortion in cattle. The fact that *Pinus ponderosa* needle consumption induces premature parturition only if cows consume it during the last half of gestation[21] may relate to the rapidly accelerating rate of fetal growth as pregnancy advances. This accelerating fetal growth may additionally contribute to the increasing efficacy and decreasing interval from the initiation of *Pinus ponderosa* needle feeding to premature parturition as pregnancy advances. The increasing nutrient demands of the fetus over the last half of gestation are predominantly met through increases in uterine blood flow resulting predominantly from a progressive reduction of caruncular arterial tone.[35,36,37] Evidence that a chronic reduction in uterine blood flow may be sufficient to initiate premature parturition was provided by Challis et al.[38] These researchers reported that the prolonged physical restriction (Teflon clamp) of blood flow to the gravid uterine horn during late gestation in the ewe resulted in fetal distress, as evidenced by a sustained elevation in fetal ACTH and cortisol. These hormones are well established initiators of the parturient process in many mammalian species including the ewe[39] and cow.[40,41,42] It has been hypothesized that the reduced caruncular arterial blood flow and thus nutrient and oxygen delivery to the fetus in association with *Pinus ponderosa* needle ingestion may result in a fetal-initiated early parturition.[43]

15.5.2 Current Concepts on Controls of Caruncular Arterial Blood Flow

Caruncular arterial blood flow is locally regulated by short-term (phasic) and long-term (tonic) contractile mechanisms during pregnancy.[44] Phasic contractile responses ($\approx$ 3–10 min) are the result of activation of α_1-adrenergic receptors on arterial smooth muscle by catecholamines released during acute maternal stress,[45] and this response remains intact during bovine pregnancy.[46] This response is vital if the female is to briefly shunt blood away from the viscera and towards the skeletal muscles and brain in response to a predator (fight or flight response). Tonic contractile responses, which regulate the baseline rate of uterine arterial blood flow, are mediated by α_2-adrenergic receptors, potential-sensitive Ca^{2+} channels, and their interactions with catechol estrogens.[47,48,49] Catecholamine activation of the α_2-adrenergic receptor opens the potential-sensitive Ca^{2+} channels, allowing a prolonged uptake of extracellular free Ca^{2+} resulting in a long-term, reduction in arterial diameter (i.e., increased tone).[37] Catechol estrogens, hydroxylated at the 2 or 4 position of the phenolic A ring of estrogens block Ca^{2+} uptake via the potential-sensitive Ca^{2+} channels.[47,48] The gravid uterus appears to be a major source of these compounds,[49] which may function locally to maintain adequate fetal-maternal exchange. Christenson et al.[50] reported that perfusion of the caruncular arterial bed of perfused placentomes from cows fed *Pinus ponderosa* needles for 3–5 days exhibited markedly increased potential-sensitive Ca^{2+} channel activity with a resultant increase in caruncular arterial tone.

15.6 ISOLATION AND CHARACTERIZATION OF THE VASOACTIVE PRINCIPLES IN *PINUS PONDEROSA*

We set out next to isolate the vasoactive factor(s) in *Pinus ponderosa* using the perfused bovine placentome to guide fractionation. When *Pinus ponderosa* needles were extracted with methylene chloride (CH_2Cl_2) and the residue was redissolved in bovine plasma, it exerted a marked tone generating activity on the bovine placentome, predominantly through an increase in potential sensitive Ca^{2+} channel activity.[51] Further, all *Pinus ponderosa* components remaining after extraction exerted no vasoactivity when redissolved in plasma. Upon further bioassay guided fractionation, a novel class of vasoactive lipids was identified.[51,52] These lipids, which contained all the vasoactivity in the CH_2Cl_2 fraction, was dominated by the presence of alkanediols esterified with myristic and/or lauric acids. Next, it was decided to evaluate the abortifacient effects of *Pinus ponderosa* components in the late pregnant guinea pig. The guinea pig was chosen as a relevant model for the cow because of its relatively long gestation length ($\approx$ 70 days), its large precocious offspring, and its fermentation-type (cecal) digestive system.[53,54] Further, Ford et al.[55] had previously demonstrated that the late pregnant guinea pig ($\geq$ 50 days of gestation) was susceptible to the abortifacient effects of *Pinus ponderosa* needle ingestion. In that study, *Pinus ponderosa*-fed females gave birth prematurely to small weak piglets and exhibited retained placentae. Further, Ford et al.[55] demonstrated that CH_2Cl_2 extracted *Pinus ponderosa* needles lost their abortifacient activity in the guinea pig. These data confirm those of James et al.,[56] who demonstrated that all abortifacient activity in cattle was removed by CH_2Cl_2 extraction of *Pinus ponderosa* needles. Further, Gardner et al.[57] reported that an 80% pure sample of isocupressic acid isolated from a CH_2Cl_2 extract of *Pinus ponderosa* needles and bark induced early parturition when gavaged into late pregnant beef cows.

More recently, Ford et al.[58] evaluated the abortifacient activity of 1-12-dodecanedioyl-dimyristate (14-12-14), the most potent member of the novel vasoactive lipids in *Pinus ponderosa* isolated by Al-Mahmoud[51] and Al-Mahmoud et al..[52] Late pregnant guinea pigs consuming CH_2Cl_2 extracted *Pinus ponderosa* needles laced with 14-12-14 experienced premature parturition of small piglets exhibiting reduced viability, and a high incidence of retained placentae. It is suggested that these vasoactive lipids may play a role in causing early parturition in pregnant beef cows ingesting *Pinus ponderosa* needles.

15.7 SUMMARY

The selectivity of novel vasoactive lipids in *Pinus ponderosa* to increase caruncular arterial tone is consistent with their effects as potential sensitive Ca^{2+} channel agonists. It must be recognized that the caruncular arterial bed is by far the most dilated of all maternal beds in the late pregnant cow. The clinical picture and available experimental results suggest that *Pinus ponderosa*-induced abortion is a reversal of the local catechol estrogen-induced progressive vasodilation of the caruncular arteries during pregnancy.

ACKNOWLEDGMENTS

Journal paper no. J-16765 of the Iowa Agric. and Home Econ. Exp. Sta., Ames, Proj. 2443 and 2444. Our work described in this review has been supported in part by grants from the U.S. Department of Agriculture, No. 89-37240-4806 and No. 92-37203-7918. The authors thank Donna Farley and Carole Hertz for their technical expertise and Donna Johnston for typing this manuscript.

REFERENCES

1. U. S. Department of Agriculture, Trees of the forest, *Forest Service*, 613, 1964.
2. Van Hooser, D. D. and Keegan, C. E., III, Symposium on Ponderosa pine: The species and its management, Sept. 28–Oct. 1, 1987, 1.
3. Bruce, E. A., Astragalus serotinae and other stock poisoning plants of British Columbia, Dominion of Canada Dept. of Agricultural Bulletin, 88, 44, 1927.
4. Gunn, W. R., British Columbia, Dept. of Agric. 43rd Annual Report, King's Printer, Victoria, British Columbia, 228 pages, 1948.
5. MacDonald, M. A., Pine needle abortion in range beef cattle, *J. Range Manage.*, 2, 150, 1952.
6. Tucker, J. S., Pine needle abortion in cattle, Proceedings, *Coll. Vet. Toxicol.*, p. 35, 1961.
7. Faulkner, L. C., Pine needle abortion, in *Abortion Diseases of Livestock*, Faulkner, L. C., Ed., Charles C. Thorne Publishing, Springfield, IL, 1969, p. 75.
8. James, L. F., Short, R. E., Panter, K. E., Molyneux, R. J., Stuart, L. D., and Bellows, R. A., Pine needle abortion in cattle: a review and report of 1973–1984 research, *Cornell Vet.*, 79, 39, 1989.
9. Allison, C. A. and Kitts, W. D., Further studies on the antiestrogenic activity of yellow pine needles, *J. Anim. Sci.*, 23, 1155, 1964.
10. Jensen, R., Pier, A. C., Kaltenbach, C. C., Murdoch, W. J., Becerra, V. M., Mills, K. W., and Robinson, J. L., Evaluation of histopathologic and physiologic changes in cows having premature births after consuming Ponderosa pine needles, *Am. J. Vet. Res.*, 50, 285, 1989.
11. Panter, K. E., James, L. F., Molyneux, R. J., Short, R. E., and Sisson, D. V., Premature bovine parturition induced by Ponderosa pine: Effects of pine needles, bark and branch tips. *Cornell Vet.*, 80, 329, 1990.
12. Allen, M. R. and Kitts, W. D., The effects of yellow pine (*Pinus ponderosa*) needles on the reproductivity of the laboratory female mouse, *Can. J. Anim. Sci.*, 41, 1, 1961.
13. Bradbury, R. B. and White, D. E., Estrogens and related substances in plants, in *Vitamin and Hormones*, Harris, R.G., Ed., 1954.
14. Cook, H. and Kitts, W. D., Antiestrogenic activity in yellow pine needles (*Pinus ponderosa*), *Acta Endocrinol.*, 45, 33, 1964.
15. Wagner, W. D. and Jackson, L. L., Phytoestrogen from *Pinus* ponderosa assayed by competitive binding with 17β-estradiol to mouse uterine cytosol, *Theriogenology*, 19, 507, 1983.
16. Magness, R. R. and Rosenfeld, C. R., The role of steroid hormones in the control of uterine blood flow, in *The Uterine Circulation*, Rosenfeld, C.R., Ed., Reproductive and Perinatal Medicine (X), Perinatology Press, Ithaca, New York, 1989, p. 239.
17. Spaziani, E., Relationship between early vascular response and growth in the rat uterus: Stimulation of cell division by estradiol and vasodilating amines, *Endocrinology*, 72, 180, 1963.
18. Resnik, R., Killam, A. P., Battaglia, F. C., Makowski, E. L., and Meschia, G., The stimulation of uterine blood flow by various estrogens, *Endocrinology*, 94, 1192, 1974.
19. Anderson, J. N., Peck, E. J., Jr., and Clark, J. H., Estrogen-induced uterine responses and growth: Relationship to receptor estrogen binding by uterine nuclei, *Endocrinology*, 96, 160, 1975.
20. Short, R. E., James, L. F., Panter, K. E., Staigmiller, R. B., Bellow, R. A., Malcolm, J., and Ford, S. P., Effects of feeding ponderosa pine needles during pregnancy: comparative studies with bison, cattle, goats, and sheep, *J. Anim. Sci.*, 70, 3498, 1992.
21. Short, R. E., Staigmiller, R. B., Bellows, R. A., Panter, K. E., and James, L. F., Effects of various factors on abortions caused by pine needles, Fort Keogh Livestock and Range Research Laboratory, Miles City, MT, *Field Day Report*, USDA-ARS and Agric. Exp. Stn., Montana State Univ., 1987.
22. Christenson, L. K., Short, R. E., and Ford, S. P., Effects of ingestion of ponderosa pine needles by late-pregnant cows on uterine blood flow and steroid secretion, *J. Anim. Sci.*, 70, 531, 1992.
23. James, L. F., Call, J. W., and Stevenson, A. H., Experimentally induced pine needle abortion in range cattle, *Cornell Vet.*, 67, 294, 1977.
24. Stevenson, A. H., James, L. F., and Call, J. W., Pine-needle (*Pinus ponderosa*) induced abortion in range cattle, *Cornell Vet.*, 62, 519, 1972.

25. Stuart, L. D., James, L. F., Panter, K. E., Call, J. W., and Short, R. E., Pine needle abortion in cattle: Pathological observations, *Cornell Vet.,* 79, 61, 1989.

26. Short, R. E., James, L. F., Staigmiller, R. B., and Panter, K. E., Pine needle abortion in cattle: Associated changes in serum cortisol, estradiol and progesterone, *Cornell Vet.,* 79, 53, 1989.

27. Chew, B. P., Erb, R. E., Randel, R. D., and Rouguette, J. M., Jr., Effect of corticoid induced parturition on lactation and on prepartum profiles of serum progesterone and the estrogens among cows retaining and not retaining fetal membranes, *Theriogenology,* 10, 13, 1978.

28. Henricks, D. M., Rawlings, N. C., and Ellicott, A. R., Plasma hormone levels in beef heifers during prostaglandin-induced parturition, *Theriogenology,* 7, 17, 1977.

29. Smith, V. G., Edgerton, L. A., Hafs, H. D., and Convey, E. M., Bovine serum estrogens, progestins and glucocorticoids during late pregnancy, parturition and early lactation, *J. Anim. Sci.,* 36, 391, 1973.

30. Chew, B. P., Erb, R. E., Fessler, J. F., Callahan, C. J., and Malven, P. V., Effects of ovariectomy during pregnancy and of prematurely induced parturition on progesterone, estrogens, and calving traits, *J. Dairy Sci.,* 62, 557, 1979.

31. Chew, B. P., Keller, H. F., Erb, R. E., and Malven, P. V., Periparturient concentrations of prolactin, progesterone and the estrogens in blood plasma of cows retaining and not retaining fetal membranes, *J. Anim. Sci.,* 44, 1055, 1977.

32. Christenson, L. K., Short, R. E., Rosazza, J. P. N., and Ford, S. P., Specific effects of blood plasma from beef cows fed pine needles during late pregnancy on increasing tone of the caruncular arteries *in vitro, J. Anim. Sci.,* 70, 525, 1992.

33. Jöchle, W., Corticosteroid-induced parturition in domestic animals, *Ann. Rev. Pharmacol.,* 16, 33, 1973.

34. Zerobin, K., Jöchle, W., and Steingruber, C., Termination of pregnancy with prostaglandins E_2 (PGE_2) and $F_{2\alpha}$ ($PGF_{2\alpha}$) in cattle, *Prostaglandins* 4, 891, 1973.

35. Rosenfeld, C. R., Consideration of the uteroplacental circulation in intrauterine growth, *Semin. Perinatol. (Phila.),* 8, 42, 1984.

36. Reynolds, L. P., Ferrell, C. L., Nienaber, J. A., and Ford, S. P., Effects of chronic environmental heat stress on blood flow and nutrient uptake of the gravid bovine uterus and foetus, *J. Agric. Sci. Camb.,* 104, 289, 1985.

37. Ford, S. P., Control of blood flow to the gravid uterus of domestic livestock species, *J. Anim. Sci.,* 73, 1852, 1995.

38. Challis, J. R. G., Fraher, L., Oosterhuis, J., White, S. E., and Bocking, A. D., Fetal and maternal endocrine responses to prolonged reductions in uterine blood flow in pregnant sheep, *Am. J. Obstet. Gynecol.,* 160, 926, 1989.

39. Liggins, G. C., Fairclough, R. J., Grieves, S. A., Kendall, J. Z., and Knox, B. S., The mechanism of initiation of parturition in the ewe, *Rec. Prog. Horm. Res.,* 29, 111, 1973.

40. Welch, R. A. S., Frost, O. L., and Bergman, M., The effect of administering ACTH directly to the fetal calf, *N.Z. Med. J.,* p. 365, 1973 (Abstr).

41. Comline, R. S., Hall, L. W., Lavelle, R. B., Nathanielsz, P. W., and Silver, M., Parturition in the cow: Endocrine changes in animals with chronically implanted catheters in the foetal and maternal circulations, *J. Endocrinol.,* 63, 451, 1974.

42. Hunter, J. T., Fairclough, R. J., Peterson, A. J., and Welch, R. A. S., Foetal and maternal hormonal changes preceding normal bovine parturition, *Acta Endocrinol.,* 84, 653, 1977.

43. Ford, S. P., Christenson, L. K., Rosazza, J. P. N., and Short, R. E., Effects of ponderosa pine needle ingestion on uterine vascular function in late-pregnant beef cows, *J. Anim. Sci.,* 70, 1609, 1992.

44. Ford, S. P., Independent steroidal modulation of the phasic and tonic contractile responses of uterine arterial smooth muscle, in *Steroid Hormones and Uterine Bleeding,* Alexander, N. J., Ed., AAAS Press, Washington, DC, 1992, p. 121.

45. Shnider, S. M., Wright, R. G., Levinson, G., Roizen, M. F., Wallis, K. L., Rolbin, S. H., and Craft, J. B., Uterine blood flow and plasma norepinephrine changes during maternal stress in the pregnant ewe, *Anesthesiology,* 50, 524, 1979.

46. Sauer, J. J., Van Orden, D. E., Farley, D. B., and Ford, S. P., Changing adrenergic sensitivity of the caruncular arterial vasculature supplying the bovine placentome, *J. Anim. Sci.,* 67, 3003, 1989.

47. Stice, S. L., Ford, S. P., Rosazza, J. P., and Van Orden, D. E., Role of 4-hydroxylated estradiol in reducing Ca^{2+} uptake of uterine arterial smooth muscle cells through potential-sensitive channels, *Biol. Reprod.,* 36, 361, 1987.

48. Stice, S. L., Ford, S. P., Rosazza, J. P., and Van Orden, D. E., Interaction of 4-hydroxylated estradiol and potential-sensitive Ca^{2+} channels in altering uterine blood flow during the estrous cycle and early pregnancy in gilts, *Biol. Reprod.,* 36, 369, 1987.

49. Ford, S. P., Farley, D. B., and Rosazza, J. P. N., Catechol estrogens and uterine vascular function, in *Local Systems in Reproduction,* Naftolin, F. and Magness, R., Eds., International Ares, Serono Symposium, Raven Press, New York, 1993, p. 225.

50. Christenson, L. K., Short, R. E., Farley, D. B., and Ford, S. P., Effects of ingestion of pine needles (Pinus ponderosa) by late-pregnant beef cows on potential sensitive Ca^{2+} channel activity of caruncular arteries, *J. Reprod. Fertil.,* 98, 301, 1993.

51. Al-Mahmoud, M. S., Isolation and characterization of vasoactive lipids of Pinus ponderosa, Ph.D. Dissertation, University of Iowa, Iowa City, 1994.

52. Al-Mahmoud, M. S., Ford, S. P., Short, R. E., Farley, D. B., and Rosazza, J. P. N., Isolation and characterization of vasoactive lipids from the needles of Pinus ponderosa, *J. Agric. Food Chem.,* 43, 2154, 1995.

53. Hagen, P. and Robinson, K. W., The production and absorption of volatile fatty acids in the intestine of the guinea pig, *Aust. J. Exp. Biol. Med. Sci.,* XXXI, 99, 1953.

54. Reid, M. E., The Guinea Pig in Research, Human Factors Research Bureau, Inc., National Press Building, Washington 4, D.C., 1958.

55. Ford, S. P., Farley, D. B., and Rosazza, J. P. N., Use of the late-pregnant guinea pig as a bioassay for the abortifacient in Ponderosa pine needles, Proceedings American Society of Animal and Dairy Science Joint Meeting, Minneapolis, MN, 1994 (Abstr. 394).

56. James, L. F., Molyneux, R. J., Panter, K. E., Gardner, D. R., and Stegelmeier, B. L., Effects of feeding Ponderosa pine needle extracts and their residues to pregnant cattle., *Cornell Vet.,* 84, 33, 1994.

57. Gardner, D. R., Molyneux, R. J., James, L. F., Panter, K. E., and Stegelmeier, B. L., Ponderosa pine needle-induced abortion in beef cattle: Identification of isocupressic acid as the principal active compound, *J. Agric. Food Chem.,* 42, 765, 1994.

58. Ford, S. P., Rosazza, J. P. N., Al-Mahmoud, M. S., Lin, S., Farley, D. B., and Short, R. E., Use of late pregnant guinea pig to confirm the abortifacient effects of a unique class of vasoactive lipids in *Pinus Ponderosa* needles, *J. Anim. Sci.,* 1996, (In press).

16 Feeding Behavior of Herbivores in Response to Plant Toxicants

F. D. Provenza

CONTENTS

16.1   INTRODUCTION

Herbivores foraging on rangelands differ in their vulnerability to poisonous plants, and not all species of deleterious plants pose an equal risk. Animals foraging in familiar environments are less likely to be poisoned than animals foraging in unfamiliar settings. Some poisonous species are eaten repeatedly despite the consequences, whereas others are sampled and then largely avoided. Some mixes of plants prompt toxicity, whereas others are likely to diminish poisoning. Currently, there is little knowledge of the causal mechanisms underlying these

0-8493-8551-2/97/$0.00+$.50

inequalities. This chapter explores some fundamental processes in the relationship between the origins of food preference and the ingestion of poisonous plants.

16.2 ORIGINS OF FOOD PREFERENCE

16.2.1 TASTE-FEEDBACK INTERACTIONS

Preference results from the interrelationship between taste and postingestive feedback, determined by an animal's physiological condition relative to a plant's chemical characteristics.[1,2,3,4] Taste (as well as smell and sight) enables animals to discriminate among foods and provides hedonic sensations. Postingestive feedback calibrates taste in accord with a food's homeostatic utility. Preference decreases when foods are deficient in nutrients, when they contain excesses of toxins, and when they are too high in rapidly digestible nutrients. Conversely, preference increases when foods are adequate in nutrients.

16.2.1.1 Increases in Preference

Animals prefer foods that meet nutritional needs.[1,2] Lambs strongly prefer flavored straw eaten during intraruminal infusions of energy (starch or glucose) or nitrogen (urea, casein, gluten) (Villalba and Provenza, unpublished).[5,6] Microbial fermentation of food produces energy (i.e., volatile fatty acids like propionate and acetate) and nitrogen (i.e., amino acids). Volatile fatty acids and amino acids are quickly absorbed and metabolized, and thus provide an immediate indication of the nutritional value of food (Villalba and Provenza, unpublished).[5,6] In the absence of energy or nitrogen, intake of flavored straw is low and variable, which suggests that flavor alone does not predict preference. Likewise, lambs offered foods differing in flavors, nutrients, and toxins prefer foods high in nutrients and low in toxins, regardless of flavor.[7,8]

Animals also prefer substances that ameliorate illness, an effect unrelated to preference for the tastes of the substances. Lambs suffering acidosis drink more of a solution containing sodium bicarbonate; otherwise, they strongly prefer plain water.[9,10,11] Thus, lambs drink the sodium bicarbonate solution because it attenuates acidosis,[12] not because they like its flavor. Rats also prefer flavors associated with recovery from threonine deficiency, but only when they are deficient in threonine.[13] Thus, preferences for substances that can rectify maladies are state dependent.

16.2.1.2 Decreases in Preference

Excesses or deficits of nutrients and excesses of toxins cause food aversions, which are manifest by a decrease in preference for the flavor of the food.[1,12,14] Lambs that receive a toxin dose after eating cinnamon-flavored rice no longer prefer cinnamon-flavored rice or wheat-flavored, both of which were formerly preferred foods.[15]

Excesses of byproducts of fermentation from energy (e.g., propionate, acetate) and nitrogen (e.g., ammonia) can condition food aversions. Lambs acquire aversions to propionate at high (>10 g) doses,[6,16] but they prefer flavored straw eaten with lower doses (5 to 7.5 g) of propionate; the same is true for acetate, combinations of propionate and acetate, and ammonia (Villalba and Provenza, unpublished).[17,18]

Preference also declines when foods are inadequate in nutrients,[1,2,19] and some plant metabolites decrease nutrient availability. For instance, phenolic compounds like tannins bind to proteins and carbohydrates in the gut of mammals and birds,[20,21] thereby reducing the availability of nitrogen and energy.[22–28] Tannins thus reduce positive postingestive effects of energy (volatile fatty acids) and nitrogen,[29] thereby causing intake and preference to decline.[1,2,30]

Preference also decreases when foods contain toxins.[1] Some tannins produce toxic effects that cannot be accounted for by digestion inhibition, but are best accounted for by stimulation of the emetic system[12,32] as a result of lesions of the gut mucosa and toxicity.[33,34] For instance, condensed tannins in the shrub blackbrush (*Coleogyne ramosissima*) cause rapid (i.e., within 1 h) and dramatic decreases (i.e., >80% reduction) in intake of current season's twigs by goats.[31]

Individual animals typically differ in their preferences for foods prior to postingestive effects. For instance, when current season's (CSG) and older growth (OG) twigs from the shrub blackbrush are offered to goats naive to blackbrush, some goats originally prefer CSG whereas others initially prefer OG. Nevertheless, all goats quickly acquire a preference for OG, even though CSG is more nutritious, because CSG contains high levels of a condensed tannin that induces a learned food aversion.[31] For nearly a decade, we mistakenly assumed that goats rejected CSG on the basis of taste alone because we had never seen goats eat CSG. Instead, goats' avoidance of CSG reflects their ability to learn quickly (goats limit intake of CSG in 1 to 4 h) based on aversive postingestive effects.

16.2.2   EXPERIENCE AND TASTE-FEEDBACK INTERACTIONS

Experiences early in life influence preference.[4,35] Animals prefer familiar foods and are reluctant to eat novel foods or familiar foods whose flavors have changed.[36,37,38,39] The greatest decreases in intake occur when lambs are offered novel foods in novel environments.[40] When lambs become ill after eating a meal of familiar and novel foods, they avoid the novel foods.[41–43] Furthermore, when they become ill after eating a meal of novel foods, they avoid the foods that are most novel.[31,44,45] Conversely, when lambs become ill after eating familiar foods, they avoid the foods eaten most frequently or in excess[2,9,31] and the foods that made them ill in the past.[46]

16.3   ACCOUNTING FOR INGESTION OF TOXIC PLANTS

Herbivores have receptor systems for detecting changes in both the external and internal milieu. Sight, smell, and taste enable animals to discern changes in plant attributes (e.g., maturity, flavor), and chemo-, osmo-, and mechano-receptors allow animals to detect changes within the body. Interactions among these receptor systems typically enable herbivores to circumvent toxicity.[1,2,47] Nevertheless, in some situations plant flavor and toxicity interact in ways that cause poisoning (Table 16.1).

TABLE 16.1
Changes in Flavor and Toxicity Affect the Likelihood that Herbivores Will Overingest Toxic Plants

	Change in Plant Toxicity	
Change in Plant Flavor	No	Yes
No	Case I	Case II
Yes	Case III	Case IV

16.3.1   CASE I: NO CHANGE IN FLAVOR OR IN TOXICITY

16.3.1.1   Lack of Alternatives

A lack of nutritious alternatives is perhaps the easiest situation in which to envision animal poisoning from ingestion of toxic plants. In this predicament, an animal has only two choices (i.e., starve or ingest toxins), and animals typically eat poisonous plants, though their intake is depressed. Sheep eat the highly toxic forb bitterweed (*Hymenoxys odorata*) if they lack alternative forages in late fall and early winter, which coincides with their greatest demand for nutrients (late gestation and early lactation).[48] Nevertheless, sheep acquire strong aversions to bitterweed, and the greater the toxin (hymenoxon) dose, the greater the reduction in intake of bitterweed.[49] Livestock are also forced to ingest toxic alkaloids when pastures are seeded to monocultures of plants like endophyte-infected tall fescue,[50] even though alkaloids in fescue depress intake.[51]

Inappropriate herding techniques also increase the likelihood of livestock eating poisonous plants. For instance, lupine (*Lupinus* spp.) is more toxic to cattle than to sheep, yet more sheep than cattle die of lupine poisoning; halogeton (*Halogeton glomeratus*) is equally toxic to sheep and cattle, yet more sheep than cattle die from this plant as well.[52] In both situations, sheep are given access to food and water only where and when allowed by the herder, whereas cattle can select times and locations to forage and drink.

16.3.1.2   Alternatives Available

In some cases, animals are poisoned when alternative foods are available. Toxicosis in this case is more difficult to explain, but may occur when the abundance and nutritional value of a poisonous plant is high relative to alternative plants. For instance, cattle eat tall larkspur (*Delphinium barbeyi*) from among a variety of alternatives, evidently because larkspur is abundant and nutritious.[53] Likewise, cattle eat locoweed (*Astragalus* and *Oxytropis* spp.) early in the spring, when it is the most nutritious alternative; they prefer grass later in the spring when locoweed matures.[54]

Cyclic patterns of intake often occur when animals eat nutritious plants that contain toxins. Positive postingestive effects from nutrients condition preferences and cause intake to increase, whereas delayed aversive effects from toxins condition aversions and cause intake to decline.[2] Cyclic patterns of intake occur with nutritious plants like larkspur, which contains alkaloids,[53] and for energy-rich foods like grain that cause acidosis.[55] The greater the positive postingestive feedback from nutrients, and the more familiar animals are with the food, the less likely they are to acquire lasting aversions.[2,5,56]

This may explain why sheep show a strong diurnal preference for highly digestible clover early in the day, and for less digestible grass later in the day.[57] Sheep acquire a mild aversion to clover as a result of toxins (i.e., cyanogenic compounds) and the rapid release of nutrients (soluble carbohydrates and protein) during the morning, which causes them to eat grass. They become averse to poorly digestible grass during the afternoon and evening, as they recuperate from the aversion to clover.[2]

Animals are also likely to be poisoned when the effects of the toxin are not experienced for some time after food ingestion. Animals acquire food aversions best when the delay between food ingestion and postingestive consequences is less than 12 hours.[43] The liver damage caused by species like *Senecio* that contain pyrrolizidine alkaloids is progressive and death may not occur for months or even years.[58] Poisoning is similarly delayed when animals consume plants like locoweed that contain indolizidine alkaloids; cellular damage occurs after 8 days and there are no clinical signs of poisoning for 3 weeks. Animals acquire aversions to these foods only after vital organs have been damaged.

Finally, some toxins may circumvent mechanisms for experiencing the consequences of food ingestion. Animals become ill and may die if the capacity of detoxification mechanisms is exceeded,[59] but this seldom occurs because animals usually acquire aversions to toxic foods.[1] Aversions involve physiological signals along gustatory and visceral nerves that converge in the brain stem, then branch to the limbic system, and thence to the cortex.[1,2,14] Excesses of toxins[12,51] and nutrients[12] and deficits of nutrients[13] all cause aversions, perhaps due to stimulation of the emetic system of the midbrain and brain stem.[1,2,12,14,51] The emetic system can be activated by excesses of nutrients or toxins from various locations in the body, including visceral afferent nerves, second-order gustatory afferent nerves, the cardiovascular system, and cerebrospinal fluid.

Herbivores are likely to eat plants containing toxins that do not stimulate the emetic system.[14,47] Thus, compounds that cause photosensitization (e.g., hypericin in *Hypericum perforatum*) are not likely to cause food aversions because they adversely affect the skin surface and not emetic receptors inside the body. Nor do animals typically avoid foods that cause allergies, bloating, or lower intestinal discomfort, again because stimulation of the emetic system is evidently essential for producing food aversions.[1,14] Likewise, compounds like gallamine, which causes neuromuscular blockage, or naloxone, which blocks the action of endogenous opiates on pain, cause strong place aversions but only weak food aversions in rats;[60] conversely, drugs like lithium chloride, which affect the emetic system,[12] cause strong food aversions but only weak place aversions in rats.[60]

16.3.2   CASE II: CHANGE IN TOXICITY BUT NOT IN FLAVOR

16.3.2.1   Plant Chemistry

Herbivores are likely to die from eating a food whose toxicity (but not flavor) has increased, especially if the toxicosis is acute. Increases in concentrations of compounds toxic in minute amounts may have little effect on flavor, but large effects on virulence.[61] Rapid increases in toxicity following frost (e.g., cyanogenic compounds in chokecherry, serviceberry, sorghum, Sudan and Johnson grasses, and bloat-producing compounds in plants like alfalfa) may be accompanied by changes in toxicity but not flavor.[62] Changes in stereochemistry can also dramatically change toxicity of many phytochemicals.[47,63] Likewise, the toxicity of a group of phytochemicals may not be associated with changes in plant flavor.[63] In these instances, animals may have little chance to learn to regulate consumption because flavor of the food does not change (see Section 16.3.3.1) and the consequences of food ingestion are rapid and lethal.

On the other hand, herbivores can regulate intake of familiar foods that are not acutely toxic.[1,47] Cattle and sheep eat little larkspur in the spring when its toxicity is high, but their intake increases later in the season as toxicity diminishes.[64] When the toxicity of a food fluctuates widely but its flavor does not, lambs limit intake in accord with the food's highest toxicity.[45] Conversely, when flavor changes constantly and is not correlated with changes in toxin or nutrient content, lambs adjust intake in accord with the concentrations of toxins and nutrients, notwithstanding the changes in flavor.[8]

16.3.2.2   Animal Physiology

An animal's physiological condition can also alter the effects of toxins.[65,66] Inadequate nutrition can impede detoxification.[59] For instance, a reduction in the sulphur-containing amino acids methionine and cysteine can reduce serum inorganic sulphate in rats and reduce acetaminophen (paractamol) conjugation with sulphate.[67] Fasting increases acetaminophen toxicity in rats, as glucuronidation and sulphation (and hence clearance) are reduced.[68] Thus, a given level of toxin may be safe for a satiated animal but hazardous for an animal that is food-deprived.

The maximum absorbed allelochemical dose depends on the supply of cosubstrate for conjugation.[69,70] Thus, animals must balance allelochemical intake rate with nutrient intake rate.[8,71] Lambs fed only one third of their *ad libitum* daily intake eat less food containing the toxin LiCl than lambs fed *ad libitum*,[72] a result consistent with the hypothesis that nutrient availability affects intake of toxins. Food deprivation affects preference, evidently by modifying interactions between energy intake and LiCl.[72] The hypoglycemia in rats and cattle produced by LiCl[73] may be aggravated by low energy intake.[72]

Many believe food-deprived animals select foods less discriminately and are more likely to overingest poisonous plants. Nonetheless, this may not occur if alternative foods are present, even in limited amounts. Food-deprived lambs avoid foods containing LiCl, no matter how much energy the foods contain, provided alternative foods are available; conversely, lambs fed *ad libitum* eat substantial amounts of LiCl-containing foods.[8] Cattle also consume less larkspur with decreasing levels of rumen fill.[64] Thus, poisonous plants may kill livestock that lack alternative foods because they are more susceptible to toxins.

16.3.2.3   Varied Diets

Little is known about the ability of animals to reduce toxicosis by ingesting various mixtures of foods. Rats decrease toxic effects by eating a mixture of foods containing tannins (most shrubs and some forbs) and saponins (many forage legumes including alfalfa) because tannins and saponins chelate in the intestinal tract, thereby reducing the aversive effects of both compounds.[66] Supplementing ruminants with polyethylene glycol, which has a high affinity for binding tannins, also diminishes the aversive effects of high-tannin foods and improves body weight gains and performance.[74,75] Lambs eat more food that causes acidosis (e.g., grain, oregano) when they have access to foods or solutions containing sodium bicarbonate, which attenuates acidosis.[10,11]

Toxins may encourage dietary diversity by limiting consumption of various kinds of nutritious foods. Different toxins (e.g., tannins, saponins, alkaloids) often differ in their effects on herbivores.[58] If the capacity of different detoxification systems is not exceeded, a mixed diet may enhance an animal's nutrient intake because animals can ingest low levels of potentially toxic compounds.[65] Interactions between toxins[66] and an animal's nutritional status[69,70] can affect an herbivore's susceptibility to toxins and can enhance dietary diversity. For instance, as the energy content of the diet increases, lambs eat a greater variety of foods containing higher concentrations of a toxin.[72]

Conversely, dietary diversity will probably decrease if the same toxin occurs in a variety of foods, as occurs when lambs are offered different foods containing the toxin LiCl.[8,71] Likewise, classes of compounds like cyanogenic glycosides act on the same sites in the body and are detoxified by the same mechanisms,[58] which should also reduce diversity. Thus, even though at least 21 different cyanogenic glycosides are found in over 1000 plant species (including amygdalin in chokecherries, wild cherries, mountain mahogany, serviceberry, and dhurrin in forage and grain sorghum, Sudan and Johnson grasses),[58] herbivores are likely to eat only the most nutritious plants containing cyanogenic glycosides. Detoxification processes associated with many kinds of toxins (e.g., terpenes, phenolics) can also lead to acidosis,[69] which may reduce diet diversity.[70]

16.3.3   CASE III: CHANGE IN FLAVOR BUT NOT IN TOXICITY

16.3.3.1   Novelty as Advantageous

A change in the flavor of a familiar food is likely to reduce the potential for poisoning because animals sample novel foods cautiously and they associate toxicosis with novel foods. Lambs initially eat little of a familiar food whose flavor has changed,[36,37] and they avoid foods with novel flavors when toxicosis ensues. For instance, lambs fed a familiar food (rolled barley), with a low and a high concentration of a novel flavor, consume small amounts of both flavored foods; if they receive a mild toxin dose after eating both foods, they avoid the barley with the highest concentration of the flavor (i.e., the barley that is most novel)[45] (see also Reference 44). Likewise, lambs avoid a novel food after eating a meal of five foods, four of which are familiar.[43] Food neophobia is part of the general phenomenon of fear exhibited in new situations.[76,77,78,79] Eating small amounts of new foods, or familiar foods with novel flavors, protects animals against plant toxins.[1,65,80]

16.3.3.2   Novelty as Disadvantageous

There may be cases in which an animal's response to novelty could increase its likelihood of toxicosis. For instance, when lambs are fed a familiar food and compound (barley and NaCl) for several days followed by barley with a novel compound (LiCl) and then by a novel food (milo), the lambs eat the familiar food with LiCl but not the novel food.[42] Lambs can discriminate between NaCl and LiCl,[71] so the response most likely reflects the greater novelty of milo. Thus, animals may be poisoned when the flavors of several foods change simultaneously (e.g., with seasonal changes or changes in weather) and when they are in unfamiliar environments containing many novel foods.

16.3.4   CASE IV: CHANGE IN FLAVOR AND IN TOXICITY

16.3.4.1   Familiar Environment

Changes in plant flavor and toxicity are not likely to be lethal when animals forage in a familiar environment due to their reluctance to eat familiar foods whose flavors (and toxicants) have changed. Lambs routinely fed elm from one location would not eat elm of the same species from another site.[36] The addition of novel flavors (e.g., onion) to familiar foods (e.g., rice) also causes dramatic declines in intake.[37]

When lambs are in a familiar environment, food preferences influence their choice of foraging location. Subgroups of lambs with different food preferences generally feed in different locations.[81] Moreover, their preference for foods is typically stronger than their preference for the companionship of other lambs, especially when the lambs were not reared together. Thus, familiarity with the physical and social environment causes lambs to restrict their foraging to foods and locations that are not likely to be harmful.

16.3.4.2   Unfamiliar Environment

On the other hand, site fidelity is less important when animals are placed in an unfamiliar area. In such cases, familiarity with companions overrides preferences for foods in choice of foraging locations. Preferences for foraging locations represent the collective preferences of the group, and lack of familiarity can lead to a broadening of food preferences and to use of a greater portion of the environment.[82,83]

Preference for the familiar is manifest by the responses of animals in unfamiliar situations. Sheep prefer to forage with companions as opposed to strangers,[82] and they prefer familiar to unfamiliar environments.[84] When introduced into unfamiliar environments, naive sheep do not forage with experienced sheep, and naive sheep stray as far as 150 km from the native herd's normal range.[85] Thus, while it is often recommended that animals naive to an environment be placed with experienced animals to learn forage and habitat selection behaviors, this is not likely to work if the animals are strangers. Sheep also prefer familiar to novel foods; they sample novel foods cautiously,[37] and they prefer familiar foods when placed in unfamiliar environments.[38] Lambs show the greatest decreases in intake when they are offered novel foods in novel environments.[40] Compared with animals reared in a familiar environment, animals in an unfamiliar environment walk more, they ingest less forage, and they suffer more from predation, malnutrition, and ingestion of toxic plants.[86,87]

16.4   IMPLICATIONS

Ruminants eat an array of plant species, varying in nutrients and toxins. This selection makes intuitive sense, but no theories adequately explain this diversity. Some maintain it reduces the likelihood of overingesting toxins,[65] whereas others contend it meets nutritional needs.[88] Nevertheless, herbivores seek variety even when toxins are not a concern and nutritional needs are met. Another explanation may be advanced for this behavior, one which encompasses the avoidance of toxins and the acquisition of nutrients.[2] A key concept in this theory is aversion, the decrease in preference for food just eaten as a result of sensory input (taste, odor) and postingestive feedback (nutritional and toxicological effects on chemo-, osmo-, and mechano-receptors) unique to each food. Aversions are pronounced when foods contain toxins or high levels of rapidly digestible nutrients. Aversions also occur when foods are deficient in specific nutrients. Aversions can even occur when animals eat nutritionally adequate foods in excess or too frequently because satiety (satisfied to the full) and surfeit (filled to nauseating or disgusting excess) are a continuum and there is a fine line between satiety and aversion. Thus, several diet-related factors can result in aversions, and encourages the consumption of varied diets.

Rangelands and pastures have often been seeded to monocultures, or have few plant species due to overuse. Providing animals with a diverse mix of nutritious plants is one of the most important means of reducing toxicosis on rangelands and in pastures, whereas overgrazing and improper pasture management increase the likelihood of poisoning.[58,89] The kinds and numbers of foods offered to ruminants are also important considerations in attempts to train animals to avoid poisonous plants.[90,91,92] In addition to creating an aversion to a target plant, providing alternative foods (or supplements) may enhance the persistence of an aversion.

Humans have planted forage species like tall fescue infected with ascomycete fungi, which produce alkaloids that adversely affect food intake and livestock performance.[50,51] The selective preference for uninfected fescue eventually leads to dominance of infected plants. Forages like white clover contain cyanogenic compounds, which also deter herbivores. However, a combination of fescue and clover may enhance intake because they contain different kinds of toxins. Many shrubs and forbs contain tannins, whereas legumes like alfalfa contain saponins. Planting these species together may reduce the aversive effects of both compounds because tannins and saponins chelate in the intestinal tract.[66] Thus, planting species with different toxins might increase livestock production. Unfortunately, little is known about how ruminants might mix their diets to reduce toxicosis.

ACKNOWLEDGMENTS

Research discussed herein was supported by grants from Cooperative States Research Service and Utah Agric. Exp. Sta. I thank E. Burritt, K. Gutknecht, J. Pfister, and M. Ralphs for suggestions to improve this manuscript.

REFERENCES

1. Provenza, F. D., Postingestive feedback as an elemental determinant of food preference and intake in ruminants, *J. Range Manage.*, 48, 2, 1995.
2. Provenza, F. D., Acquired aversions as the basis for varied diets of ruminants foraging on rangelands, *J. Anim. Sci.*, 74, 2010, 1996.
3. Provenza, F. D., A functional explanation for palatability, in *Proc. Fifth International Rangeland Congress*, West, N. E., Ed., Soc. Range Manage., Denver, CO, 1996, 123.
4. Provenza, F. D., Tracking variable environments: There is more than one kind of memory, *J. Chem. Ecol.*, 21, 911, 1995.
5. Villalba, J. J. and Provenza, F. D., Preference for flavored wheat straw by lambs conditioned with intraruminal administrations of starch, *Br. J. Nutr.*, in press, 1996.
6. Villalba, J. J. and Provenza, F. D., Preference for wheat straw by lambs conditioned with intraruminal infusions of sodium propionate, *J. Anim. Sci.*, in press, 1996.
7. Provenza, F. D., Scott, C. B., Phy, T. S., and Lynch, J. J., Preference of sheep for foods varying in flavors and nutrients, *J. Anim. Sci.*, in press, 1996.

8. Wang, J. and Provenza, F. D., Dynamics of preference by sheep offered foods varying in flavors, nutrients, and a toxin, *J. Chem. Ecol.*, in press, 1996.

9. Phy, T. S. and Provenza, F. D., Food additives and amount eaten affect preference of lambs for rolled barley, *J. Anim. Sci.*, in press, 1996.

10. Phy, T. S. and Provenza, F. D., Lambs acquire preferences for foods that attenuate lactic acidosis, *J. Anim. Sci.*, in press.

11. Phy, T. S. and Provenza, F. D., Lambs prefer fluids that attenuate lactic acidosis, *J. Anim. Sci.*, in press.

12. Provenza, F. D., Ortega-Reyes, L., Scott, C. B., Lynch, J. J., and Burritt, E. A., Antiemetic drugs attenuate food aversions in sheep, *J. Anim. Sci.*, 72, 1989, 1994.

13. Gietzen, D. W., Neural mechanisms in the responses to amino acid deficiency, *J. Nutr.*, 123, 610, 1993.

14. Garcia, J., Food for Tolman: cognition and cathexis in concert, in *Aversion, Avoidance and Anxiety*, Archer, T. and Nilsson, L., Eds., Hillsdale, New Jersey, 1989, 45.

15. Launchbaugh, K. L. and Provenza, F. D., Can plants practice mimicry to avoid grazing by mammalian herbivores? *Oikos*, 66, 501, 1993.

16. Ralphs, M. H., Provenza, F. D., Wiedmeier, W. D., and Bunderson, F. B., Effects of energy source and food flavor on conditioned preferences in sheep, *J. Anim. Sci.*, 73, 1651, 1995.

17. Farningham, D. A. H. and Whyte, C. C., The role of propionate and acetate in the control of food intake in sheep, *Br. J. Nutr.*, 70, 37, 1993.

18. Mbanya, J. N., Anil, M. H., and Forbes, J. M., The voluntary intake of hay and silage by lactating cows in response to ruminal infusion of acetate or propionate, or both, with or without distension of the rumen by a balloon, *Br. J. Nutr.*, 69, 713, 1993.

19. Provenza, F. D., Role of learning in food preferences of ruminants: Greenhalgh and Reid revisited, in *Ruminant Physiology: Digestion, Metabolism, Growth and Reproduction Proceedings VIII International Symposium on Ruminant Physiology*, Engelhardt, W. V., Leonhard-Marek, S., Breves, G., and Giesecke, D., Eds., Ferdinand Enke Verlag, Stuttgart, 1995, Chap. 11.

20. Woodford, A. and Reed, J. D., The influence of polyphenolics on the nutritive value of browse: a summary of research conducted at ILCA, *ILCA Bulletin*, 35, 2, 1989

21. Kumar, R. and Vaithiyanathan, S., Occurrence, nutritional significance and effect on animal productivity of tannins in tree leaves, *Anim. Feed Sci. Tech.*, 30, 21, 1990.

22. Barry, T. N. and Manely, R. T., The role of condensed tannins in the nutritional value of *Lotus pedunculatus* for sheep. 2. Quantitative digestion of carbohydrates and proteins, *Br. J. Nutr.*, 51, 493, 1984.

23. Nastis, A. S. and Malechek, J. C., Digestion and utilization of nutrients in oak browse by goats, *J. Anim. Sci.*, 53, 283, 1981.

24. Villena, F. and Pfister, J. A. Sand shinnery oak as forage for Angora and Spanish goats, *J. Range Manage.*, 43, 116, 1990.

25. Donnelly, E. D. and Anthony, W. B. Effect of genotype and tannin on dry matter digestibility in Sericea lespedeza, *Crop Sci.*, 10, 200, 1970.

26. Mole, S. and Waterman, P. G., A critical analysis of techniques for measuring tannins in ecological studies, *Oecologia*, 72, 137, 1987.

27. Robbins, C. T., Hanely, T. A., Hagerman, A. E., Hjeljord, O., Baker, D. L., Schwartz, C. C., and Mautz, W. W., Role of tannins in defending plants against ruminants: reduction in protein availability, *Ecology*, 68, 98, 1987.

28. Robbins, C. T., Mole, S, Hagerman, A. E. and Hanely, T. A., Role of tannins in defending plants against ruminants: reduction in dry matter digestion, *Ecology*, 68, 1606, 1987.

29. Makkar, H. P. S., Blummel, M. and Becker, K., Formation of complexes between polyvinyl pyrrolidones or polyethylene glycols and tannins, and their implications in gas production and true digestibility in *in vitro* techniques, *Br. J. Nutr.*, 73, 897, 1995.

30. Barry, T. N., The role of condensed tannins in the nutritional value of *Lotus pedunculatus* for sheep. 3. Rates of body and wool growth, *Br. J. Nutr.*, 54, 211, 1985.

31. Provenza, F. D., Lynch, J. J., Burritt, E. A. and Scott, C. B., How goats learn to distinguish between novel foods that differ in postingestive consequences, *J. Chem. Ecol.*, 20, 609, 1994.

32. Provenza, F. D., Burritt, E. A., Clausen, T. P., Bryant, J. P., Reichardt, P. B.and Distel, R. A., Conditioned flavor aversion: a mechanism for goats to avoid condensed tannins in Blackbrush, *Am. Nat.*, 136, 810, 1990.

33. Kumar, R. and Singh, M., Tannins, their adverse role in ruminant nutrition, *J. Agric. Food Chem.*, 32, 447, 1984.

34. Reed, J. D., Nutritional toxicology of tannins and related polyphenols in forage legumes, *J. Anim. Sci.*, 73, 1516, 1995.

35. Provenza, F. D., Ontogeny and social transmission of food selection in domesticated ruminants, in *Behavioral Aspects of Feeding: Basic and Applied Research in Mammals*, Galef, B. G., Jr., Mainardi, M., and Valsecchi, P., Eds., Harwood Acad. Pub., Singapore, 1994, 147.

36. Provenza, F. D., Lynch, J. J., and Nolan, J. V., The relative importance of mother and toxicosis in the selection of foods by lambs, *J. Chem. Ecol.*, 19, 313, 1993.

37. Provenza, F. D., Lynch, J. J., and Cheney, C. D., Effects of a flavor and food restriction on the intake of novel foods by sheep, *Appl. Anim. Behav. Sci.*, 43, 83, 1995.

38. Gluesing, E. A. and Balph, D. F., An aspect of feeding behavior and its importance to grazing systems, *J. Range Manage.*, 33, 426, 1980.

39. Gillingham, M. P. and Bunnell, F. L., Effects of learning on food selection and searching behavior of deer, *Can. J. Zool.*, 67, 24, 1989.

40. Burritt, E. A. and Provenza F. D., Effect of a novel environment on formation and persistence of a food aversion and on ingestion of novel foods by sheep, *Appl. Anim. Behav. Sci.*, in press, 1996.

41. Revusky, S. H. and Bedarf, E. W., Association of illness with prior ingestion of novel foods, *Science*, 155, 219, 1967.

42. Burritt, E. A. and Provenza, F. D., Food aversion learning: ability of lambs to distinguish safe from harmful foods, *J. Anim. Sci.*, 67, 1732, 1989.

43. Burritt, E. A. and Provenza, F. D., Ability of lambs to learn with a delay between food ingestion and consequences given meals containing novel and familiar foods, *Appl. Anim. Behav. Sci.*, 32, 179, 1991.

44. Kalat, J. W., Taste salience depends on novelty, not concentration, in taste-aversion learning in rats, *J. Comp. Physiol. Psych.*, 86, 47, 1974.

45. Launchbaugh, K. L., Provenza, F. D., and Burritt, E. A., How herbivores track variable environments: Response to variability of phytotoxins, *J. Chem. Ecol.*, 19, 1047, 1993.

46. Burritt, E. A. and F. D. Provenza., Amount of experience and prior illness affect the acquisition and persistence of conditioned food aversions in lambs, *Appl. Anim. Behav. Sci.*, in press, 1996.

47. Provenza, F. D., Pfister, J. A., and Cheney, C. D., Mechanisms of learning in diet selection with reference to phytotoxicosis in herbivores, *J. Range Manage.*, 45, 36, 1992.

48. Ueckert, D. N. and Calhoun, M. C., Ecology and toxicology of bitterweed (*Hymenoxys odorata*), in *The Ecology and Economic Impact of Poisonous Plants on Livestock Production*, James, L. F., Ralphs, M. H., and Nielsen, D. B., Eds., Westview Press, Boulder, 1988, Chap. 11.

49. Calhoun, M. C., Ueckert, D. N., Livingston, C. W., and Baldwin, B. C., Effects of bitterweed (Hymenoxys odorata) on voluntary feed intake and serum constituents of sheep, *Am. J. Vet. Res.*, 42, 1713, 1981.

50. Thompson, F. N. and Stuedemann, J. A., Pathophysiology of fescue toxicosis, *Agric. Ecosystems Environ.*, 44, 263, 1993.

51. Aldrich, C. G., Rhodes, M. T., Miner, J. L., Kerley, M. S., and Paterson, J. A., The effects of endophyte-infected tall fescue consumption and use of a dopamine antagonist on intake, digestibility, body temperature, and blood constituents in sheep, *J. Anim. Sci.*, 71, 158, 1993.

52. James, L. F., Introduction, in *The Ecology and Economic Impact of Poisonous Plants on Livestock Production*, James, L. F., Ralphs, M. H., and Nielsen, D. B., Eds., Westview Press, Boulder, 1988, Chap. 1.

53. Pfister, J. A., Provenza, F. D., Manners, G. D., Ralphs, M. H., and Gardner, D. R., Tall larkspur ingestion: can cattle regulate intake below toxic levels?, *J. Chem. Ecol.*, in press, 1996.

54. Ralphs, M. H., Graham, D., Molyneux, R. J., and James, L. F. Seasonal grazing of locoweeds by cattle in northeastern New Mexico, *J. Range Manage.*, 46, 416, 1993.

55. Britton, R. A. and Stock, R. A., Acidosis, rate of starch digestion and intake, in *Feed Intake by Beef Cattle*, Agric. Expt. Stn. MP 121, Oklahoma State Univ., 1987, pp 125-137.

56. Burritt E. A. and Provenza, F. D., Lambs form preferences for non-nutritive flavors paired with glucose, *J. Anim. Sci.*, 70, 1133, 1992.

57. Parsons, A. J., Newman, J. A., Penning, P. D., Harvey, A., and Orr, R. J., Diet preference of sheep: effect of recent diet, physiological state and species abundance, *J. Anim. Ecol.*, 63, 465, 1994.

58. Cheeke, P. R. and Shull L. R., *Natural Toxicants in Feeds and Poisonous Plants*. AVI Publ. Co., Westport, 1985.

59. McArthur, C., Hagerman, A. E., and Robbins C. T., Physiological strategies of mammalian herbivores against plant defenses, in *Plant Defenses Against Mammalian Herbivory*, Palo, R. T. and Robbins, C. T., Eds., CRC Press, Boca Raton, 1991, Chap. 6.

60. Lett, B.T., The pain-like effect of gallamine and naloxone differs from sickness induced by lithium chloride, *Behav. Neurosci.*, 99, 145, 1985.

61. Manners, G. D., Pfister, J. A., Ralphs, M. H., Panter, K. E., and Olsen, J. D., Larkspur chemistry: Toxic alkaloids in tall larkspur, *J. Range. Manage.*, 45, 63, 1992.

62. Osweiler, G. D., Carson, T. L., Buck, W. B., and VanGelder, G. A., *Clinical and Diagnostic Veterinary Toxicology*, Kendall/Hunt, Dubuque, 1985, 455.

63. Bryant, J. P., Reichardt, P. B., and Clausen, T. P., Chemically mediated interactions between woody plants and browsing mammals, *J. Range. Manage.*, 45, 18, 1992.

64. Pfister, J. A., Manners, G. D., Ralphs, M. H., Hong, Z. X., and Lane, M. A., Effects of phenology, site, and rumen fill on tall larkspur consumption by cattle, *J. Range Manage.* 41, 509, 1988.

65. Freeland, W. J. and Janzen, D. H., Strategies in herbivory by mammals: The role of plant secondary compounds, *Am. Nat.*, 108, 269, 1974.

66. Freeland, W. J., Calcott, P. H., and Anderson, L. R., Tannins and saponins: Interaction in herbivore diets, *Biochem. Syst. Ecol.*, 13, 189, 1985.

67. Glazenberg, E. L., Jekel-Halsema, I. N. C., Scholtens, E., Baars, A. J., and Mulder, G. J., Effects of variation in the dietary supply of cysteine and methionine on liver concentration of glutathione and active sulfate (PAPS) and serum levels of sulphate, cystine, methionine and taurine: relations to the metabolism of acetaminophen, *J. Nutr.* 113, 1363, 1983.

68. Price, V. F., Miller, M. G., and Jollow, D. J., Mechanisms of fasting-induced potentiation of acetaminophen hepato toxicity in the rat, *Biochem. Pharmacol.* 36, 427, 1987.

69. Foley, W. J., McLean, S. and Cork, S. J., Consequences of biotransformation of plant secondary metabolites on acid-base metabolism in mammals-A final common pathway? *J. Chem. Ecol.*, 21, 721, 1995.

70. Illius, A. W. and Jessop, N. S., Modeling metabolic costs of allelochemical ingestion by foraging herbivores, *J. Chem. Ecol.*, 21, 693, 1995.

71. Wang, J. and Provenza, F. D., Toxicosis affect preference of sheep for foods varying in nutrient and a toxin, *J. Chem. Ecol.*, in press, 1996.

72. Wang, J. and Provenza F.D., Food deprivation affects preference of sheep for foods varying in nutrients and toxins, *J. Chem. Ecol.*, in press, 1996.

73. Johnson, J. H., Crookshank, H. R., and Smalley, H. E., Lithium toxicity in cattle, *Vet. Hum. Toxicol.*, 22, 248, 1980.

74. Silanikove, N., Nitsan, Z., and Perevolotsky, A., Effect of a daily supplementation of polyethylene glycol on intake and digestion of tannin-containing leaves (Ceratonia siliqua) by sheep, *J. Agric. Food Chem.*, 42, 2844, 1994.

75. Silanikove, N., Gilboa, N., Nir, I., Perevolotsky, A., and Nitsan, Z., Effect of a daily supplementation of polyethylene glycol on intake and digestion of tannin-containing leaves (*Quercus calliprinos, Pistacia lentiscus and Ceratonia siliqua*) by goats, *J. Agric. Food Chem.*, 44, 199, 1996.

76. Barinaga, M., How scary things get that way, *Science*, 258, 887, 1992.

77. LeDoux, J. E., Brain mechanisms of emotion and emotional learning, *Curr. Opin. Neurobiol.*, 2, 191, 1992.

78. Davis, M., The role of the amygdala in fear and anxiety, *Ann. Rev. Neurosci.*, 15, 353, 1992.

79. LeDoux, J. E., Emotion, memory and the brain, *Sci. Am.*, 270, 50, 1994.

80. Bryant, J. P., Provenza, F. D., Pastor, J. Reichardt, P. B., Clausen, T. P., and duToit, J. T., Interactions between woody plants and browsing mammals mediated by secondary metabolites, *Annu. Rev. Ecol. Syst.*, 22, 431, 1991.

81. Scott, C. B., Provenza, F. D., and Banner R. E., Dietary habits and social interactions affect choice of feeding location by sheep, *Appl. Anim. Behav. Sci.*, 45, 225, 1995.

82. Scott, C. B., Banner, R. E., and Provenza, F. D., Observations of sheep foraging in familiar and unfamiliar environments: Familiarity with the environment influences diet selection by sheep, *Appl. Anim. Behav. Sci.*, in press, 1996.

83. Howery, L. D., Provenza, F. D., Banner, R. E., and Scott, C. B., Social and environmental manipulations by man and nature affect cattle dispersion on rangeland, *Appl. Anim. Behav. Sci.*, accepted for publication, 1996.

84. Key, C. and MacIver, R. M., The effects of maternal influences on sheep: breed differences in grazing, resting and courtship behaviour, *Appl. Anim. Ethol.*, 6, 33, 1980.

85. Warren, J. T. and Mysterud, I., Extensive ranging by sheep released onto an unfamiliar range, *Appl. Anim. Behav. Sci.*, 38, 67, 1993.

86. Provenza, F. D. and Balph, D. F., Applicability of five diet-selection models to various foraging challenges ruminants encounter, in *Behavioral Mechanisms of Food Selection*, Hughes, R. N., Ed., NATO ASI Series G: Ecological Sciences, Vol. 20. Springer-Verlag, Heidelberg, New York, 1990, 423.

87. Griffith, B., Scott, J. M., Carpenter, J. W., and Reed, C., Translocation as a species conservation tool: status and strategy, *Science*, 245, 477, 1989.

88. Westoby, M., What are the biological bases of varied diets? *Am. Nat.*, 112, 627, 1978.

89. Fusco, M., Holechek, J., Tembo, A., Daniel, A., and Cardenas, M. Grazing influences on watering point vegetation in the Chihuahuan desert, *J. Range. Manage.*, 48, 32, 1995.

90. Burritt, E. A. and Provenza F. D., Food aversion learning: conditioning lambs to avoid a palatable shrub (*Cercocarpus montanus*), *J. Anim. Sci.*, 67, 650, 1989.

91. Burritt, E. A. and Provenza, F. D., Food aversion learning in sheep: persistence of conditioned taste aversions to palatable shrubs (*Cercocarpus montanus* and *Amelanchier alnifolia*), *J. Anim. Sci.*, 68, 1003, 1990.

92. Lane, M. A., Ralphs, M. H., Olsen, J. D., Provenza, F. D., and Pfister, J. A., Conditioned taste aversion: potential for reducing cattle losses to larkspur, *J. Range Manage.*, 43, 127, 1990.

17 Modeling Animal Responses to Plant Toxicants

N. S. Jessop and A. W. Illius

CONTENTS

17.1   INTRODUCTION

Plant toxicants or secondary compounds are ingested along with other plant constituents. They act to deter feeding by herbivores although the mechanisms of deterrence are poorly understood. They can exert effects at the level of the gut (digestion inhibitors) or at a metabolic level following absorption into an animal's body at which level it seems likely that the emetic center is instrumental in the response, integrating a wide variety of signals from receptors in the oropharyngeal region, stomach, or other locations in the gastrointestinal tract with stimuli arising from the liver via the vagal nerve. The question is then: what gives rise to these stimuli? We propose that deterrence caused by such compounds works in two fundamental ways. The first and most obvious way is by a direct toxic effect, thereby damaging the animal and imposing a cost of repairing that damage ("toxic costs"). Toxic costs can vary widely from compound to compound as exemplified by large differences in the LD_{50} when expressed on a molar basis (the lower the LD_{50} the higher the toxic cost).[1] Secondly, detoxification of the absorbed secondary compound typically requires the utilization of nutrients and imposes an acid load on the animal ("Detoxification costs").[1] Both of these probably give rise to signals which are integrated by the emetic center and result in avoidance learning.[2] A third, purely hypothetical, mode of action would be where a compound has some direct neuropharmacological effect on the emetic system, and thus stimulates the food avoidance response directly and without any allied costs (see Figure 17.1). We hypothesize that plant toxicants vary in the extent to which they are toxic and impose detoxification costs, and therefore could be classified along a continuum between the two. For example, alkaloids may be extremely toxic at doses which require trivial costs of detoxification, while terpenoids such as thymol or menthol are not particularly toxic but are contained in plants at concentrations which require substantial glucuronidation capacity and acid-base restabilization.[3,4,5] Can these two classes of costs be distinguished in terms of their effects and importance for animal learning and diet selection?

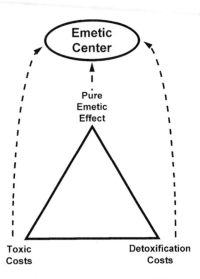

FIGURE 17.1 Schematic representation of the negative properties of foods. The triangle represents the three principle effects. Foods located at an apex present only that effect, while mixtures of two effects only are represented by places on lines between apices, and combinations of all three effects are represented as points within the triangle. The emetic centre organizes these stimuli.

Mathematical modeling is one tool that can enable such distinctions to be made. As yet, there is insufficient information available on the mechanisms of toxicity of many plant secondary compounds and thus the ability to incorporate such effects into models is limited. Sufficient information exists to allow influences of plant secondary compounds on digestion and the detoxification costs of absorbed plant secondary compounds to be predicted using this approach. This chapter reviews this work and the conclusions drawn. It describes one of the few modeling activities which has (1) provided an explanation to previously unexplained observations and (2) produced a hypothesis which has been subsequently confirmed by experimentation.

17.2   MODELING

The term modeling is generally used to refer to the quantitative descriptions of a system. The main difference between this and other scientific activities is in the rigor of the description; most science involves the raising of hypotheses (or conceptual models) which are tested experimentally, and most models start out as an attempt to integrate knowledge of a system in order to be able to explore the effects of perturbations to one part of a system on the whole. Systems under study may be simple or complex as may the quantitative description of them. As more information becomes known about particular systems, the scope for modeling at an ever more detailed mechanistic level becomes greater. Advantages of doing so are that more complete descriptions enable effects of interactions within systems to be described and to more closely represent reality.

Modeling the consequences of allelochemical impact on herbivory is in its early stages. We can distinguish two broad classes of approach to modeling the interactions of animals and allelochemicals. Policy models aim to analyze the trade-off which foraging herbivores face between positive (nutrient intake) and negative (toxin ingestion) aspects of the food plants available to them. Mechanistic models aim, broadly, to study the physiological processes and consequences of toxin ingestion, and in this sense are an adjunct to experimental investigations. Both approaches reveal the consequences of the assumptions made about the underlying processes, and are merely ways of mathematically manipulating quantitative relationships. They have the ability to generalize from the specific, and can thereby reveal a greater range of response than time and resources permit in physical experimentation.

17.2.1 POLICY MODELING

A good example of policy modeling is the study of Belovsky and Schmitz (1991),[6] who used linear programming to investigate plant defenses against herbivory. This is a technique which solves for the optimal mixture of two foods which differ in allelochemical content and which may differ in the nutrient intake rate obtained by animals eating them. The optimal diet is that mixture which achieves the animals, assumed goal of maximum nutrient intake, subject to the digestive constraints or limitation of time available for foraging. Two alternatives were examined following the classic distinction between quantitative and qualitative defenses, being defenses that, respectively, act by diluting or inhibiting nutrient absorption, and those being toxic at sufficient doses. Quantitative defenses were shown to operate by increasing the amount of food the herbivore must consume to compensate for the reduced nutrient absorption, hence reducing its profitability. In the case of qualitative defenses, herbivores were assumed to be tolerant, to a given extent, to toxins ingested in plant material. Diets predicted to maximize nutrient intake contained the defended plant, subject to the animal's fixed detoxification capacity, if the nutrient intake rate it allowed the animal was sufficiently higher than that of the undefended plant. Otherwise, none of it should be eaten. The results are interesting for contrasting positive and negative attributes of plants, contrary to the conventional assumption that the net effect of allelochemicals must be negative, and emphasizes that dietary choice always depends on the nature of alternatives available.

17.2.2 MECHANISTIC MODELING

An example of a mechanistic approach to the effect of qualitative defenses was that of Gordon and Illius (1996),[7] who used a detailed model of digestion kinetics in ruminants to analyze the consequences of the diluent effect of unabsorbed secondary compounds, such as tannins of high molecular weight, for daily nutrient intake. By reducing the quantity of rumen-fermentable substrate, secondary compounds were predicted to lead to reduced rumen microbial biomass, and hence digesta load, and thus to allow higher food intakes. The net effect, however, was to reduce daily nutrient intake. The result confirms that of Belovsky and Schmitz (1991),[6] at the expense of a more complex and sophisticated description of the underlying processes, and might thus be regarded as being less elegant, but has the advantage of allowing a more rigorous quantification of effects. The same advantages and disadvantages attend a second example, treated in detail in the remainder of this chapter, of detailed mechanistic modeling of the metabolic consequences of the detoxifying absorbable allelochemicals, with the consequent costs of maintaining acid-base homeostasis.

17.3 PLANT SECONDARY COMPOUNDS

17.3.1 DIGESTION INHIBITORS

A well-known example of such compounds are tannins which bind to proteins reducing their availability as well as reducing digestive efficiency. There is evidence to suggest that there are direct effects on gut mucosa, increasing damage and hence endogenous protein loss into the gut.[8] Counteracting this, there have been proposals that tannins reduce the load of intestinal parasites and thus may, in certain circumstances, reduce endogenous protein losses associated with such parasitism.[9] These effects have yet to be addressed using modeling. The stability of the tannin/protein complex is influenced by pH and known to be weaker at acid pH although it can vary depending on the type of tannin. Thus, tannins may exert a beneficial effect on protein supply to ruminants by protecting them from microbial degradation within the rumen, although depending on the reduction in ruminal availability of protein, such sequestration of protein may reduce the efficiency of microbial fermentation of complex carbohydrates adversely affecting intake. In addition, if rumen load is determined by dry matter, ingestion of plant material containing significant quantities of tannins will, through simple dilution effects, reduce intake by contributing to the overall dry matter load.

In order to investigate the magnitude of such effects by mathematical modeling, a suitable model that incorporates the above mechanisms needs to be used. Such models have been produced and recently used to study the above phenomena [7] and the main conclusions from this study have been discussed earlier (see Section 17.2.2). Experimentally, due to the inter-dependency of the above effects, it would be extremely difficult to design experiments to test such predictions.

17.3.2 ABSORBED PLANT TOXICANTS

The general mechanism of detoxification of absorbed allelochemicals has been well established. In the liver or in the gut, conjugation of plant secondary compounds to sulphate, glucuronic acid, or hippuric acid takes place.[1,10] This serves to increase the water solubility of the compound and aids its excretion via the urine or bile. The factor that determines which excretion route is used has been suggested to be the molecular weight of the conjugate; for compounds with molecular weights less than 200 the urine is the major route whereas for compounds with molecular weights above this then bile becomes the preferred route. For compounds excreted via bile the potential exists for recycling of plant secondary compounds in that microbial activity in the caecum or colon could hydrolyze the conjugate enabling the allelochemical to be reabsorbed across the gut wall. This possibility has not received any serious consideration as yet. A feature of such conjugation reactions is that the conjugate produced is a relatively strong acid and thus production of it will perturb acid base homeostasis.[3]

17.3.2.1 Metabolic Costs of Detoxification

The consequences of conjugation reactions are thus production of an acid load. It has been proposed that this represents a major metabolic cost to an animal and it is its ability to maintain acid base homeostasis in response to this acid load that determines the maximum intake of such compounds.[3] This hypothesis was reached from studies measuring the intake of two species of eucalyptus foliage (which differed in their terpene content) by possums. On the one with higher levels of plant secondary compounds, the classical symptoms of metabolic compensation for acidosis were observed, namely changes in the pattern of urinary nitrogen excretion. The main mechanism for neutralizing acid loads is by combination of protons with hydrogen carbonate and excretion as carbon dioxide and water.[11] Hydrogen carbonate can only be produced (without associated production of protons) from catabolism of amino acids which also produces ammonium ions.[12] Under conditions where there is no perturbation of acid base homeostasis, ureogenesis utilizes the ammonium and hydrogen carbonate produced in stoichiometric amounts. The urea that is produced is then excreted via urine. Use of hydrogen carbonate for neutralizing acid causes an excess of ammonium ions which are excreted directly in the urine. Thus, the percentage of urinary nitrogen in the form of ammonium ions increases under acidosis, and in the study of Foley (1992),[3] it was shown that this increased from 5 to 90 in animals receiving the foliage containing low or high levels of terpenes, respectively. Other studies have also shown that conditions that have caused acidosis have also resulted in an increase in rates of amino acid catabolism and protein breakdown.[13,14] Should the requirement for hydrogen carbonate exceed that produced from normal rates of amino acid catabolism, increased rates of amino acid breakdown thus occur.

Thus, one potential cost of conjugation is an increase in amino acid catabolism which, depending on the severity of the acid load, can result in decreased rates of protein synthesis (due to increasing proportion of absorbed amino acids being deaminated) or breakdown of body protein in order to supply the necessary amino acids for hydrogen carbonate production. The main conjugand used in many mammals is glucuronic acid.[1,15] This is derived from glucose. Excretion of glucuronic acid conjugates is thus a drain on glucose reserves which represents another quantifiable cost. Glucose can be derived directly from circulating supplies or from glucose stored as glycogen. In the absence of sufficient glucose from the above sources, additional glucose can be synthesized primarily from glycogenic amino acids.

Thus the two costs of plant secondary compound ingestion associated with detoxification reactions are increased amino acid use for maintenance of acid base homeostasis and increased glucose loss through loss of glucuronic acid in the urine.

17.4   MODELING DETOXIFICATION

17.4.1   CHOICE OF MODEL

Due to the nature of the cost involved, the model is required to operate at a metabolic level. Such models have been developed primarily based on those of Baldwin [16] and related ones of Gill.[17] Such models represent the flow of nutrients down known metabolic pathways using descriptions derived from enzyme kinetics. Individual reactions are not specifically represented but are grouped together as overall pathways. These are ascribed maximum rates per unit of tissue (often calculated from fluxes measured in the whole animal) and substrate effects are modeled through Michaelis Menten kinetics. The models essentially aim to balance ATP and NADPH production from catabolic pathways with that required by maintenance processes and synthetic pathways by incorporating terms for them into the Michaelis Menten descriptions, e.g., high concentrations of ATP serve to inhibit oxidative reactions which produce ATP. The maximum potential activity of any synthetic pathway, such as protein synthesis, was determined by the use of empirical data which obviously limits the scope of such models to the type and stage of maturity of the animal so described. Thus, nutrients supplied to the model in a particular ratio were used depending on their ability to support the goals of the modeled animal.

17.4.2   MODIFICATIONS

In order to describe detoxification of plant secondary compounds and the associated costs, a number of changes to such models were required. The model chosen was that of Gill et al. (1984).[17] This model considers metabolism as occurring in a single compartment represented by blood volume.

17.4.2.1   Compartmental Structure

Due to the location of detoxification being restricted to the liver and gut, the model was modified to separate these two tissues from the rest of the body with blood flow transporting nutrients between the compartments (Figure 17.2). Within the liver and the body compartments, divisions were made between blood and extracellular fluid and intracellular fluid with enzyme catalyzed reactions taking place only in the latter. Nutrients crossed the boundary between extra and intra cellular fluid either by diffusion (which depends on concentration differences, membrane permeabilities, and surface area) or by active transport depending on the type of compound. For example, all lipid soluble compounds, including absorbed plant secondary compounds, crossed membranes passively by diffusion.

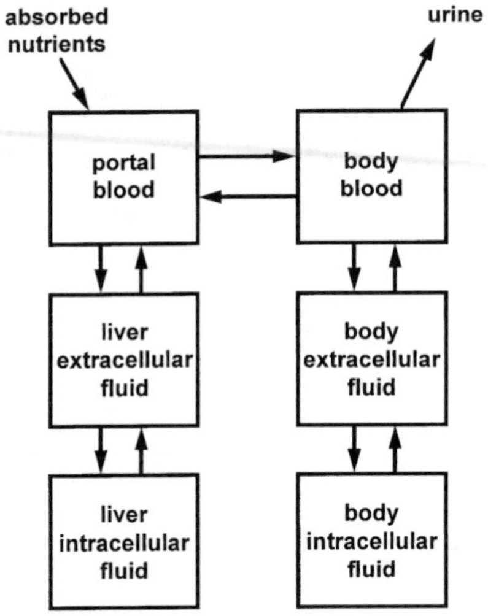

FIGURE 17.2   Illustration of the compartmental structure of the model of Illius and Jessop (1995).[21] Nutrients (including absorbable plant toxicants) enter the portal blood volume and move either through diffusion, blood flow or active transport throughout the various compartments. The plant toxicants modeled are all lipid soluble and were assumed to enter intracellular compartments by diffusion.

17.4.2.2   Acid-Base Homeostasis and Conjugation

Pools for description of hydrogen carbonate and ammonium concentrations and appropriate fluxes were added to enable acid base homeostasis to be modeled with the assumption that maintenance of this had to be ensured, i.e., if hydrogen carbonate levels were depleted by acid production, amino acid catabolism would be increased with breakdown of body protein if necessary (Figure 17.3). The metabolic pathways of plant secondary compound conjugation were added to the liver intracellular compartment with distinction being made between terpenoids and phenolics assuming that the latter did not require phase 1 reactions (Figure 17.4).[10]

17.4.2.3   Protein Turnover

The model of Gill et al. (1984)[17] only represented protein gain. In order to allow for the possibility of protein catabolism to supply amino acids for acid base homeostasis, a major change was to describe protein turnover. The model was modified to include separate descriptions of protein synthesis and protein breakdown, allowing by variation in one or the other or both, net protein gain or loss.

17.4.2.4   Animal Description

A further and linked modification was to overcome the empirical descriptions of maintenance and maximal rates of protein gain. A description of animal growth that has been developed from studies in Edinburgh was used in which the above rates are dependent on the animal's degree of protein maturity.[18,19,20] That is to say the current protein weight of an animal as a

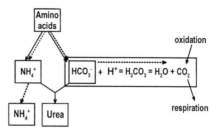

FIGURE 17.3   A summary of the metabolic trans-
actions concerned with maintenance of acid-base
homeostasis. Fluxes that increase under acidosis are
shown with a broken line.

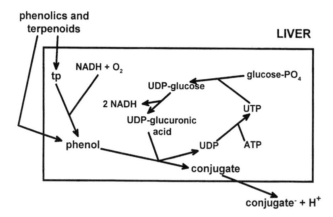

FIGURE 17.4   Representation of the pathways of conjugation of absorbed phenolics and terpenoids
within the liver.

proportion of its expected protein mass at maturity can be used to determine the energy and
amino acid requirements for maintenance and also the maximum rate of protein gain achiev-
able. Such a description is applicable to all animals and requires knowledge of the mature
protein mass of the species and of their current protein mass. These changes enable degree
of maturity effects to be investigated since a mature animal has no need to grow further
protein and thus will catabolize a greater proportion of absorbed amino acids than will a less
mature animals which is still aiming to increase its body protein mass.

17.5   MODELED EXPERIMENTS

With the above changes made, the model [21] when applied to the particular type of animal
and absorbed nutrient profile of Gill et al. (1984),[17] still gave very similar predictions. The
aim was to use the model to:

1. Explore the consequence of increasing the absorption of plant secondary compounds on metabolism within the animal,
2. Determine a maximum tolerable load of allelochemical and see how this compared
   to previously reported intakes of such chemicals, and
3. Determine if there was an interaction between the maximal tolerable load and
   nutrient intake.

It was assumed that all absorbed toxicants would be conjugated with glucuronic acid and
excreted via the urine. This is equivalent to compounds with molecular weights of around 150
such as many low molecular weight phenolics and terpenoids found in *Eucalyptus* foliage.[22]
In order to determine an appropriate range of toxicant absorption to simulate, data from the

study of Foley (1992)[3] were used. The total quantity of ingested phenolics and low molecular weight terpenoids were reported and if it is assumed that one quarter of the former and all of the latter were absorbed then dose rates of up to 0.03 mol per $W^{0.75}$ per day are given. Thus for a 37 kg animal this is equivalent to dose rates up to 0.5 mol per day.

17.5.1   SHORT-TERM EFFECTS

These were estimated from fluxes 12 hours after the start of each run. The two main costs of toxicant ingestion were separated by, for each run, replacing allelochemical absorption by a molar equivalent of protons. The response in terms of the pattern of nitrogen excretion did not differ between allelochemical or proton absorption and showed an increase in the proportion of nitrogen excreted as ammonium rather than as urea as acid load increased. The cost in terms of amino acid catabolism did differ between the two types of infusion (Figure 17.5). An absorption of 0.075 moles of allelochemical per day increased amino acid catabolism sufficiently to reduce protein gain to zero while it took 0.1 moles of protons per day to achieve the same effect. At a rate of allelochemical absorption of 0.5 moles per day, 12 g of body protein were lost each day but this reduced to 4 g of daily protein loss at the same level of proton absorption. Thus, the cost of maintaining acid base homeostasis was significantly less than that of supplying the glucose necessary for conjugation.

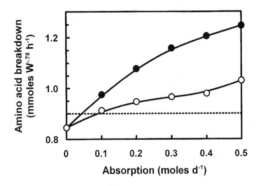

FIGURE 17.5 Predicted changes in the rate of amino acid catabolism caused by the absorption of either protons (open symbols) or allelochemical (closed symbols). Amino acid absorption (---------) was constant throughout. (From Illius, A. W. and Jessop, N. S., *J. Chem. Ecol.,* 21, 693, 1995.[21] With permission.)

17.5.2   LONG-TERM EFFECTS

Maintenance of homeostasis for a period of four days was taken as a definition of tolerance. Allelochemical absorption was sequentially increased at each of a range of intake levels (varying from zero to three times maintenance) to determine the maximal tolerable dose. The variation in predicted tolerance over the range of intakes is shown in Figure 17.6 where it can be seen that it generally increases with increased nutrient absorption although not in a linear manner. The maximal concentration of allelochemical in a food with an energy density of 10 MJ ME kg^{-1} DM can be calculated from this response and is shown in Figure 17.7. The response shown in this figure is an effective boundary which has important consequences. For a food with an allelochemical concentration of 50 g kg^{-1}, an animal has to be able to eat sufficient quantities of it to be able to obtain at least its maintenance requirement. If abundance of this food source limits intake to levels below maintenance, then the model predicts that this food cannot be tolerated and the animal would be better off not eating it at all. This may explain observations made on diet choice by animals during winter or the dry season when food availability declines. Faced with depletion of food resources at such times, animals

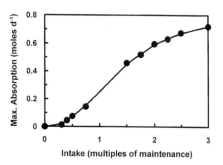

FIGURE 17.6   Predicted allelochemical tolerance over a range of intakes from 0 to 3 times maintenance. (From Illius, A. W. and Jessop, N. S., *J. Chem. Ecol.*, 21, 693, 1995.[21] With permission.)

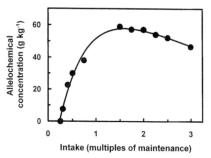

FIGURE 17.7   The maximal tolerable concentration of allelochemical in a food (with an energy density of 10 MJ ME kg[-1] DM) over a range of intakes from 0 to 3 times maintenance. (From Illius, A. W. and Jessop, N. S., *J. Chem. Ecol.*, 21, 693, 1995.[21] With permission.)

avoid current-season's growth of wood, browse, and choose more mature growth instead.[23,24,25] As the quantity of allelochemical protected feeds declines, the ability of animals to eat them is reduced since the total nutrient intake available would not enable the absorbed allelochemical load to be tolerated.

17.6   IMPLICATIONS

The modeling exercise enabled quantitation of the proposed costs of allelochemical absorption and supported the hypothesis of Foley (1992)[3] that such costs could play an important role in determining the quantity of food eaten. A somewhat surprising outcome was that for a particular allelochemical concentration in the food there was a threshold level of intake required before the animal was able to tolerate the absorbed dose and that if this could not be achieved then it would be to the animal's advantage not to eat any food at all. This, however, takes no account of the psychology of food intake, and it may be that the benefits of having some food inside the gut outweighs the costs of increased loss of body protein. Indeed, since it has been shown that animals are capable of losing up to 25% of their body protein before any serious consequences are met,[26,27] the perceived cost of loss of body protein might also vary considerably depending on the level of protein stores.

The costs identified were glucose and amino acid use and the main cause for failure of the model during tolerance tests was inability to produce sufficient glucose from amino acids.

17.7   EXPERIMENTAL TEST

The hardest test of a prediction from a mathematical model is experimental verification. As discussed earlier in this chapter, this is often difficult to do due to the interaction between various components of a system that can be controlled within a model but not within an animal. We have tested (Jessop, N. S. and Illius, A. W., unpublished) the main outcome from the modeling exercise described above. Weanling rats were acclimatized to one of two allelochemicals which were incorporated into a semi-purified diet given as a mash. The allelochemicals used were menthol and anisole, both of which are of low molecular weight and have been shown to be excreted principally via urine as conjugates with glucuronic acid. During the period of acclimatization, the dosage of each allelochemical was increased over a two week period until each animal was consuming one LD_{50} per day. Animals were then housed individually and given separately a rationed amount of protein (gelled in agar) and an unrestricted quantity of remaining nutrients (oil, starch, sucrose, vitamins, and minerals; also gelled with agar) containing two LD_{50} of the respective allelochemical per unit of maintenance energy. The voluntary consumption of both supplied feeds was measured. The quantity of protein given to each animal was varied between two thirds and six times their calculated maintenance protein requirement (at the maximum amount the supply of protein was not sufficient to meet the animal's maintenance need for energy). Control animals underwent the same treatment except that no allelochemicals were added to the diets. All animals ate the rationed quantity of protein but chose to eat variable amounts of the feed containing allelochemical. There was no difference in amount of this feed eaten between allelochemicals but this did vary depending on the quantity of protein consumed (Figure 17.8). The ability of the animals to tolerate allelochemicals was greater at higher levels of protein intake presumably due to greater supplies of amino acids for maintenance of acid base homeostasis and gluconeogenesis.

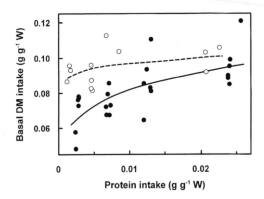

FIGURE 17.8   Intake by rats of a basal diet containing an acidosis-inducing allelochemical (solid line and symbols) when offered one of four quantities of protein supplement. Control diet (broken line and open symbols) contained no allelochemical. The slope (response of basal DM intake to protein intake) of the treatment group is greater than zero (P<0.001); the slope of the control group is not different from zero (P<0.1).

17.8   CONCLUSION

We demonstrate how modeling can be used to address the costs of ingestion of plant toxicants by herbivores. Such an approach requires information about the mechanisms of the processes concerned and integrates such knowledge in a quantitative manner to enable consequences and responses to be predicted. The most stringent test of any model is whether its predictions can be tested experimentally and we show that, in this case, this was so.

REFERENCES

1. Scheline, R. R., Handbook of mammalian metabolism of plant compounds, CRC Press, Boca Raton, 1991.
2. Provenza, F. D., Postingestional feedback as an elementary determinant of food preference and intake in ruminants, *J. Range Manage.*, 48, 2, 1995.

3. Foley, W. J., Nitrogen and energy retention and acid-base status in the common ringtail possum (*Pseudocheirus peregrinus*): Evidence of the effects of absorbed allelochemicals, *Physiol. Zool.*, 65, 403, 1992.

4. Foley, W. J. and McArthur, C., The effects and costs of allelochemicals for mammalian herbivores: An ecological perspective, in *The Digestive System in Mammals*, Chivers, D. J. and Langer, P., Eds., Cambridge University Press, Cambridge, 1994, 370.

5. Foley, W. J., McLean, S., and Cork, S. J., Consequences of biotransformation of plant secondary metabolites on acid-base metabolism in mammals-a final common pathway?, *J. Chem. Ecol.*, 21, 721, 1995.

6. Belovsky, G. E. and Schmitz, O. J., Mammalian herbivore optimal foraging and the role of plant defenses, in *Plant Defenses against Mammalian Herbivory*, Palo, R. T. and Robbins, C. T., Eds., CRC Press, Boca Raton, 1991, 1.

7. Gordon, I. J. and Illius, A. W., The nutritional ecology of ruminants: a reinterpretation, *J. Anim. Ecol.*, 65, 18, 1996.

8. Shahkhali, Y., Finot, P. A., Harnell, R., and Fern, E., Effects of foods rich in polyphenols on nitrogen excretion in rats, *J. Nutr.*, 120, 346, 1989.

9. McLeod, M. N., Plant tannins — their role in forage quality, *Nutr. Abs. Revs.*, 44, 803, 1974.

10. Häussinger, D., Meijer, A. J., Gerok, W., and Sies, H., Hepatic nitrogen metabolism and acid-base homeostasis, in *pH Homeostasis: Mechanisms and Control*, Häussinger, D., Ed., Academic Press, London, 1988, 337.

11. Welbourne, T. C., Interorgan glutamine fluxes in acid-base disturbances, in *pH Homeostasis: Mechanisms and Control*, Häussinger, D., Ed., Academic Press, London, 1988, 379.

12. Williams, B., Layward, E., and Walls, J., Skeletal muscle degradation and nitrogen wasting in rats with chronic metabolic acidosis, *Clin. Sci.*, 80, 457, 1991.

13. Reaich, D., Channon, S. M., Scrimgeour, C. M., and Goodship, T. H. J., Ammonium chloride-induced acidosis increases protein breakdown and amino acid oxidation in humans, *Am. J. Physiol.*, 263, E735, 1992.

14. Mulder, G. J., Temmink, T. J. M., and Koster, H. J., The effect of fasting on sulfation and glucuronidation in the rat *in vivo*, *Biochem. Pharmacol.*, 31, 1941, 1982.

15. Baldwin, R. L., Smith, N. E., Taylor, J., and Sharp, M., Manipulating metabolic parameters to improve growth rate and milk secretion, *J. Anim. Sci.*, 51, 1416, 1980.

16. Gill, M., Thornley, J. H. M., Black, J. L., Oldham, J. D., and Beever, D. E., Simulation of the metabolism of absorbed energy yielding nutrients in young sheep, *Br. J. Nutr.*, 52, 621, 1984.

17. Caldwell, J., Conjugation reactions in foreign-compound metabolism: Definition, consequences and species variations, *Drug Metab. Rev.*, 13, 745, 1982.

18. Taylor, St. C., S., Genetic size scaling rules in animal growth, *Anim. Prod.*, 30, 161, 1980.

19. Taylor, St. C., S., Live-weight growth from embryo to adult in domesticated mammals, *Anim. Prod.*, 31, 223, 1980.

20. Emmans, G. C. and Fisher, C., Problems in nutritional theory, in *Nutrient Requirements of Poultry and Nutritional Research*, Fisher, C. and Borman, K. N., Eds., Butterworths, London, 1986, 9.

21. Illius, A. W. and Jessop, N. S., Modeling metabolic costs of allelochemical ingestion by foraging herbivores, *J. Chem. Ecol.*, 21, 693, 1995.

22. Eberhard, I. H., McNamara, J., Pearse, R. J., and Southwell, I. A., Ingestion and excretion of Eucalyptus punctata D. C., and its essential oil by the koala, *Phascolarctos cinereus* (Goldfuss), *Aust. J. Zool.*, 23, 169, 1975.

23. Bryant, J. P. and Kuropat, P. J., Selection of winter forage by subarctic browsing vertebrates: The role of plant chemistry, *Annu. Rev. Syst. Ecol.*, 11, 261, 1980.

24. Reichardt, P. B., Bryant, J. P., Clausen, T. P., and Weiland, G. D., Defense of winter-dormant Alaska paper birch against snowshoe hares, *Oecologia (Berlin)*, 65, 58, 1984.

25. Provenza, F. D. and Malachek, J. C., Diet selection by domestic goats in relation to black-brush twig chemistry, *J. Appl. Ecol.*, 21, 831, 1984.

26. Allison, J. B. and Wannemacher, R. W., The concept and significance of labile and over-all protein reserves of the body, *Am. J. Clin. Nutr.*, 16, 445, 1965.

27. Pine, A. P., Jessop, N. S., and Oldham, J. D., Maternal protein reserves and their influence on lactational performance in rats, *Br. J. Nutr.*, 71, 13, 1994.

18 Medicinal Applications of Plant Toxicants

A. D. Kinghorn and E. J. Kennelly

CONTENTS

18.1 INTRODUCTION

Previous chapters of this volume have described a wide range of plant toxicants with detrimental effects on animals and humans. However, humans have used toxic compounds for millennia for medicinal purposes since, at subtoxic doses, some toxicants can be beneficial. By accurately regulating the dosage of plant toxicants ingested, a variety of maladies can be treated, including certain cancers as well as cardiovascular and infectious diseases.

Plant toxicants used medicinally encompass a wide variety of natural product structural types, and a number of volumes have appeared on this general subject.[1-6] While many plant drugs were first introduced after initially following up on folkloric observations, such as digitoxin, morphine, and quinine, others were obtained after laboratory study of plants used as toxins or hunting poisons, such as ouabain from African *Strophanthus* species, physostigmine from calabar beans, and *d*-tubocurarine from a South American dart poison.[7] It is primarily the plant toxicants classified as secondary metabolites that are used as drugs, and these molecules,

0-8493-8551-2/97/$0.00+$.50
© 1997 by CRC Press, Inc.

even with great structural complexity, tend to be produced naturally in the exact chiral form to exhibit maximal biological activity.[8]

While many plant toxicants have a proven track record of being used in an unmodified form as medicines, they play another important role as lead compounds for synthetic modification for the design of entirely new drug substances.[7,9–12] Though prototype bioactive compounds themselves must have significant pharmacological or biological activity, they frequently have undesirable qualities such as poor aqueous solubility or unacceptable toxicity. Among good examples of lead compounds from plants are the belladonna tropane alkaloids that have been used in the design of synthetic anticholinergic agents, and cocaine in the development of modern local anesthetics, as well as opiate alkaloids such as codeine and morphine in the synthetic improvement of less addictive analgesic agents.[7,9]

This chapter will examine three major classes of plant toxicants, namely, alkaloids, isoprenoids, and phenols and quinones, and will highlight examples of plant drugs used clinically as medicines as well as some promising compounds presently under clinical evaluation. Owing to space limitations, it will not be possible to cover many other promising biologically active plant secondary metabolites which have not yet been evaluated in humans. Only selected drugs and drug candidates from higher plants will be considered. A listing of plant-derived drugs is shown in Table 18.1, although this is not intended to be comprehensive and not all of the compounds mentioned are equally toxic on a weight for weight basis. Again in the interests of brevity, the structures of only a few of the examples mentioned in the text will be presented in this chapter.

18.2   ALKALOIDS

Alkaloids are well known and reasonably widespread plant toxicants (see Chapters 1 to 4), and it is considered that their primary ecological role is to serve as chemical defense substances to protect plants from predators.[13] These nitrogenous compounds are basic, and therefore can be to made into alkaloidal salts, thus facilitating both their laboratory evaluation as well as their therapeutic administration.[14] In general, alkaloids are, to a varying degree, toxic substances that exert effects on the central nervous system.[15] Alkaloids are structurally and biologically diverse, and thus difficult to define, but can be broadly classified based on their biogenetic origin and heterocyclic ring type.[5,16] Alkaloidal toxicants that have particular medicinal importance include those derived from the amino acids ornithine, phenylalanine/tyrosine, tryptophan, and histidine, the isoprenoid-derived diterpenoid alkaloids, and a protoalkaloid that will be categorized herein as a miscellaneous alkaloid.

18.2.1   ORNITHINE-DERIVED

The tropane alkaloids atropine, hyoscyamine, and scopolamine constitute a group of pharmaceutically important anticholinergic compounds. These alkaloids are found commonly in the plant family Solanaceae, especially from three commercial sources, *Atropa belladonna* (deadly nightshade), *Datura stramonium* (jimson weed), and *Hyoscyamus niger* (henbane).[6] Historically, many tropane-containing plants are well-known poisons, with henbane being used in European witchcraft in the Middle Ages.[5] Cases of poisoning by these compounds are rare in livestock due to the strong unpleasant odors of the plants producing these toxins.[17,18] Since these solanaceous alkaloids penetrate the central nervous system, they are sometimes abused by young adults in the U.S. in order to experience a euphoric effect. The toxic symptoms of abusing jimson weed in this manner include blurred vision, coma, confusion, difficulty in swallowing, hallucinations, hyperthermia, seizures, thirst, and urinary retention.[19] Solanaceous alkaloids may also cause anticholinergic poisoning by contaminating herbal teas, as in the case of several people in the U.S. who became ill recently as a result of drinking an adulterated stimulant beverage from South America.[20] Atropine is competitive

TABLE 18.1
Examples of Active Plant Toxicants with Drug Use

Compound	Compound Class	Biological Activity/Medicinal Use(s)
Acetylditigoxin	Steroidal glycoside	Cardiotonic
Ajmalicine	Indole alkaloid	Circulatory disorders
Anisodamine	Tropane alkaloid	Antidysenteric
Anisodine	Tropane alkaloid	Antimigraine
Artemisinin	Sesquiterpenoid	Antimalarial
Atropine	Tropane alkaloid	Anticholinergic, mydriatic
Caffeine	Purine base	CNS stimulant
Camptothecin	Quinoline alkaloid	Antineoplastic
Cocaine	Tropane alkaloid	Local anesthetic
Colchicine	Alkaloidal amine	Antigout
Deserpidine	Indole alkaloid	Antihypertensive, tranquilizer
Deslanoside	Steroidal glycoside	Cardiotonic
Digitoxin	Steroidal glycoside	Cardiotonic
Digoxin	Steroidal glycoside	Cardiotonic
Emetine	Isoquinoline alkaloid	Antiamebic, antiemetic
Ephedrine	Protoalkaloid	Bronchodilator
Gossypol	Sesquiterpenoid	Male contraceptive
Hyoscyamine	Tropane alkaloid	Anticholinergic
Lanatosides A, B, and C	Steroidal glycosides	Cardiotonics
Morphine	Morphinan alkaloid	Analgesic
Noscapine	Isoquinoline alkaloid	Antitussive
Ouabain	Steroidal glycoside	Cardiotonic
Paclitaxel	Diterpenoid	Antineoplastic
Papaverine	Isoquinoline alkaloid	Smooth muscle relaxant
Physostigmine	Indole alkaloid	Cholinergic
Picrotoxin	Sesquiterpenoid	Central and respiratory stimulant
Pilocarpine	Imidazole alkaloid	Cholinergic
Podophyllotoxin	Lignan	Caustic for papillomas
Proscillaridin	Steroidal glycoside	Cardiotonic
Protoveratrines A and B	Steroidal alkaloids	Hypotensives
Pseudoephedrine	Protoalkaloid	Decongestive
Quinidine	Quinoline alkaloid	Cardiac depressant
Quinine	Quinoline alkaloid	Antimalarial
Rescinnamine	Indole alkaloid	Antihypertensive
Reserpine	Indole alkaloid	Antihypertensive, tranquilizer
Scillarens A and B	Steroidal glycosides	Cardiotonics
Scopolamine	Tropane alkaloid	Anticholinergic
Sparteine	Quinolizidine alkaloid	Oxytocic
Strophanthidin	Steroidal glycoside	Cardiotonic
Strychnine	Indole alkaloid	CNS stimulant
Δ^9-Tetrahydrocannabinol	Cannabinoid	Antiemetic
dl-Tetrahydropalmatine	Isoquinoline alkaloid	Sedative
Theophylline	Purine base	Smooth muscle relaxant
d-Tubocurarine	Alkaloid	Skeletal muscle relaxant
Vinblastine	Bisindole alkaloid	Antineoplastic
Vincristine	Bisindole alkaloid	Antineoplastic
Yohimbine	Indole alkaloid	α-Adrenergic antagonist

Source: Information compiled from Farnsworth and Soejarto (1991),[64] Bruneton (1995),[5] and Robbers et al. (1996).[6]

with acetylcholine at the postganglionic synapse of the parasympathetic nervous system, and it is because of its antimuscarinic effects that atropine and the other alkaloids of belladonna are used as spasmolytic drugs to control gastrointestinal motility.[6] In addition, atropine is an antidote in cases of poisoning caused by cholinesterase inhibitors such as physostigmine and organophosphate insecticides.[6]

Cocaine, perhaps the most highly renowned tropane alkaloid, is obtained from the leaves of *Erythroxylum coca* and *Erythroxylum truxillense*, and is commonly used illicitly to induce euphoria and hyperactivity, with the drug being introduced to the body by the intranasal route. "Crack cocaine" is the free base of the alkaloid obtained by treating the hydrochloride salt with bicarbonate, and is extremely addictive.[6] Cocaine hydrochloride is available by prescription in the U.S. for use as a topical local anesthetic solution, and is also an ingredient in Brompton's cocktail which is used to curtail pain in terminal cancer patients.[3] Cocaine hydrochloride is regarded as being too toxic for internal use therapeutically as a local anesthetic since it has multiple adverse central and peripheral nervous system effects.[6]

Scopolia tangutica, a plant used in traditional Chinese medicine, biosynthesizes the tropane alkaloids anisodamine (Figure 18.1 [1]) and anisodine (Figure 18.1 [2]). Anisodamine is used in the People's Republic of China for the treatment of septic shock resulting from toxic bacillary dysentery, fulminant epidermic meningitis, and hemorrhagic enteritis. This compound has been shown to stimulate microcirculation.[21] Anisodine is used in China for the treatment of migraine headaches.[5] Anisodamine and anisodine therefore extend the range of clinical uses typically associated with the anticholinergic and mydriatic solanaceous alkaloids.

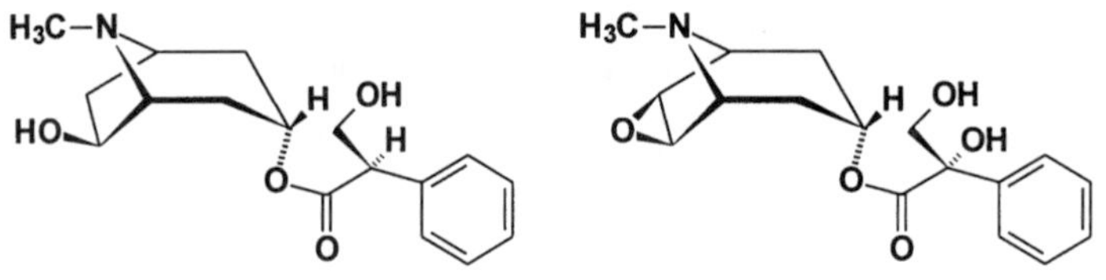

FIGURE 18.1   Structures of the tropane alkaloids anisodamine [1] and anisodine [2].

18.2.2   PHENYLALANINE-/TYROSINE-DERIVED

One of the most well-known drugs discovered as a result of the ethnobotanical uses of a plant toxicant by indigenous peoples is the skeletal muscle relaxant, *d*-tubocurarine. The compound is contained in a preparation from the South American plant *Chondodendron tomentosum* which is applied to the tips of blow gun darts to aid in hunting small game. *d*-Tubocurarine, an acetylcholine antagonist, became an important anesthetic used in surgery to attain full muscular flaccidity without deep anesthesia.[9] The use of *d*-tubocurarine has curtailed recently as a result of the introduction into therapy of synthetic neuromuscular blocking agents such as atracurium and vecuronium.[22] However, *d*-tubocurarine is still used to control convulsions of strychnine.[6]

The isoquinoline alkaloid emetine is an ingredient in many over-the-counter emetics, such as ipecac syrup and balsam syrup, being used to rid the stomach of poisonous substances, such as in accidental drug overdoses in children.[5] While the compound is still isolated from *Cephaelis ipecacuanha* and *Cephaelis acuminata* roots, the semisynthetic derivative, dehydroemetine, is also employed now due to decreased toxicity.[5] Emetine induces emesis by irritation of the gastrointestinal mucosa and a central medullary effect by stimulation of the chemoreceptor trigger zone.[6] Syrup of ipecac can be given orally to induce emesis in about 15 to 30 min.[23] There are a number of subacute and chronic toxicity effects associated with emetine ingestion, including arrhythmias, gastrointestinal distress, hypotension, and muscular weakness.[5] Emetine can also be employed as an antiamebic by administration intramuscularly or subcutaneously.[6]

The alkaloidal amine colchicine is obtained from the meadow saffron, *Colchicum autumnale*. Colchicine was officially employed as early as the 19th century as an anti-inflammatory to treat gout.[5] This substance has been found to exert a wide range of cellular effects, and there has been an increase in its clinical use in the treatment of numerous inflammatory diseases such as Behcet's syndrome, psoriasis, scleroderma, and sclerosis. It is also used for the prevention of attacks in patients with familial Mediterranean fever.[24] Acute poisoning by colchicine produces ascending paralysis of the CNS, hemorrhagic gastroenteritis, and nephrotoxicity, among other symptoms.[25]

18.2.3 TRYPTOPHAN-DERIVED

The quinoline alkaloid camptothecin (Figure 18.2 [3]) and its analogs have shown great potential as antineoplastic agents. Camptothecin was discovered in the 1960s by Wall and Wani as a constituent of the Chinese tree *Camptotheca acuminata*, and found to be a potent antileukemic and antitumor agent. However, initial clinical trials on camptothecin sodium salt were halted in the 1970s in the U.S. due to the toxicity observed, including myelosuppression at unacceptable incidence and severity levels.[26,27] Interest in camptothecin and its analogs was revived in the 1980s when these compounds were found to target the enzyme DNA topoisomerase I specifically, and then some of those compounds showed the ability to inhibit human colon adenocarcinomas in marine xenograft studies.[28] Clinical trials are now being carried out on camptothecin, and several of its more water-soluble analogs, including 9-amino-20(S)-camptothecin (Figure 18.2 [4]), topotecan (Figure 18.2 [5]), and irinotecan (Figure 18.2 [6]). Several of these compounds have shown excellent activity in patients with colorectal cancer, adenocarcinoma, gynecological cancers, and lung cancer.[28] Irinotecan has been approved as an antineoplastic agent in Japan and France.[27] In addition to myelosuppression, other toxic effects reported in clinical trials to date on the camptothecins include gastrointestinal disturbances and neutropenia.[28]

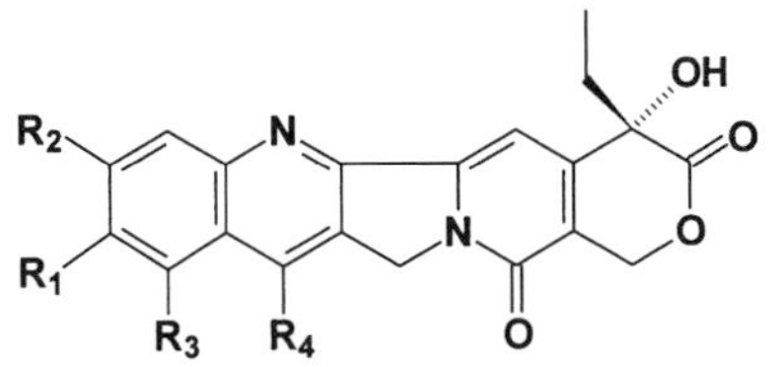

	R_1	R_2	R_3	R_4
3	H	H	H	H
4	H	H	NH_2	H
5	OH	H	$CH_2N(CH_3)_2$	H
6	$C_5H_{10}N-C_5H_9N-COO$	H	H	CH_2CH_3

FIGURE 18.2 Structures of the quinoline alkaloid 20(S)-camptothecin [3], and its analogs 9-amino-20(S)-camptothecin [4], topotecan [5], and irinotecan [6].

Two other quinoline alkaloids with extensive clinical use are quinine and its stereoisomer, quinidine, which are used as an antimalarial and as a cardiac antiarrhythmic, respectively. These compounds are obtained from several *Cinchona* species and their hybrids, and from *Remijia pedunculata*.[6] Of the 300 to 500 metric tons of quinine extracted and purified each year, about half is used in the food industry as a bitter principle in soft drinks, and most of the rest is used to semisynthesize quinidine.[5] In overdose, the cinchona alkaloids produce the symptoms of cinchonism, involving temporary loss of hearing with ringing in the ears.[6]

The indole alkaloids represent perhaps the largest group of therapeutically useful alkaloids, and include the *Rauvolfia* alkaloids (inclusive of reserpine, deserpidine, and rescinnamine), physostigmine (derived from the Calabar bean, *Physostigma venenosum*), and yohimbine (from *Pausinystalia yohimbe*). It should also be pointed out that the ergot alkaloids are also therapeutically important indoles, although these are not obtained from a higher plant source, but rather are of fungal origin (*Claviceps purpurea* and *Claviceps paspali*). The newest group of clinically useful indole alkaloids, however, are the vinca (*Catharanthus*) alkaloids, which are employed as cancer chemotherapeutic agents.[6] The bisindole alkaloids, vincristine (Figure 18.3 [7]) and vinblastine (Figure 18.3 [8]), obtained from the leaves of *Catharanthus roseus*, have been used clinically for over 30 years for the treatment of several forms of cancer, usually in combination chemotherapy.[6] Vinblastine sulfate is employed to treat advanced testicular cancer, Hodgkin's disease, Kaposi's syndrome, and lymphocytic leukemia, while vincristine sulfate is a drug of choice for acute lymphocytic leukemia, and is also used for rhabdomyosarcoma and Wilms' tumor.[6] As with most antineoplastic agents, the vinca alkaloids have numerous neurotoxic side effects, including headache, myalgia, neuralgia, and paresthesia.[5] A number of semisynthetic derivatives of the vinca alkaloids have been developed in an effort to decrease the toxicity of these compounds.[29] Recently, the semisynthetic compound vinorelbine (5′-noranhydrovinblastine, Navelbine®) (Figure 18.3 [9]), a drug developed in France, has been approved by the FDA in the U.S. as a first-line therapy for non-small lung cancer either alone or in combination with cisplatin.[30]

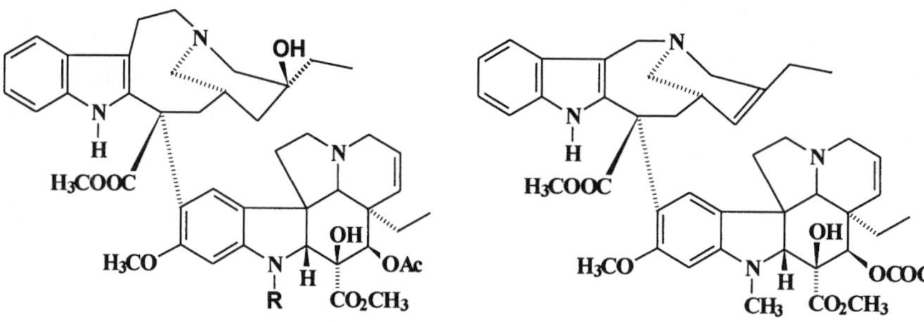

	R	
7	CHO	
8	CH$_3$	**9**

FIGURE 18.3 Structures of the bisindole alkaloids vincristine [7], vinblastine [8], and the semisynthetic derivative, vinorelbine [9].

18.2.4 HISTIDINE-DERIVED

The imidazole alkaloid pilocarpine is obtained from *Pilocarpus microphyllus* and other *Pilocarpus* species indigenous to South America, and acts as a parasympathomimetic. This ophthalmic cholinergic agent is used clinically for the treatment of glaucoma, and because it directly stimulates the muscarinic receptors of the eye, it causes the ciliary muscle to contract and the pupil to constrict, thereby resulting in a decrease of intraocular pressure.[6] Salts of pilocarpine are used in topical solutions and in the form of a long-acting ocular delivery system. Toxic effects include myosis and transient headaches.[5,6]

18.2.5   ISOPRENOID-DERIVED

From the roots of *Aconitum napellus*, a number of toxic diterpene alkaloids have been isolated. Extracts of this plant have been used for centuries as arrow poisons in many part of the world, with the principal alkaloid aconitine being highly toxic.[5] There have been numerous reported cases of aconite poisoning in both humans and animals, resulting in restlessness, salivation, weakness, irregularity of heartbeat, and occasionally death.[18] Tincture of aconite is used in over-the-counter decongestion and cold medicines in some countries.[5] Lappaconitine, isolated from *Aconitum sinomontanum*, is used clinically as a local anesthetic in the People's Republic of China, being much more potent than cocaine.[21] Other toxic diterpenoid alkaloids have been isolated from the larkspur, *Delphinium* spp.[31] Methyllycaconitine (LD_{50} 3.2 mg kg^{-1} i.v. mice) is the most toxic of the larkspur alkaloids.[32] Instances of cattle poisoning by larkspur are common in the western U.S. resulting in significant economic loss.[18] These alkaloids have a curare-like effect on skeletal muscles, and act by blocking postsynaptic conduction by acetylcholine.[31]

18.2.6   MISCELLANEOUS

Ephedrine is a phenethylamine derivative (protoalkaloid) produced by several species of the genus *Ephedra*, including *Ephedra distachya*, *Ephedra quisetina*, and *Ephedra sinica* ("Ma Huang"), and is a potent sympathamimetic that stimulates alpha, beta$_1$, and beta$_2$ adrenergic receptors.[6] Ephedrine hydrochloride is used to treat acute cases of asthma.[5] "Ma Huang" has been use in traditional Chinese medicine for over 5000 years, and the use of herbal products containing "Ma Huang" has increased greatly in the last several years in the U.S., with these products being promoted as an herbal alternative to the illegal hallucinogenic drug "ecstasy."[33] Side-effects of ephedrine ingestion include increase of both systolic and diastolic blood pressure, increased heart rate, and possible palpitations as well as dizziness, headaches, insomnia, and nervousness.[34] Since 1993, the U.S. Food and Drug Administration (FDA) has noted 395 adverse reactions to products containing "Ma Huang," and 15 deaths that may be related to ingestion of ephedrine from these products.[33] The FDA has recently issued a warning about the potentially serious side-effects of ephedrine-containing herbal products.[35]

18.3   ISOPRENOIDS

The isoprenoids, represented by both the terpenoid and steroids, are mevalonic acid-derived compounds which are widespread in the plant kingdom. While many terpenes and steroids are innocuous, being found in certain food plants for example, some are potent toxins. In the following sections, compounds in this group that are classified as sesquiterpenoids, diterpenoids, and steroidal glycosides will be discussed in turn.

18.3.1   SESQUITERPENOIDS

The sesquiterpene lactone artemisinin (also known as "Qinghaosu") (Figure 18.4 [**10**]) obtained from the Chinese medicinal plant *Artemisia annua*, has become an important prototype antimalarial agent. Like many other terpenoids, artemisinin is poorly soluble in water, but reduction to the secondary alcohol and salt formation to produce sodium artesunate (Figure 18.4 [**11**]), results in a more highly water-soluble compound. In the People's Republic of China, and other countries in southeast Asia, sodium artesunate is manufactured in parenteral and oral formulations to treat falciparum malaria. Artemether (Figure 18.4 [**12**]), the methyl ester of artemisinin, is also used clinically in Asian countries.[36,37] Artemether is given intramuscularly, which provides greater bioavailability compared to oral dosing, and it is intended that this be introduced to Western countries as an antimalarial in the near future.[37] Other derivatives of artemisinin include arteether (Figure 18.4 [**13**]), which has undergone clinical trials supported by the World Health Organization, and artelinate (Figure 18.4 [**14**]), a water-soluble ether derivative of artemisinin which is more stable than sodium artesunate.[36] However, there has recently been a report on the fatal neurotoxicity of high doses of artemether and arteether in rats,[38] which is of some concern to the future development of these agents as clinically effective antimalarials.[37]

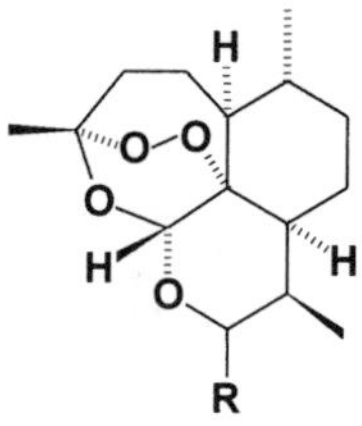

	R
10	=O
11	β-OCOCH$_2$CH$_2$CO$_2$Na
12	β-OCH$_3$
13	β-OCH$_2$CH$_3$
14	β-OCH$_2$C$_6$H$_4$CO$_2$Na

FIGURE 18.4 Structures of the sesquiterpene lactone artemisinin [**10**], and analogs, sodium artesunate [**11**], artemether [**12**], arteether [**13**], and sodium artelinate [**14**].

The toxicity of the phenolic sesquiterpene gossypol has been been described in depth in Chapter 7 of this volume. This constituent of cotton oil (*Gossypium* species) was extensively evaluated in the 1980s for its antifertility properties in the People's Republic of China, since it causes a decrease in the sperm count in human males.[39] The (-)-enantiomer of this compound has also been found to inhibit the replication of the HIV virus.[40] Gossypol went into human clinical trials in the People's Republic of China as a male contraceptive. Although this was effective, limiting side-effects were apparent, including hypokalemia, lassitude, and occasional paralysis.[41]

18.3.2 DITERPENOIDS

In late 1992, the diterpene paclitaxel (Taxol®) (Figure 18.5 [**15**]), previously known in the scientific literature as taxol, was approved by the U.S. Food and Drug Agency for the treatment of refractory ovarian cancer. In April of 1994, paclitaxel was also approved for the therapy of metastatic breast cancer. This has marked the first chemically unmodified plant secondary metabolite to be approved for use in the U.S. drug market for nearly 30 years, and the first diterpenoid to be introduced into clinical therapy.[42,43] This compound was discovered as a constituent of the bark of the North American plant, *Taxus brevifolia* as an antitumor agent in the 1960s by Wall and Wani.[42,43] The compound's unique mechanism of action, in being an antimitotic agent which stabilizes microtubules and prevents depolymerization, was demonstrated by Horwitz and co-workers.[42,43]

In the lengthy development of paclitaxel as an approved antineoplastic agent, many problems were encountered and had to be solved. Since the compound is not water soluble, early clinical trials were hampered due to the lack of adequate drug delivery, and the drug needed to be formulated in an oily vehicle.[43] Furthermore, the amount of paclitaxel necessary for clinical trials initially produced a severe supply crisis for the drug. It proved necessary to extract 15,000 pounds of *T. brevifolia* bark to obtain 1 kg of pure paclitaxel, and because in one particular year over 1.6 million pounds of the bark of this tree were extracted to produce the drug in the states of Oregon and Washington alone, this has led to a considerable environmental

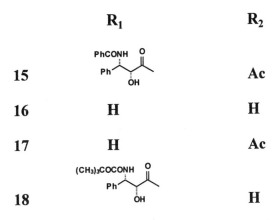

	R_1	R_2
15	PhCONH—CH(Ph)—CH(OH)—C(=O)	Ac
16	H	H
17	H	Ac
18	$(CH_3)_3COCONH$—CH(Ph)—CH(OH)—C(=O)	H

FIGURE 18.5 Structures of paclitaxel [**15**] from *Taxus brevifolia*, and its analogs 10-deacetylbaccatin III [**16**], baccatin III [**17**], and docetaxel [**18**].

concern.[44] Much scientific effort and ingenuity has been required to afford alternative solutions to the commercial production of paclitaxel for clinical use, including the development of semisynthetic routes from diterpenoid building blocks, with independent reports of the total synthesis of this complex molecule by both the Holton and Nicolaou groups appearing in 1994.[42,43] Current production of the drug involves partial synthesis from diterpenoid moieties such as 10-deacetylbaccatin III (Figure 18.5 [**16**]) and baccatin III (Figure 18.5 [**17**]) that are abundant in ornamental *Taxus* species.[42,43] As a result of extensive structure-activity relationship studies on paclitaxel analogues, a second taxane anticancer agent has been developed, docetaxel (Taxotere®) (Figure 18.5 [**18**]).[42,43]

Poisonings by *Taxus* species occur in both humans and animals, with symptoms including gastroenteritis, dyspnea, and, at high doses, sudden death.[2,18] Clinical toxicity which has been shown by paclitaxel includes cardiac toxicity, hyposensitivity reactions, leukopenia, and neurotoxicity.[43] In turn, docetaxol (Taxotere®), which has been approved in several countries for treating advanced or metastatic breast cancer resistant to adriamycin therapy, includes similar types of toxicity and a high incidence of fluid retention.[45]

18.3.3 STEROIDAL GLYCOSIDES

There are a number of pharmaceutically important steroidal glycosides of plant origin. These compounds which are also referred to as cardiac glycosides display a stimulatory effect on heart function. Digoxin, obtained from the white foxglove (*Digitalis lanata*) is the most widely prescribed of the steroidal glycosides, and results in an increase in the force of systemic contraction.[6] Over 200 years ago, the English physician William Withering employed preparations of the purple foxglove (*Digitalis purpurea*) for the treatment of myocardial insufficiency, avoiding the toxic effects of the plant through careful dose-optimization experimentation.[46]

The digitalis glycosides are still of major importance in the treatment of atrial flutter, congestive heart failure, and fibrillation. However, the drugs have a low therapeutic index, so renal function, especially in elderly patients, must be carefully monitored to prevent toxicity.[46] There are many cases of human intoxication with digitalis glycosides, especially due to people unwittingly drinking teas from the leaves of *Digitalis* spp.[47] In toxic doses, the cardiac glycosides may increase automaticity and lead to ectopic tachyarrythmia.[6] Recently, life-threatening digitoxin intoxications have been successfully treated using digitalis-specific antibody fragments that bind and inactivate the steroidal glycosides.[48]

18.4   PHENOLS AND QUINONES

18.4.1   LIGNANS

From the rhizomes of *Podophyllum peltatum* (Mayapple), the antineoplastic lignan podophyllotoxin has been isolated. The Mayapple is well known as being poisonous, especially its resinous rhizomes, but there are few documented cases of human or animal fatalities.[18] Podophyllotoxin (Figure 18.6 [**19**]) is a principal lignan constituent of *P. peltatum*, and in the form of Podophyllum Resin has been used clinically for the treatment of various papillomas.[6] Several semisynthetic derivatives have been developed as less toxic and more potent antineoplastic agents than the parent compound, podophyllotoxin, including etoposide (Figure 18.6 [**20**]), for the treatment of small-cell lung and testicular cancers.[49] More recently, a second epipodophyllotoxin derivative, teniposide (Figure 18.6 [**21**]), has been introduced into therapy in the U.S. for use in combination chemotherapy for patients with refractory childhood acute lymphoblastic leukemia.[50] Both of these semisynthetic analogs exhibit a hematological toxicity, and thus regular blood tests are needed in the course of using such drug therapy.[5] For both etoposide and teniposide, toxicity is increased in patients with decreased serum albumin, due to decreased protein binding of the drug.[51]

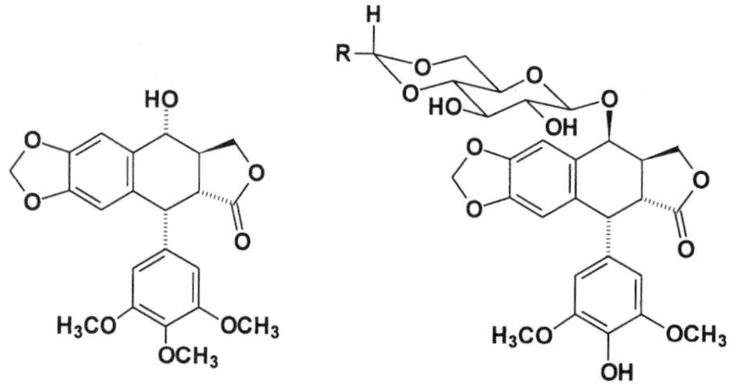

FIGURE 18.6   Structures of the lignan podophyllotoxin [**19**], and its semi-synthetic derivatives etoposide [**20**] and teniposide [**21**].

18.4.2   COUMARINS

Psoralens are photosensitizing furocoumarins (see Chapter 12) which occur in various plants in the families Apiaceae and Rutaceae, including *Ammi majus* and *Psoralea corylifera*.[6,52] These compounds have been responsible for outbreaks of photosensitization in livestock.[52] Four psoralens are used presently in the U.S. and Europe in so-called "PUVA" (psoralen and UVA light) therapy for severe psoriasis, namely, psoralen, bergapten, methoxsalen, and trioxsalen.[53] However, known side-effects of the psoralens include nausea and painful erythema.[53]

18.4.3 CANNABINOIDS

Cannabis sativa, marijuana, produces a series of biologically active terpenophenolic constituents known as cannabinoids, with the major euphoriant principle being (-)-Δ^9-*trans*-tetrahydrocannabinol (Δ^9-THC). Δ^9-THC (dronabinol), has been approved by the FDA in the U.S. to decrease nausea and vomiting in persons undergoing cancer chemotherapy.[54] Also, these compounds stimulate appetite, and thus dronabinol may be used for the treatment of wasting syndrome associated with AIDS.[6] Symptoms of toxicity associated with dronabinol include headache, malaise, and minor digestive and respiratory effects.[55] Since the marijuana plant has a bitter taste, it is rarely ingested by animals.[18] Recently, there has been considerable interest expressed in the medical and pharmaceutical literature in developing further drugs from marijuana with analgesic, anticonvulsant, antiglucoma, and other effects.[56,57]

18.4.4 QUINONES

There has been much interest in those plant secondary metabolites which have exhibited anti-HIV activity thus far in the laboratory as novel chemotype non-nucleoside leads.[58] Two of the most promising compounds in this regard are the polycyclic aromatic dimers, hypericin (Figure 18.7 [**22**]) and pseudohypericin (Figure 18.7 [**23**]), isolated from *Hypericum triquetrifolium*.[59,60] These compounds inhibit kinase C during HIV-induced CD4 phosphorylation.[61] As indicated in Chapter 13, *Hypericum* spp. are responsible for many cases of livestock poisonings, including some fatalities with hypericin being a primary photosensitizing agent.[18,52] However, Meruelo et al.[60] were able to show no toxicity of this type for mice at the low doses needed to produce antiviral effects. A small clinical trial on hypericin on human patients has been undertaken.[58]

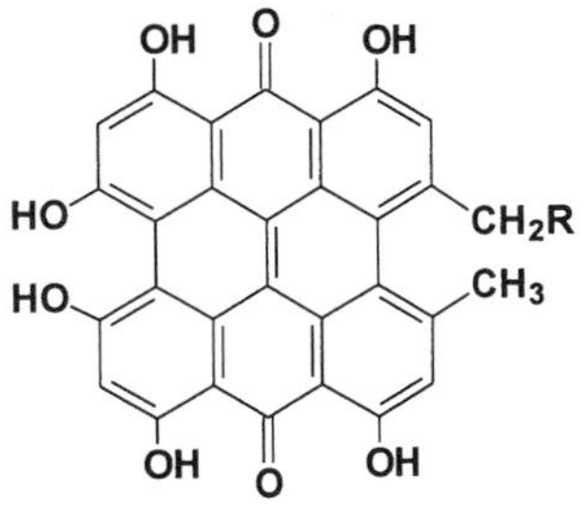

	R
22	H
23	OH

FIGURE 18.7 Structures of the quinones hypericin [**22**] and pseudohypericin [**23**].

18.5 CONCLUSIONS

Higher plants have had an important role historically as sources of prescription drugs in western medicine, and over the years there has remained a great interest in this area.[8,62–65] Although pursuing toxic principles from plants is not the only manner in which plant selection in drug discovery can be based,[11] this approach has indeed led to many plant drugs presently on the market that represent a wide range of chemical classes. As evidenced by some of the examples in this chapter, it can be seen that it has often taken many years for toxic plant secondary metabolites to be introduced to the market after their initial discovery or structure elucidation. Also, it appears that in certain cases there is a clinical role for well-known plant toxicants, so their laboratory reinvestigation may prove to be fruitful in terms of drug development. Finally, several examples have been provided herein on analogs of toxic natural product leads that have greater efficacy, more desirable physical properties, and/or less toxicity than the prototype biologically active molecule concerned.

REFERENCES

1. Lewis, W. H. and Elvin-Lewis, M. P. F., *Medical Botany: Plants Affecting Man's Health*, John Wiley and Sons, New York, 1977.
2. Der Marderosian, A. and Liberti, L. E., *Natural Product Medicine: A Scientific Guide to Foods Drugs Cosmetics*, George F. Stickley Co., Philadelphia, 1988.
3. Tyler, V. E., Brady, L. R., and Robbers, J. E., *Pharmacognosy*, 9th Edition, Lea and Febiger, Philadelphia, 1988.
4. Kinghorn, A. D. and Balandrin, M. F., Eds., *Human Medicinal Agents from Plants*, Symp. Ser. 534, American Chemical Society, Washington, DC, 1993.
5. Bruneton, J., *Pharmacognosy, Phytochemistry, Medicinal Plants*, Lavoisier Technique and Documentation, New York, 1995.
6. Robbers, J. E., Speedie, M. K., and Tyler, V. E., *Pharmacognosy and Pharmacobiotechnology*, Williams and Wilkins, Philadelphia, 1996.
7. Balandrin, M. F., Kinghorn, A. D., and Farnsworth, N. R., Plant-derived natural products in drug discovery and development: An overview, in *Human Medicinal Agents from Plants*, Symp. Ser. 534, Kinghorn, A. D. and Balandrin, M. F., Eds., American Chemical Society, Washington, DC, 1993, 2.
8. Balandrin, M. F., Klocke, J. A., Wurtele, E. S., and Bollinger, W. H., Natural plant chemicals: Sources of industrial and medicinal materials, *Science*, 228, 1154, 1985.
9. Sneader, W., *Drug Discovery: The Evolution of Modern Medicines*, John Wiley and Sons, Chichester, U.K., 1985, 435.
10. Spilker, B., 1989, *Multinational Drug Companies: Issues in Drug Discovery and Development*, Raven Press, New York, 1989, 27.
11. Kinghorn, A. D., The discovery of drugs from higher plants, in *The Discovery of Natural Products with Therapeutic Potential*, Gullo, V. P., Ed., Butterworth-Heinemann, Boston, 1994, 81.
12. Foye, W. O., Lemke, T. L., and Williams, D. A., Eds., *Principles of Medicinal Chemistry*, 4th Ed., Williams and Wilkins, Baltimore, 1995.
13. Cheeke, P. R., *Toxicants of Plant Origin. Vol. I. Alkaloids*, CRC Press, Boca Raton, Florida 1989.
14. Kinghorn, A. D., Plants as sources of medicinally and pharmaceutically important compounds, in *Phytochemical Resources for Medicine and Agriculture*, Nigg, H. N. and Seigler, D., Eds., Plenum Press, New York, 1992, 75.
15. Pelletier, S. W., *Alkaloids: Chemical and Biological Perspectives*, Vol. 1, Pelletier, S. W., Ed., John Wiley and Sons, New York, 1983.
16. Cordell, G. A., *Introduction to Alkaloids: A Biogenetic Approach*, John Wiley and Sons, New York, 1981.
17. Pammel, L. H., *Manual of Poisonous Plants*, The Torch Press, Cedar Rapids, Iowa, 1911.
18. Kingsbury, J. M., *Poisonous Plants of the United States and Canada*, Prentice-Hall, Englewood Cliffs, NJ, 1964.
19. Anonymous, Jimson weed poisoning—Texas, New York, and California, 1994, *J. Am. Med. Assoc.*, 273, 532, 1995a.
20. Anonymous, Anticholinergic poisoning associated with an herbal tea—New York City, 1994, *Morbid. Mortal. Weekly Rep.*, 44, 193, 1995b.
21. Pei-Gen, X. and Shan-Lin, F., Pharmacologically active substances of Chinese traditional and herbal medicines, in *Herbs, Spices, and Medicinal Plants: Recent Advances in Botany, Horticulture, and Pharmacology*, Vol. 2, Craker, L. E. and Simon, J. E., Eds., Oryx Press, Phoenix, 1987, 2.
22. Lewis, W. H., Plants used medically by indigenous peoples, in *Phytochemical Resources for Medicine and Agriculture*, Nigg, H. N. and Seigler, D., Eds., Plenum Press, New York, 1992, 75.

23. Klaassen, C. D., Principles of toxicology and treatment of poisoning, in *Goodman & Gilman's The Pharmacological Basis of Therapeutics*, 9th Ed., Hardman, J. G., Limbird, L. E., Molinoff, P. B., Ruddon, R. W., and Gilman, A. G., Eds., McGraw-Hill, New York, NY, 1996, 63.

24. Malkinson, F. D., Colchicine: New uses of an old, old drug, *Arch. Dermatol.*, 118, 453, 1982.

25. Insel, P. A., Analgesic-antipyretic and anti-inflammatory agents and drugs employed in the treatment of gout, in *Goodman & Gilman's The Pharmacological Basis of Therapeutics*, 9th Ed., Hardman, J. G., Limbird, L. E., Molinoff, P. B., Ruddon, R. W., and Gilman, A. G., Eds., McGraw-Hill, New York, 1996, 617.

26. Wall, M. E. and Wani, M. C., Camptothecin and analogs: From discovery to clinic, in *Camptothecins: New Anticancer Agents*, Potmesil, M. and Pinedo, H., Eds., CRC Press, Boca Raton, FL, 1995, 21.

27. Wall, M. E. and Wani, M. C., Private communication to A. D. Kinghorn.

28. Potmesil, M. and Pinedo, H., Eds., *Camptothecins: New Clinical Agents*, CRC Press, Boca Raton, FL, 1995.

29. Zhou, X. -J. and Rahmani, R., Preclinical and clinical pharmacology of Vinca alkaloids, *Drugs*, 44, 1, 1992.

30. Mancano, M. A., New drugs of 1994, *Pharmacy Times*, March, 23, 1994.

31. Olsen, J. D. and Manners, G. D., Toxicology of diterpenoid alkaloids in rangeland larkspur (*Delphinium* spp.), in *Toxicants of Plant Origin*. Vol. I. *Alkaloids*, Cheeke, P. R., Ed., CRC Press, Boca Raton, FL, 1989, 291.

32. Benn, M. H. and Jacyno, J. M., The toxicology and pharmacology of diterpenoid alkaloids, in *Alkaloids, Chemical and Biological Perspectives*, Vol. 1, Pelletier, S. W., Ed., John Wiley and Sons, New York, 1983, 153.

33. Burros, M. and Jay, S., Concern is growing over an herb that promises a legal high, *New York Times*, April 10, 1996, B1.

34. Tyler, V. E., *The Honest Herbal: A Sensible Guide to the Use of Herbs and Related Remedies*, 3rd Ed., Pharmaceutical Products Press, New York, 1993.

35. Anonymous, U.S. sees risk is "legal highs," *New York Times*, April 11, 1996, A9.

36. Klayman, D. L., *Artemisia annua*: From weed to respectable antimalarial plant, In *Human Medicinal Agents from Plants*, Symp. Ser. 534, Kinghorn, A. D. and Balandrin, M. F., Eds., American Chemical Society, Washington, DC, 1993, 242.

37. White, N. J., Artemisinin: Current status, *Trans. R. Soc. Trop. Med. Hygeine*, 88, S1/3, 1994.

38. Brewer, T. G., Grate, S. J., Peggins, J. O., Weina, P. J., Petras, J. M., Levine, B. S., Heiffer, M. H., and Schuster, B. G., Fatal neurotoxicity of arteether and artemether, *Am. J. Trop. Med. Hyg.*, 51, 251, 1994.

39. Waller, D. P., Zaneveld, L. J. D., and Farnsworth, N. R., Gossypol: Pharmacology and current status as a male contraceptive, in *Economic and Medicinal Plants Research, Vol. 1,* Wagner, H., Hikino, H., and Farnsworth, N. R., Eds., Academic Press, London, 1985, 87.

40. Lin, T. S., Schinazi, R., Griffith, B. P., August, E. M., and Eriksson, B. F. H., Selective inhibition of human immunodeficiency virus type 1 replication by the (-) but not the (+) enantiomer of gossypol. *Antimicrob. Agents Chemother.* 33, 2149, 1989.

41. Bingel, A. S. and Fong, H. H. S., Potential fertility-regulating agents from plants, in *Economic and Medicinal Plants Research*, Vol. 2, Wagner, H., Hikino, H., and Farnsworth, N. R., Eds., Academic Press, London, 1988, 33.

42. Georg, G. I., Chen, T. T., Ojima, I., and Vyas, D. M., Eds., *Taxane Anticancer Agents: Basic Science and Current Status*, Symp. Ser. 583. American Chemical Society, Washington, DC, 1995.

43. Suffness, M., Ed., *TAXOL®: Science and Applications*, CRC Press, Boca Raton, FL 1995.

44. Cragg, G. M., Schepartz, S. A., Suffness, M., and Grever, M. R., The taxol supply crisis. New NCI policies for handling the large-scale production of novel natural product anticancer and anti-HIV agents, *J. Nat. Prod.*, 56, 1657, 1993a.

45. Summerhayes, M., New drugs: docetaxel, *Pharm. J.*, 256, 125, 1996.

46. Rietbrock, N. and Woodcock, B. G., Two hundred years of foxglove therapy: *Digitalis purpurea* 1785-1985, *Trends Pharmacol. Sci.*, 71, 267, 1985.

47. Bain, R. J., Accidental digitalis poisoning due to drinking herbal tea, *Br. Med. J.*, 290, 1624, 1985.

48. Kurowski, V., Iven, H., and Djonlagic, H., Treatment of a patient with severe digitoxin intoxication by Fab fragments of anti-digitalis antibodies, *Intern. Care Med.*, 18, 439, 1992.

49. Cragg, G. M., Boyd, M. R., Cardellina II, J. H., Grever, M. R., Schepartz, S. A., Snader, K. M., and Suffness, M., Role of plants in the National Cancer Institiute drug discovery and development program, in *Human Medicinal Agents from Plants*, Symp. Ser. 534, Kinghorn, A. D. and Balandrin, M. F., Eds., American Chemical Society, Washington, DC, 1993b, 80.

50. Anonymous, *Physician's Desk Reference*, 49th Ed., Medical Economics Data Production Co., Montvale, 1995c, 694.

51. Chabner, B. A., Allegra, C. J., Curt, G. A., and Calabresi, P., Antineoplastic agents, in *Goodman & Gilman's The Pharmacological Basis of Therapeutics*, 9th Ed., Hardman, J. G., Limbird, L. E., Molinoff, P. B., Ruddon, R. W., and Gilman, A. G., Eds., McGraw-Hill, New York, 1996, 1233.

52. Johnson, A. E., Photosensitizing toxins from plants and their biologic effects, in *Handbook of Natural Toxins: Plant and Fungal Toxins*, Vol. 1, Keeler, R. F. and Tu, A. T., Eds., Marcel Dekker, Inc., New York, 1983, 345.

53. Guzzo, C. A., Lazarus, G. S., and Werth, V. P., Dermatological pharmacology, in *Goodman & Gilman's The Pharmacological Basis of Therapeutics*, 9th Ed., Hardman, J. G., Limbird, L. E., Molinoff, P. B., Ruddon, R. W., and Gilman, A. G., Eds., McGraw-Hill, New York, 1996, 1593.

54. Nieforth, K.A. and Gianutsos, G., Central nervous system stimulants, in *Principles of Medicinal Chemistry*, 4th Ed., Foye, W. O., Lemke, T. L., and Williams, D. A., Eds., Williams and Wilkins, Baltimore, 1995, 270.

55. Brunton, L. L., Agents affecting gastrointestinal water flux and motility; emesis and antiemetics; bile acids and pancreatic enzymes, in *Goodman & Gilman's The Pharmacological Basis of Therapeutics*, 9th Ed., Hardman, J. G., Limbird, L. E., Molinoff, P. B., Ruddon, R. W., and Gilman, A. G., Eds., McGraw-Hill, New York, NY, 1996, 917.

56. Gray, C., Cannabis—the therapeutic potential, *Pharm J.*, 254, 771, 1995.

57. Grinspoon, L. and Bakalar, J.B., Marihuana as medicine: A plea for reconsideration, *J. Am. Med. Assoc.*, 273, 1875, 1995.

58. Kinghorn, A. D., Plant-derived anti-HIV agents, in *Anti-AIDS Drug Development: Challenges, Strategies, and Prospects*, Mohan, P. and Baba, M. Eds., Harwood Academic Publishers, Chur, Switzerland, 1995, 211.

59. Lavie, G., Valentine, F., Levin, B., Mazur, Y., Gallo, G., Lavie, D., Weiner, D., and Meruelo, D., Studies of the mechanisms of action of the antiretroviral agents hypericin and pseudohypericin. *Proc. Natl. Acad. Sci. U.S.A.*, 85, 5230, 1989.

60. Meruelo, D., Lavie, G., and Lavie, D., Therapeutic agents with dramatic antiretroviral activity and little toxicity at effective doses: Aromatic polycyclic diones hypericin and pseudohypericin, *Proc. Natl. Acad. Sci. U.S.A.*, 85, 5230, 1988.

61. Takahashi, I., Nakanishi, S., Kobayashi, E., Nakano, H., Suzuki, K., and Tamaoki, T., Hypericin and pseudohypericin specifically inhibit protein kinase C: Possible relation to their antiretroviral activity, *Biochem. Biophys. Res. Commun.*, 165, 1207, 1989.

62. Lewis, W. H. and Elvin-Lewis, M. P. F., Contributions of herbology to modern medicine and dentistry, in *Handbook of Natural Toxins Plant and Fungal Toxins*, Vol. 1, Keeler, R. F. and Tu, A. T., Eds., Marcel Dekker, Inc., New York, 1983, 785.

63. Phillipson, J. D. and Anderson, L. A., Plants as sources of new medicines, *Pharm. J.*, November 28, 662, 1987.

64. Farnsworth, N. R. and Soejarto, D. D., Global importance of medicinal plants, in *The Conservation of Medicinal Plants*, Cambridge University Press, Cambridge, U.K., 1991, 26.

65. Anonymous, Pharmaceuticals from plants: Great potential, few funds, *Lancet*, 393, 1513, 1994.

19 Aflatoxins

J. E. Smith

CONTENTS

19.1 INTRODUCTION

The aflatoxins are undoubtedly the most documented of all mycotoxins and have a wide product presence. While the aflatoxins have at times been detected in most agricultural commodities, their presence is of particular significance in corn, cottonseed, groundnuts, and treenuts.[1] Aflatoxin occurrence is significantly higher in warm humid climates although formation can occur in temperate and cooler climatic areas, particularly when there is on-farm feed storage.

Production of the aflatoxins is confined to certain strains of *Aspergillus flavus* and *Aspergillus parasiticus* as well as the newly identified species *Aspergillus nominus* . These species are ubiquitous in tropical and subtropical countries with hot and humid climates.[2] Contamination of crops by aflatoxin can occur in the field or during subsequent storage as a consequence of inadequate drying. Whereas in tropical countries such as Thailand inadequate storage conditions and a warm humid environment lead to extensive toxigenic mold growth and aflatoxin production, in the U.S. where good post-harvest storage facilities are regularly available, the problem of aflatoxin occurrence in corn largely occurs in the field.[3,4] Predominantly in the U.S., *A. flavus* contamination of corn arises in two basic ways, viz. (1) airborne and insect transmitted conidia contaminate the silks and grow into the ear; (2) insect or bird damaged kernels can become colonized with the fungus leading to high levels of aflatoxin presence.[2] Also it has been shown that drought, nutrient or temperature stressed plants are more susceptible to colonization by *A. flavus*.[5] *A. flavus* and *A. parasiticus* are weak pathogens of the reproductive organs of the susceptible plant species and are particularly aggressive in

0-8493-8551-2/97/$0.00+$.50
© 1997 by CRC Press, Inc.

the mature seed where high concentrations of oil are often present. The ecology of *A. flavus* and *A. parasiticus* in North and South America and subsequent contamination with aflatoxin has been comprehensively reviewed.[2]

However, the presence of the aflatoxins is a world-wide problem since the majority of agricultural crops in which they are naturally produced are in international commerce. In Europe, most aflatoxin presence in foods and feeds derives from the importation of contaminated agricultural produce.[6] The economics of many developing countries rely heavily on the export of specific agricultural raw materials and due to insufficient drying equipment, coupled with generally humid atmospheric conditions can lead to unacceptable levels of aflatoxins in harvested groundnuts, palm kernels, and corn. As the import regulations of most developed countries becomes increasingly stringent, this has resulted in restriction of export potential and concomitant damage to producer countries.[6]

19.2   ECONOMIC IMPACT OF THE AFLATOXINS

Each year a significant proportion of the world's grain (especially corn) and oil seed supply will be contaminated with the aflatoxins.[1,6] While much of this presence and impact will go undetected in developing countries, most developed economies now have in operation suitable analytical methods to identify and quantify the presence of the aflatoxins in a wide range of foods and feeds. The extent of aflatoxin presence and concomitant economic loss is now becoming more generally recognized.[7]

Direct economic losses due to the presence of aflatoxin-producing fungi in agricultural crops can be detected in reduced crop yields and lower quality, reduced animal performance and reproductivity capabilities, and increased disease incidence.[2,6] Indirect losses are vastly greater than had previously been considered.[7] Crop producers with aflatoxin contaminated products will incur downgrading of crops, reduced markets, increased handling and processing, and increased costs of detoxification or dilution. Food processors must also recognize increased costs related to further processing requirements, in particular, analyses and monitoring for presence of the aflatoxins. Animal producers will identify increased production costs related to veterinary requirements, seeking new feed supplies and reduced output.[1]

While to some extent all of these effects of aflatoxin occurrence have been well documented, it has been and continues to be very problematic to quantify the true financial implications. However, serious attempts are now being made to estimate the impact on net revenue of aflatoxin contamination in grain and milk production.

An approach to the prediction of the impact of mycotoxins on net revenues for grain producers has recently been explored by Canadian scientists[7] and prediction formulae devised, viz:

$$NCR = [((100-X)/100) \times AP \times PP] + [(x/100) \times AP \times PC]$$

where NCR is net cash receipts, X is percentage of crop contaminated, AP is annual grain production in metric tonnes per year, PP is average prices of premium grain in dollars per metric tonne, PC is the average price of contaminated grain in dollars per metric tonne. The net loss of cash income is estimated by calculating NCR when X is O and then subtracting the actual NCR. When such economic analyses are carried out it is clear that when aflatoxin contamination does occur at levels above the legal limit it can lead to multimillion dollar losses. In 1983, the Texas corn crop had its worst aflatoxin problem in a decade.[8] Associated with this, members of the South Region of Associated Milk Producers Inc. lost a considerable quantity of milk over a two week period due to aflatoxin-contaminated feed.

There is an extensive literature documenting the adverse effects on animal health and productivity when aflatoxin is present in the diet.[6,9] Although mycotoxins, especially the aflatoxins, can significantly influence the health of most animal species, the toxic effects are especially noticeable in assembled groups of farmed animals since their normal feeding regimes invariably include a high intake of concentrated plant-derived feed ingredients. Animals can demonstrate variable susceptibilities to the aflatoxins, especially AFB_1 depending on genetic factors (species, breed, and strain) physiological factors (age, nutrition, etc.), and environmental factors (climatic conditions, husbandry, and management).[6] Furthermore, it is becoming increasingly evident that compound animal feeds due to their range of sourcing are exposing animals to multiple mycotoxins assault.[2]

In most developed economies, it is now realized that natural contamination levels of the aflatoxins and indeed other mycotoxins are not normally occurring at levels likely to cause overt toxic syndromes but rather to induce symptoms of chronic primary mycotoxicoses and immune suppression.[1,6,7] Diagnosis of these manifestations of disease are extremely difficult and often go undetected but undoubtedly are the most common forms of mycotoxicoses in farmed animals. It is in these areas that large economic losses are occurring, e.g., reduced reproductive efficiency, reduced feed conversion efficiency, slow growth rates, etc.

With respect to the aflatoxins, poultry have undoubtedly been the most studied group of animals and observed effects include liver damage, reduced productivity and reproductive efficiency, decreased egg production and inferior egg-shell quality, downgraded carcasses, and enhanced susceptibility to disease.[10]

In swine, the aflatoxins can cause liver damage while in cattle there can be reduced weight gain, liver and kidney damage, as well as reduced milk production. Increasing efforts are now being applied to reduce and possibly eliminate the aflatoxins and other mycotoxins from feedstuffs.[11]

Recognizing the economic and health problems created by mycotoxins in many countries is clearly the first step towards implementing appropriate programs for their prevention and control. These must include not only the prevention of mycotoxin formation in agricultural products but also, where possible, their removal through decontamination or detoxification.

19.3 CHEMISTRY AND MOLECULAR BIOLOGY OF AFLATOXIN BIOSYNTHESIS

19.3.1 CHEMICAL STRUCTURES

The discovery of the aflatoxins in the 1960s had a major impact on the science of mycotoxicology leading to a much wider international awareness of toxic fungal metabolites and their deleterious effects especially on farmed animals and man. There are four naturally occurring aflatoxins B_1, B_2, G_1, and G_2 together with other aflatoxins which occur as metabolic products of microbial, animal, or human metabolic systems. Chemically, the aflatoxins are fluorescent heterocyclic compounds characterized by dihydrofuran or tetrahydrofuran moieties fused to a substituted coumarin moiety (Figure 19.1). The G aflatoxins differ chemically from the B group in that they contain a δ- lactone ring instead of a cyclopentenone ring. The occurrence of an 8,9 double bond in the form of a vinyl ether at the terminal furan ring in aflatoxin B_1 and G_1 but not B_2 and G_2 is associated with a very significant change in activity. Resulting from this small structural difference, AFB_1 and AFG_1 are carcinogenic and considerably more toxic than AFB_2 and AFG_2. Aflatoxin B_1 is by far the most toxic and carcinogenic of the aflatoxins.[12] The strong fluorescence associated with the aflatoxins has been harnessed and utilized in their detection, analysis, and regulation.[1]

19.3.2 THE AFLATOXIN BIOSYNTHETIC PATHWAY

In recent years, there have been extensive studies on the biosynthetic pathways of the aflatoxins involving the isolation and characterization of several mutants blocked in aflatoxin biosynthesis.[13] By the use of these mutants together with metabolic inhibitors and stable isotope or

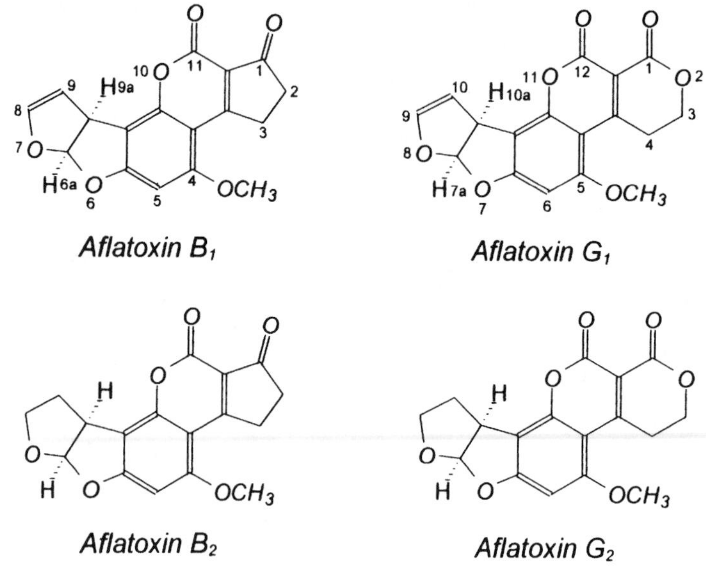

FIGURE 19.1   Chemical structures of aflatoxins B_1, B_2, G_1, and G_2.

radioisotope-labeled precursors or pathway intermediates, it has been possible to achieve the current detailed understanding of the sequence and mechanisms controlling the pathways.[14,15,16] At least 17 different enzymes are now known to be involved with several being purified to homogeneity. The main components of the biosynthetic pathway are summarized in Figure 19.2.

19.3.3   GENETIC MECHANISMS CONTROLLING AFLATOXIN BIOSYNTHESIS

The ability to clone many of the genes involved in the biosynthesis of the aflatoxins has rapidly lead to an increased understanding of the molecular biology of the pathway. This complex area of molecular biology will only be briefly summarized here and reference should be made to recent comprehensive reviews together with the 1995 Aflatoxin Elimination Workshop.[17]

Several different approaches have been used to isolate and clone aflatoxin biosynthetic genes. A genetic complementation approach isolated genes encoding three enzymes in the pathway *nor-1*, *ver-1*, and *uvm8* and one regulatory gene *aflR* and transformation systems developed.[18,19] The role of these genes in the biosynthesis of aflatoxin were confirmed by gene disruption or recombinational inactivation.[20] The use of reverse genetics to isolate genes relied on the presence of purified pathway enzymes and the production of specific antibodies to these enzymes. The *omt-1* gene encoding *O*-methylsterigmatocystin activity has been isolated from a cDNA expression library of the aflatoxin pathway in *Escherichia coli*.[21]

The genes involved in aflatoxin biosynthesis would appear to be physically clustered on one chromosome and the physical role of the genes appears to be similar to the order of the enzyme reactions catalyzed by their gene products.[22,23] The clustering of genes involved in other types of secondary metabolism would appear to be quite common.

Current ongoing research on the molecular biology of aflatoxin synthesis in the U.S. is now strongly directed to two main areas — (1) structure, function, organization, and comparative mapping of the aflatoxin genes and gene clusters in *A. parasiticus*, *A. flavus*, and *Aspergillus nidulans*, and (2) identification of molecular mechanisms which regulate pathway genes (regulatory genes; aflatoxin promoter structures and function).[14]

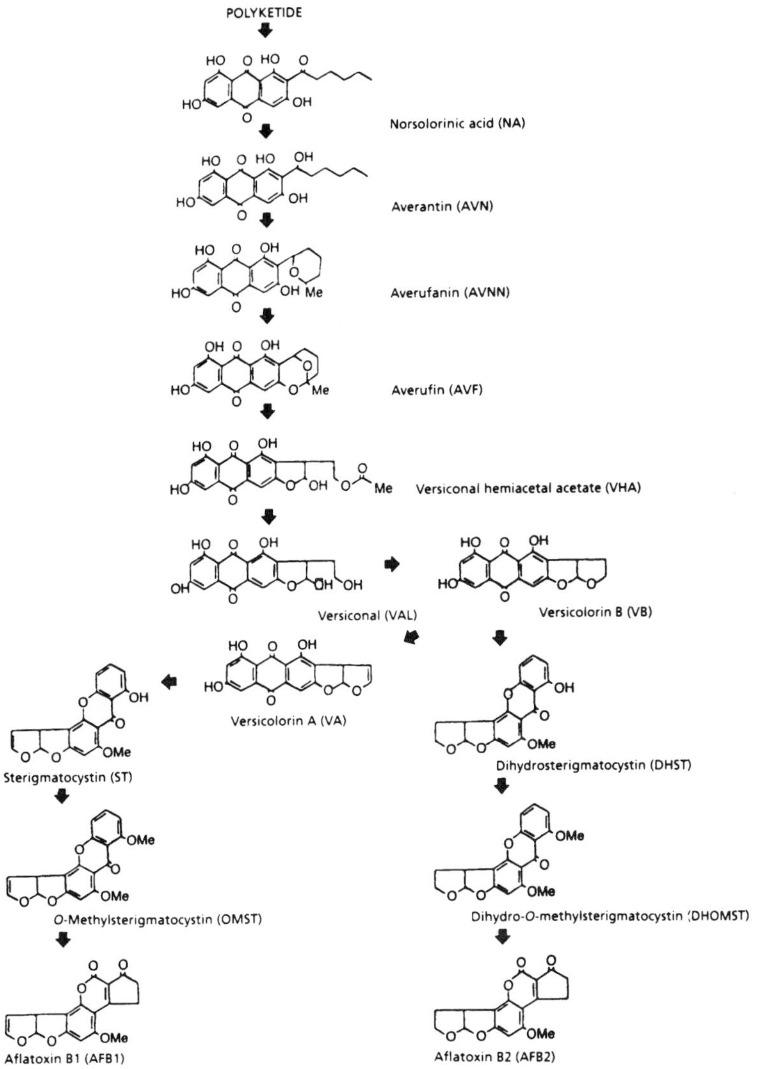

FIGURE 19.2 Aflatoxin B_1 and B_2 biosynthetic pathway. (From Trail, F., Mahanti, N. and Linz, J., Molecular biology of aflatoxin biosynthesis, *Microbiology*, 141, 755, 1995. With permission.)

19.4. AFLATOXIN METABOLISM AND DISEASE

The aflatoxins primarily enter the mammalian system by ingestion of contaminated food and feedstuffs, and absorption from the alimentary tract is considered to be the primary route of absorption.[24] Humans and animals can be exposed to aflatoxins in multiple ways.[25] While the principle point of entry will be by direct consumption of aflatoxin contaminated food or feed, other increasingly recognized ways of intake include inhalation in the form of fungal spores or mycelial fragments or grain dust, mother's milk, and finally transplacental transfer.

Aflatoxins can be acutely toxic, carcinogenic, mutagenic, teratogenic, and immunosuppressive to most mammalian species. The rank order of toxicity, carcinogenicity, etc., is $AFB_1 > AFG_1 > AFB_2 > AFG_2$ implying that the unsaturated terminal furan of AFB_1 is critical for determining the level of biological activity of the aflatoxins. The aflatoxins are mainly metabolized in the liver by microsomal mixed-function oxidases and cytosolic enzymes and most of the primary metabolites produced by these enzymes are further detoxified by conjugation with amino acids, glucuronic acid, or sulfate or bile salts and eliminated by way of urine and feces.[26] As a consequence of metabolic detoxification or hydroxylation of aflatoxin B_1, aflatoxin M_1 is formed and excreted in the milk of lactating animals and humans that have consumed aflatoxin B_1 in contaminated food and feed intake.[25,26]

Carcinogenicity and mutagenicity of AFB_1 are believed to arise from the formation of a reactive epoxide at the 8,9 position of the terminal furan (Figure 19.3). Epoxide formation permits the covalent binding of AFB_1 to cellular macromolecules such as DNA by way of isoforms of cytochrome P_{450}. Carcinogenicity is a biological problem of increasing relevance and concern to humans but of little practical significance in agricultural livestock production because of the short life span of these animals. The confirmation of aflatoxin carcinogenicity to animals *and* man is shown in Table 19.1.[12]

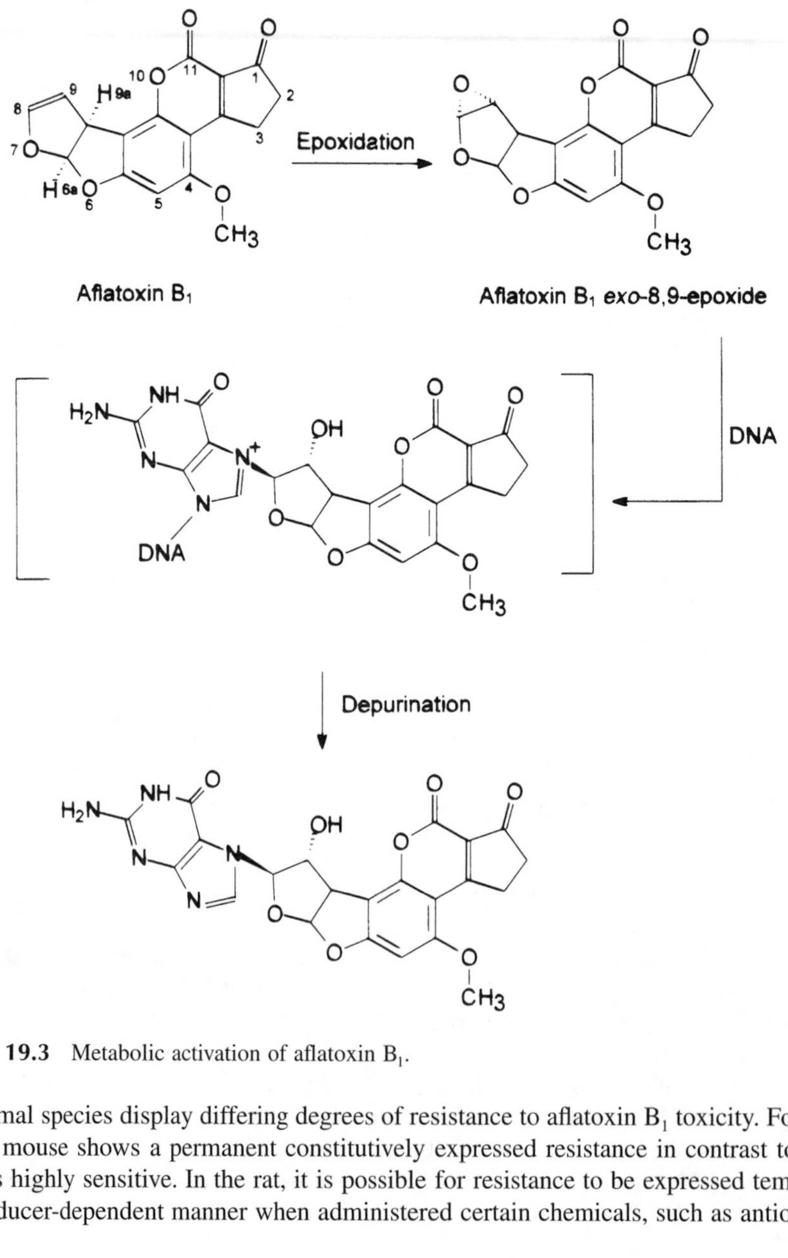

FIGURE 19.3 Metabolic activation of aflatoxin B_1.

Animal species display differing degrees of resistance to aflatoxin B_1 toxicity. For example, the mouse shows a permanent constitutively expressed resistance in contrast to the rat which is highly sensitive. In the rat, it is possible for resistance to be expressed temporarily in an inducer-dependent manner when administered certain chemicals, such as antioxidants.

TABLE 19.1
Summary of IARC 1993 Evaluation Concerning the Aflatoxins

Product	Degree of evidence of carcinogenicity		
	In man	In animals	Global evaluation
Aflatoxins (naturally occurring mixtures of)	S	S	1
Aflatoxin B_1	S	S	
Aflatoxin B_2		L	
Aflatoxin G_1		S	
Aflatoxin G_2		I	
Aflatoxin M_1	I	S	2B

Note: I, insufficient evidence; L, limited evidence; S, sufficient evidence. 1, carcinogenic to humans; 2B, possibly carcinogenic to humans.

Source: Adapted from IARC 1993.[12]

Recent studies at the MRC Toxicology Unit at Leicester, U.K., have identified enzymes involved in the resistance to aflatoxin toxicity.[27] It has been shown that glutathione S-transferase isomers containing the novel subunit Y_{C2} possess high catalytic activity towards the aflatoxin B_1-expoxide which is virtually absent from adult male rats but expressed in livers of aflatoxin-resistant antioxidant treated rats. An aldehyde reductase enzyme has also been identified in the livers of antioxidant treated rats. It is considered that the glutathione S-transferase is chemoprotective by removing the aflatoxin B_1-epoxide and the aldehyde reductase by metabolizing the aldehydic form of aflatoxin B_1-dihydrodiol to a non-protein binding metabolite[28,29] (Figure 19.4).

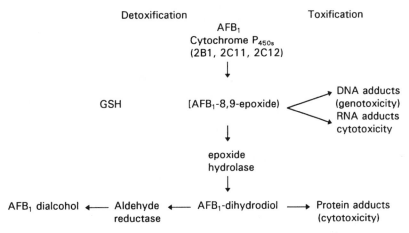

FIGURE 19.4 Toxifying and detoxifying metabolism of aflatoxin B_1. (From Neal, G. E., *MRC Toxicology Unit Biennial Report*, 20, 1993/1995. With permission.)

It is now well established that lactating animals and humans fed with aflatoxin B_1 and B_2 secrete hydrolyzed metabolites of these toxins, aflatoxin M_1 and M_2, in the milk. The percentage of conversion of the B-toxins to M-toxins varies with the animal species. It has been calculated that with a contamination concentration of 50 pg aflatoxin M_1 ml^{-1} milk, a baby of 2.5 kg at one week of age will be exposed to about 10 ng aflatoxin M_1 kg^{-1} body weight on the basis of average milk consumption. In countries where breastfeeding extends well beyond one year, e.g., in the tropics, where aflatoxin contamination is highest, this must constitute a serious health impediment for the developing child.[30]

The possibility of aflatoxin accumulating in the fetus when exposed to such toxins *in vitro* must now be beyond doubt, since the M aflatoxins have been measured in cord blood and maternal blood samples.[31,32] Furthermore, since it is generally recognized that fetuses are usually low in levels of hepatic microsomal monooxygenase activity, it could well imply that a toxic insult by aflatoxin could lead to metabolic disfunctions such as carcinogenesis and immune disturbances. In tropical areas, there may be considerable damage to child health as a result of high aflatoxin exposure prenatal and in early childhood.

Establishing a causal relationship between aflatoxin exposure and human disease continues to be a complex phenomenon because of the many uncertainties associated with human epidemiological studies. Assessing the extent of human exposure continues to be a major impediment, but confirmation of aflatoxin exposure can now be better followed through the development of rapid, sensitive, and highly accurate methods for the identification of adducts formed with aflatoxin. In this way the presence of such adducts in blood and urine now makes it possible to assess actual dietary intake of aflatoxin.[34]

The immune system of mammals plays an indispensable role in the defense against infection and tumor formation. The main effects of the aflatoxins would appear to involve cellular immune phenomenon and nonspecific humoral factors associated with immunity causing thymic aplasia, inhibitory phagocytosis by macrophages, delayed cutaneous hypersensitivity, lymphocyte proliferation, and leukocyte migration.[26]

While aflatoxins are now recognized as animal and more recently human carcinogens,[12,35] they have also been shown to (1) increase neonatal susceptibility to infection and jaundice, (2) to be involved in the etiology of kwashiorkor, (3) compromise the immune response to prophylactic immunization of animals, (4) increase childhood susceptibility to infection and malignant diseases, and (5) play a role in the pathogenesis of diseases in heroin addicts.[26,36-39]

19.5. THE CONTROL OF AFLATOXIN IN FOODS AND FEEDS

19.5.1 STRATEGY

Any strategy to control the aflatoxins and indeed other mycotoxins in foods and feeds should incorporate the following:[39]

1. An understanding of the ecology and etiology of aflatoxin production within a range of pre-and post-harvest ecosystems.
2. The collection and interpolation of surveillance data that identifies the nature and extent of the aflatoxin contamination in foods and feeds.
3. The introduction of treatments to reduce the incidence of mycotoxins.

In the developed world and especially in the U.S. and EC the aflatoxins are the only mycotoxins that are formally and jurisdictionally regulated. This has arisen from their regular occurrence and now accepted animal and human carcinogenicity.[12] Yet, there are still no satisfactorily presentable data based on toxicological judgment of the aflatoxins to form an experimentally valid basis for a regulation for a maximum permitted level in foods and feeds. The need for legislation imposing limits to the concentration of mycotoxins in foods and feeds have largely derived from studies on the aflatoxins and many overviews of regulations and their rationales have been published.[6,40,41]

Aflatoxin has been viewed as an unavoidable contaminant in food and feed when good manufacturing practices have been performed. In the U.S., the FDA has regulated aflatoxin under the Food Drug and Cosmetic Act Section 402(a)(1), viz. "The FDA has established more specific guidance on acceptable levels of aflatoxin in human foods and animal feeds by establishing action levels permitting the removal of a violative lot from interstate commerce (Table 19.2)." Within the EC, individual member states have already adopted regulations and recommendations respecting maximum permitted levels of aflatoxin B_1 and M_1. Because of some variation between states, EC efforts are now being directed towards harmonizing these existing levels (Table 19.3).[42]

TABLE 19.2
FDA Guidance Levels for Total Aflatoxins in Livestock Feeds and Human Foods

Class of Animal	Feed	Total AF Level mg kg^{-1}
Finishing beef cattle	Corn and peanut products	0.3
Beef cattle, swine, poultry	Cotton seed meal	0.3
Finishing swine over 100 lbs	Corn or peanut products	0.2
Breeding cattle, breeding swine and mature poultry	Corn and peanut products	0.1
Immature animals	Animal feeds and ingredients excluding cottonseed meal	0.02
Dairy animals and others not listed above	Animal feeds and ingredients	0.02
Humans	Milk	0.0005 (AFM$_1$)
	Any food except milk	0.02

TABLE 19.3
EC Guidance Levels for Aflatoxins in Feeding Stuffs and Human Foods

Substances, Products	Feedingstuffs	Maximum Contents in mg kg^{-1} of Unadulterated Matter at Moisture Content of 12%
Aflatoxin B$_1$	Straight feedingstuffs with the exception of: Groundnut,	0.05
	copra, palm kernel, cotton seed, babassu, maize, and products derived from the processing thereof	0.02
	Complete feeding stuffs for cattle, sheep, and goats (with the exception of complete feedingstuffs for calves, lambs, and kids)	0.05
	Complete feedingstuffs for pigs and poultry (with the exception of young animals)	0.02
	Other complete feedingstuffs	0.01
	Complementary feedingstuffs for cattle, sheep and goats (with the exception of complementary feedingstuffs for dairy animals, calves, and lambs)	0.05
	Complementary feedingstuffs for pigs and poultry (with the exception of young animals)	0.03
	Other complementary feedingstuffs	0.005

Human Foods

Substances, Products	Feedingstuffs	Maximum Contents
Aflatoxin B$_1$	Cereals, nuts, nut products, and dried fruits (mostly imported)	0.002–0.004
	Babyfood and food for infants	0.001–0.002
Aflatoxin M$_1$	Milk/milk products	0.0005
	Baby foods and food for infants	0.0005

19.5.2   INNOVATIONS IN AFLATOXIN ANALYTICAL PROCEDURES

Efficient sampling, sample preparation, and methods of analysis have been the continuing goals for identifying the nature and extent of aflatoxin presence in raw agricultural products, feeds, and foods and to form the basis of quality control procedures. The analytical procedures for isolating the aflatoxins from complex biological matrices and subsequent separation and purification follows well established flow patterns, viz. sampling, extraction, clean-up, separation, detection, quantification, and finally confirmation. While there are now well established protocols for each stage of the analytical procedures which have been adopted and incorporated into legislation, modifications are continuously being devised permitting more accurate and reproducible identification of the aflatoxins in natural products and mammalian body fluids.[6,43]

Central to any analytical process for the aflatoxins is the design and efficacy of the original sampling plan since this is invariably impeded by the highly positively skewed distribution of the aflatoxins in the raw organic products to be analyzed. All product sampling steps must be carried out as accurately as possible so that the final samples to be chemically analyzed are truly representative of the batch under investigation. When inappropriate or biased samples are used this will obviously invalidate the resultant analyzed data. The complexities associated with the design of acceptable sampling protocols for the aflatoxins have recently been comprehensibly reviewed by Coker et al.[39] who have examined these differing but complementary approaches, viz.

1. The use of theoretical probability models
2. The use of statistical models to express the variance of the sampling tests results
3. Sampling plans for problematic commodities

They recommend that single, composite samples of unprocessed kernels (e.g., groundnuts, corn, oil palm) should be within the range 10–30 kg (composed of approximately 100 incremental samples), whereas samples of processed commodities (e.g., oilseed cakes and meals) need samples of approximately 5 kg (composed of approximately 50 incremental samples). There is also a considerable need to develop sampling plans for finished foods to measure the distribution of aflatoxin among individual retail packages and so give added consumer safety. Repackaging of foods derived from temperate countries in warm humid climates can lead to secondary *A. flavus* infection and aflatoxin formation in previously clean cereal baby products (personal observation).

Aflatoxin purification from commodities has been improved in recent years by the use of solid phase extraction columns which are quick, solvent-efficient and very economical.[6] The analyte can be eluted into a small solvent volume to then be injected into a liquid chromatograph with possible automation for large scale sample handling. Reverse phase liquid chromatography is increasingly being used and precolumn derivatization with trifluoroacetic acid has been used in several commodities after clean-up using multifunctional column filtration. These multifunctional cleanup columns provide an extremely rapid removal of interferences from food extracts by lysophilic and charged sites present in the proprietary packing material.[44] The aflatoxins pass through the column and can be quantified by HPLC with precolumn or postcolumn derivatization.[46-50]

The production of well-characterized monoclonal antibodies has played an important role in establishing immunochemical methods for aflatoxin analysis. The use of such specific antibodies in competitive enzyme immunoassays either as microtiter plate assays or membrane-based quick tests as well as for the production of immunoaffinity columns is now widely practiced for total aflatoxins, aflatoxin B_1 and aflatoxin M_1. The range of immunochemical techniques now available are increasingly used at different stages of an integrated analytical system for aflatoxin determination in feed and food systems. Monoclonal antibodies for the aflatoxins have provided the advantage of a permanent supply of immunoreagents with identical properties. Their current wide use in immunoaffinity columns and in test strips arises from their high capacity on the one hand and their low background color development on the other hand. Aflatoxin-specific antibodies in clean-up columns are now widely used as a substitute for conventional physico-chemical methods of clean-up, allowing them a direct quantification by TLC or HPLC.[6]

The development and improvement of analytical methodology and sampling plans for the aflatoxins has been greatly improved by the increased availability of matrix matched certified reference materials (CRMs). The type of matrix CRMs and concentration of the mycotoxin is based on the natural occurrence pattern of the toxin in specified foods and feeds. Pure reference aflatoxins rather than CRMs are used when calibrating rapid procedures and/or sensors for the detection of the aflatoxins.

CRMs certified for their aflatoxin B_1, B_2, G_1, G_2 and total aflatoxin content in peanut butter (CRM No 282 and 401) aflatoxin M_1 content in full-cream milk powders (CRM No. 282–285) and aflatoxin B_1 content in feed CRMs (CRM No., 262–264, 375, and 376) have been prepared and certified (Table 19.4). The recent availability of suitable CRMs, while being a prerequisite for the implementation of regulation and standards, will also be invaluable in many ways for the validation of new methods, solving trade disputes and for harmonizing proficiency schemes.[53,53]

Throughout the world, there are statutory limits or in some cases advisory guidelines for the maximum levels of the aflatoxins permitted in foods and feeds. Such limits have often been set at surprisingly low levels when there is ample evidence of the difficulties encountered in aflatoxin sampling and analysis. It is increasingly recognized that aflatoxin analysis requires considerable skill and attention to detail to achieve consistent results (for example 1 µg kg^{-1} aflatoxin is equivalent to less than the weight of one kernel in 1000 tons of nuts).[54,55]

While consistency can be more easily achieved within an individual laboratory, there still exists considerable interlaboratory variability when analyzing at threshold limits.[56] In many cases, current limits are set close to the limits of performance of available analytical methods.[56]

19.5.3 PRE-HARVEST PREVENTION OF AFLATOXIN FORMATION

Historically, there have been extensive efforts to prevent aflatoxin formation in agricultural commodities at farm level incorporating better management protocols such as improved irrigation techniques and fertilizer additions, pest control, fungus resistant varieties of crop plants, and crop rotation. While many approaches are now available to control fungal plant pathogens in the field, they have mostly not been particularly successful in the field control of aflatoxin-producing fungi infecting peanuts, cottonseed, and corn. Novel approaches are now increasingly being revised, and of particular significance is the attempt to replace afla-toxigenic strains of *A. flavus* and *A. parasiticus* in the field with nonaflatoxigenic strains. A competitive, nonaflatoxin producing strain of *A. parasiticus* has been quite successful in replacing wild toxigenic strains of this fungus in peanut test plots without significantly increas-ing the level of the fungus present in the soil but with a marked reduction in aflatoxin levels. In a similar way, the aflatoxin producing capability of endemic *A. flavus* populations infecting cotton plant has been reduced through application of nontoxigenic strains.[17,57] How well such toxigenic strains will continue to survive in nature is currently under study.

19.5.4 POST-HARVEST CONTROL OF AFLATOXIN FORMATION

There are continued improvements of post-harvest techniques such as drying practices and storage conditions to ensure safe levels of moisture content of agricultural commodities. Again in the tropics this still continues to be a source of continued problems where inadequate drying and humid conditions facilitate aflatoxin formation.

TABLE 19.4

Overview on the Different CRMs and RMs for Aflatoxin Analysis and Corresponding Certified Values

CRM No. or RM No.	Matrix	Mycotoxin	Certified value mass fraction or mass concentration ($\mu g\ kg^{-1}$)	Uncertainty mass fraction or mass concentration ($\mu g\ kg^{-1}$)
CRM385	Peanut butter	Aflatoxin B_1	7.0[d]	±0.8
		Aflatoxin B_2	1.1[d]	±0.2
		Aflatoxin G_1	1.7[d]	±0.3
		Aflatoxin G_2	0.3[d]	±0.2
		Total aflatoxins	10.1[b]	±1.5[c]
CRM401	Peanut butter	Aflatoxin B_1	<0.2[d]	—[e]
		Aflatoxin B_2	<0.2[d]	—
		Aflatoxin G_1	<0.3[d]	—
		Aflatoxin G_2	<0.2[d]	—
		Total aflatoxins	<0.9[d]	—
CRM282	Full-cream	Aflatoxin M_1	<0.05[a]	—
CRM283	Full-cream	Aflatoxin M_1	0.09[d]	+0.04
CRM284	Milk powder	Aflatoxin M_1	0.31[d]	±0.06
CRM285	Milk powder	Aflatoxin M_1	0.76[d]	±0.05
CRM262	Defatted peanut meal	Aflatoxin M_1	<3[d]	—
CRM263			43.3[d]	±2.8
CRM264			206[d]	±13
CRM375	Compound feed		<1[d]	—
			9.3[d]	±0.5
RM423	Chloroform	Aflatoxin M_1	Study planned	

[a] probable content is in the range of 0.01 to 0.02 $\mu g\ kg^{-1}$.

[b] calculated by linear addition of the mean of means obtained for the four aflatoxins.

[c] calculated by quadratic addition of the variances obtained for the four aflatoxins and additional application of a safety factor.

[d] includes allowances for incomplete recovery during certification measurements.

[e] not given due to the certification of a "less than" value.

Source: From Boenke, A., *Nat. Toxins*, 2, 243, 1995. With permission.

When physical methods of preservation are not suitable, increased uses of chemical preservatives becomes essential. When such compounds are used at correct internationally agreed levels, these compounds can successfully restrict the growth of aflatoxin producing fungi. Propionic acid and sorbic acid have the greatest commercial use.[6] However, it must be noted that inadequate levels of these fungistatic preservatives can lead to increased levels of aflatoxin in products.[58]

19.5.4.1 Decontamination and Detoxification

There are as yet no totally safe and reliable methods available that can guarantee aflatoxin-free agricultural crops. Thus, when aflatoxin contamination is demonstrated in a crop, e.g., peanuts, cottonseed, or corn, what can be done?

There have been many strategies directed to the reduction or detoxification of aflatoxins in grains and oilseeds with varying levels of effectiveness. In practice there are five main approaches, viz: (1) food and feed processing; (2) biocontrol and microbial inactivation;

(3) dietary modification and chemoprotection; (4) chemical degradation, and (5) reduction in toxin bioavailability by selective chemisorption with clays.[6]

In practice, detoxification processes or decontamination will involve degradation, destruction, and/or inactivation of aflatoxin.[59] Detoxification while reducing the concentration of aflatoxin to safe levels must not produce toxic degradation products or reduce the nutritional or palatability properties of the commodities. Furthermore, the method should be simple, inexpensive, and possibly use existing technology. While many methods have been developed to detoxify aflatoxins, only a relatively small number offer realistic hope for practical application. As yet no FDA or EC fully approved methods exist for aflatoxin detoxification in human foods.[1,6] To date, ammonia either in the gaseous form or as an ammonium hydroxide solution has been used at various temperatures, pressures, moisture contents, and reaction times to degrade aflatoxins in various animal feedstuffs. There have been extensive studies using two processes, viz. (1) high pressure and high temperature, and (2) atmospheric pressure and ambient temperature, and it has been possible to reduce aflatoxin levels in the contaminated products by up to 99% (Table 19.5).[60,61]

TABLE 19.5
Parameters and Application of Ammonia/Aflatoxin Decontamination Procedures

Temperature	Process High Temperature High Pressure	Ambient Temperature Atmospheric Pressure
Ammonia level (%)	1–4	1.8–2
Pressure (PSI)	45 (3 bars)	Atmospheric
Temperature (°C)	80–120	Ambient
Duration	0.5–3.0h	14–21 days
Moisture (%)	14–20	17
Commodities	Whole cottonseed, cottonseed meal, peanut meal, peanut cakes	Whole cottonseed, Corn
Application	Feed mill	Farm

The exact chemical nature of the reaction products resulting from the ammoniation detoxification of aflatoxin has yet to be elucidated.

A major EC program is currently in progress to hopefully provide definitive information on the value, usefulness, and safety of the ammoniation process for aflatoxin degradation. In a multicenter operation, the following aspects are being covered:

1. Extraction and identification of decomposition products and the mechanisms involved.
2. Assessment of the toxicity of decontaminated feed for larger animals.
3. Assessment of the risk for the consumer of products derived from animals fed with decontaminated feed.

Biological detoxification or the biotransformation or degradation of aflatoxin by microbial systems to a metabolite(s) that is either nontoxic when ingested by animals or less toxic than the original toxin and readily excreted from the body is being studied in several laboratories.[59] As yet, such methods do not as yet constitute a realistic practical approach.

Of particular recent interest has been the application of phyllosilicate clays (hydrated sodium calcium aluminosilicate, or HSCAS).[62] Phyllosilicate clays are made up of layer-lattice silicates and chain silicates with repeating layers or chains of divalent or trivalent cations such as aluminums held in octahedral coordination with oxygens and hydroxyls and silicas tetrahedrally coordinated with oxygens and hydroxyls. HSCAS clays have been shown to chemisorb aflatoxins in aqueous suspensions, including milk, reduce the uptake of aflatoxin by blood and its distribution to body organs such as the liver, reduce the transmission of aflatoxin M_1 to milk in lactating dairy goats and cattle, and to decrease the toxic effect of aflatoxin to many animal species. It is considered that the mechanism of action of HSCAS when added to the feed intake is the selective chemisorption of the aflatoxins in the gastrointestinal tract of animals resulting in a noticeable reduction in the bioavailability of the aflatoxins. However, it will be essential to ensure that HSCAS does not cause adsorption and therefore loss of trace minerals and water soluble vitamins from the animal diet.[6]

19.5.5 APPLICATION OF MOLECULAR BIOLOGY TO AFLATOXIN ELIMINATION

An understanding of the molecular factors controlling aflatoxin biosynthesis in host plant cells will certainly be the prelude to the development of biocontrol strategies and to the development of crop lines which could be highly resistant to aflatoxin-producing fungi.

Preharvest reduction or elimination of aflatoxin presence in food plants could be accomplished at two levels, viz. (1) by blocking the molecular interaction of the toxigenic fungus with the host plant, and (2) inhibiting the synthesis of aflatoxin by the producer fungus.[14]

The identification of constitutive and *de novo*-produced anti-fungal compounds in crops subject to aflatoxin contamination strongly suggests that endogenous resistance mechanisms exist that could be enhanced by conventional plant breeding or by utilizing genetic manipulation approaches. Compounds of current interest include sesquiterpenoids such as cadalens and their oxidized products stilbenes and lacililenes. Genes for potential antifungal hydrolases such as glucanases and chitinases and for biosynthetic enzymes catalyzing the synthesis of specific photodexins have been cloned and could be utilized to introduce resistance into susceptible crop plants. Future strategies will almost certainly involve cloning and amplifying native antifungal genes or the incorporation of new, non-host inhibitor genes for antifungal substances into important agricultural crops such as corn, peanut, and cotton seed by genetic manipulation methods. Aflatoxin gene/GUS reporter constructs have proved to be valuable tools for identifying plant compounds which inhibit or stimulate fungal infection, growth, or toxin formation.[63]

Aflatoxin synthesis has now been shown to be carried out by enzymes and regulatory proteins (transacting factors, TAF), encoded by at least 16 genes. Most of these genes have now been localized to gene clusters in *A. flavus* and *A. parasiticus* and their functions determined by molecular biology techniques. These genes will now be utilized to identify possible targets suitable for inhibition of fungal growth and aflatoxin biosynthesis. Expression of genes within a cell is a multistep process which includes transcription, RNA processing, and transport, translation, post-translational processing, and localization. Thus, blocking the expression of a gene at any stage will prevent its functioning by blocking the expression of genes involved in AFB_1 synthesis which would lead to critical reduction in essential toxic pathway intermediates.

These concepts on the genetic mechanism controlling aflatoxin biosynthesis and regulation have recently been authoritatively reviewed[14,16] and further expanded at the Aflatoxin Elimination Workshop 1995.[17]

REFERENCES

1. CAST (Council for Agricultural Science and Technology), *Mycotoxins: Economic and Health Risks*, Ames: 1A. Task Force Report No. 116, 1-91, 1989.

2. Millar, J. D., Fungi and mycotoxins in grain: implications for stored product research. *J. Stored Prod. Res.*, 31, 1, 1995.

3. Payne, G. A., Aflatoxin in maize, *Curr. Rev. Plant Sci.*, 10, 423, 1992.

4. Siriacho, P., Tamboon-Ek, P., and Buanzguwon, D., Aflatoxin in maize in Thailand, *ACIAR Proceedings*, 36, 187, 1991.

5. Diener, U. L., Cole, R. J. and Hill, R. A., Epidemiology of aflatoxin formation by *Aspergillus flavus*, *Ann. Rev. Phytopathol.*, 25, 249, 1987.

6. Smith, J. E., Lewis, C. W., Anderson, J. G. and Solomons, G. L., *Mycotoxins in Human Nutrition and Health*, EUR Report No. 16048, EN 1994, Luxembourg, European Commission.

7. Charmley, L. L., Trenholm, H. L., Prelusky, D. B., and Rosenberg, A., Economic losses and decontamination, *Nat. Toxins*, 3, 199, 1995.

8. Nichols, T. E., Economic effects of aflatoxin in corn, in *Aflatoxin* and *Aspergillus flavus* in corn, Diener, U., Asquith, R., and Dickens, J., Eds., Southern Coop Service Bulletin, Auburn AL, 279, 67, 1983.

9. Smith, J. E. and Henderson, R. S. (Eds.), *Mycotoxins and Animal Feeds*, CRC Press, Boca Raton, FL.

10. Wyatt, R. D., Poultry, in *Mycotoxins and Animal Foods*, Smith, J. E. and Henderson, R. S., Eds., 1991, 553, CRC Press, Boca Raton, FL.

11. Raisbeck, M. F., Rottinhaus, G. E., and Kendell, J. D., Effects of naturally occurring mycotoxins on ruminants, in, *Mycotoxins and Animal Foods,* Smith, J. E. and Henderson, R. S., Eds., 1991, 553, CRC Press, Boca Raton, FL.

12. IARC, IARC Monographs on the Evaluation of Carcinogenic Risks to Humans, V.56, *Some Naturally Occurring Substances: Food Items and Constituents, Heterocyclic Aromatic Amines and Mycotoxins*, Lyon, France, 1993.

13. Bennett, J. W. and Papa, K. E., The aflatoxigenic *Aspergillus* spp., *Adv. Plant Pathol.*, 6, 265, 1988.

14. Trail, F., Mahanti, N., and Linz, J., Molecular biology of aflatoxin biosynthesis, *Microbiology*, 141, 755, 1995.

15. Yebe, K., Matsuyama, Y., Ando, Y., Nakajuma, H., and Hamasaki, T., Stereochemistry during aflatoxin biosynthesis: conversion of norsolorinic acid to averngin. *Appl. Environ. Microbiol.*, 59, 2486, 1993.

16. Bhatnager, D., Payne, G., Linz, J. E., and Cleveland, T. E., Molecular biology to eliminate aflatoxins, *INFORM*, 6, 262, 1995.

17. Aflatoxin Elimination Workshop. Robins, J. F. (Ed.). Atlanta, Georgia, October 23–24, 1995.

18. Spory, C. D., Chang, P.-K., and Linz, J. E., Regulated expression of the *nor-1* and *ver-1* genes associated with aflatoxin biosynthesis, *Appl. Environ. Microbiol.*, 59, 1642, 1993.

19. Woloshuk, C., Serp E., Payne, G., and Adkins, C., Gene transformation system for the aflatoxin-producing fungus *Aspergillus flavus*, *Appl. Environ. Microbiol.*, 55, 86, 1989.

20. Liang, S.-H., and Linz, J. E., Structural and functional characterisation of the *ver-1* and proteins from *Aspergillus parasiticus* associated with the conversion of versicolorin A to sterigmatocystin in aflatoxin biosynthesis, *Proceedings of Current Issues in Food Safety*, National Food Safety Toxicology Centre, Michigan State University, 1994.

21. Yu, J., Cary, J. W., Bhatnager, D., Cleveland, T. E., Keller, N. P., and Chu, F. S., Cloning and characterisation of a *c*DNA from *Aspergillus parasiticus* encoding on *O*-methyltransferase involved in aflatoxin biosynthesis, *Appl. Environ. Microbiol.*, 59, 3564, 1993.

22. Trail, F., Mahanti, N., Rarick, M., Mehigh, R., Liang, S-H., Zhou, R., and Linz, J. E., A physical and transcriptional map of an aflatoxin gene cluster in *Aspergillus parasiticus* and the functional disruption of a gene involved early in the aflatoxin pathway, *Appl. Environ. Microbiol.*, 61, 2665, 1995.

23. Trail, F., Mahanti, N., Mehigh, R., Rarick, M., Liang, S.-H., Zhou, R., and Lunz, J. E., A physical and transcriptional map of the aflatoxin gene cluster and the functional disruption of a gene involved in the early part of the pathway, *Appl. Environ. Microbiol.*, 61, in press.

24. Smith, J. E., Solomons, G., Lewis, C., and Anderson, J. G., Role of mycotoxins in human and animal nutrition and health, *Nat. Toxins*, 3, 187, 1995.

25. Raisuddin, S., Toxic responses to aflatoxins in a developing host, *J. Toxicol. Toxin Rev.*, 12, 178, 1993.

26. Eaton, D. L. and Gallagher, E. P., Mechanisms of aflatoxin carcinogenesis, *Annu. Rev. Pharmacol. Toxicol.*, 34, 135, 1994.

27. Neal, G. E., An investigation of the biochemical pathways involved in resistance to aflatoxin toxicity, *MRC Toxicol. Unit Biennial Rep.*, 20, 1993/1995.

28. McLellan, L. I., Judah, D. J., Neal, G. E., and Hayes, J. D., Regulation of aflatoxin B₁-metabolising aldehyde reductase and glutathione S-transferase by chemoprotectors, *Biochem. J.*, 300, 17, 1994.

29. Ellis, E. M., Judah, D. J., Neal, G. E., and Hayes, J. D., An ethoxyquin-inducible aldehyde reductase from rat liver that metabolises aflatoxin B₁ defines a subfamily of aldo-keto reductases, *Proc. Nat. Acad. Sci. U.S.A.*, 90, 350, 1993.

30. Wild, C. P., Pineau, F. A., Montesano, R., Mutino, C. F., and Chesanga, C. J., Aflatoxin detected in human breast milk by immunoassay, *Int. J. Cancer*, 40, 328, 1987.

31. Maxwell, S. M., Apeagyei, F., de Vries, H. R., Muranmut, D., and Hendrickse, R. G., Aflatoxins in breast milk, neonatal cord blood and sera of pregnant women, *J. Toxicol. Toxin Rev.*, 8, 19, 1989.

32. Denning, D. W., Allen, R., Wilkinson, A. P., and Morgan, M. R. A., Transplacental transfer of aflatoxin in humans, *Carcinogenesis*, 11, 1033, 1990.

33. Tredger, J. M., Chabra, R. S., and Foults, J. R., Postnatal development of mixed function oxidation as measured in microsomes from small intestine and liver of rabbits, *Drug Metab. Disposal*, 4, 17, 1976.

34. Groopman, J. D., Molecular dosimetry methods for assessing human aflatoxin exposure, in *The Toxicology of Aflatoxin, Human Health, Veterinary and Agricultural Significance*, Eaton, D. L. and Groopman, J. D., Eds., Academic Press, New York, 1995, Chapt. 10.

35. Castegnaro, M. and Wild, G. P., IARC activities in mycotoxin research, *Nat. Toxins*, 3, 327, 1995.

36. El Nezami, H. S., Nicoletti, G., Neal, G. E., Donohue, D. C., and Ahokes, J. T., Aflatoxin M₁ in human breast milk samples from Victoria, Australia and Thailand, *Food Chem. Toxicol.*, 33, 173, 1995.

37. Denning, D. W., Quiepo, S. C., Altman, D. G., Makarananda, K., Neal, G. E., Camallere, E. L., Morgan, H. R. A., and Tupasi, T. E., Aflatoxin and outcome from acute lower respiratory infection in children in The Philippines, *Annu. Trop. Paed.*, 15, 209, 1995.

38. Kirby, G. M., Wolf, C. R., Neal, G. E., Judah, D. J., Henderson, C. J., Srivatanakul, P., and Wild, C. P., *In vitro* metabolism of aflatoxin B₁ by normal and tumorous liver tissue from Thailand, *Carcinogenesis*, 14, 2613, 1993.

39. Coker, R. D., Nagler, M. J., Blunden, G., Sharkey, A. J., Defize, P. R., Derksen, G. B., and Whitaker, T. B., Design of sampling plan for mycotoxins in foods and feeds, *Nat. Toxins*, 3, 257, 1995.

40. Stoloff, L., van Egmond, H. P., and Park D. L., Rationales for the establishment of limits and regulations for mycotoxins, *Food Add. Contam.*, 8, 213, 1991.

41. van Egmond, H. P. and Detcher, W. H., Worldwide regulations for mycotoxins in 1994. *Nat. Toxins*, 3, 332, 1995.

42. Commission of the European Communities, Commission Directive 92/95/EEC and Commission Directive 94/14/EC amending the annex of the Seventh Commission Directive 76/372/EEC establishing community methods of analysis for the official control of feeding stuffs. *Off. J. Eur. Comm.*, L, 154, 1992.

43. Gilbert, J., Recent advances in analytical methods for mycotoxins. *Food Add. Contam.*, 10, 37, 1993.

44. Wilson, T. J. and Romer, T. R., Use of the Mycosep multifunctional cleanup column for liquid chromatographic determination of aflatoxins in agricultural products, *J. Assoc. Analy. Chem.*, 74, G51, 1991.

46. Dietrich, R., Schneider, E., Usleber, E., and Martlbauer, E., Use of monoclonal antibodies for the analysis of mycotoxins, *Nat. Toxins*, 3, 288, 1995.

47. Groopman, J. D., Trudel, L. J., Donarhuez, P., Marshak-Rothstein, A., and Wogan, G. N., High-affinity monoclonal antibodies for aflatoxins and their application to solid-phase immunoassays, *Proc. Natl. Acad. Sci. U.S.A.*, 81, 7728, 1984.

48. Marlbauer, E. and Babler, H. M. S., Immunoaffinitatssaulen Zur Propenaufbereiteing von Lebens-und Futtermittehn fur die Aflatoxinbestimmung, *Bioforum*, 17, 19, 1994.

49. Waychik, N. A., Hinsdill, R. D., and Chu, F. S., Production and characterisation of monoclonal antibodies against aflatoxin M_1, *App. Environ. Microbiol.*, 48, 1096, 1984.
50. Candlish, A. A. G., The determination of mycotoxins in animal feeds by biological methods, in *Mycotoxins and Animal Foods*, Smith, J. E. and Henderson, R. S., Eds., Boca Raton, CRC Press 1991, 223.
51. Gilbert, J., Sherman, M., Wood, G. M., Bocuke, A., and Wagstaffe, J. P., The preparation, validation and certification of the aflatoxin content of two peanut butter reference materials. *Food Add. Contam.*, 8, 305, 1991,
52. Boenke, A., van Egmond, H. P., and Wagstaffe, P. J., Certification of the aflatoxin B_1 content in animal feed reference materials and intercomparison with supercortical fluid extraction (SFE). *Fresenius, J. Anal. Chem.*, 345, 1993.
53. Boenke, A., BCR and MBT-activities in the area of mycotoxin analysis in food and feedstuffs, *Nat. Toxins*, 3, 243, 1995.
54. Friesen, M. D. and Garren, L., International mycotoxins check sample programme,. Part II, Report on laboratory performance for determination of aflatoxin M_1 in milk. *J. Assoc. Off. Anal. Chem.*, 65, 864, 1982.
55. McKinney, J. D., Analyst performance with aflatoxin methods as determined from AOCS Sundley check sample program: short-term and long-term views. *J. Assoc. Off. Anal. Chem.*, 67, 25, 1984.
56. Horwitz, W., Albert, R., and Nesheim, S., Reliability of mycotoxin assays: an update. *J. Assoc. Off. Anal. Chem. Int.*, 76, 461, 1993.
57. Cotty, P. J., Bayman, P., Egel, D. S., and Elias, D. S., Agriculture, aflatoxins and *Aspergillus*, in *The Genus Aspergillus*, Powell, K. A., Fenwick, A., and Peberdy, J. F., Eds. Plenum Press, New York, 1994, 1.
58. Al-Hilli, A. L. and Smith, J. E., Influence of propionic acid on growth and aflatoxin production by *Aspergillus flavus* in liquid submerged and solid substrate conditions, *J. Environ. Pathol., Toxicol. Oncol.*, 11, 57, 1992.
59. Smith, J. E. and Bol, J., Biological detoxification of aflatoxin, *Food Biotechnol.*, 3, 127, 1989.
60. Jemmali, M., Decontamination and detoxification of mycotoxins, *J. Environ. Pathol., Toxicol. Oncol.*, 10, 154, 1990.
61. Coker, R. D., Jewers, K., and Jones, N. E., The detoxification of aflatoxin by ammonia: practical possibilities, *Trop. Sci.*, 25, 139, 1985.
62. Phillips, T. D., Sarr, A. B., and Grant P. G., Selective chemisorption and detoxification of aflatoxins in phyllosilicate clay, *Nat. Toxins*, 3, 204, 1995.
63. Flaherty, J. E., Weaver, M. A., Payne, G. A., and Wotoshuk, C. P., A beta-glucuronidase reporter gene construct for monitoring aflatoxin biosynthesis in *Aspergillus flavus*, *Appl. Environ. Microbiol.*, 61, 2482, 1995.

20 *Fusarium* Mycotoxins

J. P. F. D'Mello, J. K. Porter, A.M.C. Macdonald, and C. M. Placinta

CONTENTS

20.1 INTRODUCTION

Fusarium species of fungi produce a wide range of mycotoxins, of which the most important from the standpoint of animal and human health are the trichothecenes, zearalenone and its derivatives, fumonisins, moniliformin, fusarochromanones, fusaric acid, fusarins, cyclic peptides, and amino acid esters (beauvericin type). In excess of 100 trichothecenes have been isolated, characterized, and classified into four types. However, most attention has focused on Type A trichothecenes, including T-2 toxin, HT-2 toxin, neosolaniol (NEO) and diacetoxyscirpenol (DAS), and on Type B trichothecenes, comprising deoxynivalenol (DON; vomitoxin) and its 3-acetyl and 15-acetyl derivatives (3-ADON and 15-ADON, respectively), nivalenol (NIV), and fusarenon-X. The structures of the major trichothecenes and other *Fusarium* mycotoxins are presented by Flannigan.[1] In this chapter, we review the production of mycotoxins by the common species of *Fusarium* and the toxicity of a number of these fungal metabolites.

20.2 PRODUCTION OF *FUSARIUM* MYCOTOXINS

The production of mycotoxins by four *Fusarium* species is summarized in Table 20.1, compiled from investigations over the past 20 years or so.[2-28] The table is not designed to be exhaustive

TABLE 20.1
Production of Mycotoxins by Four Species of *Fusarium* Molds[a]

Fusarium species	T-2	HT-2	NEO	DAS	Ref.
		(1) Type A trichothecenes			
F. sporotrichioides	+	+	+		2
	+			+	3
	+			+	4
	+	+	+		5
	+	+	+	+	6
	+	+	+	+	7
	+	+	+	+	8
	+				9
	+	+			10
F. culmorum		+			8
F. graminearum	+	+		+	8
F. poae	+	+	+		2
	+	+		–	27
	–	+	–	+	8
	–		–	+	28
	+	+			10

	DON	3-ADON	15-ADON	NIV	ZEN	Ref.
		(2) Type B trichothecenes and ZEN				
F. sporotrichioides					+	2
	–		–		–	3
					+	5
	+		+	+		8
F. culmorum					+	11
	+				+	6
		+				12
	+	+		–	+	13
	+		+	+		8
	+				+	14
				+		15
	+		+			10
F. graminearum					+	16
	+	+			+	17
	+		+		+	3
	+					18
	+		+			19
	+		+		–	20
					+	11
	+			+	+	6
					+	21
					+	22
	+		+		+	23
	+					24
	+		+	+		8
					+	25
	+				+	14
	+					26
	+		+	+		10
F. poae				+	+	27
	+		+	+		8
	–	–		+	–	28
				+		10

[a] Key: + = confirmed presence; – = confirmed absence (below detection limit); empty cell = no data.

but rather illustrative of certain distinguishing features in the production of the commonly occurring mycotoxins. All four species synthesize a variable mixture of mycotoxins. However, it is clear that the production of Type A trichothecenes predominates in *Fusarium sporotrichioides*. Similar comments apply to *Fusarium poae*, but the evidence is less convincing in view of the confirmed absence of T-2 and NEO in two studies.[8,28] Production of Type B trichothecenes occurs principally in *Fusarium culmorum* and *Fusarium graminearum*, but *F. poae* is also a consistent producer of one of the Type B trichothecenes, namely NIV. A unifying feature is that all four species possess at least some capacity to synthesize zearalenone (ZEN). Similarly, the studies of Abramson et al.[8] indicate that fusarenon-X (FX) production occurs in all four species of *Fusarium* as well as in other species of this genus. Among the latter group, *Fusarium equiseti* is a consistent producer of FX and other Type B trichothecenes[8] as well as ZEN.[6,23] In addition to 3-ADON and 15-ADON, other metabolites of trichothecenes are also produced by various *Fusarium* species. For example, T-2 triol and T-2 tetraol are synthesized by *F. sporotrichioides*[5,7] and 4-acetyl NIV by *F. poae*.[28] A distinctive feature of trichothecene biosynthesis in *F. tricinctum* is the production of a wide array of compounds including T-2 toxin, DAS, NEO, as well as metabolites such as 15-acetoxyscirpenol and acetyl T-2 toxin.[22,29,30]

Other *Fusarium* species are also endowed with the capacity to elaborate trichothecenes and ZEN. Thus, *F. equiseti* produces HT-2, DAS, DON, 15-ADON, and FX,[8] while ZEN is a consistent metabolite.[6,8,23,31] *Fusarium avenaceum* and *Fusarium crookwellense* can synthesize DON and 15-ADON, and in addition, the latter species has been shown to produce FX, a property it shares with *Fusarium acuminatum*.[8] *Fusarium nivale* is capable of synthesizing 3-ADON from DON.[32]

At least seven *Fusarium* isolates producing ZEN may also synthesize closely related derivatives, including cis-ZEN, α- and β-zearalenol and α- and β-zearalanol.[11]

Mycotoxins other than the trichothecenes and ZEN may be synthesized by *Fusarium* species. *Fusarium moniliforme* is capable of producing the fumonisins, moniliformin and fusarin C.[1] The fumonisins comprise 6 structurally related metabolites, of which fumonisins B_1 and B_2 (FB_1 and FB_2 respectively) have been implicated in human and animal disorders.[33] Moniliformin is also produced by *F. avenaceum*,[34] *F. sporotrichioides*, *F. culmorum*,[35] and *Fusarium oxysporum*.[36] In addition, *F. oxysporum* is an established source of the hemorrhagic factor, wortmannin[37] and fusaric acid.[38]

20.3 NATURAL OCCURRENCE OF *FUSARIUM* MYCOTOXINS

The ubiquitous distribution of *Fusarium* species particularly in association with cereal plants predisposes to at least some mycotoxin contamination of grain when appropriate environmental conditions prevail. Scott[39] has provided an exhaustive survey of the global occurrence of the trichothecenes in foods, mainly cereal grains. Since that review in 1989, more evidence has emerged of world-wide contamination of foods and feeds with trichothecenes and other *Fusarium* mycotoxins.

20.3.1 Trichothecenes and Zearalenone

Scott[39] recorded values for DON in the range 0.01 to 20 µg g^{-1} in cereal grains and their by-products in Germany. Values for DAS, NIV, T-2 toxin, and HT-2 toxin varied from 0.05 to 32, 0.01 to 0.94, 0.065 to 14, and 0.1 to 10 µg g^{-1} respectively in positive samples. A more recent study with wheat in Germany[40] yielded similar levels for DON (0.004 to 20.5 µg g^{-1}), but values were markedly lower for NIV (0.003 to 0.032 µg g^{-1}), T-2 (0.003 to 0.25 µg g^{-1}) and HT-2 (0.003 to 0.02 µg g^{-1}). Levels of DAS were below the detection limit, but values for ZEN ranged from 0.001 to 8.04 µg g[-1]. As might be expected (Table 20.1), ZEN occurred with DON in a significant number of samples (20%).

In Polish wheats, Scott[39] recorded values of 0.007 to 30 µg g[-1] for DON but recent data[41] suggest higher levels in samples from central Poland, at 2 to 40 µg g[-1], while NIV was lower at 0.01 µg g[-1], compared with 0.003 to 0.35 µg g[-1] previously.[39] In addition, the co-occurrence of 3-ADON with 15-ADON was observed in the recent Polish study[41] while ZEN levels of 0.01 to 2 µg g[-1] were also reported.

In the survey of Scott,[39] concentrations of DON in feeds and grains in Finland ranged from 0.001 to 0.12 µg g[-1], but a recent study in that country[42] indicated higher values in the range 0.007 to 0.3 µg g[-1], with 3-ADON at 0.013 to 0.12 µg g[-1] and ZEN at 0.022 to 0.095 µg g[-1]. Six lots of oats contained toxic levels of DON (1.3 to 2.6 µg g[-1]).

Values up to 18 µg g[-1] for DON in corn samples in Italy were listed by Scott,[39] while ranges for other trichothecenes were: NIV, 0.08 to 0.2 µg g[-1]; T-2, 0.05 to 0.30 µg g[-1]; and DAS, 0.15 to 0.30 µg g[-1]. The occurrence of DON, ZEN, and zearalenols in corn plants infected with stalk rot caused by *Fusarium* species has also been reported recently in southern Italy.[43]

A study in the Netherlands[44] revealed a relatively high frequency of contamination of cereal grains with *Fusarium* mycotoxins, including the natural co-occurrence of DON, NIV, and ZEN. DON concentrations (µg g[-1]) varied from 0.020 to 0.231 in wheat, 0.004 to 0.152 in barley, 0.056 to 0.147 in oats, and 0.008 to 0.384 in rye. Levels (µg g[-1]) of NIV were 0.007 to 0.203 for wheat, 0.030 to 0.145 for barley, 0.017 to 0.039 for oats, and 0.010 to 0.034 for rye. For ZEN, data (µg g[-1]) ranged from 0.002 to 0.174 in wheat, 0.004 to 0.009 in barley, and 0.016 to 0.029 in oats, with a single value for rye at 0.011 µg g[-1].

In Japan, there is evidence of consistent and occasionally severe contamination of cereal grains with DON and NIV. Values in positive samples of up to 50 and 37 µg g[-1] respectively were listed by Scott.[39] More recently, Sugiura et al.[28] reported contamination of wheat in seven locations in Japan, with DON at 0.03 to 1.28 µg g[-1], NIV at 0.04 to 1.22 µg g[-1], and ZEN at 0.002 to 0.025 µg g[-1]. In a subsequent study,[45] ZEN levels of 0.010 to 0.658 µg g[-1] were reported for 14 barley samples.

Data relating to trichothecene contamination of grains and feeds in North America is voluminous, as indicated by the survey of Scott.[39] The impetus has been maintained, particularly in relation to DON,[46] with values ranging from nondetectable to 9.3 µg g[-1] in wheat grains harvested in the U.S. in 1991. Highest levels were seen in wheat from Missouri, North Dakota and Tennessee. In another survey of the 1991 harvest in the U.S.,[47] values for DON ranged from <0.1 to 4.9 µg g[-1] in winter wheat and <0.1 to 0.9 µg g[-1] in spring wheat. In the 1993 harvest, 483 wheat samples had DON concentrations ranging from <0.5 to 18 µg g[-1], with 86% of samples from Minnesota and up to 78% of samples from North and South Dakota containing levels in excess of 2 µg g[-1]. In 147 samples of barley from that harvest,[47] DON concentrations ranged from <0.5 to 26 µg g[-1]. A new advisory was issued by the FDA, recommending a maximum DON level of 1 µg g[-1] in finished wheat products destined for human consumption, and 5 to 10 µg g[-1] in animal feeds. In a study of mycotoxin contamination of grain in Atlantic Canada,[48] DON and ZEN were detected in respectively 53 to 62% and 25 to 29% of samples tested. Most of the values for DON and ZEN were equal to or less than 0.5 and 0.3 µg g[-1], respectively. Of 55 samples tested, five had levels of T-2 toxin ranging from 0.16 to 0.31 µg g[-1], two contained HT-2 toxin (0.12 and 0.44 µg g[-1]), and two were contaminated with DAS (both at 0.11 µg g[-1]).

20.3.2   FUMONISINS

World-wide contamination of corn with fumonisins has been reported,[33] with samples from Italy, Portugal, Zambia, and Benin containing $FB_1 + FB_2$ levels of up to 2.85, 4.45, 1.71, and

3.31 μg g^{-1}, respectively, and with incidence rates of 82 to 100%. In India, FB$_1$ levels of 300 to 366 μg g^{-1} have been reported in corn infected with *F. moniliforme*.[49]

20.4 FACTORS AFFECTING PRODUCTION OF *FUSARIUM* MYCOTOXINS

The production of *Fusarium* mycotoxins is affected by a diverse array of factors, broadly divisible into biological, physical, and chemical. It is recognized, however, that complex interactions among these factors may exert significant effects on mycotoxin synthesis.

20.4.1 BIOLOGICAL

Wide variations exist in toxigenic potential among different strains of a particular *Fusarium* species.[13] Nevertheless, many of the toxigenic species of *Fusarium* are also pathogenic towards cereal plants. For example, head blight of cereals may be caused by *F. graminearum* and *F. culmorum*, both of which are recognized DON producers (Table 20.1). A direct quantitative assessment of the link between head blight and DON contamination of wheat grains has recently been elucidated in the Netherlands,[13] using genotypes of differing resistance to the disease. A striking correlation was evident between the severity of head blight and DON concentrations in wheat kernels. Work in Canada also supports the view that wheat grains from cultivars susceptible to head blight contain more DON than those from Chinese genotypes which are recognized to be resistant.[10]

At harvest, seeds are likely to be colonized by different species of fungi[50] and mycotoxin production may be affected by fungal interactions, for example during storage.[21]

20.4.2 PHYSICAL

A variety of interacting physical factors may affect mycotoxin production under field conditions and during storage.[51] In a number of countries, *Fusarium* head blight has been associated with years of high rainfall. It is also now believed that rain impact may play an important role in the dispersal of *Fusarium* inoculum and in the development of head blight epidemics.[52] It is of interest that the higher than average levels of DON in wheat samples from Missouri, North Dakota, and Tennessee were tentatively attributed to increased rainfall in these states.[46]

Laboratory studies indicate that both time and water activity affect FB$_1$ production in corn kernels.[53] Maximum synthesis occurred at 21 days of incubation at a water activity of 1, but on lowering water activity to 0.95, maximum production of FB$_1$ did not occur until 47 days had elapsed. However, both laboratory and field studies[14,18] provide evidence that DON production declines with time. There is some controversy regarding the effects of temperature on ZEN synthesis in laboratory cultures, with Merino et al.[54] maintaining the need for thermic shock while others[14,31,55] have reported ZEN production at constant temperatures.

20.4.3 CHEMICAL

The chemical definition of media which promote mycotoxin production from field isolates of *Fusarium* species has been attempted. Depletion of carbohydrate appears to be an important stimulus for DON production,[3,19] while T-2 formation is enhanced by the presence of sorbic acid.[56]

As might be expected, fungicides can influence mycotoxin production, but the effects are variable and dose-dependent (Table 20.2). Insecticides may inhibit mycotoxin production in *Fusarium* molds,[16,58] but type and dose level of the insecticide are important determinants of efficacy.

TABLE 20.2
Effects of Fungicides on Production of *Fusarium* Mycotoxins

Fungicide	Methods	Effects	Ref.
Dicloran, iprodione vinclozolin	Added separately at levels of up to 500 μg ml⁻¹ potato-dextrose broth; static culture of *F. graminearum*	Dose-related inhibition of growth and production of DAS and ZEN; total inhibition of DAS and ZEN production at higher levels of each fungicide	25
Propiconazole, thiabendazole	Field trial with wheat; heads sprayed with *F. graminearum* inoculum 2 days after anthesis; fungicides applied separately prior to, during and after inoculation; propiconazole applied at 120 g ha⁻¹, thiabendazole at 360 g ha⁻¹	Propiconazole reduced *Fusarium* infection by 39–55%; DON levels reduced by 34–78%. Thiabendazole had no effect on infection level, but DON levels reduced by up to 83%	24
Propiconazole, thiabendazole, tebuconazole	Field trial with wheat inoculated with *F. graminearum* at 3 stages from beginning of flowering; fungicide rates: propiconazole, 140; thiabendazole, 280; tebuconazole, 140 g ha⁻¹	Head blight incidence ranged from 83 to 84% irrespective of fungicide (control, 87%); DON levels ranged from 12.9 to 16.7 mg kg⁻¹ (control, 12.0 mg kg⁻¹)	26
Tebuconazole with triadimenol (Matador)	Field trial with wheat; heads inoculated with *F. culmorum*; Matador applied at 11 ha⁻¹	Fungicide reduced head blight; 16-fold increase in NIV content of grain from fungicide-treated plants	15
Tridemorph	Shake-flask cultures of *F. sporotrichioides* in a defined medium; fungicide added at 6 and 36 μg ml⁻¹	At 6 μg ml⁻¹, growth enhanced but T-2 toxin and DAS production inhibited; at 36 μg ml⁻¹, growth inhibited but T-2 production stimulated	4
Carbendazim	Cultures of *F. sporotrichioides* on potato-dextrose agar; fungicide added at 0.1 and 1.0 μg ml⁻¹	Growth unaffected by fungicide; 6-fold increase in T-2 toxin production with fungicide at 0.1 μg ml⁻¹ and small increases in ZEN and NEO production	57

20.5   TOXICOLOGY

20.5.1   Risk Assessment

Food safety, animal health and productivity, and human health problems identified with fungal-contaminated grains most recently have concentrated research on *Fusarium* species and their toxic metabolites. Major world-wide agricultural problems with *Fusarium* have been recognized since: *F. moniliforme* and the fumonisins have been associated with equine leukoencephalomalacia;[59–61] the human oncologic implications with esophageal cancer in certain areas of South Africa, China, northern Italy, and possibly Iran;[62–69] and the routine occurrence of the fumonisins and other *Fusarium* toxins in corn, wheat, barley, rice, and other cereal grains.[70–73] *F. moniliforme* infection of corn can be asymptomatic and presents its greatest concerns since fungal toxins surreptitiously enter animal and human foods.[74] Although *Fusarium* species and their toxins appear to be most prevalent in corn, wheat, and barley, they are also found in nuts, fruits and vegetables, and in non-food items of economic importance (e.g., tobacco, cotton, forage grasses, alfalfa, red clover, and flax).[71,75] Corn, wheat, and barley comprise two thirds of the world cereal production and there are over 24 *Fusarium* species associated with animal and human health problems.[76] Toxigenic *Fusarium* species, therefore, are a major agricultural problem not only because of human and animal health, but also because of losses due to plant diseases, losses incurred by grain and livestock producers, and effects of contaminated grain on export/import markets.[77]

20.5.2 STRUCTURAL DIVERSITY AND BIOLOGICAL ACTIVITY

The spectrum of *Fusarium* metabolites that affect animal and human health are not only the most divergent structurally unrelated, but also are the most divergent biologically acting group of compounds known.[76,78] The fumonisins are pentahydroxyicosanes, structurally similar to sphingosine and dihydrosphingosine, the interference of which in ceramide synthesis is detrimental to cell maintenance.[60,73,79–81] DON is a 12,13-epoxytrichothecene associated with feed refusal and emetic responses primarily in swine.[75,78,82] Zearalenone is a β-resorcylic acid lactone associated with hyperestrogenic activity also in swine,[83] while the fusarins are pyrrolo-polyketides of which fusarin C is mutagenic.[84,85] Beauvericin, a cyclic hexadepsipeptide consisting of three N-methylphenylalanyl- and three 2-hydroxy-3-methyl-butyric acid residues in a continuous-alternating sequence, is reported toxic to both mammalian and insect cell lines.[86,87] Fusaproliferin is a new sesterterpene toxic to both brine shrimp larvae and human B-lymphocytes cell line.[88]

Several fumonisins have been identified from *F. moniliforme*-infected grains.[60,72,73] FB_1 is 2-amino-12, 16-dimethyl-3,5,10,14,15-pentahydroxyicosane with a propane-1,2,3-tricarboxylate substituent at C-14 and C-15. FB_2 and FB_3 are the C-10 and the C-5 deoxy analogues, respectively, whereas FB_4 lacks the hydroxyl moiety at both C-5 and C-10. In addition, fumonisin AB_1 and AB_2 have been defined as the corresponding N-acetyl-analogues of FB_1 and FB_2, respectively, and the N-acetyl-C15-keto form of FB_1 has also been reported.[89]

The C-22-aminopentol "backbone" of FB_1, commonly referred to as hydrolyzed FB_1, structurally resembles the free sphingoid bases, sphingosine and sphiganine (dihydrosphingosine).[60,73,79–81] These bases are necessary for ceramide synthesis, which is important to both mammalian and plant cell maintenance and viability.[79,80,90,91] FB_1, FB_2, and hydrolyzed FB_1 (i.e., the C-22-aminopentol backbone) are specific inhibitors of *de novo* sphingolipid biosynthesis, the primary target being sphinganine N-acetyltransferase.[70,79,81] These effects cause an abnormal accumulation of the free sphingoid bases, which are extremely detrimental to cellular function.[80] Analogous inhibitions have been reported with the structurally similar AAL toxins from *Alternaria alternata* f. sp. *lycopersici*.[90,91]

The carcinogenic potential of FB_1 in laboratory animals[62,65] and the statistical correlation of *F. moniliforme*-infected corn and FB_1 with human esophageal cancer has been established.[61–69] Additionally, the possible mechanisms of fumonisin-induced disruption of sphingolipid metabolism and carcinogenesis has been discussed in detail.[61,69] Disruption of *de novo* sphingolipid metabolism also has been correlated with *F. moniliforme* and FB_1-induced hepato- and nephrotoxicity in rats,[92,93] porcine pulmonary edema,[94] and dietary exposure to FB_1 in equidae.[95] These mechanisms are indeed consistent with FB_1-induced chronic hepatotoxicity in primates with an inference to atherosclerosis.[96] The inhibition of sphingolipid synthesis by FB_1 also reduces ganglioside synthesis and affects axonal growth in cultured hippocampal neurons.[97] Whether disruption of sphingolipid metabolism is directly related to liquefactive necrosis in the brain of equines, pulmonary edema in swine, and/or certain syndromes in poultry remains to be determined. However, increases in the sphinganine to sphingosine ratio occur long before any indication of toxicity,[92,98] and the changes in the relative amounts of free sphinganine and sphingosine may be used as a biomarker for animals consuming fumonisins.[94,95,98] Similar results have been reported in plants exposed to both fumonisin(s) and the AAL toxins.[91] Interestingly, FB_1, FB_2, and the AAL toxins have similar effects in cultured mammalian cells.[99]

Although DON is only one of several mycotoxins produced by *Fusarium* species, it is among the most frequent trichothecenes analyzed in cereal crops.[68,78] Animal and public health concerns stem from its occurrence primarily in wheat, corn, and barley and in feed and food commodities derived from these crops. Swine appear to be the most sensitive to the effects of DON; however, several cases of human toxicity have been attributed to the chronic consumption of flour/bread contaminated with DON.[68] The hazard assessments of DON (and other trichothecenes), which are based on toxicology data (i.e., inhibition of protein and DNA synthesis; reproductive, embryotoxic, and possible teratogenic effects), estimate a minimum consumption level at 0.6 mg kg^{-1} body weight (i.e., that which produces no adverse effect).[100] Current guidelines for DON in Canadian wheat are 2 mg kg^{-1} in uncleaned soft wheat used for nonstaple foods, including bran, and 1 mg kg^{-1} in soft wheat destined for infant food.[100]

ZEN and related metabolites are nonsteroidal estrogenic mycotoxins produced by several species of *Fusarium*[78,83] (Table 20.1) and associated with hyperestrogenism in swine, and reproductive problems in cattle and sheep.[83,101] In swine, hyperestrogenism is a well defined syndrome,[83] and although these animals appear to be more sensitive to ZEN and DON, poultry and ruminants seem more tolerant.[78]

20.5.3   INTERACTIONS

Guidelines regulating the amount of mycotoxins found in agricultural commodities used for both animal and human consumption are generally based on toxicologic investigations with pure compounds. Fungal contamination of food and feed, however, rarely involve exposure to a single toxin. *Fusarium* species are common pathogens of a wide variety of crop plants and there is high potential for mycotoxin contamination of many foodstuffs. Routinely, animal toxicity problems occur in which the quantity of the individual mycotoxins found in the suspected feed(s) does not explain the observed syndromes.[68,77,78,100] Currently, the combined effects of mycotoxins on animal and human health have aroused concern because synergistic activities present a unique set of problems in defining both toxicity and food safety guidelines.[77,78]

Dowd[102] has discussed in detail the benefits fungi derive from producing simple molecules that act as synergists with other more complex molecules and the probability that this synergism is widespread in nature. Some of the benefits proposed are energy efficient mechanisms for host invasion, protective mechanisms for survival, and competitive advantage. The endophytic association of *F. moniliforme* and *F. proliferatum* to the corn plant and the external and systemic association to the kernel[71,74] suggest some symbiotic relationships within the *Fusarium*-host associations. Other fungal endophytic-plant interactions that confer insect resistance, drought tolerance, and increased competition with other plants (i.e., allelochemicals) have been described in Chapter 4.

Fusaric acid (5-butylpicolinic acid) is a common metabolite of several *Fusarium* species including *F. moniliforme*, *F. oxysporum*, *F. subglutinans*, and *F. crookwellense*,[75,78,103] and its co-occurrence with other *Fusarium* toxins (e.g., ZEN, fumonisins, DON, and other trichothecenes has been reported.[104] The 10,11-dehydro; 10,11-dihydroxy-; 10-hydroxy-; and the 11-carboxy-analogues of fusaric acid have also been identified from cultures of *F. moniliforme*,[105,106] but fusaric acid appears to be the major analogue occurring in *Fusarium*-contaminated plants and grains. Although of minor toxicity at the levels detected in nature,[104,107,108] there is increasing evidence that fusaric acid may enhance the activity of other *Fusarium* toxins.[71,102-104,107,109,110] This was recognized in tests with certain insects in which DAS, DON, and T-2 toxicity were potentiated by fusaric acid.[102] Furthermore, fusaric acid enhanced the activities of certain insecticides by inhibiting oxidative enzymes responsible for toxin metabolism by insects.[102]

Bacon et al.[103] reported a toxic interaction between fusaric acid and FB_1 in the fertile chicken egg. Each mycotoxin at 5 μg per egg in sterile water produced virtually no effects on the eggs (pip) incubated for 21 days. However, a combination of fusaric acid and FB_1 at 5 μg each per egg was lethal to 46% of eggs. The lethal effects of this combination were dose-dependent and enhanced in a phosphate buffer solution.

Fusaric acid, DON, and ZEN were the major mycotoxins isolated from suspect duck and ostrich feed, and individually were below levels generally considered to cause problems in poultry.[104,108] However, in a 3-week feeding trial using day-old broiler chicks, turkey poults, and ducklings, the feed caused a significant dose-dependent growth depression and mortality, consistent with observed toxicities on both duck and ostrich farms.[104]

The lactational transfer of fusaric acid from the feed of nursing dams to the neonate rat[107] and the potentiation by this mycotoxin of the adverse effects of FB_1 in the fertile chicken egg[103] has raised concern about their synergistic effects. Porter et al.[107] observed fusaric acid concentrations in the stomach colostrum taken from 4-day old rats were directly proportional to the quantity of the mycotoxin in the feed of the nursing dams; these observations were consistent in both F_1 and F_2 generations. Additionally, in both male and female weanlings (21-day old rats), serotonin and tyrosine were decreased in the pineal gland with concurrent increases in serum and pineal melatonin in these animals.[107,111] In adult rats, parental administration of fusaric acid increased tyrosine, serotonin, and dopamine in the brain with concomitant increases of N-acetylserotonin in the pineal gland.[104] Altered neurochemical effects in the brain and pineal gland were partially attributed to fusaric acid inhibiting dopamine-β-hydroxylase and tyrosine-hydroxylase activities and indirectly on its peripheral effects on adrenal catecholamines and serotonin.[104,107] Fusaric acid also elevated serotonin in the brain of swine,[109] and the potential synergistic significance of co-occurring fusaric acid and DON has been proposed in feed refusal and emesis in these animals.[110] Aberrant changes with serotonin and melatonin and/or other neurotransmitters in the brain and pineal gland have been related to adverse effects on seasonal physiological and endocrine changes in both animals and humans.[104,107,112] Therefore, in addition to its synergistic potential, fusaric acid has diverse neurochemical effects that may negatively influence animal productivity, maturation and behavior.

Miller et al.[84] have reported the production of the fumonisins with the fusarins by *F. moniliforme* and suggested that the variable toxicity of fungal extracts may be due to this combination. Although the fusarins A through F have been isolated and characterized, fusarin C has been primarily implicated in the mutagenic activity of *F. moniliforme*.[84,85,113] Following the correlation of FB_1 (and FB_2 and FB_3) with cancer induction and promotion,[114] it is not too inconceivable that the fusarins (e.g., fusarin C) could enhance these activities.

Madahyastha et al.[113] have shown synergistic responses with a combination of T-2 and HT-2; T-2 and T-2-4ol; DON and NIV; or DON and T-2 in a yeast bioassay. Harvey et al.[115] described the synergistic effects of FB_1 and DON on reduced bodyweight gains and serum biochemical indices in growing barrows. To complicate matters even more, varying degrees of antagonism have been observed with other trichothecenes.[113]

Beauvericin is produced by *F. subglutinans*, *F. proliferatum*, and *F. moniliforme* and fusaproliferin is produced by *F. proliferatum*.[86,88] Although their role in animal and human pathology has yet to be determined, both have been reported to co-occur with FB_1 and moniliformin.[86–88]

20.6   CONCLUSIONS

Fusarium molds may produce a diverse range of mycotoxins. Since these fungi are also pathogenic to cereal plants, it is likely that grain from infected plants may become contaminated with one or more of the *Fusarium* mycotoxins. Although fungicides are generally perceived as effective agents in the control of *Fusarium* diseases, there is limited evidence to indicate that at sublethal levels some fungicides may enhance mycotoxin production. This effect may be significant in view of proposals to reduce fungicide applications for environmental reasons and also in the context of the development of fungicide resistance in *Fusarium* molds. The co-occurrence of *Fusarium* mycotoxins[116] in cereals should be considered in any assessment of risk for human and animal health since potentiating effects have been observed in mammalian systems. The co-occurrence of the fumonisins, fusarins, moniliformin, the trichothecenes (primarily DON), ZEN, and fusaric acid not only complicates toxicologic investigations but also creates major difficulties in defining tolerance limits and regulatory guidelines.

REFERENCES

1. Flannigan, B., Mycotoxins, in *Toxic Substances in Crop Plants*, D'Mello, J. P. F., Duffus, C. M., and Duffus, J. H., Eds., Royal Society of Chemistry, Cambridge, 1991, Chap. 10.
2. Szathmary, C. I., Mirocha, C. J., Palyusik, M., and Pathre, S. V., Identification of mycotoxins produced by species of *Fusarium* and *Stachybotrys* obtained from Eastern Europe, *Appl. Environ. Microbiol.*, 32, 579, 1976.
3. Miller, J. D., Taylor, A., and Greenhalgh, R., Production of deoxynivalenol and related compounds in liquid culture by *Fusarium graminearum*, *Can. J. Microbiol.*, 29, 1171, 1983.
4. Moss, M. O. and Frank, J. M., Influence of the fungicide tridemorph on T-2 toxin production by *Fusarium sporotrichioides*, *Trans. Br. Mycol. Soc.*, 84, 585, 1985.
5. Richardson, K. E., Hagler, W. M., Haney, C. A., and Hamilton, P.B., Zearalenone and trichothecene production in soybeans by toxigenic *Fusarium*, *J. Food Prot.*, 48, 240, 1985.
6. Thrane, U., Detection of toxigenic *Fusarium* isolates by thin layer chromatography, *Lett. Appl. Microbiol.*, 3, 93, 1986.
7. Marasas, W. F. O., Yagen, B., Sydenham, E., Combrinck, S., and Thiel, P. G., Comparative yields of T-2 toxin and related trichothecenes from five toxicologically important strains of *Fusarium sporotrichioides*, *Appl. Environ. Microbiol.*, 53, 693, 1987.
8. Abramson, D., Clear, R. M., and Smith, D. M., Trichothecene production by *Fusarium* spp. isolated from Manitoba grain, *Can. J. Plant Pathol.*, 15, 147, 1993.
9. Proctor, R. H., Hohn, T. M., McCormick, S. P., and Desjardins, A. E., *Tri 6* encodes an unusual zinc finger protein involved in regulation of trichothecene biosynthesis in *Fusarium sporotrichioides*, *Appl. Environ. Microbiol.*, 61, 1923, 1995.
10. Wong, L. S. L., Abramson, D., Tekauz, A., Leisle, D., and McKenzie, R. I. H., Pathogenicity and mycotoxin production of *Fusarium* species causing head blight in wheat cultivars varying in resistance, *Can. J. Plant Sci.*, 75, 261, 1995.
11. Richardson, K. E., Hagler, W. M., and Mirocha, C. J., Production of zearalenone, α- and β-zearalenol, and α- and β-zearalanol by *Fusarium* spp. in rice culture, *J. Agric. Food Chem.*, 33, 862, 1985.
12. Greenhalgh, R., Levandier, D., Adams, W., Miller, J. D., Blackwell, B. A., McAlees, A. J., and Taylor, A., Production and characterization of deoxynivalenol and other secondary metabolites of *Fusarium culmorum* (CMI 14764, HLX 1503), *J. Agric. Food Chem.*, 34, 98, 1986.
13. Snijders, C. H. A. and Perkowski, J., Effects of head blight caused by *Fusarium culmorum* on toxin content and weight of wheat kernels, *Phytopathology*, 80, 566, 1990.
14. O'Neill, K., Damoglou, A. P., and Patterson, M. F., Toxin production by *Fusarium culmorum* IMI 309344 and *F. graminearum* NRRL 5883 on grain substrates, *J. App. Bacteriol.*, 74, 625, 1993.
15. Gareis, M. and Ceynowa, J., Influence of the fungicide Matador (tebuconazole/triadimenol) on mycotoxin production by *Fusarium culmorum*, *Zeitschr. Lebensmittel-Untersuchung-Forsch.*, 198, 244, 1994.
16. Berisford, Y. C. and Ayres, J. C., Effect of insecticides on growth and zearalenone (F-2) production by the fungus *Fusarium graminearum*, *Environ. Entomol.*, 5, 644, 1976.

17. Greenhalgh, R., Neish, G. A., and Miller, J. D., Deoxynivalenol, acetyl deoxynivalenol, and zearalenone formation by Canadian isolates of *Fusarium graminearum* on solid substrates, *Appl. Environ. Microbiol.*, 46, 625, 1983.

18. Scott, P. M., Nelson, K., Kanhere, S. R., Karpinski, K. F., Hayward, S., Neish, G. A., and Teich, A. H., Decline in deoxynivalenol (vomitoxin) concentrations in 1983 Ontario winter wheat before harvest, *Appl. Environ. Microbiol.*, 48, 884, 1984.

19. El-Bahrawy, A., Hart, L. P., and Pestka, J. J., Comparison of deoxynivalenol (vomitoxin) production by *Fusarium graminearum* isolates in corn steep-supplemented Fries medium, *J. Food Prot.*, 48, 705, 1985.

20. Miller, J. D., Young, J. C., and Sampson, D. R., Deoxynivalenol and Fusarium head blight resistance in spring cereals, *Phytopathol. Zeitschr.*, 113, 359, 1985.

21. Cuero, R., Smith, J. E. and Lacey, J., Mycotoxin formation by *Aspergillus flavus* and *Fusarium graminearum* in irradiated maize grains in the presence of other fungi, *J. Food Prot.*, 51, 452, 1988.

22. Halasz, A., Badaway, A., Sawinsky, J., Kozma-Kovacs, E., and Beczner, J., Effect of γ-irradiation on F-2 and T-2 toxin production in corn and rice, *Folia Microbiol.*, 34, 228, 1989.

23. Bosch, U., Mirocha, C. J., and Wen, Y., Production of zearalenone, moniliformin and trichothecenes in intact sugar beets under laboratory conditions, *Mycopathologia*, 119, 167, 1992.

24. Boyacioglu, D., Hettiarachchy, N. S., and Stack, R. W., Effect of three systemic fungicides on deoxynivalenol (vomitoxin) production by *Fusarium graminearum* in wheat, *Can. J. Plant Sci.*, 72, 93, 1992.

25. Hasan, H. A. H., Fungicide inhibition of aflatoxins, diacetoxyscirpenol and zearalenone production, *Folia Microbiol.*, 38, 295, 1993.

26. Milus, E. A. and Parsons, C. E., Evaluation of foliar fungicides for controlling Fusarium head blight of wheat, *Plant Dis.*, 78, 697, 1994.

27. Pettersson, H., Nivalenol production by *Fusarium poae*, *Mycotoxin Res.*, 7, 26, 1991.

28. Sugiura, Y., Fukasaku, K., Tanaka, T., Matsui, Y., and Ueno, Y., *Fusarium poae* and *Fusarium crookwellense*, fungi responsible for the natural occurrence of nivalenol in Hokkaido, *Appl. Environ. Microbiol.*, 59, 3334, 1993.

29. Roinestad, K. S., Montville, T. J., and Rosen, J. D., Inhibition of trichothecene biosynthesis in *Fusarium tricinctum* by sodium bicarbonate, *J. Agric. Food Chem.*, 41, 2344, 1993.

30. Paster, N., Barkai-Golan, R., and Calderon, M., Control of T-2 toxin production using atmospheric gases, *J. Food Prot.*, 49, 615, 1986.

31. Paster, N., Blumenthal-Yonassi, J., Barkai-Golan, R., and Menasherov, M., Production of zearalenone in vitro and in corn grains stored under modified atmospheres, *Int. J. Food Microbiol.*, 12, 157, 1991.

32. Yoshizawa, T. and Morooka, N., Biological modification of trichothecene mycotoxins: acetylation and deacetylation of deoxynivalenols by *Fusarium* spp., *Appl. Microbiol.*, 29, 54, 1975.

33. Doko, M. B., Rapior, S., Visconti, A., and Schjoth, J. E., Incidence and levels of fumonisin contamination in maize genotypes grown in Europe and Africa, *J. Agric. Food Chem.*, 43, 429, 1995.

34. Bosch, U., Mirocha, C. J., Abbas, H. K., and di Menna, M., Toxicity and toxin production by *Fusarium* isolates from New Zealand, *Mycopathologia*, 108, 73, 1989.

35. Scott, P. M., Abbas, H. K., Mirocha, C. J., Lawrence, G. A., and Weber, D., Formation of moniliformin by *Fusarium sporotrichioides* and *Fusarium culmorum*, *Appl. Environ. Microbiol.*, 53, 196, 1987.

36. Rabie, C. J., Marasas, W. F. O., Thiel, P. G., Lubben, A., and Vleggaar, R., Moniliformin production and toxicity of different *Fusarium* species from Southern Africa, *Appl. Environ. Microbiol.*, 43, 517, 1982.

37. Abbas, H. K. and Mirocha, C. J., Isolation and purification of a hemorrhagic factor (Wortmannin) from *Fusarium oxysporum* (N17B), *Appl. Environ. Microbiol.*, 54, 1268, 1988.

38. Luz, J. M., Paterson, R. R. M., and Brayford, D., Fusaric acid and other metabolite production in *Fusarium oxysporum* f. sp. *vasinfectum*, *Lett. Appl. Microbiol.*, 11, 141, 1990.

39. Scott, P. M., The natural occurrence of trichothecenes, in *Trichothecene Mycotoxicosis: Pathophysiologic Effects*, Volume I, Beasley, V. R., Ed., CRC Press Inc., Boca Raton, 1989, Chap. 1.

40. Muller, H. M. and Schwadorf, K., A survey of the natural occurrence of *Fusarium* toxins in wheat grown in southwestern area of Germany, *Mycopathologia*, 121, 115, 1993.

41. Perkowski, J., Plattner, R. D., Golinski, P., Vesonder, R. F., and Chelkowski, J., Natural occurrence of deoxynivalenol, 3-acetyl-deoxynivalenol, 15-acetyl-deoxynivalenol, nivalenol, 4,7-dideoxynivalenol and zearalenone in Polish wheat, *Mycotoxin Res.*, 6, 7, 1990.

42. Hietaniemi, V. and Kumpulainen, J., Contents of *Fusarium* toxins in Finnish and imported grains and feeds, *Food Add. Contam.*, 8, 171, 1991.

43. Bottalico, A., Logrieco, A., and Visconti, A., *Fusarium* species and their mycotoxins in infected corn in Italy, *Mycopathologia*, 107, 85, 1989.

44. Tanaka, T., Yamamoto, S., Hasegawa, A., Aoki, N., Besling, J. R., Sugiura, Y., and Ueno, Y., A survey of the natural occurrence of *Fusarium* mycotoxins in the Netherlands, *Mycopathologia*, 110, 19, 1990.

45. Tanaka, T., Teshima, R., Ikebuchi, H., Sawada, J., and Ichinoe, M., Sensitive enzyme-linked immunosorbent assay for the mycotoxin zearalenone in barley and Job's-tears, *J. Agric. Food Chem.*, 43, 946, 1995.

46. Fernandez, C., Stack, M. E., and Musser, S. M., Determination of deoxynivalenol in 1991 U.S. winter and spring wheat by high-performance thin-layer chromatography, *J. AOAC Int.*, 77, 628, 1994.

47. Trucksess, M. W., Thomas, F., Young, K., Stack, M. E., Fulgueras, W. J., and Page, S. W., Survey of deoxynivalenol in U.S. 1993 wheat and barley crops by enzyme-linked immunosorbent assay, *J. AOAC Int.*, 78, 631, 1995.

48. Stratton, G. W., Robinson, A. R., Smith, H. C., Kittilsen, L., and Barbour, M., Levels of five mycotoxins in grains harvested in Atlantic Canada as measured by high performance liquid chromatography, *Arch. Environ. Contam. Toxicol.*, 24, 399, 1993.

49. Chatterjee, D. and Mukherjee, S. K., Contamination of Indian maize with fumonisin B_1 and its effects on chicken macrophage, *Lett. Appl. Microbiol.*, 18, 251, 1994.

50. D'Mello, J. P. F., Macdonald, A. M. C., and Cochrane, M. P., A preliminary study of the potential for mycotoxin production in barley grain, *Aspects Appl. Biol. Cereal Qual.*, 36, 375, 1993.

51. Lacey, J., Prevention of mould growth and mycotoxin production through control of environmental factors, in *Proceedings of the Seventh International IUPAC Symposium on Mycotoxins and Phycotoxins*, Natori, S., Hashimoto, K. and Ueno, Y., Eds., Elsevier Science Publishers, Amsterdam, 1989, 161.

52. Jenkinson, P. and Parry, D. W., Splash dispersal of conidia of *Fusarium culmorum* and *Fusarium avenaceum*, *Mycological Res.*, 98, 506, 1994.

53. Cahagnier, B., Melcion, D., and Richard-Molard, D., Growth of *Fusarium moniliforme* and its biosynthesis of fumonisin B_1 on maize grain as a function of different water activities, *Lett. Appl. Microbiol.*, 20, 247, 1995.

54. Merino, M., Ramos, A. J., and Hernandez, E., A rapid HPLC assay for zearalenone in laboratory cultures of *Fusarium graminearum*, *Mycopathologia*, 121, 27, 1993.

55. Cuero, R. G., Smith, J. E., and Lacey, J., Interaction of water activity, temperature and substrate on mycotoxin production by *Aspergillus flavus*, *Penicillium viridicatum* and *Fusarium graminearum* in irradiated grains, *Trans. Br. Mycol. Soc.*, 89, 221, 1987.

56. Gareis, M., Bauer, J., Von Montgelas, A., and Gedek, B., Stimulation of aflatoxin B_1 and T-2 toxin production by sorbic acid, *Appl. Environ. Microbiol.*, 47, 416, 1984.

57. Placinta, C. M., D'Mello, J. P. F., and Macdonald, A. M. C., unpublished data, 1995.

58. Draughon, F. A. and Churchville, D. C., Effect of pesticides on zearalenone production in culture and in corn plants, *Phytopathology*, 75, 553, 1985.

59. Kellerman, T. S., Marasas, W. F. O., Thiel, P. G., Gelderblom, W. C. A., Cawood, M., and Coetzer, J. A. W., Leukoencephalomalacia in two horses induced by oral dosing of fumonisin B_1, *Onderstepoort J. Vet. Res.*, 57, 269, 1990.

60. Bezuidenhout, S. C., Gelderblom, W. C. A., Gorst-Allman, C. P., Horak, R. M., Marasas, W. F. O., Spiteller, G., and Vleggaar, R., Structure elucidation of the fumonisins, mycotoxins from *Fusarium moniliforme*, *J. Chem. Soc. Chem. Comm.*, 743, 1988.

61. Norred, W. P. and Voss, K. A., Toxicity and role of fumonisins in animal diseases and human esophageal cancer, *J. Food Prot.*, 57, 522, 1994.

62. Gelderblom, W. C. A., Jaskiewicz, K., Marasas, W. F. O., Thiel, P. G., Horak, R. M., Vleggaar, R., and Kriek, N. P. J., Fumonisins - novel mycotoxins with cancer-promoting activity produced by *Fusarium moniliforme, Appl. Environ. Microbiol.*, 54, 1806, 1988.

63. Yoshizawa, T., Yamashita, A., and Luo, Y., Fumonisin occurrence in corn from high- and low-risk areas for human esophageal cancer in China, *Appl. Environ. Microbiol.*, 60, 1626, 1994.

64. Franceschi, S., Bidoli, E., Baron, A. E., and La Vecchia, C., Maize and risk of cancers of the oral cavity, pharynx and esophagus in northern Italy, *J. Cancer Inst.*, 82, 1407, 1990.

65. Gelderblom, W. C. A., Kriek, N. P. J., Marasas, W. F. O., and Thiel, P. G., Toxicity and carcinogenicity of the *Fusarium moniliforme* metabolite, fumonisin B_1, in rats, *Carcinogenesis*, 12, 1247, 1991.

66. Marasas, W. F. O., Jaskiewicz, K., Venter, F. S., and van Schalkwyk, D. J., *Fusarium moniliforme* contamination of maize in oesophageal cancer areas in Transkei, *S. Afr. Med. J.*, 74, 110, 1988.

67. Thiel, P. G., Marasas, W. F. O., Sydenham, E. W., Shephard, G. S., and Gelderblom, W. C. A., The implications of naturally-occurring levels of fumonisins in corn for human and animal health, *Mycopathologia*, 117, 3, 1992.

68. Beardall, J. M. and Miller, J. D., Diseases in humans with mycotoxins as possible causes, in *Mycotoxins in Grain: Compounds other than Aflatoxin*, Miller, J. D. and Trenholm, H. L., Eds., Egan Press, St. Paul, 1994, Chap. 14.

69. Riley, R. T., Voss, K. A., Yoo, H., Gelderblom, W. C. A., and Merill, A. H., Mechanism of fumonisin toxicity and carcinogenesis, *J. Food Prot.*, 57, 638, 1994.

70. Riley, R. T., Norred, W. P., and Bacon, C. W., Fungal toxins in foods: recent concerns, *Annu. Rev. Nutr.*, 13, 167, 1993.

71. Bacon, C. W. and Nelson, P. E., Fumonisin production in corn by toxigenic strains of *Fusarium moniliforme* and *Fusarium proliferatum*, *J. Food Prot.*, 57, 514, 1994.

72. Sydenham, E. W., Shephard, G. S., Thiel, P. G., Marasas, W. F. O., Rheeder, J. P., Sanhueza, C. E. P., Gonzalez, H. L., and Resnik, S. L., Fumonisins in Argentinian field-trial corn, *J. Agric. Food Chem.*, 41, 891, 1993.

73. Plattner, R. D., Norred, W. P., Bacon, C. W., Voss, K. A., Peterson, R., Shackelford, D. D., and Weisleder, D., A method of detection of fumonisins in corn samples associated with field cases of leukoencephalomalacia, *Mycopathologia*, 82, 698, 1990.

74. Bacon, C. W., Bennett, R. M., Hinton, D. M., and Voss, K. A., Scanning electron microscopy of *Fusarium moniliforme* within asymptomatic corn kernels and kernels associated with equine leukoencephalomalacia, *Plant Dis.*, 76, 144, 1992.

75. Drysdale, R. B., The production and significance in phytopathology of toxins produced by species of *Fusarium*, in *The Applied Mycology of Fusarium*, Moss, M. O. and Smith, J. E., Eds., Cambridge University Press, New York, 1984, Chap. 5.

76. ApSimon, J. W., The biosynthetic diversity of secondary metabolites, in *Mycotoxins in Grain: Compounds other than Aflatoxin*, Miller, J. D. and Trenholm, H. L., Eds., Egan Press, St. Paul, 1994, Chap. 1.

77. Charmley, L. L., Rosenberg, A., and Trenholm, H. L., Factors responsible for economic losses due to *Fusarium* mycotoxin contamination of grains, foods and feedstuffs, in *Mycotoxins in Grain: Compounds other than Aflatoxin*, Miller, J. D. and Trenholm, H. L., Eds., Egan Press, St. Paul, 1994, Chap.13.

78. Prelusky, D. B., Rotter, B. A., and Rotter, R. G., Toxicology of mycotoxins, in *Mycotoxins in Grain: Compounds other than Aflatoxin*, Miller, J. D. and Trenholm, H. L., Eds., Egan Press, St. Paul, 1994, Chap. 9.

79. Wang, E., Norred, W. P., Bacon, C. W., Riley, R. T., and Merrill, A. H. Jr., Inhibition of sphingolipid biosynthesis by fumonisins: implications for diseases associated with *Fusarium moniliforme, J. Biol. Chem.*, 266, 14486, 1991.

80. Hannun, Y. A. and Bell, R. M., Functions of sphingolipids and sphingolipid breakdown products in cellular regulation, *Science*, 243, 500, 1989.

81. Merrill, A. H., van Echten, G., Wang, E., and Sandhoff, K., Fumonisin B_1 inhibits sphingosine (sphinganine) N-acetyltransferase and *de novo* sphingolipid biosynthesis in cultured neurons *in situ, J. Biol. Chem.*, 268, 27299, 1993.

82. Thiel, P. G., Meyer, C. J., and Marasas, W. F. O., Natural occurrence of moniliformin together with deoxynivalenol and zearalenone in Transkeian corn, *J. Agric. Food Chem.*, 30, 308, 1982.

83. Haschek, W. M. and Haliburton, J. C., *Fusarium moniliforme* and zearalenone toxicoses in domestic animals: a review, in *Diagnosis of Mycotoxicoses*, Richards, J. L. and Thurston, J. R., Eds., Martinus Nijhoff Publishers, Dordrecht, 1986, p.213.

84. Miller, J. D., Savard, M. E., Rapior, S., Hocking, A. D., and Pitt, J. I., Production of fumonisins and fusarins by *Fusarium moniliforme* from Southeast Asia, *Mycologia*, 85, 10458, 1993.

85. Savard, M. E. and Miller, J. D., Characterization of fusarin F, a new fusarin from *Fusarium moniliforme*, *J. Nat. Prod.*, 55, 64, 1992.

86. Macchia, L., DiPaola, R., Fornelli, F., Nenna, S., Moretti, A., Napoletano, R., Logrieco, A., Caiaffa, M. F., Tursi, A., and Bottalico, A., Cytotoxicity of beauvericin to mammalian cells, *International Seminar on Fusarium Mycotoxins, Taxonomy and Pathogenicity*, May 9–13, 1995, Martina Franca, Italy, Book of Abstracts, p. 72.

87. Ritieni, A., Logrieco, A., Fogliano, V., Moretti, A., Ferracane, R., Randazzo, G., and Bottalico, A., Natural occurrence of fusaproliferin in pre-harvest infected maize ear rot in Italy, *International Seminar on Fusarium Mycotoxins, Taxonomy and Pathogenicity*, May 9–13, 1995, Martina Franca, Italy, Book of Abstracts, p. 23.

88. Ritieni, A., Fogliano, V., Randazzo, G., Scarallo, A., Logrieco, A., Moretti, A., Mannina, L., and Bottalico, A., Isolation and characterizaton of fusaproliferin, a new toxic metabolite from *Fusarium proliferatum*, *Nat. Toxins*, 3, 17, 1995.

89. Musser, S. M., Eppley, R. M., Mazzola, E. P., Hadden, C. E., Shockor, J. P., Crouch, R. C., and Martin, G. E., Identification of an N-acetylketo derivative of fumonisin B_1 in corn cultures of *Fusarium proliferatum*, *J. Nat. Prod.*, 58, 1392, 1995.

90. Kaneshiro, T., Vesonder, R. F., and Peterson, R. E., Fumonisin-stimulated N-acetyldihydrosphingosine, N-acetylphytosphingosine and phytosphingosine products of *Pichia (Hansenula) ciferri*, NRRL Y-1031, *Curr. Microbiol.*, 24, 319, 1992.

91. Abbas, H. K., Tanaka, T., Duke, S. O., Porter, J. K., Wray, E. M., Hodges, L., Sessions, A. E., Wang, E., Merrill, A. H., Jr., and Riley, R. T., Fumonisin- and AAL-toxin-induced disruption of sphingolipid metabolism with the accumulation of free sphingoid bases, *Plant Physiol.*, 106, 1085, 1994.

92. Riley, R. T., Hinton, D. M., Chamberlain, W. J., Bacon, C. W., Wang, E., Merrill, A. H., Jr., and Voss, K. A., Dietary fumonisin B_1 induces disruption of sphingolipid metabolism in Sprague-Dawley rats: a new mechanism of nephrotoxicity, *J. Nutr.*, 124, 594, 1994.

93. Voss, K. A., Chamberlain, W. J., Bacon, C. W., and Norred, W. P., A preliminary investigation on renal and hepatic toxicity in rats fed purified fumonisin B_1, *Nat. Toxins*, 1,1,1993.

94. Haschek, W. M., Kim, H. Y., Motelin, E. L., Stair, E. L., Beasley, V. R., Chamberlain, W. J., and Riley, R. T., Pure fumonisin B_1 as well as fumonisin-contaminated feed, alters swine serum and tissue sphinganine and sphingosine levels, biomarkers of exposure, *Toxicologist*, 13, 232, 1993.

95. Wang, E., Ross, P. F., Wilson, T. M., Riley, R. T., and Merrill, A. H., Jr., Increases in serum sphingosine and sphinganine and decreases in complex sphingolipids in ponies given feed containing fumonisins, mycotoxins produced by *Fusarium moniliforme*, *J. Nutr.*, 122, 1706, 1992.

96. Fincham, J. E., Marasas, W. F. O., Taljaard, J. J.F., Kriek, N. P. J., Badenhorst, C. J., Gelderblom, W. C. A., Seier, J. V., Smuts, C. M., Faber, M., Weight, M. J., Slazus, W., Woodroof, C. W., van Wyk, M. J., Kruger, M., and Theil, P. G., Athrogenic effects in non-human primates of *Fusarium moniliforme* cultures added to a carbohydrate diet, *Atherosclerosis*, 94, 13, 1992.

97. Harel, R. and Futerman, A. H., Inhibition of sphingolipid synthesis affects axonal outgrowth in cultured hippocampal neurons, *J. Biol. Chem.*, 268, 14476, 1993.

98. Riley, R. T., Wang, E., and Merrill, A. H., Jr., Liquid chromatographic determination of sphinganine and sphingosine: use of the free sphinganine-to-sphingosine ratio as a biomarker for the consumption of fumonisins, *J. AOAC Int.*, 77, 533, 1994.

99. Shier, W. T., Abbas, H. K., and Mirocha, C. J., Toxicity of mycotoxins fumonisins B_1 and B_2 and *Alternaria alternata* f.sp. *lycopersici* toxin (AAL) in cultured mammalian cells, *Mycopathologia*, 116, 97, 1991.

100. Kuipe-Goodman, T., Prevention of human mycotoxicoses through risk assesssment and risk management, in *Mycotoxins in Grain: Compounds other than Aflatoxin*, Miller, J. D. and Trenholm, H. L., Eds., Egan Press, St. Paul, 1994, Chap. 12.

101. Towers, N. R. and Sprosen, J., High zearalenone residues in grazing cows with fertility problems: cause or coincidence? *International Seminar on Fusarium Mycotoxins, Taxonomy and Pathogenicity*, May 9–13, 1995, Martina Franca, Italy, Book of Abstracts, p.6.

102. Dowd, P. F., Toxicology and biochemical interactions of fungal metabolites fusaric acid and kojic acid with xenobiotics in *Heliothis zea* (F.) and *Spodoptera frugiperda* (J. E. Smith), *Pesticide Biochem.*, 32, 123, 1988.

103. Bacon, C. W., Porter, J. K., and Norred, W. P., Toxic interaction of fumonisin B_1 and fusaric acid measured by injection into fertile chicken egg, *Mycopathologia*, 129, 29, 1995.

104. Porter, J. K., Bacon, C. W., Wray, E. M., and Hagler, W. M., Jr., Fusaric acid in *Fusarium moniliforme* cultures, corn and feeds toxic to livestock and the neurochemical effects in the brain and pineal gland of rats, *Nat. Toxins*, 3, 91, 1995.

105. Burmeister, H. R., Grove, M. D., Peterson, R. E., Weisleder, D., and Plattner, R. D., Isolation and characterization of two new fusaric acid analogs from *Fusarium moniliforme* NRRL 13163, *Appl. Environ. Microbiol.*, 50, 311, 1985.

106. Pitel, D. W. and Vining, L. C., Accumulation of dehydrofusaric acid and its conversion to fusaric acid and 10-hydroxyfusaric acid in cultures of *Giberella fujikuroi*, *Can. J. Biochem.*, 48, 623, 1970.

107. Porter, J. K., Wray, E. M., Rimando, A. M., Stancel, P. C., Bacon, C. W., and Voss, K. A., Lactational passage of fusaric acid from the feed of nursing dams to the neonate rat and effects on pineal neurochemistry in the F_1 and F_2 generations at weaning, *J. Toxicol. Environ. Health*, 49, 101, 1996.

108. Ogunbo, S., Ledoux, D. R., Bermudez, A. J., and Rottinghaus, G. E., Effects of fusaric acid on broiler chicks and turkey poults, *Poultry Sci.*, 73, 154 (S102), 1994.

109. Smith, T. K. and MacDonald, E. J., Effects of fusaric acid on brain regional neurochemistry and vomiting behavior in swine, *J. Anim. Sci.*, 69, 2044, 1991.

110. Smith, T. K., Recent advances in the understanding of *Fusarium* trichothecene mycotoxicoses, *J. Anim. Sci.*, 70, 3989, 1992.

111. Rimando, A. M., Porter, J. K., and Stancel, P. C., Enzyme linked immunosorbent assay (ELISA) of melatonin (MEL) in serum of weanling rats: effects of fusaric acid, *210th National Meeting of the American Chemical Society*, August 20–24, 1995, Chicago, Book of Abstracts, AGFD 080.

112. Reiter, R. J., The melatonin rhythm: both clock and calendar, *Experientia*, 49, 654, 1993.

113. Madhyastha, M. S., Marquardt, R. R., and Abramson, D., Structure-activity relationships and interactions among trichothecene mycotoxins as assessed by the yeast bioassay, *Toxicon*, 32, 1147, 1994.

114. Gelderblom, W. C. A., Marasas, W. F. O., Thiel, P. G., Veggaar, R., and Cawood, M. E., Fumonisins, chemical characterization and toxicological effects, *Mycopathologia*, 117, 11, 1992.

115. Harvey, R. B., Edrington, T. S., and Kubena, L. F., Interactive toxicity of fumonisin B_1 and deoxynivalenol contaminated diets fed in combination to growing swine, *International Seminar on Fusarium Mycotoxins, Taxonomy and Pathogenicity*, May 9–13, 1995, Martina Franca, Italy, Book of Abstracts, p.67.

116. Logrieco, A., Moretti, A., Ritieni, A., Bottalico, A., and Corda, P., Occurrence and toxigenicity of *Fusarium proliferatum* from preharvest maize ear rot, and associated mycotoxins, in Italy, *Plant Dis.*, 79, 727, 1995.

21 Toxicants of the Genus *Penicillium*

D. Abramson

CONTENTS

21.1   GENERAL CONSIDERATIONS

Toxic secondary metabolites from *Penicillium*, and mycotoxins from other fungal genera, have been discussed in several reviews since 1990. These works have covered several aspects of the topic including fungal production,[1,2] chemical properties,[3] occurrence in commodities,[4] biological effects,[5] and analysis procedures. Continuing coverage of detection methods and commodity surveys for major mycotoxins is provided by the Association of Official Analytical Chemists International, which publishes an annual summary in the January issue of its journal.[6]

In this chapter, toxicants from *Penicillium* species are classified to reflect their economic and health significance. The classes are cumulative (Table 21.1), in the sense that characteristics of lower categories usually apply to the categories above them. For example, although ochratoxin A is a "Class A" toxicant because of the compound's association with human illness, attributes of the lower categories also apply. Ochratoxin A has been implicated in livestock toxicosis (Class B), has been found in foods and feeds (Class C), and, of course, demonstrates significant oral toxicity and can be produced in fungal cultures (Class D). Some secondary metabolites from *Penicillium* are not discussed in this chapter; culture-produced metabolites without proven oral toxicity, such as erythroskyrin and rugulosin, are treated elsewhere.[7] These compounds are of limited interest in an economic or health sense, but may be of value in bioassay development or in structure-activity studies.

TABLE 21.1
Classification of *Penicillium* Toxins

Attribute	Class			
	A	**B**	**C**	**D**
Implicated in human disease	■			
Implicated in animal disease	■	■		
In foods or feeds; oral toxicity	■	■	■	
From fungal cultures; oral toxicity	■	■	■	■

The toxicants listed below are produced mainly by *Penicillium* fungi, although other genera, such as *Aspergillus*, *Chaetomium*, and *Claviceps* may produce some of these same compounds as well.[7] Classification of toxigenic *Penicillium* species has recently undergone re-examination on the basis of several new taxonomic criteria. One criterion currently used is based on secondary metabolite studies of pure *Penicillium* cultures[8] by liquid chromatography and diode-array spectrometry.[9,10] Although there is some question about the correct identification[11] of several secondary metabolites, the scheme is gaining acceptance.[3] This current taxonomic classification[12] will be used wherever possible in this review.

21.2 CLASS A: TOXICANTS IMPLICATED IN HUMAN DISEASE

The mycotoxins implicated in human toxicosis, or found in human blood, are the most significant in terms of agro-economics and human health, and are listed in Class A. Ochratoxin A, citreoviridin, and cyclopiazonic acid are included in this classification.

21.2.1 OCHRATOXIN A

The ochratoxins are a group of isocoumarin derivatives linked with an amide bond to the amino group of L-β-phenylalanine. Ochratoxin A (Figure 21.1) is the most abundantly produced and most toxic member of this group. In tropical and semi-tropical zones, the most significant producers are Aspergilli, particularly *Aspergillus ochraceus*, while in temperate zones, Penicillia are the prime sources. The ochratoxin-producing Penicillia formerly included several species, but in the current taxonomic scheme,[12] only *Penicillium verrucosum* and its chemotypes are regarded as producers.

The ochratoxins act primarily as nephrotoxins, but ochratoxin A is also a potent teratogen in rats, hamsters, and mice. The oral LD_{50} for ochratoxin A in rats is 22 mg kg^{-1}. In day-old chicks, the LD_{50} (oral) is 3.6 mg kg^{-1} for ochratoxin A and 54 mg kg^{-1} for ochratoxin B, with acute kidney damage observed; ochratoxin C manifested similar effects. The carcinogenicity of ochratoxin A in mice has been documented.[13] A study at the National Institutes of Health in the U.S. also showed ochratoxin A to be a carcinogen in rats.[14]

Ochratoxin A contaminates a variety of plant and animal products, and is particularly likely to appear in stored cereal grains. Although ochratoxin A is a worldwide problem, its impact is greatest in temperate climates where much of the world's grain is produced and stored. In commodity surveys, ochratoxin A has been found in Australian feed grains, Japanese rice, Canadian cereals and feed grains, American corn and wheat, British cereals and feeds, German barley, wheat, corn, and oats, Italian corn, Norwegian cereals, Polish wheat, barley, oats, and rye, Bulgarian corn, and various grains from the former state of Yugoslavia.[4] Cheese is evidently also a good substrate for ochratoxin production, and this toxin has been discovered in moldy cheese from Britain.[15] Ochratoxin A is found in certain meat products from monogastric animals, and has been detected in pork sausages in Germany[16] and Switzerland.[17] In

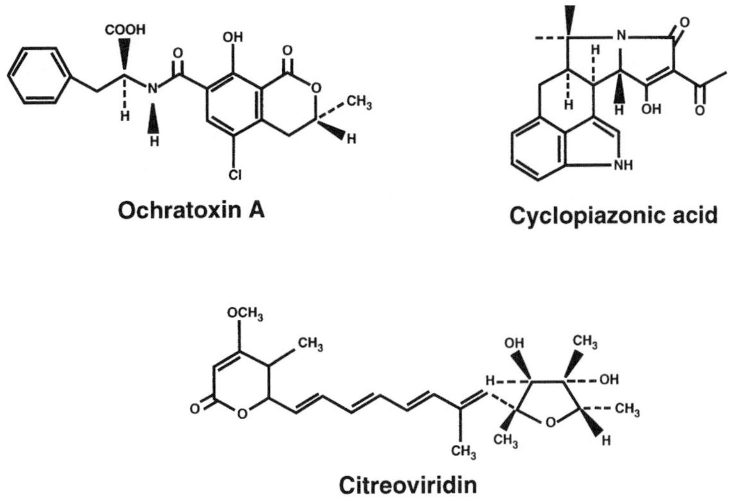

Ochratoxin A

Cyclopiazonic acid

Citreoviridin

FIGURE 21.1 Chemical structures of *Penicillium* toxicants implicated in human disease.

chickens, this toxin can be carried over[18,19] from contaminated feed into the muscle tissue and eggs.

In terms of animal and human health, ochratoxin A is the most important of the *Penicillium* mycotoxins. It has long been of particular significance to the poultry[20] and swine industries[21] because monogastric animals lack the ability to hydrolyze ochratoxin A rapidly, as compared to ruminants. Monogastric livestock are far more susceptible to the nephrotoxic effects of ochratoxin A than ruminants, and residues can enter the human food chain via their organ and meat products.[13] In contrast, the adverse effects on ruminant livestock appear to be limited mainly to suppression of milk production, which is transient.[22]

Ochratoxin A is thus a major concern to livestock producers, especially in European countries. Ochratoxin A introduced into the feed of monogastric livestock can contaminate eggs, organs, fat, muscle tissue, and biological fluids. Although limited data is available for poultry, swine are quite susceptible owing to a rather long serum half-life of 72–120 h.[23,24] Recent surveys have detected ochratoxin A as a natural contaminant of swine blood in Canada, and in European countries, namely Germany, Norway, Poland, Sweden, and the former state of Yugoslavia. In addition, ochratoxin A has been found in swine kidneys in the U.S., Austria, Belgium, Denmark, Finland, Germany, Poland, Switzerland, Britain, and the former state of Yugoslavia.[4]

Ochratoxin A is the mycotoxin most frequently found in the blood of humans, and has been associated with the high incidence of a nephropathic illness in eastern Europe.[25] This disease is encountered in Bulgaria[26] and in the former state of Yugoslavia, and is known as Balkan endemic nephropathy. Ochratoxin A residues have been compared in food and blood samples from nephropathic and non-nephropathic regions of this geographical area.[27] Ochratoxin A has also been found in human blood samples from other parts of Europe,[28] including Germany, Poland, and the Czech Republic. This mycotoxin has recently been detected in human blood in Canada,[29] Japan,[30] and Italy.[31] The threat of ochratoxin A in terms of human and animal health has been reviewed and assessed.[13,32]

21.2.2 CITREOVIRIDIN

Citreoviridin (Figure 21.1) was among the first mycotoxins studied in modern times in a formal scientific manner.[33] In early 20th century Japan, this compound became associated with acute cardiac disease arising from consumption of rice infected with *Penicillium citreo-viride*.[34] In subsequent studies, it was discovered that rice is an ideal substrate for *P. citreo-viride*, and in a 1958 survey this fungus was found on 7.4% of rice surveyed in Italy, Spain, Thailand, Burma, and other countries.[35] Citreoviridin is also produced by *Penicillium pulvullorum*,[36] *Penicillium ochrasalmoneum* (=*Eupenicillium ochrasalmoneum*),[37] and *P. charlesii*.[38]

The acute toxicity of citreoviridin has been described and reviewed by Uraguchi.[33] Citreoviridin administered to mice results in progressive paralysis in the limbs, vomiting, convulsions, cardiovascular damage and respiratory arrest. In cats and dogs, the symptoms are the same, with vomiting preceding the progressive paralysis. The toxin appears to attack motor neurons and interneurons of the spinal chord, medulla oblongata, and central nervous system. Kymography indicates dilation on the right side of the heart and paralysis of the diaphragm.

Modern chromatographic methods and standards have been available for citreoviridin analysis since about 1970. This mycotoxin has been found in the U.S. in moldy pecans[38] and in standing corn.[37] Identification of citreoviridin in the moldy yellow rice associated with human cardiac disease epidemics in early 20th century Japan is based on retrospective evidence.[33] This documentation is fairly complete and includes published accounts of the epidemic, isolation of the fungus, toxicology studies with fungal extracts, and purified fractions, and structural elucidation of the toxin.

The particular cardiac disease associated with citreoviridin was first thought to be identical with the thiamin deficiency syndrome arising from eating polished rice. Between 1890 and 1925, Sakaki, Ogata, and their colleagues in Japan showed that a different type of beri-beri called "Shoshin-kakke," due to toxicosis rather than avitaminosis, was on the increase in Japan, particularly in urban centers such as Tokyo. From 1909 onwards, increasingly stringent government regulations to restrict trade in uninspected rice reduced the Tokyo death rate from >11/10,000 to approx 2/10,000. By 1929 the epidemic was effectively controlled.[33]

Some clinical manifestations of Shoshin-kakke are described in Uraguchi's historical summary.[33] During the course of the disease, the afflicted patient initially experiences cardiac distress and tachypnea followed by nausea and vomiting. Later, the patient undergoes painful seizures and rolls from side to side. As blood pressure falls, the patient experiences tachycardia, dyspnea, pupillary dilation, and presents cold and cyanotic extremities. Increasing paralysis of the respiratory muscles impair pulmonary circulation, overloading the right ventricle of the heart, and finally causing cardiac failure.

21.2.3 Cyclopiazonic Acid

Cyclopiazonic acid was isolated from, and named for, *Penicillium cyclopium* in 1968.[39] In the current taxonomic scheme,[12] this mycotoxin appears to be produced mainly by *Penicillium camemberti*, *Penicillium commune*, *Penicillium griseofulvum*, and several Aspergilli. The chemical and biological aspects of cyclopiazonic acid have been reviewed.[40]

Oral administration to male and female rats proved fatal after one to five days, and gave LD_{50} values of 36 and 63 mg kg^{-1}, respectively.[41] In rats, cyclopiazonic acid produces degenerative changes and necrosis in the liver, pancreas, spleen, kidney, salivary glands, myocardium, and skeletal muscle. Although the teratogenic potential proved to be low in rats,[42] significant retardation in embryonic skeletal development was evident after administration of 5–10 mg of cyclopiazonic acid during pregnancy. When rations containing 100 mg kg^{-1} cyclopiazonic acid were fed to chickens for seven weeks, the test group experienced decreased weight gain, poor feed conversion, and a sixfold increase in mortality compared to chickens receiving toxin-free feed.[43] Postmortem examination revealed proventricular lesions characterized by mucosal erosion and hyperemia, and yellow foci in the livers and spleens. The birds also experienced mucosal necrosis in the gizzard, and hepatic and splenetic necrosis and inflammation.

Cyclopiazonic acid is often co-produced along with aflatoxins by *A. flavus* in corn[44] and peanuts[45] in the U.S. In a survey of the 1990 Georgia corn and peanut crops,[46] 51% of the

45 corn samples contained this mycotoxin at levels up to 2.8 mg kg⁻¹ (average 0.47 mg kg⁻¹); 90% of the 50 peanut samples contained up to 2.9 mg kg⁻¹ (average 0.46 mg kg⁻¹). Cyclopiazonic acid can also be produced by *P. camemberti* in Camembert-type cheeses.[47] A study showed that this mycotoxin did not appear during the 9-day ripening period at 14–18°C, nor during 12 days of storage at 8°C, but cheese kept at 25°C for 5 days accumulated this toxin at levels up to 4 mg kg⁻¹. A feeding study has demonstrated the carry-over of cyclopiazonic acid from feed into muscle tissue in chickens,[48] and the potential for contamination of human diets by chicken meat containing this toxin. Analysis of chicken meat following oral dosing with 10 mg cyclopiazonic acid kg⁻¹ body weight indicated that 14.5% of the dose was in the meat 48 h later.

In India, cyclopiazonic acid has been found in millet associated with human toxicosis.[49] Toxicosis from moldy millet, known as "kodua poisoning,"[50] is a nonlethal illness characterized in humans by fatigue, tremors, slurring of speech, and nausea. Although known since antiquity, two recent cases of human toxicosis arising from millet contaminated with cyclopiazonic acid have been reported.[49] In both cases, characteristic symptoms, including fatigue and nausea were reported. Cyclopiazonic acid was isolated from the suspect millet by extraction and preparative thin-layer chromatography, and identified from infrared and mass spectrum comparisons.

Cattle experience a type of neurological disorder characterized by prostration and head nodding after ingesting *Paspalum* grass infected by *Claviceps paspali*. In India *Paspalum* grass toxicosis is often referred to as "mona poisoning" or "kodua poisoning."[51,52] Several indole-diterpene tremorgens, namely paspalinine and the paspalitrems, have been implicated in this *Paspalum-Claviceps* mycotoxicosis.[53] This type of "kodua poisoning" should not be confused with the human toxicosis associated with cyclopiazonic acid described above.

21.3   CLASS B: TOXICANTS IMPLICATED IN ANIMAL DISEASE

In Class B are mycotoxins associated with animal poisoning incidents, but which are not implicated in human disease. These toxins have been found in the feed or stomach contents of the affected animals, producing symptoms consistent with those observed in experimental laboratory toxicosis studies. This group includes roquefortine, penitrem A, and citrinin (Figure 21.2).

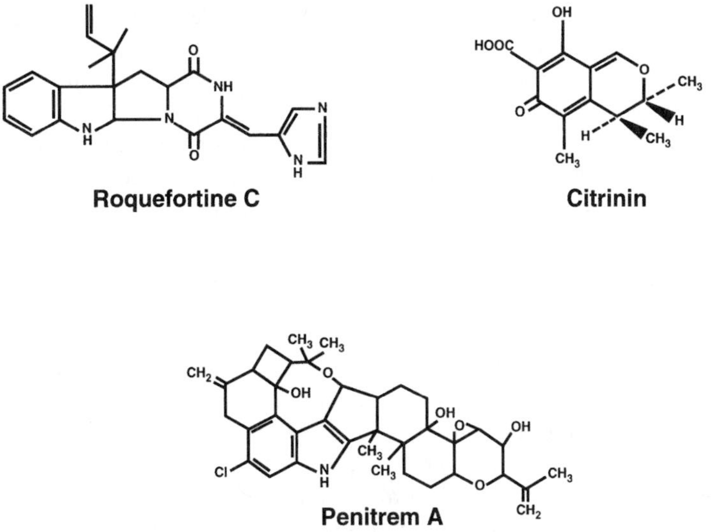

Roquefortine C   **Citrinin**

Penitrem A

FIGURE 21.2   Chemical structures of *Penicillium* toxicants implicated in animal disease.

21.3.1 Roquefortine C

This toxin is produced by a several Penicillia[12] including *Penicillium chrysogenum*, *Penicillium crustosum*, *Penicillium expansum*, *Penicillium griseofulvum*, *Penicillium hirsutum*, *Penicillium hordei*, and *Penicillium melanoconidium*, but most notably by *Penicillium roqueforti*, which is used in cheese production. According to Pitt and Leistner,[1] attempts to find nontoxigenic strains for use in cheese production have so far been unsuccessful. The production, analysis, and toxicity of roquefortine C have been reviewed.[54]

Although oral LD_{50} figures have not been established for roquefortine C, day-old chicks, orally dosed via crop intubation, have been used as a bioassay for this toxin.[55] The chicks initially lost their balance, and remained in sitting and leaning postures. The chicks died in characteristic postures with head and neck extended backwards, and legs and feet extended outwards from the body.

Roquefortine C has been found mainly in blue cheese and blue cheese products. In a survey[56] of 16 blue cheese samples from Denmark, Finland, Germany, France, Britain, Italy, and Canada, all samples contained this mycotoxin at levels up to 6.8 mg kg^{-1} (average 0.95 mg kg^{-1}). In a later survey,[57] 12/12 American blue cheese samples contained roquefortine C at average levels of 0.42 mg kg^{-1}; for American blue cheese salad dressing, 2/2 samples averaged 0.045 mg kg^{-1}.

In a cattle poisoning incident, roquefortine C was found in barley-based feed at 25 mg kg^{-1} on a Swedish farm.[58] The grain was heavily infected by *P. roqueforti*, with a large accumulation of mycelium. Cattle developed extensive paralysis which did not respond to calcium treatment. The disease symptoms disappeared when the cattle were fed sound grain.

Roquefortine was recently found in the stomach contents of two out of six dogs showing strychnine-like symptoms of fatal food poisoning.[59] These particular animals had evidently eaten garbage which had become moldy during warm local weather conditions. Fatal canine poisoning due to ingestion of roquefortine from moldy blue cheese has also been reported.[60] Because of their scavenging habits, dogs are more likely than other animals to seek out and consume moldy foods from garbage, and thus become affected by food-borne mycotoxins.[61]

21.3.2 Penitrem A

Penitrem A, previously known as tremortin A, is a tremorgenic toxin produced by *P. crustosum* and *P. melanoconidium*,[12] and was reported earlier in cultures of *P. commune*.[55] Due to the high incidence of *P. crustosum* isolates producing significant levels of this toxin,[1] the presence of this fungal species in food or feed should be regarded as a warning signal of penitrem contamination. Penitrem A and other tremorgenic mycotoxins have recently been reviewed.[62]

Of the tremorgenic mycotoxins produced by Penicillia, penitrem A is the most toxic.[63] In low doses, the penitrems cause sustained tremors in animals which are otherwise able to feed and function normally. 5- to 20-fold increases in dosage are rapidly fatal with discernible pathological effects, particularly in the hepatocytes.[64] The amount of toxin producing a given response varies from species to species.[65] The oral LD_{50} for penitrem A in mice is 10 mg kg^{-1};[66] in chickens, it is 42 mg kg^{-1}.[67]

Penitrem A has been implicated in intoxications of dogs manifesting tremors after eating moldy cream cheese[68] and moldy walnuts.[69] In the former case, isolation of *P. crustosum* from moldy cream cheese and inoculation onto normal cream cheese resulted in production of penitrem A by the fungus.[70] In the latter case, a chloroform extract of the walnuts gave an lipid residue which produced tremors in mice after oral administration in olive oil. A case of canine poisoning by penitrem A from moldy bread indicated that levels of 35 mg penitrem A kg^{-1} in bread were achieved by *P. crustosum*, and that approximately 0.175 mg penitrem A kg$^-$ body weight was sufficient to produce severe muscle tremors in dogs.[61]

21.3.3 CITRININ

This nephrotoxin is produced by, and named for, *Penicillium citrinum*. Citrinin is also produced by *P. expansum* and *P. verrucosum*.[12] *P. citrinum* and *P. verrucosum* are common storage fungi found on cereal and oilseed crops in the U.S., Canada, and other countries.

Citrinin often occurs together with ochratoxin A, and has been reported in North America, Europe, and Asia in cereals, oilseeds, and their products. Citrinin is a contaminant of American corn, Canadian wheat, rye, barley, and oats, Danish feed grains, Swedish barley, and British wheat flour;[4] it has also been found in Japanese rice,[71] and in peanuts,[72] rice, and coconut products from India.[73] If convenient screening tests for citrinin were available, this listing would likely be much longer.

The toxicity of citrinin has recently been reviewed.[74] Oral toxicity varies with duration and species; LD_{50} values have been reported for rabbits (134 mg kg^{-1}, 72 h), guinea pigs (43 mg kg^{-1} d^{-1}, 14 d), mice (105–112 mg kg^{-1}, 72 h), chicks (95 mg kg^{-1}, 72 h), turkey poults (56 mg kg^{-1}, 72 h) and ducklings (57 mg kg^{-1}). Citrinin was implicated in the deaths of 200/9600 feeder cattle manifesting uremia and fibrosing nephritis.[75] *P. citrinum* and 2–3 mg kg^{-1} citrinin were found in the feed. Citrinin has also been reported as a possible cause of livestock poisoning in Brazil.[76]

21.4 CLASS C: ORAL TOXICANTS OCCURRING IN FOODS OR FEEDS

Table 21.2 lists some *Penicillium* mycotoxins having known oral toxic effects, and which are found in foods or feeds. These toxins, depicted in Figure 21.3, have significant potential for

TABLE 21.2
Oral Toxicants from *Penicillium* Occurring in Foods or Feeds

	Producers[a]	Food or Feed Contamination	Oral LD_{50}, mg kg^{-1}	Ref.
Patulin	*P. expansum*	Apples	35, mouse	79
	P. griseofulvum			
Penicillic Acid	*P. aurantiogriseum*	Corn	90, chicken	80
	P. aurantiovirens	Beans	600, mouse	81
	P. cyclopium			
	P. freii			
	P. viridicatum			
Rubratoxin B	*P. purpurogenum*	Corn	63, duckling	78
	P. rubrum		83, chick	
			120, mouse	
Secalonic Acid D	*P. oxalicum*	Corn (dust)	25, rat	82, 83
Viomellein	*P. freii*	Barley	b	4
	P. cyclopium			
	P. viridicatum			
Xanthomegnin	*P. freii*	Corn	c	4
	P. cyclopium	Cereal grains		
	P. viridicatum			

[a] According to Frisvad and Thrane (1995).[12]
[b] Jaundice in male mice at 456 mg viomellein kg^{-1} g feed.
[c] Jaundice in male mice at 448 mg xanthomegnin kg^{-1} feed.

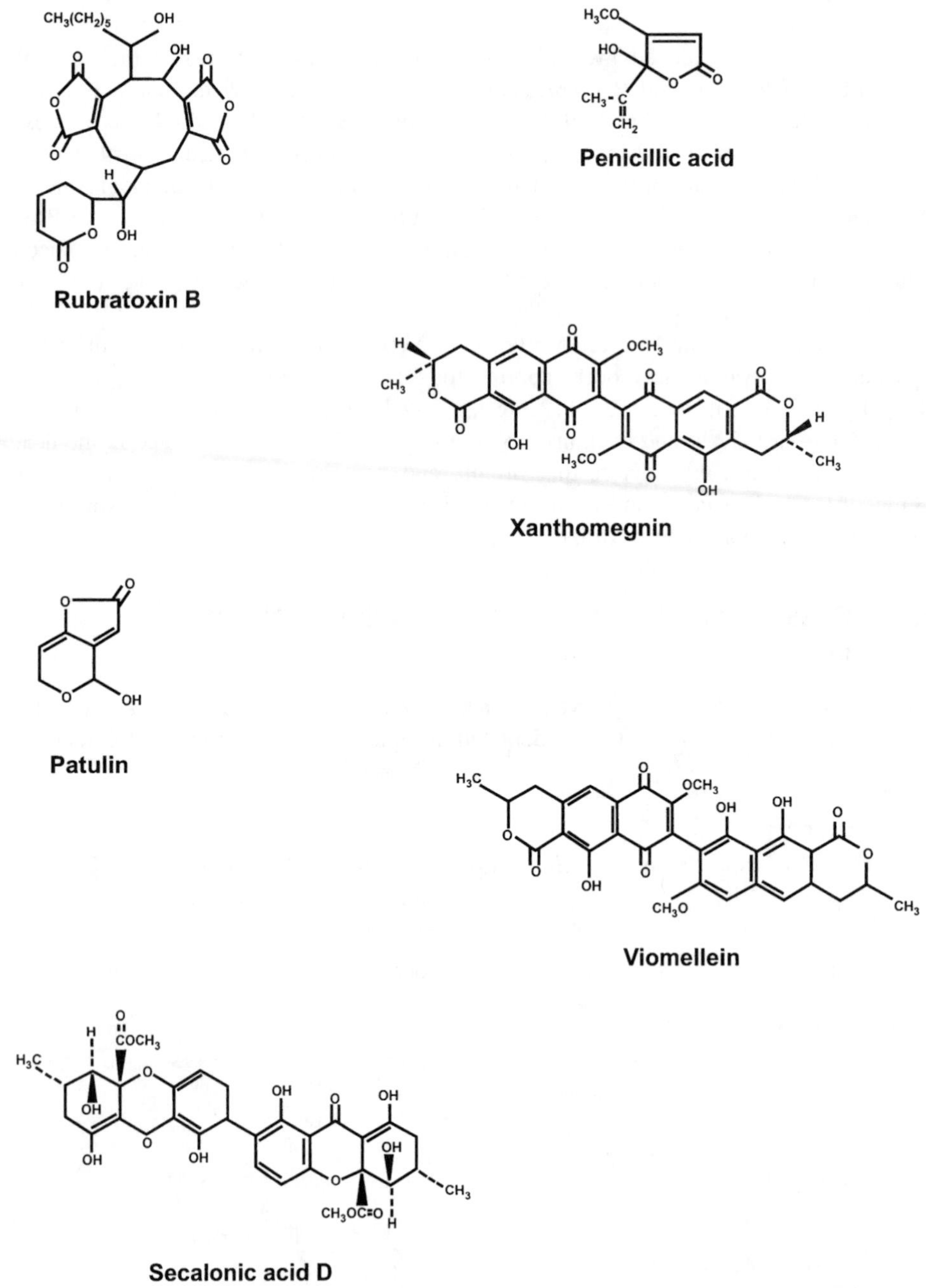

Rubratoxin B

Penicillic acid

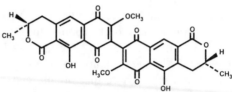

Xanthomegnin

Patulin

Viomellein

Secalonic acid D

FIGURE 21.3 Chemical structures of oral toxicants from *Penicillium* occurring in foods or feeds.

TABLE 21.3
Oral Toxicants from *Penicillium* Cultures

	Producers[a]	Oral LD$_{50}$, mg kg^{-1}	Ref.
8-chlororugulovasine A	*P. islandicum*	75–125, chick	84
brefeldin A	*P. brefeldianum*[b]	275, rat	7
		62, chick	7
cyclochlorotine	*P. islandicum*	6.6, mouse	7
emodin	*P. islandicum*	3.7, chick	7
gliotoxin	*P. bilaii*	67, mouse	7
griseofulvin	*P. griseofulvum*	400, rat	85
luteoskyrin	*P. islandicum*	221, mouse	7
mycophenolic acid	*P. brevicompactum*	700, rat	7
	P. roqueforti		
oosporein	*P. phoenicium*[b]	6.1, chick	7
PR toxin	*P. roqueforti*	115, rat	7
verruculogen	*P. simplicissimum*	127, mouse	7
verruculotoxin	*P. verruculosum*[b]	20, chick	7
viridicatum toxin	*P. brasilianum*	122, rat	7

[a] According to Frisvad and Thrane (1995),[12] except where indicated as [b].

causing human and animal health problems, and are occasionally the subject of food contaminant surveys. Most of these compounds are available as standards from commercial suppliers. Mycotoxins such as PR toxin, which have been produced in cultures of *Penicillium* species isolated from foods, but which have not been detected as natural contaminants in actual commodities,[77] are listed in Table 21.3 rather than in Table 21.2.

21.5   CLASS D: ORAL TOXICANTS FROM *PENICILLIUM* CULTURES

Table 21.3 lists several *Penicillium* metabolites demonstrating oral toxicity, but which are presently found only in fungal cultures. These toxins, shown structurally in Figure 21.4, have some potential for causing animal health problems, but generally are not subject to monitoring in foods or feeds. Most of these compounds are not yet available as standards from commercial suppliers. These Class D toxins should not be dismissed as mere "laboratory curiosities;" many mycotoxins which today pose problems in human and animal health, such as ochratoxin A[13] and citrinin,[73] were originally isolated from fungal cultures.

ACKNOWLEDGMENTS

The author thanks Diane Smith, Mike Malyk, Gaye Miller, Reg Sims, Amber Bole, and Dr. Noel White, all of the Winnipeg Research Center, for their help in the preparation, illustration, and proofreading of this work.

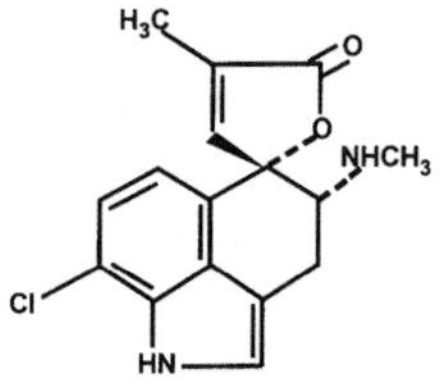

8-Chlororugulovasine A

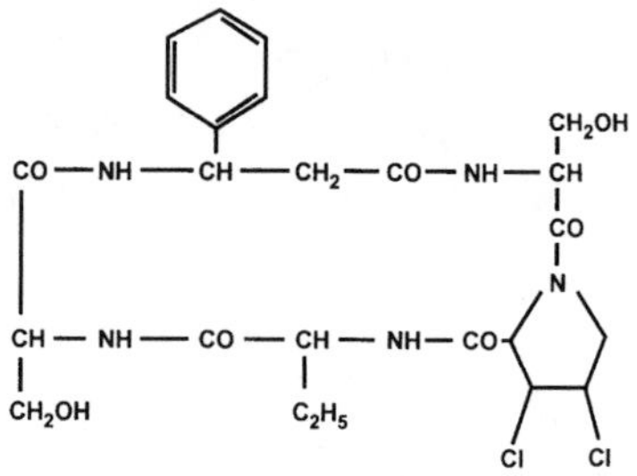

Cyclochlorotine

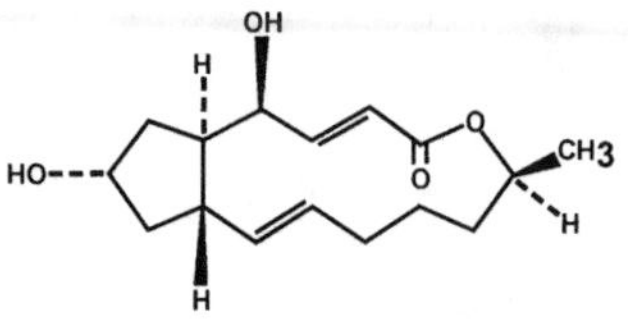

Brefeldin A

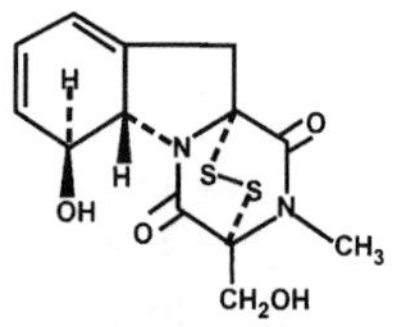

Gliotoxin

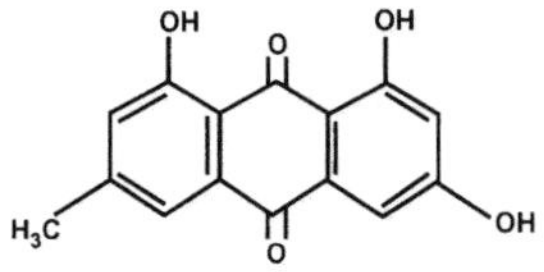

Emodin

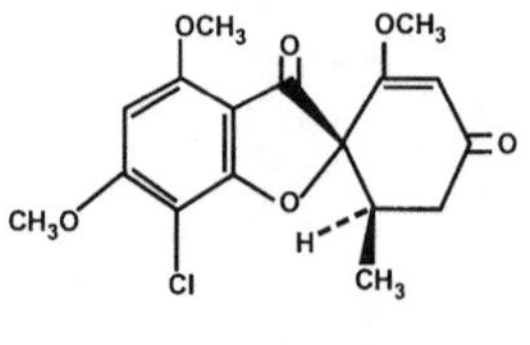

Griseofulvin

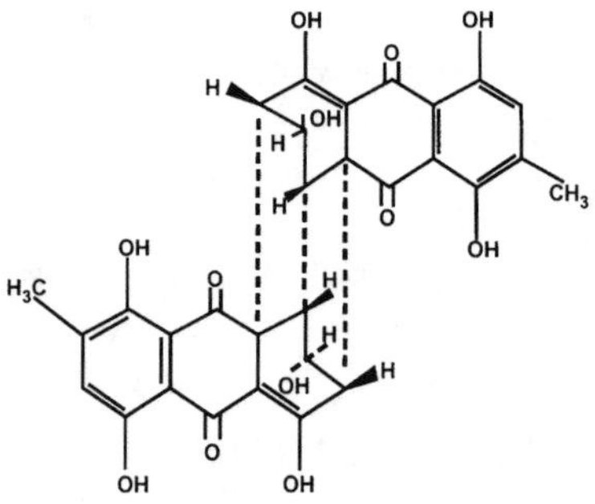

Luteoskyrin

FIGURE 21.4 Chemical structures of oral toxicants from *Penicillium* cultures.

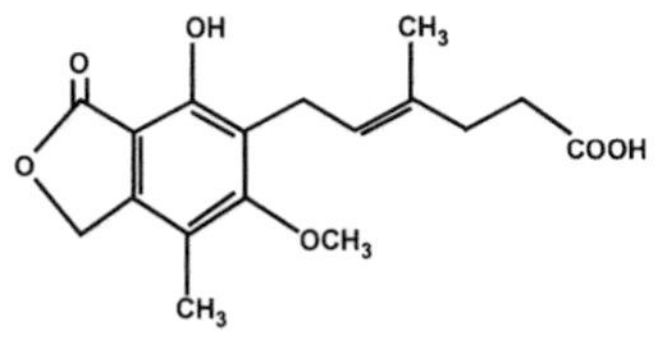

Mycophenolic acid

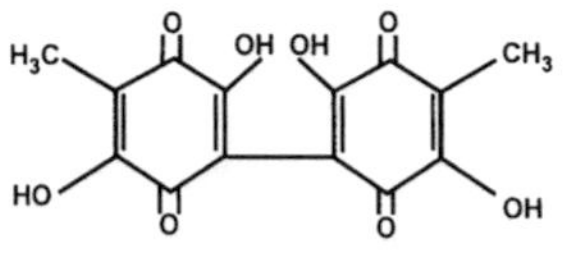

Oosporein

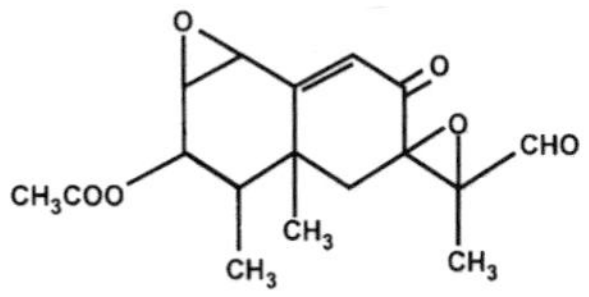

PR toxin

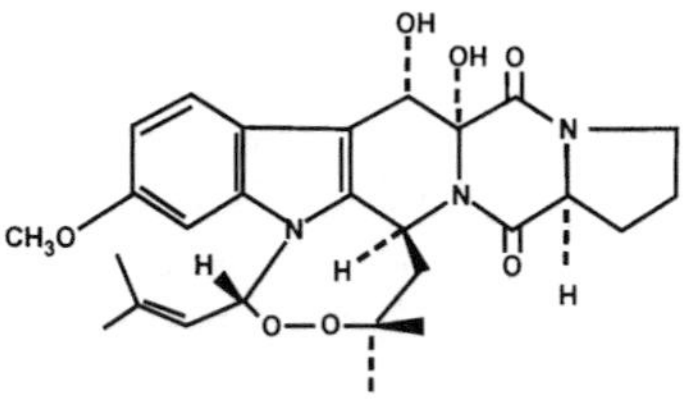

Verruculogen

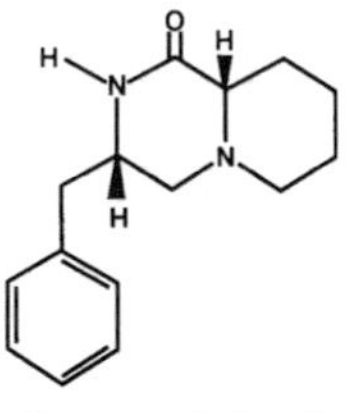

Verruculotoxin

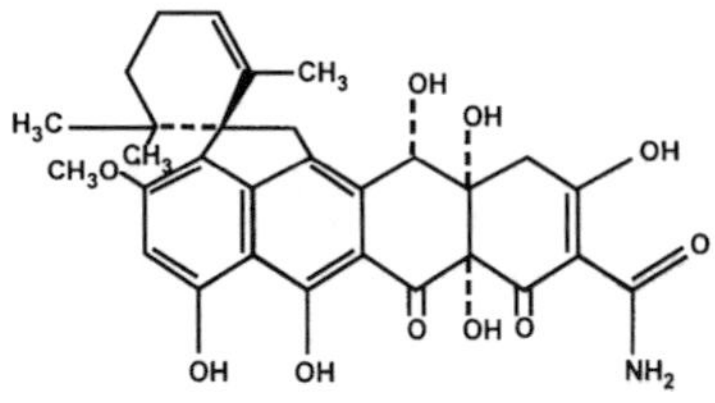

Viridicatum toxin

FIGURE 21.4 (continued)

REFERENCES

1. Pitt, J. I. and Leistner, L., Toxigenic *Penicillium* species, in *Mycotoxins and Animal Foods*, Smith, J. E. and Henderson, R. S., Eds., CRC Press, Boca Raton, 1991, 81.

2. Frisvad, J. C., Mycotoxins and mycotoxigenic fungi in storage, in *Stored-Grain Ecosystems*, Jayas, D. S., White, N. D. G., and Muir, W. E., Eds., Marcel Dekker Inc., New York, 1995, 251.

3. Scott, P. M., *Penicillium* and *Aspergillus* toxins, in *Mycotoxins in Grain: Compounds other than Aflatoxin*, Miller, J. D. and Trenholm, H. L., Eds., Eagan Press, St. Paul, 1994, 261.

4. Wilson, D. M. and Abramson, D., Mycotoxins, in *Storage of Cereal Grains and Their Products*, 4th Edition, Sauer, D. B., Ed., American Association of Cereals Chemists, St. Paul, 1992, 341.

5. Smith, J. E. and Henderson, R. S., Eds., *Mycotoxins and Animal Foods*, CRC Press, Boca Raton, 1991.

6. Trucksess, M. W., Mycotoxins, *J. AOAC Int.*, 78, 135, 1995.

7. Cole, R. J. and Cox, R. H., *Handbook of Toxic Fungal Metabolites*, Academic Press, New York, 1981.

8. Frisvad, J. C., Fungal species and their specific production of mycotoxins, in *Introduction to Food-Borne Fungi*, 3rd Edition, Samson, R. A. and van Reenen-Hoekstra, E. S., Eds., Centraalbureau voor Schimmelcultures, Baarn, 1988, 239.

9. Frisvad, J. C. and Thrane, U., Standardized high-performance liquid chromatography of 182 mycotoxins and other fungal metabolites based on alkylphenone retention indices and UV-VIS spectra (diode array detection), *J. Chromatogr.*, 404, 195, 1987.

10. Frisvad, J. C., The use of high-performance liquid chromatography and diode array detection in fungal chemotaxonomy based on profiles of secondary metabolites, *Bot. J. Linn. Soc.*, 99, 81, 1989.

11. Paterson, R. R. M. and Kemmelmeier, C., Neutral alkaline and difference ultraviolet spectra of secondary metabolites from *Penicillium* and other fungi and comparisons to published maxima from gradient high-performance liquid chromatography with diode-array detection, *J. Chromatogr.*, 511, 195, 1990.

12. Frisvad, J. C. and Thrane, U., Mycotoxin production by food-borne fungi, in *Introduction to Food-Borne Fungi*, 4th Edition, Samson, R. A., Hoekstra, E. S., Frisvad, J. C., and Thrane, U., Eds., Centraalbureau voor Schimmelcultures, Baarn, 1995, 251.

13. Kuiper-Goodman, T. and Scott, P. M., Risk assessment of the mycotoxin ochratoxin A. *Biomed. Environ. Sci.*, 2, 179, 1989.

14. National Toxicology Program, *NTP Technical Report on the Toxicology and Carcinogenesis Studies of Ochratoxin A (CAS No. 303-47-9) in F334/N Rats (Gavage Studies), NIH Publication No. 89-2813*, National Institutes of Health, Washington, DC, 1989.

15. Chapman, W. B., Cooper, S. J., Williams, A. R., and Jarvis, B., Mycotoxins in molded cheeses, in *Proc. Int. Symp. Mycotoxins* Sept. 6–8, 1981, Cairo, Egypt, Naguib, K., Naguib, M. M., Park, D. L., and Pohland, A. E., Eds., National Research Center, Cairo, 1983, 363.

16. Scheuer, R. and Leistner, L., Occurrence of ochratoxin A in pork and pork products, *Proc. 32nd Eur. Mtg. Meat Res. Workers*, Ghent, August 24 to 29, 1986, 191.

17. Baumann, U. and Zimmerli, B., Einfache ochratoxin-A-bestimmung in lebensmitteln, *Mitt. Gebiete Lebensm. Hyg.*, 79, 151, 1988.

18. Juszkiewicz, T., Piskorska-Pliszczynska, J., and Wisniewska, H., Ochratoxin A in laying hens: tissue disposition and passage into eggs, in *Proc. V Int. IUPAC Symp. Mycotoxins Phycotoxins*, Austrian Chem. Soc., Vienna, 1982, 122.

19. Bauer, J., Niemiec, J., and Scholtyssek, S., Ochratoxin A im Legehennenfutter. 2. Mitteilung: Rückstände in Serum, Leber und Ei, *Arch. Geflügelk.*, 52, 71, 1988.

20. Elling, F., Hald, B., Jacobsen, C., and Krogh, P., Spontaneous toxic nephropathy in poultry associated with ochratoxin A, *Acta Pathol. Microbiol. Scand. A*, 83, 739, 1975.

21. Rutqvist, L., Bjorklund, N. E., Hult, K., and Gatenbeck, S., Spontaneous occurrence of ochratoxin residues in kidneys of fattening pigs, *Zbl. Vet. Med. Reihe A*, 24, 402, 1977.

22. Ribelin, W. E., Ochratoxicosis in cattle, in *Mycotoxic Fungi, Mycotoxins, Mycotoxicoses*, Vol. 2, Wyllie, T. D. and Morehouse, L. G., Eds., John Wiley and Sons, New York, 1978, 28.

23. Galtier, P., Alvinerie, M., and Charpenteau, J. L., The pharmacokinetic profiles of ochratoxin A in pigs, rabbits and chickens, *Food Cosmet. Toxicol.*, 19, 735, 1981.

24. Mortensen, H. P., Hald, B., Larsen, A. E., and Madsen A., Ochratoxin A contaminated barley for sows and piglets: pig performance and residues in milk and pigs, *Acta Agric. Scand.*, 33, 235, 1983.

25. Pavlovic, M., Plestina, R., and Krogh, P., Ochratoxin A contamination of foodstuffs in an area with Balkan (endemic) nephropathy, *Acta. Path. Microbiol. Scand. B*, 87, 243, 1979.

26. Petkova-Bocharova, T., Chernozemsky, I. N., and Castegnaro, M., Ochratoxin A in human blood in relation to Balkan endemic nephropathy and urinary system tumors in Bulgaria, *Food Addit. Contam.*, 5, 229, 1988.

27. Pepeljnjak, S., and Cvetnic, Z., The mycotoxicological chain and contamination of food by ochratoxin A in the nephropathic and non-nephropathic areas of Yugoslavia. *Mycopathologia*, 90, 147, 1985.

28. Hald, B., Human exposure to ochratoxin A, in *Mycotoxins and Phycotoxins '88*, Natori, S., Hashimoto, K., and Ueno, Y., Eds., Elsevier, Amsterdam, 1989, 57.

29. Frohlich, A. A., Marquardt, R. R., and Ominsky, K. H., Ochratoxin A as a contaminant of the human food chain: a Canadian perspective, in *Mycotoxins, Endemic Nephropathy and Urinary Tract Tumors*, Castegnaro, M., Plestina, R., Dirheimer, G., Chernozemsky, I. N., and Bartsch, H., Eds., Oxford University Press, Oxford, 1991, 139.

30. Kawamura, O., Maki, S., Sato, S., and Ueno, Y., Ochratoxin A in livestock and human sera in Japan quantified by a sensitive ELISA, in *Human Ochratoxicosis and its Pathologies*, Creppy, E. E., Castegnaro, M., and Dirheimer, G., Eds., Editions INSERM, Paris, 1993, 167.

31. Breitholtz-Emanuelsson, A., Minervini, F., Hult, K., and Visconti, A., Ochratoxin A in human serum samples collected in southern Italy from healthy individuals and individuals suffering from different kidney disorders, *Nat. Toxins*, 2, 366, 1994.

32. Creppy, E. E., Castegnaro, M., Dirheimer, G., Eds., *Human Ochratoxicosis and its Pathologies*, John Libbey Eurotext, Montrouge, France, 1993.

33. Uraguchi, K., Mycotoxic origin of cardiac beriberi, *J. Stored Prod. Res.*, 5, 227, 1969.

34. Ueno, Y., The toxicology of mycotoxins, *Crit. Rev. Toxicol.*, 14, 99, 1985.

35. Tsunoda, H., Tsuruta, O., and Takahashi, M., Research for the microorganisms which deteriorate the stored cereals. XVII. Parasites of imported rice, *Bull. Food Res. Inst.*, 13, 29, 1958.

36. Nagel, D.W. and Steyn, P., Production of citreoviridin by *Penicillium pulvullorum*, *Phytochemistry*, 11, 627, 1972.

37. Wicklow, D. T. and Cole, R. J. Citreoviridin in standing corn infested by *Eupenicillium ochrosalmonium* Scott and Stolk, *Mycologia*, 76, 959, 1984.

38. Cole, R. J., Dorner, J. W., Cox, R. H., Hill, R. A., Cluter, H. G., and Wells, J. M., Isolation of citreoviridin from *Penicillium charlesii* cultures and molded pecan fragments, *Appl. Environ. Microbiol.*, 42, 677, 1981.

39. Holzapfel, C. W., The isolation and structure of cyclopiazonic acid, a toxic metabolite of *Penicillium cyclopium* Westling, *Tetrahedron*, 24, 2101, 1968.

40. Cole, R. J., Cyclopiazonic acid and related toxins, in *Mycotoxins: Production, Isolation, Separation, and Purification*, Betina, V., Ed., Elsevier, Amsterdam, 1984, 405.

41. Purchase, I. F. H., The acute toxicity of the mycotoxin cyclopiazonic acid to rats, *Toxicol. Appl. Pharmacol.*, 18, 114, 1971.

42. Morrissey, R. E., Cole, R. J., and Dorner, J. W., The effects of cyclopiazonic acid on pregnancy and fetal development of Fischer rats, *J. Toxicol. Environ. Health*, 14, 585, 1984.

43. Dorner, J. W., Cole, R. J., Lomax, L. G., Gosser, H. S., and Diener, U. L., Cyclopiazonic acid production by *Aspergillus flavus* and its effects on broiler chickens, *Appl. Environ. Microbiol.*, 46, 698, 1983.

44. Gallagher, R. T., Richard, J. L., Stahr, H. M., and Cole, R. J., Cyclopiazonic acid production by aflatoxigenic and non-aflatoxigenic strains of *Aspergillus flavus*, *Mycopathologia*, 66, 31, 1978.

45. Lansden, J. A. and Davidson, J. I., Occurrence of cyclopiazonic acid in peanuts, *Appl. Environ. Microbiol.*, 45, 766, 1983.

46. Urano, T., Trucksess, M. W., Beaver, R. W., Wilson, D. M., Dorner, J. W., Dowell, and F. E., Co-occurrence of cyclopiazonic acid and aflatoxins in corn and peanuts, *J. AOAC Int.*, 75, 838, 1992.

47. Still, P., Eckardt, C., and Leistner, L., Production of cyclopiazonic acid by *Penicillium camemberti* isolates from cheese, *Fleischwirtsch.*, 58, 876, 1978.

48. Norred, W. P., Cole, R. J., Dorner, J. W., and Lansden, J. A., Liquid chromatographic determination of cyclopiazonic acid in poultry meat, *J. Assoc. Off. Anal. Chem.*, 70, 121, 1987.

49. Rao, B. L. and Husain, A., Presence of cyclopiazonic acid in kodo millet (*Paspalum scorbiculatum*) causing "kodua poisoning" in man and its production by associated fungi, *Mycopathologia*, 89, 177, 1985.

50. Bhide, N. K., Pharmacological study and fractionation of *Paspalum scrobiculatum* extract, *Br. J. Pharmacol.*, 18, 7, 1962.

51. Bazlur, M., Probable mona grass (*Paspalum commersoni*) poisoning, *Indian Vet. J.*, 37, 31, 1960.

52. Nayak, N. C. and Misra, D. B., Cattle poisoning by *Paspalum scrobiculatum* (Kodua Poisoning), *Indian Vet. J.*, 39, 501, 1962.

53. Raisbeck, M. F., Rottinghaus, G. E., and Kendall, J. D., Effects of naturally occurring mycotoxins on ruminants, in *Mycotoxins and Animal Foods*, Smith, J. E. and Henderson, R. S., Eds., CRC Press, Boca Raton, 1991, 647.

54. Scott, P. M., Roquefortine, in *Mycotoxins: Production, Isolation, Separation, and Purification*, Betina, V., Ed., Elsevier, Amsterdam, 1984, 463.

55. Wagener, R. E., Davis, N. D., and Diener, U. L., Penitrem A and roquefortine production by *Penicillium commune, Appl. Environ. Microbiol.*, 39, 882, 1980.

56. Scott, P. M. and Kennedy, B. P. C., Analysis of blue cheese for roquefortine and other alkaloids of *Penicillium roqueforti, J. Agric. Food Chem.*, 24, 865, 1976.

57. Ware, G. M., Thorpe, C. W., and Pohland, A. E., Determination of roquefortine in blue cheese and blue cheese dressing by high pressure liquid chromatography with ultraviolet and electrochemical detectors, *J. Assoc. Off. Anal. Chem.*, 63, 637, 1980.

58. Häggblom, P., Isolation of roquefortine C from feed grain, *Appl. Environ. Microbiol.*, 56, 2924, 1990.

59. Lowes, N. R., Smith, R. A., and Beck, B. E., Roquefortine in the stomach contents of dogs suspected of strychnine poisoning in Alberta, *Can. Vet. J.*, 33, 535, 1992.

60. Puls, R. and Ladyman, E., Roquefortine toxicity in a dog, *Can. Vet. J.*, 29, 568, 1988.

61. Hocking, A. D., Holds, K., and Tobin, N. F., Intoxication by tremorgenic mycotoxin (penitrem A) in a dog. *Austral. Vet. J.*, 65, 82, 1988.

62. Moreau, C., Les mycotoxines a effets trémorgéniques, *Cryptogamie Mycol.*, 11, 89, 1990.

63. Cole, R. J., Fungal tremorgens, *J. Food Prot.*, 44, 715, 1981.

64. Hayes, A. W., Presley, D. B., and Neville, J. A., Acute toxicity of penitrem A in dogs, *Toxicol. Appl. Pharmacol.*, 35, 311, 1976.

65. Richard, J. L., Peden, W. M., and Thurston, J. R., Occurrence of penitrem mycotoxin and clinical manifestations of penitrem intoxications, *Curr. Topics Vet. Med. Anim. Sci.*, 33, 51, 1986.

66. Ueno, Y. and Ueno, I., Toxicology and biochemistry of mycotoxins, in *Toxicology Biochemistry and Pathology of Mycotoxins*, Uraguchi, K. and Yamazaki, M. (Eds.), John Wiley and Sons, New York, 107, 1978.

67. Wyatt, R. D., Hamilton, P. B., Colwell, W. M., and Ciegler, A., The effect of tremortin A on chickens, *Avian Dis.*, 16, 461, 1972.

68. Arp, L. H. and Richard, J. L., Intoxication of dogs with the mycotoxin penitrem A, *J. Am. Vet. Med. Assoc.*, 175, 565, 1979.

69. Richard, J. L., Bacchetti, P., and Arp, L. H., Moldy walnut toxicosis in a dog caused by the mycotoxin penitrem A, *Mycopathologia*, 76, 55, 1981.

70. Richard, J. L. and Arp, L. H., Natural occurrence of the mycotoxin penitrem A in moldy cream cheese. *Mycopathologia*, 67, 107, 1979.

71. Sugimoto, T., Minamisawa, M., Takano, K., Sasamura, Y., and Tsuruta, O., Detection of ochratoxin A, citrinin and sterigmatocystin from storaged rice by natural occurrence of *Penicillium viridicatum* and *Aspergillus versicolor, J. Food Hyg. Soc. Jpn.*, 18, 176, 1977.

72. Subrahmanyan, P. and Rao, A. S., Occurrence of aflatoxins and citrinin in groundnut (*Archis hypogaea* L.) at harvest in relation to pod condition and kernel moisture content, *Curr. Sci.*, 43, 707, 1974.

73. Wilson, D. M., Citrinin: analysis and occurrence, in *Biodeterioration Research 4*, Llewellyn, G. C., Dashek, W. V., and O'Rear, C. E. (Eds.), Plenum Press, New York, 1994, 65.

74. Hanika, C. and Carlton, W. W., Toxicology and pathology of citrinin, in *Biodeterioration Research 4*, Llewellyn, G. C., Dashek, W. V., and O'Rear, C. E. (Eds.), Plenum Press, New York, 1994, 41.

75. Lloyd, W. E., Holter, J. A., Wohlgemuth, K., and Classick, L., A case of probable citrinin toxicosis in a herd of Iowa feeder cattle, *Proc. Amer. Assoc. Vet. Lab. Diag. 21st Ann. Mtg.*, 147, 1978.

76. Rosa, C. A. da R., da Cruz, L. C. H., Chagas, W. A., and Veiga, C. E. de O., Ocurrencia natural de nefropatia micotoxica suina causada pela ingestao de ceveda contaminada con citrina. *Rev. Bras. Med. Vet.*, 7, 87, 1985.

77. Scott, P. M. and Kanhere, S. R., Instability of PR toxin, a metabolite of *Penicillium roqueforti* in blue cheese, *J. Assoc. Off. Anal. Chem.*, 62, 141, 1979.

78. Davis, R. M. and Richard, J. L., Rubratoxins and related substances, in *Mycotoxins: Production, Isolation, Separation, and Purification*, Betina, V., Ed., Elsevier, Amsterdam, 1984, 315.

79. Engel, G. and Teuber, M., Patulin and other small lactones, in *Mycotoxins: Production, Isolation, Separation, and Purification*, Betina, V., Ed., Elsevier, Amsterdam, 1984, 291.

80. Murnaghan, M. F., The pharmacology of penicillic acid, *J. Pharmacol. Exp. Ther.*, 88, 119, 1946.

81. Huff, W. E., Hamilton, P. B., and Ciegler, A., Evaluation of penicillic acid for toxicity in broiler chickens, *Poult. Sci.*, 59, 1203, 1980.

82. Reddy, C. S., Hayes, A. W., Williams, W. L., and Ciegler, A., Toxicity of secalonic acid D, *J. Toxicol. Environ. Health*, 5, 1159, 1979.

83. Ehrlich, K. C., Lee, L. S., Ciegler, A., and Palmgren, M. S., Secalonic acid D: natural contaminant of corn, *Appl. Environ. Microbiol.*, 44, 1007, 1982.

84. Cole, R. J., Kirksey, J. W., Cutler, H. G., Wilson, D. M., and Morgan-Jones, G., Two toxic indole alkaloids from *Penicillium islandicum*, *Can. J. Microbiol.*, 22, 741, 1976.

85. Paget, G. E. and Wampole, A. L., Some cytological effects of griseofulvin, *Nature*, 182, 1320, 1958.

22 *Alternaria* Toxins

S. Panigrahi

CONTENTS

Abbreviations used: $LD_{25,50,100}$ (as mg of toxin kg^{-1} body weight) — lethal dose for 25, 50, or 100% mortality; similarly LC, EC, ED, ID = lethal, effective, inhibitory concentration or dose; *i.p.*—intraperitoneal, *s.c.*—subcutaneous, or *i.v.*—intravenous; TA—tenuazonic acid, AOH—alternariol, AME—alternariol monomethyl ether; ATX—altertoxin; and ALT—altenuene.

22.1   INTRODUCTION

The genus *Alternaria* includes some of the commonest fungi, which occur ubiquitously in temperate and tropical climates. They are found in the soil, in household dust, in decaying organic matter, and in cultivated and noncultivated plants, and have been isolated from pulpy and citrus fruits, vegetables, cereals, oilseeds, edible nuts, beans, flowers, tobacco, grass, silage, hay, and weeds. The genus may comprise up to 100 species (with much nomenclatural confusion[1]), but the most important species is *Alternaria alternata*, a common collective fungus that is aggressive as a saprophyte and causes pathogenic or indefinite opportunistic diseases in many crops. *Alternaria* fungi cause spoilage of most agricultural commodities both in the field and during storage, but due to its high water requirement for growth, the most susceptible commodities are fruits and vegetables. Heavy infestation of cereal leaves, stem, and grains can, however, occur if maturation and harvest coincide with wet and warm weather. *Alternaria* are weakly invasive so that they may not penetrate unblemished commodities, but invade through imperfections such as a thin layer of skin, or damaged parts that are weakened due to over-ripening or chilling during cold storage.

Alternaria fungi damage plants through secondary metabolites, substances that apparently play no function in the primary metabolism of the organism. There is much concern about the harmful effects of these compounds on agricultural productivity, spoilage of commodities during transport and storage, and acute and chronic toxicity to humans and animals consuming contaminated food. Approximately 125 secondary metabolites of *Alternaria* are known (Table 22.1 collated from literature), three-quarters of which are reported as phytotoxic, and a quarter mycotoxic. Mycotoxicity studies of individual compounds have been rather limited due perhaps to low natural incidence of the major *Alternaria* toxins reported for the important food crops,[2] and because it is assumed that damaged fruits and vegetables will be discarded by appearance and thus, did not constitute a hazard. More studies are, however, warranted in view of the widespread occurrence of the genus, and the high toxicity reported for crude *Alternaria* culture extracts. For example, one study showed 78% of 83 *Alternaria* isolates to be orally toxic to rats, in comparison with 61% of 152 *Fusarium* isolates, 43% of 197 *Aspergillus* isolates, 47% of 196 *Penicillium* isolates, and 47% of all 943 fungal isolates tested; in another study, extracts of some *Alternaria* isolates were as or more toxic to brine shrimps, chick embryos, and rats than 13 other genera of fungi.[2] Further, of 176 *Alternaria* isolates examined from samples of barley in Spain, 89% produced TA, whereas only 6% of the 190 isolates of *Aspergillus flavus* produced aflatoxin.[3]

Although this paper focuses on mycotoxicity, an overview of *Alternaria* phytotoxins is necessary because of the possible correlation between plant and animal toxicity.[4] Further, lack of toxicity studies on numerous secondary metabolites necessitates the preparation of a listing of all compounds to consider overall hazards. First, however, the pitfalls in assessing toxic hazards are discussed, which also serves to provide some insights into the nature of *Alternaria* toxins and the methodologies appropriate for studying them. The acute toxicity associated with *Alternaria* fungi are than described in relation to studies in which contaminated commodities or culture extracts were tested. Following this, individual compounds are assessed by reference to these reports and examining others in which the pure compounds were tested. Mutagenicity and carcinogenicity, and inhalational and dermal toxicity are treated separately in view of their chronic toxicity implications and hazards of particular concern to humans. A section then considers synergism between fungal metabolites and, finally, a brief account of phytotoxicity is presented.

22.2   METHODOLOGICAL CONSIDERATIONS

Fungi toxicity is assessed on the basis of reports of bioassays and feeding trials with small animals conducted by different groups of workers using either the contaminated raw material, crude extracts of it, crude extracts of cultures, or purified mycotoxins. Several reports need to be examined taking particular note of bioassay design,[5] in particular the following aspects.

First, a single *Alternaria* species may produce a number of mycotoxins, known and unknown, which can act synergistically to increase toxicity, as demonstrated for other mycotoxins.[5] Thus, culture extracts, which may contain several toxins, have generally been reported to be more toxic than the compounds assayed.[6,7,8] Second, the toxins and quantities produced in the field or in the culture will vary with conditions such as the fungal species and strain, substrate, wavelength of light, temperature, and pH[9,10,11,12,13] with the possibility of antagonistic interactions taking place between different fungal species leading to suppression or degradation of a metabolite or the production of a new metabolite;[14,15] field samples generally contain a host of different fungal species. Third, the chemical form of a compound

TABLE 22.1
Classification of *Alternaria* Mycotoxins According to Species and Nature of Toxicity

Fungal Metabolite	Source of *Alternaria* and Other Fungi	Mycotoxic Effects	Phytotoxic Effects	Host-specificity
Tenuazonic acid	*alternata* (tenuis) (*longipes*), *mali, citri, cheiranthi, kikuchiana, raphani, tomato, brassicicola, consortiale, japonica, tenuissima, solani;* wild *Alternaria; Pyricularia oryzae, Aspergillus* spp., *Pleospora infectoria* ("*Alternaria*" state), *Phoma sorghina, Sphaeropsidales* sp.	+	C,G,S,N,P	–
Alternariol	*alternata* (tenuis) (*longipes*), *solani, dauci, cucumerina, cheiranthi, tomato, kikuchiana, raphani, brassicicola, consortiale* (=*Ulocladium consortiale*), *hemicota, tenuissima,* wild *Alternaria, Pleospora infectoria* ("*Alternaria*" state)	+	C	–
Alternariol methyl ether	*alternata* (tenuis) (*longipes*), *solani, cucumerina, raphani, brassicicola, dauci, chartarum, consortiale, hemicota, tomato, kikuchiana, tenuissima,* wild *Alternaria, Pleospora infectoria* ("*Alternaria*" state)	+	C	–
Altenuene	*alternata* (tenuis), *raphani,* wild *Alternaria*	+	C	?
5-epialtenuene	*alternata*	?	+	–
Neoaltenuene	*alternata*	?	?	?
Isoaltenuene	*alternata*	?	N	?
Altenuisol (=Altertenuol)	*alternata* (tenuis)	+	?	–
Altenusin	*alternata* (tenuis)	?	?	?
Dehydroaltenusin	*alternata* (tenuis)	?	?	?
Altertoxin I (=Dihydroalterperylenol)	*alternata* (tenuis), *mali, tenuissima, raphani, chartarum, consortiale, cassiae, tomato,* wild *Alternaria*	+	+	–
Altertoxin II (=Stemphyltoxin II)	*alternata, mali, cassiae, tomato, Stemphylium botryosum*	+	+	?
Altertoxin III	*alternata, tomato*	+	?	?
Stemphyltoxin III	*alternata, Stemphylium botryosum* var. *lactucum*	+	?	?
Altenuic acids, I, II & III	*alternata* (tenuis)	?	C	?
Alternarian acid	*mali, porri*	?	+	–
Alternaric acid	*solani, mali, oryzae*	?	C,N,M,F,R	potato, tomato

TABLE 22.1 (CONTINUED)
Classification of *Alternaria* Mycotoxins According to Species and Nature of Toxicity

Fungal Metabolite	Source of *Alternaria* and Other Fungi	Mycotoxic Effects	Phytotoxic Effects	Host-specificity
Alternarine	*solani*	+	+	–
Phytoalternarins A, B, and C	*kikuchiana*	?	+	–
Tentoxin	*alternata (tenuis), porri, mali, brassicae, brassicicola*	+	C	–
Dihydrotentoxin	*alternata*	?	?	?
Dihydrotentoxin dipeptide	*alternata*	?	?	?
3,6,8-trihydroxy, 3-methyl-3,4 dihydroisocoumarin	*kikuchiana*	–	+	–
Ergostatetraenone	*alternata, brassicicola, kikuchiana*, other genera	+	?	–
Ergosterol	*alternata, brassicicola, kikuchiana*, other genera	+	?	?
Bostrycin	*eichhorniae*	+	+	–
4-deoxybostrycin	*eichhorniae*	+	+	–
Brefeldin A (=Ascotoxin, Ducumbin, Cyanein)	*carthami, Ascoshyta imperfecta, Penicillium simplicissimum, P. brefeldianum, P. decumbens, P. cyaneum, Curvularia lunata, C. subulata, Nectoria radiciola*	+	W	?(safflower ?)
Dehydrobrefeldin A	*carthami*	?	N	?
Curvularin α (& β?)	*cinerariae, macrospora, cucumerina, tomato, Curvularia* sp., *Penicillium* sp.	+	+	–
αβ-dehydrocurvularin	*cinerariae, macrospora, cucumerina, radicina, tomato, Curvularia* sp., *Alternaria* state of *Pleospora scirpicola* (=*Cercospora scirpicola*), *Aspergillus aerofulgens*	+	+	–
β-hydroxycurvularin	*cinerariae, tomato*	?	+	
Destruxin A	*brassicae, Metarrhizium anisopliae* (=*Oospora destructor*)	+	+	(*Brassica* sp ?)
Destruxin B	*brassicae, Aspergillus ochraceus, Metarrhizium anisopliae*	+	C,N,S	(*Brassica campestris* ?)
Destruxin B2	*brassicae*	?	?	?
Homodestruxin B	*brassicae*	?	?	?
Desmethyldestruxin B	*brassicae*	?	?	?
Brassicicolin A	*brassicicola*	+	+	?
Altersolanol A (=Stemphylin)	*porri, solani, Phomopsis juniperovora, Dactylaria lutea*	+	?	?(stoneleek?)
Altersolanol B (=Dactylarin)	*porri, solani, Phomopsis juniperovora, Dactylaria lutea*	+	+	?(stoneleek?)
Altersolanol C (=Dactylariol)	*porri, solani, Dactylaria lutea*	+	+	?(stoneleek?)
Altersolanols D, E & F	*porri*	+	?	?
Tetrahydroaltersolanol B	*solani*	?	N	–
Hexahydroanthronol	*solani*	?	+	–

TABLE 22.1 (CONTINUED)
Classification of *Alternaria* Mycotoxins According to Species and Nature of Toxicity

Fungal Metabolite	Source of *Alternaria* and Other Fungi	Mycotoxic Effects	Phytotoxic Effects	Host-specificity
Solanapyrones A, B & C	*solani*	?	N	–
Anthraquinone A (=2-methylxanthopurpurin-7-methylether)	*porri, solani* (?)	?	N	?(stoneleek)
Anthraquinone B (=3,4,5, tri-hydroxy-7-methoxy-2-methyl-anthraquinone)	*porri, solani*	?	N	?(stoneleek?)
Anthraquinone C (=Macrosporin)	*solani, porri* (=*Macrosporium porri*), *bataticola, cucumerina, Phomopsis juniperovora*	?	N	?(stoneleek?)
3-methyl-xanthopurpurin-6-methyl-ether)	*porri, solani*	?	?	?
6-methyl-xanthopurpurin-3-methyl-ether)	*bataticola*	?	?	?
1,2,8-trihydroxy-6-methoxy-3-methyl anthraquinone	*porri, solani*	?	?	?
5-hydroxy-7-methoxy-2-methyl anthraquinone	*porri, bataticola, cucumerina*	?	?	?(stoneleek?)
Erythroglaucin (=1,4,5-tri-hydroxy-7-methoxy-2-methyl-anthraquinone	*porri, Aspergillus chevalieri, A. ruber*	+	?	?(stoneleek?)
Physcion (=4,5-dihydroxy-7-methoxy-2-methyl-anthraquinone)	*porri, Aspergillus chevalieri*	+	?	?(stoneleek?)
Alterporriols A, B & C	*porri*	?	+	?(stoneleek?)
Alterporriols D & E	*porri, solani*	?	+	?
Stemphyperylenol	*cassiae, Stemphylium botryosum*	?	+	?(crab grass?)
Porritoxin	*porri*	?	S	?
6-(3,3-dimethylallyloxy,4-methoxy-5-methylphthalide	*porri*	?	?	?
3,5,dihydroxy-7-methoxy-2-2-methyl anthraquinone	*porri*	?	?	?
Porriolide	*porri*	?	RL	?
Zinniol	*zinniae, dauci, cichorii, porri, solani, tagetica, carthami*	+	W,C,N	–
Zinnolide	*solani*	?	S,RL	?
Zinnimidine	*cichorii*	?	N	?
Z-hydroxyzinnimidine	*cichorii*	?	N	?
Zinnol	*cichorii*	?	N	?
Zinndiol	*cichorii*	?	N	?
Cichorine	*cichorii*	?	N	?
Radicinin (=Stemphylone)	*helianthi, chrysanthemi, radicina* (=*Stemphylium radicinum*), *Cochliobolus lunatus*	+	N	–
Radicinol	*radicina, chrysanthemi, Cochliobolus lunatus*	?	N	?
3-epiradicinol	pair culture: *chrysanthemi + longipes*	?	?	?

TABLE 22.1 (CONTINUED)
Classification of *Alternaria* Mycotoxins According to Species and Nature of Toxicity

Fungal Metabolite	Source of *Alternaria* and Other Fungi	Mycotoxic Effects	Phytotoxic Effects	Host-specificity
3-epideoxyradicinol	*helianthi*	?	+	?
Deoxyradicinol	*helianthi*	?	+	?
4-deoxyradicinin	*helianthi*	?	+	–
Radianthin	*helianthi*	?	+	+
Pyrenphorin	*radicina, Pyrenophora avenae*	?	+	+
Pyrenocine A (=Citreopyrone)	*helianthi, Pyrenochaeta terrestris, Penicillium citreoviride*	?	+	–
Pyrenocine B	*helianthi, Pyrenochaeta terrestris*	?	+	–
Altenin	*kikuchiana*	?	+	?
Alteichin (=Alterperylenol)	*eichhorniae, cassiae alternata*	?	+	–
Dihydrosporogenin AO-1	*citri*	?	?	?
Abscisic acid	*brassicae, Botrytis cinerea, Cercospora rosicola*, higher plants	?	?	?
AL toxin	*eichhorniae*	?	N	?
AT toxin	*alternata (longipes)*	?	N	tobacco
Maculosin	*alternata*	?	N	knapweed
Alterlosin I & II	*alternata*	?	N	?(knapweed?)
ACR (or ACRL) toxins I-IV	*citri (alternata)*	?	N,O	rough lemon
ACR compound A	*citri (alternata)*	?	?	(inactive on rough lemon)
ACT (or ACTG) toxins A-H	*citri (alternata)*	?	N	tangerine, mandarin
ACT toxin Ib, II, c (?)	*citri (alternata)*	?	VN,E,M	tangerine, mandarin
AF toxin I & III	*alternata*	?	N,E,M	strawberry, Japanese pear
AF toxin II	*alternata*	?	N,E,M	Japanese pear
AK toxin I & II	*kikuchiana (=alternata)*	?	VN,P,E,M	Japanese pear, apple
AM toxin I (=alternariolide)	*mali*	?	VN	apple, pear
AM toxin II & III	*mali*	?	N,E	apple, pear
AAL-toxin (s?)	*A. alternata* f. sp. *lycopersici*	+	N	tomato
1-aminodimethylheptadecapentol	*A. alternata* f. sp. *lycopersici*	?	+	–
Fumonisin B$_1$	*A. alternata* f. sp. *lycopersici, Fusarium moniliforme, F. proliferatum*	+	+	–
9-10-epoxy-8-hydroxy-9-methyl-deca-trienoic acid	*kikuchiana*	?	?	?

Note: Species names for different morphologically-identical pathotypes are shown in brackets. + toxic; – not toxic; ? - toxicity cannot be ruled out. Mycotoxicity refers to effects in whole animal, cell culture, protozoa, bacteria and/or other *in vitro* bioassays, while phytotoxicity refers to activity in fungi, yeast, and algae, and all types of plant bioassays and metabolic investigations. Phytotoxicity key: W—wilting; C—chlorosis; G—germination inhibited; S—stunted growth; RL—root elongation reduced; F—anti-fungal activity, R—cell respiration inhibited; E—electolyte loss; P—protein synthesis inhibited; N—necrosis; VN—veinal necrosis; O—oxidative phosphorylation inhibited, M—plasma membrane potential or integrity changed.

may affect toxicity. TA is optically active and converts to iso-TA, a mixture of D-*allo*-TA and TA. It is therefore stored as a dibenzylethylenediamine or copper salt,[16] the latter as a mono- or tri-hydrate. Each salt has a different molecular weight so that the concentration of the biologically active component should be reported. More importantly, TA exhibits stereospecificity in some biological systems. D-*allo*-potassium-TA was no more than 25% as anti-tumorogenic against human adenocarcinoma (Hadl) as sodium-L-TA, the D-isomer being inactive; similarly, its cytotoxicity to Eagle's KB cell carcinoma were 40–90, 1300, and 300 µg ml^{-1} for the L-, D-, and D-*allo* isomers.[17] However, its toxicity to *Bacillus megaterium* was not stereospecific. The dibenzylethylenediamine-TA was more toxic than the sodium salt to rats and mice *i.v.*, but it was less toxic orally.[18] The solubility and stability of toxins in different carrier solvents frequently give problems in bioassays. ATX-II and III were unstable in dimethylsulfoxide, thus affecting mutagenicity tests.[19] ATX-I and II, and AOH and AME had low solubility in methanol that limited their testing in brine shrimp bioassay[20] to 20 and 200 µg ml^{-1}, respectively. AME also posed a solubility problem in chloroform.[21] Only 58 to 74% of the AOH, AME, ATX-1, and TA were recovered after 28 days, AME being prone to loss on exposure to air or ultraviolet light.[16] Fourth, some toxins may be difficult to quantify due to lack of standard compounds of reliable purity and stability.[16] Many methods require standardization; for example, AME and AOH interfered with the analysis of zearalenone and aflatoxin,[22] the ACT-toxin 1c might be an artefact of ACT-toxin 1b exposed to light during purification,[23] more than one compound has been referred to as ATX-II, and the structure of altenuisol has been questioned.[16] Fifth, toxicity varies with animal species, sex, age, and the route of administration. The gastrointestinal tract represents a barrier to absorption, and in ruminants, the rumen may reduce toxicity; thus, compounds active in *in vitro* bioassays and *in vivo* studies involving *i.p.*, *s.c.*, or *i.v.* administration may not pose a hazard orally, Further, oral studies should be interpreted in different ways depending on whether feeding trials were conducted, or by whether toxins were given by stomach intubation (gavage); the gavage method is useful because it precludes feed refusal by animals and yields data that are directly comparable with those from *i.v.* or *i.p.* routes, while the dietary method allows depression of feed intake to be used as a measure of toxicity, and it represents the practical situation. The chick embryo bioassay data is particularly difficult to interpret[24] due to uncertainties on how well different toxins are absorbed into the embryo by the air-sac and yolk methods of administration at different stages of incubation.

22.3 TOXICITY OF CONTAMINATED FOODS AND CULTURE EXTRACTS

22.3.1 ORAL ADMINISTRATION TO ANIMALS

The following reports demonstrate the acute oral toxicity of *Alternaria* contamination in mammals. Of 85 isolates of *A. alternata* cultured on corn-rice substrate and fed to rats, 67% were lethal within 10 days, with symptoms of anorexia, weight loss, and intestinal hemorrhages.[25] Other studies showed that culture extracts of 78% of 83,[2] and 95% of 21 *Alternaria* isolates were lethal to rats within 7 days when given orally (see Reference 26). Female mice fed *Alternaria mali* culture lost weight and died within 10 weeks;[27] males were less severely affected than females, the LD$_{50}$ being 367 and 1100 mg of culture extract (consumed by 20–25 g mice), respectively. Symptoms included hyperkeratosis of the forestomach, pulmonary hemorrhage, and vacuolated liver cord cells.

Weather-damaged sorghum grains containing 10.8 mg kg^{-1} of AOH, 5.4 mg kg^{-1} of AME, and 1.26 mg kg^{-1} of zearalenone depressed pig growth without reducing digestibility, indicating that the poorer performance was due to the mycotoxins.[28] *A. alternata* culture given in drinking water decreased selenium-glutathione peroxidase and superoxide dismutase activities and increased lipid peroxidation, symptoms resembling the oxygen radicals-induced damage in humans with Kaschin-Beck disease.[29] *A. alternata* f. sp *lycopersici* caused 100% mortality when cultures were fed to rats in a diet or given by gavage, with symptoms of intestinal hemorrhage and hematuria.[8] Fodder on which an *Alternaria* sp. (possibly *A. alternata*) was propagated may have caused the death of 39 out of 40 rabbits, which had symptoms of catarrhallic inflammation of the small intestines.[30]

In avian species, *Alternaria tenuissima* culture extracts produced 100% mortality within 14 days in day-old cockerels by a single gavage dose.[31] Two of four isolates of *A. alternata* were toxic when *Alternaria* culture was fed to chicks at 500 g kg^{-1} diet.[32] Culture extracts of 31 out of 96 isolates of *Alternaria longipes* fed to chicks produced death, with a further 12 of isolates depressing weight gain;[33] affected chicks displayed poor appetite, lethargy, diarrhea, loss of muscular control, followed by coma and death in 4–8 days, and *post mortem* examination showed hemorrhages in the proventriculus and gizzard erosion. *Alternaria*-contaminated feed caused diarrhea, depression, prostation, and death of 19 out of 20 young chicks within 3 days, and it was also toxic to 6 week-old chicks,[34] which became emaciated and developed hemorrhages and congestion in the skin, skeletal muscle, heart, thymus, proventriculus, gizzard (which were also eroded), intestines, caeca, lungs, liver, and kidneys; in some experiments, congestion, vacuolation, fatty infiltration, and paleness of the liver and kidneys, kidney tubule necrosis, and round cell infiltration, hyalinization of the spleen and decreased formation of blood elements in bone marrow were also observed. Moldy sorghum naturally contaminated with 10.0 mg kg^{-1} of AOH and 3.6 mg kg^{-1} of AME, or 7.2 mg kg^{-1} of each reduced weight gain and feed conversion efficiency, with increased liver, spleen, and pancreas weights and regressed Bursa of Fabricius.[35] A culture extract of *Alternaria tenuis* given in the diet produced mortality in 10 out of 15 goslings within 15 days, with symptoms of subcutaneous hemorrhages and diffusion of blood.[36]

22.3.2   Injection of Culture Extracts to Animals

To determine *in vivo* toxicity as distinct from assessing oral hazards, test materials are injected. The tetrahydrofuran culture extracts of 71% of 11 *Alternaria* isolates were lethal to mice at 300 mg kg^{-1} *i.p.*[2] Culture extracts (of most ?) of 27 *A. alternata* isolates given to male mice *i.p.* were highly toxic within 24 hours, the LD$_{50}$ (mg kg^{-1}) ranging between 14.9 and 26.8.[37] *A. mali* culture extract at 300 mg kg^{-1} *i.p.* produced death of three out of six mice (both sex), with symptoms of hyperkeratosis of the forestomach.[27]

22.3.3   Bench-Scale Bioassay of Culture Extracts

Toxicity, rather than oral hazards, may also be assessed by conducting bioassays using organisms that can be easily handled in general purpose laboratories. Of 33 fungal isolates from apples, the culture extracts of which were toxic to brine shrimp larvae, 31 were *Alternaria* spp., with another study showing *A. alternata* and *A. tenuissima* to be toxic.[2] Zajkowski et al.[6] reported that 10 out of 22 culture extracts of *A. alternata*, *A. raphani*, *A. consortiale*, and *A. chartarum* were toxic in the brine shrimp motility test, the EC$_{50}$ ranging between 0.65 to 144 mg ml^{-1}; there was no correlation between toxicity and the quantity of AME, AOH, ALT, ATX-I, and TA present in the extracts, and synergy between toxins could not be demonstrated. Some isolates even stimulated larvae motility. Similarly, rice culture extract of 76 out of 101 isolates belonging to several identified *Alternaria* spp. were toxic to brine shrimp larvae,[7] with several toxic extracts not containing TA, AOH, or AME, while some nontoxic extracts contained all three compounds.

In bacterial bioassays, *A. mali* mycotoxins appear particularly toxic to gram-positive bacteria such as *Staphylococcus albus*, *S. aureus*, *Sarcina lutea*, and *Gaffkya* sp.[27] In one study, out of 127 *Alternaria* species tested, 86 produced antibiotics, 75 were active against gram-positive bacteria, 24 against gram-negative, 43 against acid-fast, 19 against gram-positive and gram-negative, and 10 against all three.[38] Interestingly, there were combinations of species that were anti-bacterial and anti-yeast; anti-bacterial and anti-fungal; or anti-bacterial, anti-yeast, and anti-fungal; five were active in all tests including being anti-algal, while 41 species were inactive. Of particular interest was that most isolates of *Alternaria brassicicola* exhibited strong broad-spectrum anti-microbial activity, but *Alternaria brassicae* was considerably less active. Culture filtrate of *A. alternata* isolated from aubergine inhibited the growth of *Bacillus megaterium* and *Bacillus subtilis*.[39]

In the chick embryo bioassay, reports include the finding that the culture extracts of the majority of 27 isolates of *A. alternata* tested were toxic to the embryos by the air-sac method, but with marked variation in toxicity.[37] *A. tenuissima* was shown to be toxic in another study.[31] In cytotoxicity tests, culture extracts of *A. brassicae* and *A. brassicicola* (containing compounds that were not tentoxin, AOH, or AME) were toxic to human epithelial (HEp2) cells.[40] *A. mali* culture extracts at 5, 10, and 20 μg ml^{-1} caused 80–100% inhibition of HeLa cell growth by 72 hours, the highest dose producing partial cell death in 24 hours.[27]

22.4 TOXICITY OF PURIFIED MYCOTOXINS

22.4.1 TENUAZONIC ACID

The study of Meronuck et al.[25] implicated TA as the primary compound in *Alternaria* toxicity in rats, with 20 of 23 toxigenic isolates producing it and hematological symptoms being of particular significance. In another study,[35] absence of TA did not produce any hematological symptoms in chicks even though growth was depressed; and while it is possible that pig performance was depressed by a combination of AOH, AME, and zearalenone in the study of Williams et al.,[28] TA was not assayed and its presence in the diet could not be ruled out. Of particular interest is the report of Sauer et al.[32] that diets containing as high as 39 mg kg^{-1} of AOH, 24 mg kg^{-1} of AME, and 10 mg kg^{-1} of ALT (three to five times greater than the concentrations in the aforementioned studies) were nontoxic, while those containing TA and ATX were lethal to chicks and rats. These studies indicate the high acute toxicity of TA, which is corroborated by studies using the purified compound. In male mice, single doses of sodium-TA gave LD$_{50}$ values (in mg kg^{-1}) of 125 *i.v.*, 150 *i.p.*, 145 *s.c.*, and 225 orally; whereas monkeys tolerated a single gavage dose of 50 mg kg^{-1}, and at 300 mg kg^{-1} five out of six died.[41] Another study[18] determined the LD$_{50}$ (in mg kg^{-1}) of sodium-TA by oral and *i.v.* routes, respectively, as follows: in rats, males 180 and 146, females 168 and 157; in mice, males 186 and 162, and females, 81 and 115. These data may suggest that, in female mice, sodium-TA might be at least as toxic (if not more) by oral as by *i.v.* route. Dogs were much more susceptible to TA, 60 mg kg^{-1} in three quick equal doses *i.p.* producing death with severe hemorrhagic gastroenteropathy, and two out of two dogs becoming moribund by daily doses of 10 mg kg^{-1} in 7 days. Oral toxicity of dibenzylethylenediamine-TA was 2–4 times lower in monkeys than in dogs. Symptoms of salivation, emesis, bloody diarrhea, and hemorrhagic gastroenteropathy led the authors to suggest that TA has an emetic and a cardiovascular effect, the latter causing circulatory failure, leading to tachycardia, hemconcentration, elevated bromosulphonphthalein retention and serum glutamic oxaloacetic transaminase activity, and hemorrhagic gastroenteropathy. TA may be more toxic to day-old chicks than to mammals, the gavage LD$_{50}$ being 38 mg kg^{-1}, and the majority of birds dying within 4 days with hemorrhages of the musculature, heart, and subcutaneous tissues.[42] Dietary concentrations of TA at 10 mg kg^{-1} or daily gavage doses of 1.25 or 2.50 mg kg^{-1} to three-week old chicks depressed weight gain and feed efficiency; gross symptoms included enlarged and mottled spleen, gizzard erosion, hemorrhages of the intestinal lumen, thigh muscle, and heart, and myocardial edema, while microscopic examination showed that the livers and kidneys has also developed congestion of blood vessels and hemorrhage. Liver hematoma has been observed by this author in chicks fed diets with 15 mg kg^{-1} of TA.

There may be a link between these hematological symptoms and Onyalai disease in humans in southern Africa. Onyalai is a noninfectious, acute, purpuric disease in localities where people consume stored sorghum and millets. This disease is characterized by distinctive hemorrhagic bullae in the mouth similar to symptoms of idiopathic thrombocytopenia. Studies revealed a correlation with the incidence in cereals of the TA-producing *Phoma sorghina*, cultures of which produced hemorrhages in the gastrointestinal and respiratory tracts and rapid death when fed to chickens. Other isolates caused severe weight loss and death in rats, with symptoms of hematuria and epistasis, and blood in the feces; *post mortem* observations included blood in the mouth and nose, hematuria, blood in the stomach and intestines, echymoses, petechiae in stomach, internal organs, and muscles. Steyn and Rabie[43] speculated that TA's toxicity might in part be due to its ability to complex with trace metals *in vivo*. The hematological symptoms have also been related to poultry hemorrhagic disease (PHD).[34] PHD has a range of characteristics and causes (including avitaminosis K), but fungi such as *Penicillium rubrum, P. purpurogenum*, and *Alternaria* sp. that are normally found in the litter of broiler houses, were especially implicated. PHD may appear in birds between 3–12 weeks of age, reaching a peak at 4–7 weeks, and declining thereafter to be absent at 12 weeks; the appearance perhaps related to dampness of the litter.

TA is toxic to a diverse range of biological systems, its LC_{50} being 120 µg ml^{-1} in *Lucilla sericata* insect larvae[44] and 75 µg ml^{-1} in brine shrimp larvae.[21] In 7-day chick embryos (yolk-sac method), the LD_{50} of TA was 548 µg per egg (without malformations);[45] however, in another study the LD_{50} and 5 mg kg^{-1} (300 µg per egg) for 10 and 18-day embryos, respectively.[11] Since compounds are generally less toxic in older organisms, it may be possible that this bioassay is influenced by absorption-related problems, with increasing age perhaps resulting in greater absorption, and hence, toxicity. TA was toxic to *Bacillus megaterium* at more than 100 µg per disc,[17] and to rumen micro-organisms as detected by an *in vitro* fermentation bioassay in which gas production was recorded as the end-point (author's unpublished work). TA was cytopathic to several types of viruses,[41] with its LC_{50} in herpes simplex (HF) being 160 µg ml^{-1}. Sodium L-TA was anti-tumorogenic *in vitro* to human adenocarcinoma, the ED_{60} being 126 µg per egg,[17] and to bronchogenic carcinoma A-42.[2] TA also arrested metaphase in Yoshida sarcoma cells.[46] TA inhibited the incorporation of glycine, leucine, phenylalanine, lysine, and valine into cellular proteins of intact Ehlrich Ascites tumor cells, and of glycine and lysine *in vivo* in the liver, spleen, thymus, and intestinal mucosa of rats.[47] These and other studies using rat liver and cell-free systems indicated that TA inhibits protein synthesis by suppressing the release of newly-formed microsomal proteins. However, another study showed that TA blocked peptide bond formation in ribosomes *in vitro*, strongly in human tonsil and pig liver, to a limited extent in *Euglena gracilis* and *Phaseolus vulgaris*, and not significantly in a yeast species.[48]

22.4.2   ALTERTOXINS

As a group, the altertoxins may be the next most toxic of the *Alternaria* mycotoxins. The LC_{50} of ATX-I in brine shrimp larvae[21] was 200 µg ml^{-1}. ATX-I and ATX-II produced 100% mortality in female mice at 200 mg kg^{-1} *i.p.*;[49] the toxicity of ATX-I was characterized by inactivity, subendocardial and subarachnoid hemorrhage, and blood in cerebral ventricles. Both compounds were toxic to *B. mycoides* at 250 µg per disc, whereas, the ID_{50} of ATX-I in HeLa cells was 20 µg ml^{-1}, and for ATX-II it was only 0.5 µg ml^{-1}; the latter was the most toxic of the seven *Alternaria* mycotoxins tested in this bioassay. In a comparative cytotoxicity test and a test to determine inhibition of gap junction communication in the Chinese hamster lung metabolic cooperation assay (V79), ATX II was also found to be the most cytotoxic (0.02–0.0008 µg ml^{-1}) followed by ATX-III (0.2–0.04 µg ml^{-1}) and ATX-I (1–5 µg ml^{-1});[50] however, of the three, only ATX-I disrupted metabolic communication at 4 µg ml^{-1}.

22.4.3 ALTERNARIOL

AOH was lethal to 3 out of 10 female mice at 400 mg kg^{-1} *i.p.*[49] The LC$_{50}$ in brine shrimp larvae[21] was 100 μg ml^{-1}, and it was nontoxic to chick embryos at 1 mg per egg.[45] In different studies, AOH was anti-bacterial to *Staphylococcus aureus* at 25 μg ml^{-1}, *Escherichia coli* at 50 μg ml^{-1}, and *Bacillus mycoides* at 60 μg per disc; AOH might have different effects on gram-positive and gram-negative bacteria, since in one study (see Reference 2) inhibition took place against *S. aureus* at 5 μg ml^{-1}, for a *Corynebacterium* sp. at 10 μg ml^{-1}, and for *E. coli* at 80 μg ml^{-1}. AOH was toxic to HeLa and mouse lymphoma cells,[49] the ID$_{50}$ being 6 μg ml^{-1}, and at 15 μg ml^{-1}; it inhibited cell replication completely.[51] AOH also exhibited competitive anticholinesterase activity *in vitro*.[52]

22.4.4 ALTERNARIOL MONOMETHYL ETHER

In female mice, AME was only slightly toxic at 400 mg kg^{-1} *i.p.*[49] In male rats, daily gavage dose of 3.75 mg AME for 30 days produced no significant effects on weight gain, serum enzyme levels, and gross morphological changes in the major body organs;[53] however, in pregnant Syrian golden hamsters, at 200 mg kg^{-1} *i.p.* AME was maternally toxic. Dietary AME at 100 mg kg^{-1} produced no ill effects in day-old chicks, and at 500 μg per egg in the 7-day embryo it produced no ill effects.[45] AME also showed relatively low toxicity to *Bacillus mycoides* at 500 μg per disc, although it was more toxic to HeLa and mouse lymphoma cells,[49] the ID$_{50}$ being between 8–14 μg ml^{-1}. The low toxicity indicated for AME may perhaps be due to its poor absorption from the gastrointestinal tract;[2] however, AME's poor solubility also raises doubt about its suitability for testing in certain bioassays.[21]

22.4.5 ALTENUENE

ALT given *i.p.* at 50 mg kg^{-1} was lethal to one out of three female mice tested.[49] The LC$_{50}$ of ALT to brine shrimp larvae[21] was 375 μg ml^{-1}. It was toxic to 5 gram-positive and gram-negative bacteria at greater than 250 μg per disc,[2] to *Bacillus mycoides* at 125 μg per disc,[49] and HeLa cells with an ID$_{25}$ of 28 μg ml^{-1}. ALT did not effect the development of 7-day-old chick embryos at 1 mg per egg.[45]

22.4.6 OTHER MYCOTOXINS

Of nine rat hepatoma cell lines tested, seven were sensitive to fumonisin B$_1$ and AAL toxin,[54] the IC$_{50}$ for the most sensitive line being 4 and 10 μg ml^{-1}, respectively. However, in rats 21 mg of fumonisin B$_1$ or 3.6 mg of AAL-toxin by gavage did not produce any acute toxicity, although the culture extracts from which the compounds were isolated were highly toxic.[8] This may indicate either that the compounds are of low toxicity or that they interact with other toxins to produce their effects. It should be noted however that fumonisin B$_1$ is implicated in equine leukoencephalamalacia in horses, and is a hepatotoxin and a carcinogen in rats.[55]

In male mice, the LD$_{50}$ of physcion *i.p.* was 10 mg kg^{-1}, but it appeared to be nontoxic by gavage to day-old cockerels and mice at about 100 mg kg^{-1}, and to chick embryos at 1 mg per egg;[24] erythroglaucin may be more toxic than physcion to chick embryos, but less toxic to mice. Alternarine displays broad-spectrum anti-microbial activity, including to a number of gram-positive and gram-negative bacteria (see Reference 27). Altenuisol was toxic to *Bacillus mycoides* at 5 μg per disc,[49] and its ID$_{50}$ in HeLa cells was 8 μg ml^{-1}. Dactylarin was slightly toxic to gram-positive bacteria, but strongly inhibitory to the protozoans, *Leishmania braziliensis* and *Entamoeba invadens* (but nontoxic to some others); and it was cytotoxic to HeLa cells.[56] Injection *i.p.* of destruxins A and B caused rapid convulsions and death by a lowest dose, respectively, of 1.4 and 16.9 mg kg^{-1} in mice, and 0.14 and 0.17 μg ml^{-1} in 5th

Instar silkworm larvae, while dietary presentation to the potato beetle *Epilachna sparsa* caused reduction in feed intake and death within 24 hours;[57] mosquito larvae are also susceptible to these toxins. Radicinin was insecticidal, and anti-bacterial to *Staphylococcus aureus* and a *Clostridium* sp.[58] At between 12.5 and 100 µg ml[-1], most altersolanols (except D and F) were toxic to *Staphylococcus aureus*, *Micrococcus luteus*, *Bacillus subtilis*, and *Pseudomonas aeruginosa*, but not to *Escherichia coli*, one yeast, and two species of fungi.[59] Curvularin and αβ-dehydrocurvularin were toxic to *Bacillus subtilis*, the latter compound also being toxic to *Staphylococcus aureus* and *Escherichia coli*;[60] it is also noteworthy that extracts of *Curvularia* species (which may also produce these compounds) were highly toxic to chick embryos.[31] Brassicicolin A was toxic[61] to *Corynebacterium fascians* and *Bordetella bronchiseptica* at 12.5 µg ml[-1], and to *Bacillus subtilis* at 100 µg ml[-1]. Brefeldin A was antibacterial at 1 µM and inhibited transport of protein from endoplasmic reticulum to the Golgi apparatus in rats;[62] it was also active against mammalian tumor cells and was antiviral, antifungal, antimitotic, and toxic to nematodes (see Reference 63). Ergostatetraenone, which occurs widely in fungi as one of the earliest metabolites, may be identical to a primary toxin of *Alternaria* sp. isolated from apples, which was more toxic towards female animals, perhaps because of its similarity to ergosterol.[12]

22.5   FETOTOXICITY AND TERATOGENICITY

At 200 mg kg[-1] given *i.p.* to pregnant Syrian golden hamsters on day 8 of gestation, AME was fetotoxic, causing an increased number of resorptions and decrease in fetal weight, but no malformations.[53] AOH was fetotoxic at 100 mg kg[-1] *s.c.* when it was given to mice at 9–12 days of gestation, the percentage of malformed fetuses appearing to increase when administered at 13–16 days.[49] Feeding pregnant mice from day one of gestation on a diet contaminated with a culture extract of *A. alternata* led to a high death rate in mothers but no fetotoxicity;[26] however, daily *s.c.* doses from day 1–6 of an extract that contained ALT, AME, and ATX-I caused some fetotoxicity, while three other extracts containing different toxins proved nontoxic. Feeding *A. solani*-blighted potatoes to hamsters caused no teratogenic effects; however, in pigs and rabbits it increased the number of fetuses, and in rabbits, there were some indications of organogenesis problems in the vertebral column and brain of a few fetuses.[64]

22.6   MUTAGENICITY AND CARCINOGENICITY

All five cultures of *Alternaria* species and *A. alternata* tested proved mutagenic with or without liver microsomal activation in *Salmonella* strains at 0.1 ml ethyl acetate extracts per plate,[65] and cultured spore suspensions of *A. brassicicola* given *i.p.* to mice affected metaphase in bone marrow cells and produced chromosomal aberrations such as acentric fragments, translocation, polyploidy, aneuploidy, and centromeric separation.[66] For purified compounds, TA and AOH were nonmutagenic in *Salmonella typhimurium* at 1000 and 500 µg per plate, respectively, while AME was mildly mutagenic at the lower dose.[67] However, the carcinogenicity of AOH needs to be reconsidered in view of (1) the report that it is a phototoxic DNA-intercalating agent in the dark and a DNA cross-linking mycotoxin in near ultraviolet light,[68] (2) its teratogenicity (with the apparent synergism with AME),[49] (3) the appearance of micronuclei in HeLa and lymphoma cells after AOH treatment, suggesting that it affects chromosomal distribution during mitosis,[51] and (4) the similarity of its structure with that of furanocoumarins.[68]

The altertoxins appear to be strongly mutagenic with or without metabolic activation, ATX-I being active at 3.4 µg per plate[67] in *Salmonella typhimurium* strains. However, ATX-III is the most potent followed by ATX II and ATX-I, the number of revertants produced per ρmole being 0.7, 0.5, and less than 0.03, respectively[19] (mutagenicity was still, at most, one tenth that of aflatoxin B_1, one of the most potent biologically-produced genotoxic compounds known). ATX-I and ATX-III may also be tumorogenic, as indicated by their activation of Raji cell Epstein-Barr virus-early antigen expression, and enhancement of murine fibroblast cell transformation, in which ATX-I was more active.[69] Stemphyltoxin III was also mutagenic in three strains of *Salmonella typhimurium* with or without metabolic activation,[70] and fumonisin B_1 produced primary hepatocellular carcinoma *in vivo* in rats.[55] Many anthraquinones (which often represent considerable quantities in organic extracts of cultures) also possess "DNA attacking ability," for example, physcion was active at 100 μg per plate in one *Salmonella typhimurium* strain with metabolic activation; with erythroglaucin being more potent.[24]

The mutagenic and cell-transforming activities of *Alternaria* mycotoxins may be a causative factor in the etiology of esophageal cancer in the Linxuan Province in China (and parts of southern Africa), where the incidence of this disease was positively correlated with that of *A. alternata* in staple cereal grains.[71] One isolate damaged human cell DNA *in vitro*, while another induced 6-thioguanine-resistant mutants in V79 cells and caused transformation of NIH/373 mouse fibroblast cells. Other studies in this series showed that, with or without metabolic activation, AME induced reverse mutation in *Escherichia coli* ND160 strain, unscheduled DNA synthesis in cultured human amnion FL cells, lymphocytic chromosomal abberation and sister chromatid exchange in human peripheral blood, mutation of V79 cells, and transformation of NIH/373 cells. The transformed cells grew well in agar and were tumorogenic to "nude" mice. Thus, compared with the low mutagenicity reported in *Salmonella typhimurium*,[67] AME was more potent, indicating, perhaps that it is selective to specific genomic regions or DNA sequences.

22.7   INHALATIONAL AND DERMAL TOXICITY

Airborne *Alternaria* spores and conidia can cause inhalational allergies in humans, such as bronchial asthma and hypersensitivity pneumonitis.[72] In view of the high incidence of hay fever and other allergies in Europe during the summer months, it may be relevant that *Alternaria* spores reached a peak in Danish homes during August.[73] The fungi have also been implicated in dermatophytes in cat, dogs, cows, horses, and deers, with symptoms of skin lesions, alopecia, scaling, encrusted lesions, and skin nodules.[74] *A. infectoria* caused phaeohyphomycosis in a cat that was characterized by ulcerated nodular lesions on the nose.[75] In humans, too, *Alternaria* spp. may produce opportunistic cutaneous infections (superficial alternariosis) with severe skin lesions characterized by multiple nonhealing ulcers covered with dry crusts.[76] It seems likely that these effects are mediated through mycotoxins, particularly in view of the reports that a culture extract of *Alternaria* spp.[30] and a PHD factor[34] produced skin inflammatory reactions in rabbits. Since cigarette, cigar, and pipe tobaccos contain *Alternaria* spores, Forgacs and Carll[77] also determined the inhalational effects in mice of smoke from hay on which *Alternaria* had been grown. It caused pulmonary congestion, edema and emphysema, fatty infiltration or vacuolation, necrosis of hepatic cells, hemorrhage, congestion, tubular dilation, and glomerular interstitial nephritis in kidneys; control mice on uninoculated hay were clinically normal, with histological changes being limited to pulmonary chronic inflammation.

22.8   SYNERGISM BETWEEN TOXINS

Alternaria mycotoxins may act synergistically, as indicated by the high acute toxicity for culture extracts in relation to the major pure compounds noted in this review. The most direct synergism reported is that between AOH and AME in the *Bacillus mycoides*,[49] a combination of AOH and AME elicited a zone of inhibition at only 0.25 μg per disc in contrast to the 60 and 500 μg per disc that were required with the individual toxins. In the same study, there were indications of synergism in fetotoxicity and teratogenicity in mice, since a combination of AOH and AME at 25 mg kg^{-1} *s.c.* increased the numbers of dead and resorbed fetuses, runts per litter, and malformed fetuses in excess of that expected on the basis of the potency of the individual compounds. However, no synergism was apparent on maternal mortality in mice, nor on toxicity to HeLa cells.

Aflatoxin B$_1$ can co-occur with TA (and other mycotoxins) in poultry feeds (author's unpublished findings) which raises concern of synergistic toxicity in view of the former's hepatic necrosis and TA's inhibition of protein synthesis. With regard to mutagenicity and carcinogenicity, it is particularly noteworthy that AME and AOH are the most frequently reported *Alternaria* mycotoxins in natural occurrence, and are also generally produced in the largest quantities in culture. It is not clear whether these toxins can interact synergistically with the other food-borne mutagens and/or carcinogens, such as the altertoxins, Stemphyl-toxin III, the anthraquinones, fumonisin B$_1$, and others such as the *Fusarium* and *Aspergillus* mycotoxins and food N-nitrosamines that may be consumed by humans.

22.9   PHYTOTOXINS

Alternaria species are implicated in numerous crop diseases, among which are: leaf spot of cotton by *A. longipes*; black spot on leaves of Japanese pear by *A. kikuchiana* and *A. alternata*; black spot and leaf spot of the oilseed brassicas by *A. brassicae* and *A. brassicicola*; early blight of potato by *A. solani*, collar rot and early blight on tomatoes by *A. solani*; leaf spot of pigeon pea by *A. tenuissima*; and "*Berkak ungu*" of shallots by *A. porri*. The diseases are mediated through secondary metabolites, the effects of which can be so strong that some toxins and the pathogens have been suggested as potential herbicides for weed control, for example, *A. macrospora* for spurred anoda and *A. eichhorniae* for water hyacinth. The compounds (Table 22.1) are of diverse chemical structures, a characteristic that is believed to improve the chances of producing an effective phytotoxin to facilitate invasion of the plant. The mechanisms of phytotoxin production are not well-understood, although factors believed to trigger their production are nutritional imbalance in the fungi[10,12] and chemical compounds produced by plants, in some cases, operating through complex host-parasite interactions. The phytotoxins may be classified under "general toxins" which affect a range of hosts as well as nonhosts, and "host-specific toxins" which are toxic only to the host susceptible to the pathogen and are capable of producing all the disease symptoms. Equally, lack of specificity may indicate a role in symptom development as a secondary rather than a primary determinant of pathogenicity. However, toxins may also act synergistically to produce all the disease symptoms.

Among the general phytotoxins are zinniol, tentoxin, anthraquinones, solanapyrones, alter-porriols, curvularin and its derivatives, radicinin and its derivatives, AOH, AME, ALT, altenuic acid, and TA. These cause effects such as leaf chlorosis (by disrupting chloroplast development in the case of tentoxin), necrotic lesions on leaves, and also activities with uncertain ecological implications, such as inhibitions of seed germination and growth,[20,39] and pollen germination and tube length growth. Most host-specific *Alternaria* phytotoxins such as ACT toxin-1b appear to act by causing veinal necrosis and electrolyte-release through the disruption of plasma membrane integrity, an effect that facilitates infection by the pathogen.[23] Actual degeneration of the membrane may be caused by activation of phospholipase A2 enzyme, in some cases. Other metabolic effects may include inhibition of endogenous respiration by uncoupling of oxidative phosphorylation and alteration of mitochondrial membrane potential.

22.10 CONCLUSIONS

This review has been concerned specifically with the mycotoxicity, and to a lesser extent, the phytotoxicity of *Alternaria* secondary metabolites. The chemistry of these compounds has been presented elsewhere.[62] This chapter has also not discussed the important issue of the hazards arising from natural contamination. This is because very few investigations have been undertaken, especially in tropical developing countries, in which commodities are screened for the presence of mycotoxins produced by several genera of fungi using a multi-mycotoxin assay procedure.

An attempt has been made in this review to compile a list of known and possible *Alternaria* toxins as a guide to identifying areas for future research. It is evident that there are considerable gaps in the literature on the mycotoxicity of *Alternaria* metabolites. A particular problem encountered is that most studies of the lesser known mycotoxins have been restricted to bioassays using microorganisms, such as bacteria, yeast, virus, fungi, and cell cultures without simultaneously conducting animal studies *in vivo*, the crucial data that is needed for assessing hazards. Even the important study of Lindenfelser and Ciegler[38] tested *Alternaria* culture extracts in species belonging to the Monera and Prostita, ignoring the viruses and the plant and animal kingdoms. Furthermore, few studies have tested toxins singly and in combination to examine synergism. Thus, some of the analysis presented in this chapter has been speculative in nature.

Assessment of the toxicities reported for certain *Alternaria* species and the toxins already identified from them also indicates that other phytotoxins and mycotoxins may still be discovered. Species worthy of particular study are as follows: for compounds with phytotoxic and/or mycotoxic activities—*A. anagallidis* var. *lineariae*, *A. angustiovoidea*, *A. battaticola*, *A. brassicicola*, *A. crass*, *A. cucumerina* var. *cyamopsidis*, *A. eichhorniae*, *A. gossypina*, *A. helianthi*, *A. infectoria*, *A. iridis*, *A. macrospora*, *A. mali*, *A. passifloriae*, *A. porri*, *A. radicina*, *A. solani*, *A. tenuissima* and *A. zinniae*; and for compounds that are only phytotoxic—*A. anagallidis*, *A. brassicae*, *A. carthami*, *A. dauci*, *A. dianthi*, *A. dianthicola*, *A. humicola*, *A. lallemantiae*, *A. linicola*, *A. oleracea*, *A. raphani*, *A. ricini*, *A. sesami*, *A. tomato*, *A. tabacina* and *A. triticina*.

Finally, it should be stated that while food quality standards in modern developed countries are generally high so that humans may not be exposed to the acute toxicity of *Alternaria* mycotoxins on a significant scale, in developing countries, downgraded foods are frequently purchased by the poor (also the most vulnerable group), while subsistence farmers and parastatals may store sorghum and millet for several years for famine relief. Since high levels of natural contamination of cereals with TA, AME, and AOH have been reported,[2,9] the dangers to human of disorders such as Onyalai, Kashin-Beck, and esophageal cancer from *Alternaria* toxins must be considered real, pending evidence of alternative etiology. Particular attention is drawn to the need for research on synergism between fungal and other food-associated mutagens and carcinogens.

REFERENCES

1. Kwasna, H. A., Ecology and nomenclature of *Alternaria*, in *Alternaria Biology, Plant Diseases and Metabolites, Topics in Secondary Metabolism—Volume 3*, Chelkowski, J. and Visconti, A., Eds., Elsevier, 63–100, 1992.
2. King, A. D., Jr. and Schade, J. E., *Alternaria* toxins and their importance in food, *J. Food Prot.*, 47, 886, 1984.
3. Sanchis, V., Sanclemente, A., Usall, J., and Vinas, I., Incidence of mycotoxigenic *Alternaria alternata* and *Aspergillus flavus* in barley, *J. Food Prot.*, 56, 246, 1993.

4. Sobers, E. K. and Doupnik, B., Jr., Relationship of pathogenicity to tobacco leaves and toxicity to chicks of isolates of *Alternaria longipes*, *Appl. Microbiol.*, 23, 313, 1972.

5. Panigrahi, S., Bioassay of mycotoxins using terrestrial and aquatic, animal and plant species, *Food Chem. Toxicol.*, 31, 767, 1993.

6. Zajkowski, P., Grabarkiewicz-Szcesna, J., and Schmidt, R., Toxicity of mycotoxins produced by four *Alternaria* species to *Artemia salina* larvae, *Mycotoxin Res.*, 7, 11, 1991.

7. Bruce, V. R., Stack, M. E., and Mislivec, P. B., Incidence of toxic *Alternaria* species in small grains from the USA, *J. Food Sci.*, 49, 1626, 1984.

8. Mirocha, C. J., Gilchrist, D. G., Shier, W. T., Abbas, H. K., Wen, Y., and Vesonder, R. F., AAL toxins, fumonisins (biology and chemistry) and host-specificity concepts, *Mycopathologia*, 117, 47, 1992.

9. Ansari, A. A. and Shrivastava, A. K., Natural occurrence of *Alternaria* mycotoxins in sorghum and ragi from North Bihar, India, *Food Add. Contam.*, 7, 815, 1990.

10. Wei, C. I. and Swartz, D. D., Growth and production of mycotoxins by *Alternaria alternata* in synthetic, semisynthetic and rice media, *J. Food Prot.*, 48, 306, 1985.

11. Davis, N. D., Diener, U. L., and Morgan-Jones, G., Tenuazonic acid production by *Alternaria alternata* and *Alternaria tenuissima* isolated from cotton, *Appl. Environ. Microbiol.*, 34, 155, 1977.

12. Stinson, E. E., Mycotoxins—their biosynthesis in *Alternaria*, *J. Food Prot.*, 48, 80, 1985.

13. Ozcelik, S. and Ozcelik, N., Interacting effects of time, temperature, pH, and simple sugars on biomass and toxic metabolite production by three *Alternaria* spp., *Mycopathologia*, 109, 171, 1990.

14. Dalcero, A., Chulze, S., Etcheverry, M., Farnochi, C., and Varsavsky, E., Aflatoxins in sunflower seeds: influence of *Alternaria alternata* on aflatoxin production by *Aspergillus parasiticus*, *Mycopathologia*, 108, 31, 1989.

15. Sheridan, H., Smyth, C., Canning, A. M., and James, J. P., Stereoselective reduction of radicinin by liquid cultures of *Alternaria longipes*, *J. Nat. Prod.*, 55, 986, 1992.

16. Schade, J. E. and King, A. D., Jr., Analysis of the major *Alternaria* toxins, *J. Food Prot.*, 47, 978, 1984.

17. Gitterman, C. O., Antitumor, cytotoxic and antibacterial activities of tenuazonic acid and cogeneric tetramic acids, *J. Med. Chem.*, 8, 483, 1965.

18. Smith, E. R., Fredrickson, T. N., and Hadidian, Z., Toxic effects of the sodium and the N,N'-dibenzylethylenediamine salts of tenuazonic acid (NSC-525816 and NSC-82260), *Cancer Chemother. Rep.*, 52, 579, 1968.

19. Stack, M. E. and Prival, M. J., Mutagenicity of the *Alternaria* metabolites altertoxin I, II, and III, *Appl. Environ. Microbiol.*, 52, 718, 1986.

20. Visconti, A., Sibilia, A., and Sabia, C., *Alternaria alternata* from oilseed rape: mycotoxin production, and toxicity to *Artemia salina* larvae and rape seedlings, *Mycotoxin Res.*, 8, 9, 1992.

21. Panigrahi, S. and Dallin, S., Toxicity of the *Alternaria* spp. metabolites tenuazonic acid, alternariol, altertoxin-I and alternariol monomethyl ether to brine shrimp (*Artemia salina* L.) larvae, *J. Sci. Food Agric.*, 66, 493, 1994.

22. Seitz, L. M., Sauer, D. B., Mohr, H. E., Burroughs, R., and Paukstelis, J. V., Metabolites of *Alternaria* in grain sorghum. Compounds which could be mistaken for zearalenone and aflatoxin, *J. Agric. Food Chem.*, 23, 1, 1975.

23. Kohmoto, K., Itoh, Y., Shimomura, N., Kondoh, Y., Otani, H., Kodama, M., Nishimura, S., and Nakatsuka, S., Isolation and biological activities of two-host specific toxins from the tangerine pathotype of *Alternaria alternata*, *Phytopathology*, 83, 495, 1993.

24. Bachman, M., Luthy, J., and Schlatter, C., Toxicity and mutagenicity of molds of the *Aspergillus glaucus* group. Identification of physcion and three related anthraquinones as main toxic constituents from *Aspergillus chevalieri*, *J. Agric. Food Chem.*, 27, 1342, 1979.

25. Meronuck, R. A., Steele, J. A., Mirocha, C. J., and Christensen, C. M., Tenuazonic acid, a toxin produced by *Alternaria alternata*, *Appl. Microbiol.*, 23, 613, 1972.

26. Younis, S. A. and Al-Rawi, F. I., Studies on the fetotoxic effect of *Alternaria alternata* metabolites in mice, *J. Biol. Sci. Res.*, 19, 245, 1988.

27. Slifkin, M. K. and Spalding, J., Studies of the toxicity of *Alternaria mali*, *Toxicol. Appl. Pharmacol.*, 17, 375, 1970.

28. Williams, K. C., Blaney, B. J., and Peters, R. T., Nutritive value of weather-damaged sorghum grain for pigs, *Proc. Austr. Soc. Anim. Prod.*, 16, 395, 1986.

29. Jiajun, T. and Xintong, H., Studies on metabolic extract of *Alternaria alternata* and toxicity, *J. Environ. Sci. (China)*, 3, 29, 1991.

30. Wawrzkiewicz, K., Gluch, A., Rubaj, B., and Wrobel, M., *Alternaria* sp., an opportunistic-pathogenic fungus, *Medycyna Weterynaryjna*, 45, 27, 1989.

31. Diener, U. L., Morgan-Jones, G., Wagener, R. E., and Davis, N. D., Toxigenicity of fungi from grain sorghum, *Mycopathologia*, 75, 23, 1981.

32. Sauer, D. B., Seitz, L. M., Burroughs, R., Mohr, H. E., West, J. L., Milleret, R. J., and Anthony, H. D., Toxicity of *Alternaria* metabolites found in weathered grain at harvest, *J. Agric. Food Chem.*, 26, 1380, 1978.

33. Doupnik, B., Jr. and Sobers, E. K., Mycotoxicoses: toxicity to chicks of *Alternaria longipes* isolated from tobacco, *Appl. Microbiol.*, 16, 1596, 1968.

34. Forgacs, J., Koch, H., Carll, W. T., and White-Stevens, R. H., Mycotoxicoses, I. Relationship of toxic fungi to mouldy-feed toxicosis in poultry, *Avian Dis.*, 6, 363, 1962.

35. Bryden, W. L., Suter, D. A. I., and Jackson, C. A. W., Response of chickens to sorghum contaminated with *Alternaria*, *Proc. Nutr. Soc. Austr.*, 9, 109, 1984.

36. Palyusik, M., Szep, I., and Szoke, F., Data on susceptibility to mycotoxins of day-old goslings, *Acta Vet. Acad. Sci. Hungaricae Tomus*, 18, 363, 1968.

37. Jawad, A. L. M., Hussain, M. I., Imad, H. R., Razzak, A. A. W., and Khulud, F. A., Evaluation of mycotoxins produced by *Alternaria alternata* isolated from tomato in Iraq, *Symbiosis*, 2, 347, 1986.

38. Lindenfelser, L. A. and Ciegler, A., Production of antibiotics by *Alternaria* species, *Dev. Ind. Microbiol.*, 10, 271, 1969.

39. Vijayalakshmi, M. and Rao, A. S., Toxin production by *Alternaria alternata* pathogenic to brinjal (*Solanum melongena* L.), *Curr. Sci.*, 57, 150, 1988.

40. Mckenzie, K. J., Robb, J., and Lennard, J. H., Toxin production by *Alternaria* pathogens of oilseed rape (*Brassica napus*), *Crop Res. (Hort. Res.)*, 28, 67, 1988.

41. Miller, F. A., Righstel, W. A., Sloan, B. J., Ehrlich, J., French, J. C., and Bartz, Q. R., Antiviral activity of tenuazonic acid, *Nature*, 200, 1338, 1963.

42. Giambrone, J. J., Davis, N. D., and Diener, U. L., Effect of tenuazonic acid on young chickens, *Poultry Sci.*, 57, 1554, 1978.

43. Steyn, P. S. and Rabie, C. J., Characterization of magnesium and calcium tenuazonate from *Phoma sorghina*, *Phytochemistry*, 15, 1977, 1976.

44. Cole, M. and Rolinson, G. N., Microbial metabolites with insecticidal properties, *Appl. Microbiol.*, 24, 660, 1972.

45. Griffin, G. F. and Chu, F. S., Toxicity of the *Alternaria* metabolites alternariol, alternariol methyl ether, altenuene and tenuazonic acid in the chicken embryo assay, *Appl. Environ. Microbiol.*, 46, 1420, 1983.

46. Hashimoto, Y., Oshima, H., and Yuki, H., Metaphase arresting action of carcinostatic tenuazonic acid, *Gann*, 63, 79, 1972.

47. Shigeura, H. T. and Gordon, C. N., The biological activity of tenuazonic acid, *Biochemistry*, 2, 1132, 1963.

48. Carrasco, L. and Vasquez, D., Differences in eukaryotic ribosomes detected by selective action of an antibiotic, *Biochem. Biophys. Acta*, 319, 209, 1973.

49. Pero, R. W., Posner, H., Blois, M., Harvan, D., and Spalding, J. W., Toxicity of metabolites produced by the "*Alternaria*." *Enrivon. Health Perspect.*, June, 87, 1973.

50. Boutin, B. K., Peeler, J. T., and Twedt, R. M., Effects of purified altertoxins I, II, and III in the metabolic communication V79 system, *J. Toxicol. Environ. Health*, 26, 75, 1989.

51. Spalding, J. W., Pero, R. W., and Owens, R. G., Inhibition of the G2 phase of the mammalian cell cycle by the mycotoxin alternariol, in: Abstracts of the Tenth Annual Meeting of the American Society of Cell Biology, Abstract No. 527, *J. Cell Biol.*, 47, 199a, 1970.

52. Mohammed, Y. S., Osman, M., and Gabr, Y., Alternariol: a new fungal anticholinesterase drug, Part I, *Arzneimittelforsch (Drug Res.)*, 24, 121, 1974.

53. Pollock, G. A., DiSabatino, C. E., Heimsch, R. C., and Hilbelink, D. R., The subchronic toxicity and teratogenicity of alternariol monomethyl ether produced by *Alternaria solani*, *Food Chem. Toxicol.*, 20, 899, 1982.

54. Shier, W. T., Abbas, H. K., and Mirocha, C. J., Toxicity of mycotoxins fumonisin B1 and B2 and *Alternaria alternata* F. sp. *lycopersici* toxin (AAL) in cultured mammalian cells, *Mycopathologia*, 116, 97, 1991.

55. Gelderblom, W. C. A., Kriek, N. P. J., Marasas, W. F. O., and Thiel, P. G., Toxicity and carcinogenicity of the *Fusarium moniliforme* metabolite fumonisin in rats, *Carcinogenesis*, 12, 1247, 1991.

56. Horakova, K., Navarova, J., Nemec, P., and Kettner, M., Effect of dactylarin on HeLa cells, *J. Antibiotics*, 27, 408, 1974.

57. Kodiara, Y., Studies on the new toxic substances to insects, Destruxin A and B, produced by *Oospora destructor*. Part I. Isolation and purification of Destruxin A and B, *Agric. Biol. Chem.*, 26, 36, 1962.

58. Robeson, D. J., Gray, G. R., and Strobel, G. A., Production of the phytotoxins radicinin and radicinol by *Alternaria chrysanthemi*, *Phytochemisty*, 21, 2359, 1982.

59. Yagi, A., Okamura, N., Haraguchi, H., Abo, T., and Hashimoto, K., Antimicrobial tetrahydroanthraquinones from a strain of *Alternaria solani*, *Phytochemistry*, 33, 87, 1993.

60. Robeson, D. J. and Strobel, G. A., αβ-dehydrocurvularin and curvularin from *Alternaria cinerariae*, *Z. Naturforsch. Teil. C.*, 36c, 1081, 1981.

61. Ciegler, A. and Lindenfelser, L. A., An antibiotic complex from *Alternaria brassicicola*, *Experientia*, 25, 719, 1969.

62. Montemurro, N. and Visconti, A., *Alternaria* metabolites—chemical and biological data, in *Alternaria Biology, Plant Diseases and Metabolites, Topics in Secondary Metabolism—Volume 3*, Chelkowski, J. and Visconti, A., Eds., Elsevier, 449–558, 1992.

63. Hayashi, T., Takatsuki, A., and Tamura, G., Effect of brefeldin A on synthesis of cellular components in *Candida albicans*, *Agric. Biol. Chem.*, 46, 2241, 1982.

64. Sharma, R. P., Willhite, C. C., Wu, M. T., and Salunkhe, D. K., Teratogenic potential of blighted potato concentrate in rabbits, hamsters, and miniature swine, *Teratology*, 18, 55, 1978.

65. Harwig, J., Scott, P. M., Stoltz, D. R., and Blanchfield, B. J., Toxins of molds from decaying tomato fruit, *Appl. Environ. Microbiol.*, 38, 267, 1979.

66. Manna, G. K. and Banerjee, M., Spores of the fungus *Alternaria brassicicola* as a clastogen in treated mice, *Curr. Sci.*, 56, 1052, 1987.

67. Scott, P. M. and Stoltz, D. R., Mutagens produced by *Alternaria alternata*, *Mutat. Res.*, 78, 33, 1980.

68. DiCosmo, F. and Straus, N. A., Alternariol, a dibenzopyrone mycotoxin of *Alternaria* spp., is a new photosensitizing and DNA-cross linking agent, *Experentia*, 41, 1188, 1985.

69. Osborne, L. C., Jones, V. I., Peeler, J. T., and Larkin, E. P., Transformation of C3H/10T$_{1/2}$ cells and induction of EBV-early antigen in raji cells by altertoxins I and III, *Toxicol. In Vitro*, 2, 97, 1988.

70. Davis. V. M. and Stack, M. E., Mutagenicity of stemphyltoxin III, a metabolite of *Alternaria alternata*, *Appl. Environ. Microbiol.*, 57, 180, 1991.

71. Liu, G. T., Qian, Y. Z., Zhang, P., Dong, Z. M., Shi, Z. Y., Zhen, Y. Z., Miao, J., and Xu, Y. M., Relationships between *Alternaria alternata* and oesophageal cancer, in *Relevance to Human Cancer of N-nitroso Compounds, Tobacco Smoke and Mycotoxins*, O'Neill, I. K., Chen, J., and Bartsch, H., Eds., International Agency for Research on Cancer Scientific Publications No. 105, Lyon, France, 258–262, 1991.

72. Schlueter, D. P., Fink, J. N., and Hensley, G. T., Wood-pulp worker's disease: a hypersensitivity pneumonitis caused by *Alternaria*, *Ann. Int. Med.*, 77, 907, 1972.

73. Graveson, S., Identification and quantification of indoor airborne micro-fungi during 12 months from 44 Danish homes, *Acta Allergol.*, 27, 337, 1972.

74. Aho, R., Saprophytic fungi isolated from the hair of domestic and laboratory animals with suspected dermatophytosis, *Mycopathologia*, 83, 65, 1983.

75. Roosje, P. J., De Hoog, G. S., Koeman, J. P., and Willemse, T., Phaeohyphomycosis in a cat caused by *Alternaria infectoria* E. G. Simmons, *Mycoses*, 36, 451, 1993.

76. Pedersen, N. G., Mardh, P. A., Hallberg, T., and Jonsson, N., Cutaneous alternariosis, *Br. J. Dermatol.*, 94, 201, 1976.
77. Forgacs, J. and Carl, W. T., Mycotoxicoses: toxic fungi in tobaccos, *Science*, 152, 1634, 1966.

Index

A

A58365A, 15
Abortion, 219
Absorption
    isoflavonoids, 130–131
    modeling detoxification, 247–249
    oxalate and calcium, 212
    plant toxicants, 246
    pyrimidine glycosides, 144
Acacetin, 104, 115
Acacia spp., 82
Acclimatization, 251–252, see also Modeling
ACE, see Angiotensin-converting enzyme
Acetate, 232
Acetate-malonate pathway, 78
Acetone, 24
Acetyl bromide, 82
Acetyl derivatives, 126–128, see also Isoflavonoids
Acetylcholine, 255
Acetylcholinesterase, 8
Acetylleptinidine, 8, 9
N-Acetylloline, 55
Acid catalysis, 209
Acid detergent fiber (ADF), 81–83, see also
    Proanthocyanidins
Acid detergent lignin (ADL), 82, 83, see also
    Proanthocyanidins
Acid fogs, 181
Acidosis, 232, 246–247, see also Feeding, preferences
Acid-base homeostasis, 246–247, 248, 250, see also
    Modeling
Aconitine, 261
Aconitum napellus, 261
Acremonium spp., 52–57
Actinidine, 168
Activated charcoal, 200
Addictions, 258
Adenocarcinoma, 66
ADF, see Acid detergent fiber
ADL, see Acid detergent lignin
Administration route, 325, 326, 327, see also *Alternaria*
    toxins
Adrenal corticoids, 145
Adrenergic receptors, 225
Aflatoxins
    chemistry and molecular biology, 271–273
    control in foods and feed, 276–282
    coproduction with cyclopiazonic acid, 306–307
    economic impact, 270–271
    metabolism and disease, 273–276
    synergistic actions of *Alternaria* toxins, 332

AFO, see Algar-Flynn-Oyamada reaction
Agalactia, 51
Agave spp., 194, 196
Age, 328, see also *Alternaria* toxins
Age groups, 197
Aggregation, 149
Agriotes obscurus, 26
Agroclavine, 52
AIDS, see Human immunodeficiency virus
Alexa spp., 4
Algar-Flynn-Oyamada reaction (AFO), 103
Alimentary carcinoma, 66, see also
    Carcinogenicity
Alkaloids, see also Individual entries
    alkenoid, 38, 40, 45, 46
    bisindole, 260
    detoxification costs, 244
    dienoid, 38, 39, 40, 43, 45
    endophyte
        biological activity, 57–58
        factors affecting production, 55–57
        feeding preferences, 234
        natural occurrence, 52–56
    Elaeocarpus, 11–12
    ergot
        biological activity, 57
        medicinal applications, 260
        natural occurrence, 52–54
        production, 56
    ergot peptide, 52
    Erythrina
        pharmacology and toxicology, 46
        spectral features, 43–46
        structural features and distribution, 37–43
    Glochidion, 14
    glycodienoid, 38
    histidine-derived, 260
    indole, 38, 260
    ingestion by herbivores and feeding preferences,
        234, 238
    indolizidine, see Indolizidine alkaloids
    isoprenoid-derived, 261
    lactonic, 38, 41, 43, 45
    lolitrem, 54–57
    lysergic acid amides, 52, 53
    medicinal applications, 256, 258–261
    ornithine-derived, 256, 258
    phenylalanine-derived, 258–259
    Rauvolfia, 260
    tryptophan-derived, 259–260
Allan-Robinson reaction, see Kostanecki-Robinson
    reaction